MICROBIOLOGY
for the Health Sciences

Sixth Edition

MICROBIOLOGY
for the Health Sciences

Gwendolyn R. W. Burton, Ph.D.

Professor Emeritus
Science Department
Front Range Community College
Westminster, Colorado

Paul G. Engelkirk, Ph.D.

Faculty
Science Department
Central Texas College
Killeen, Texas

LIPPINCOTT WILLIAMS & WILKINS
A **Wolters Kluwer** Company
Philadelphia · Baltimore · New York · London
Buenos Aires · Hong Kong · Sydney · Tokyo

Editor: Lawrence McGrew
Editorial Assistant: Frank Musick
Marketing Manager: Deborah Hartman
Production Editor: Robert D. Magee
Cover Image: Custom Medical Stock Photo, Inc.

Printed in the United States of America

First Edition, 1979 　　Third Edition, 1988 　　Fifth Edition, 1996
Second Edition, 1983 　Fourth Edition, 1992

Library of Congress Cataloging-in-Publication Data

Burton, Gwendolyn R. W. (Gwendolyn R. Wilson)
　　Microbiology for the health sciences / Gwendolyn R.W. Burton, Paul G. Engelkirk.—6th ed.
　　　　p.　　　cm.
　　Includes bibliographical references and index.
　　ISBN 0-7817-1844-9
　　1. Microbiology.　2. Medical microbiology.　I. Engelkirk, Paul G.　II. Title.
　QR41.2.B88　1999
　616′.01—dc21　　　　　　　　　　　　　　　　　　　　　99–19388
　　　　　　　　　　　　　　　　　　　　　　　　　　　　　　CIP

The material contained in this volume was submitted as previously unpublished material, except in the instances in which credit has been given to the source from which some of the illustrative material was derived.

Any procedure or practice described in this book should be applied by the health-care practitioner under appropriate supervision in accordance with professional standards of care used with regard to the unique circumstances that apply in each practice situation. Care has been taken to confirm the accuracy of information presented and to describe generally accepted practices. However, the authors, editors, and publisher cannot accept any responsibility for errors or omissions or for any consequences from application of the information in this book and make no warranty, express or implied, with respect to the contents of the book.

The authors and publisher have exerted every effort to ensure that drug selection and dosage set forth in this text are in accordance with current recommendations and practice at the time of publication. However, in view of ongoing research, changes in government regulations, and the constant flow of information relating to drug therapy and drug reactions, the reader is urged to check the package insert for each drug for any change in indications and dosage and for added warnings and precautions. This is particularly important when the recommended agent is a new or infrequently employed drug.

Materials appearing in this book prepared by individuals as part of their official duties as U.S. Government employees are not covered by the above-mentioned copyright.

The publishers have made every effort to trace the copyright holders for borrowed material. If they have inadvertently overlooked any, they will be pleased to make the necessary arrangements at the first opportunity.

To purchase additional copies of this book, call our customer service department at **(800) 638-3030** or fax orders to **(301) 824-7390**. International customers should call **(301) 714-2324**.

99 00 01 02 03
6 7 8 9 10

For my husband, Lynn; children, Cindy, Gary, Alan, and Earl; and grandchildren, Jennifer, Penny, RaeLynn, Austin, Kelsey, and Emily: for their caring, respect and love.

-GRWB

For my parents, Paul and Phyllis, who set me on the right path, and for my wife, Janet, who shares with me each and every step along that path.

-PGE

About the Authors

Gwendolyn R. Wilson Burton, Ph.D., is retired Chairperson and Professor Emeritus of the Science Department at Front Range Community College, Westminster, Colorado, where she taught microbiology and human biology for 20 years. She also taught microbiology and immunology at the University of Denver and lectured at many colleges and high schools in the Denver area on sexually transmitted diseases. Dr. Burton received her bachelors degree in Chemistry from Colorado State University, took graduate studies at the University of Oklahoma, and completed masters and doctoral degrees in Microbiology and Higher Education at the University of Denver. She developed 39 computer-interfaced video-taped microbiology lectures for individual study, and a series of self-paced learning materials for human biology students. Dr. Burton served as a state and international high school Science Fair Judge. As a delegate with the People-to-People Microbiology Delegation to the Peoples' Republic of China, she lectured on giardiasis at the medical schools in Beijing, Nanchang, and Guangzhow. Dr. Burton developed and began the widely recognized program for Hazardous Materials Technology Training at Front Range Community College, one of the first such programs at a community college in the nation. She is a member of the American Society for Microbiology and served as President of the Rocky Mountain Branch of that organization. Some of the honors presented to Dr. Burton include the Academic Excellence Science Award from the American Association of Community and Junior Colleges; Outstanding Educators of America; the World Safety Organization Special Recognition Award; and Distinguished Leadership Award for her many accomplishments in the educational fields of hazardous materials and microbiology.

Paul G. Engelkirk, Ph.D., is a faculty member in the Science Department at Central Texas College in Killeen, Texas, where he teaches introductory microbiology to approximately 200 students per year. Before joining Central Texas College, he was an Associate Professor at the University of Texas Health Science Center in Houston, Texas, where he taught diagnostic microbiology to medical technology students for 7 years. Before that, Dr. Engelkirk served 22 years as an officer in the U.S. Army Medical Department, supervising a variety of clinical pathology, diagnostic microbiology, and clinically oriented microbiology research laboratories in Germany, Vietnam, and the United States; he retired with the rank of Lieutenant Colonel. Dr. Engelkirk received his bachelors degree (in Biology) from New York State University and his masters and doctoral degrees (both in Microbiology and

Public Health) from Michigan State University. He received additional medical technology and tropical medicine training at Walter Reed Army Hospital in Washington, D.C., and specialized training in anaerobic bacteriology, mycobacteriology, and virology at the Centers for Disease Control and Prevention in Atlanta, GA. Dr. Engelkirk is the author or co-author of 3 microbiology textbooks, 10 additional book chapters, 4 medical laboratory-oriented self-study courses, and many scientific articles. Together with his wife, Dr. Janet Duben-Engelkirk, he writes and publishes three quarterly newsletters, entitled *Anaerobe Abstracts, Clinical Parasitology Abstracts,* and *Medical Mycology Abstracts.* Dr. Engelkirk has been engaged in various aspects of clinical microbiology for over 35 years and is a Past President of the Rocky Mountain Branch of the American Society for Microbiology.

Preface

Microbiology, the study of microorganisms, is a fascinating topic to those of us who feel its importance in and impact on our daily lives. Others find it necessary to learn microbiological concepts and vocabulary in order to function well in their chosen vocations. For example, those who plan to work as nurses, in allied health professions, or in any other area of healthcare must be aware of the principles of sterilization, infectious disease causation, and infectious disease prevention.

Microbiology for the Health Sciences will aid those who want to learn the basic microbiological concepts that apply to nursing, the allied health professions, and many other fields involving the care of patients and protection against infectious diseases. It is sufficiently comprehensive for use by students having little or no science background and by mature students returning to school after an absence of several years.

There is a need for a fundamental microbiology text that presents major concepts clearly and concisely for people entering healthcare occupations. This book is appropriate for use in a one-term, allied health microbiology class or as one unit in a basic science class for all healthcare-oriented students. It contains all the core themes and concepts for an introductory microbiology course, as described by the American Society for Microbiology.

Microbiology is an enormous and complex subject with many interrelated facets and hundreds of scientific terms. The authors have attempted a very fundamental approach to the subject matter by presenting, at the beginning of each chapter, the basic information necessary to understand the more complex concepts with which the chapters conclude. New terms are listed at the beginning of each chapter for easy reference and are highlighted and defined in the text. A complete glossary, with pronunciation keys, can be found at the end of the book. A brief outline and objectives are presented at the beginning of each chapter to enable the student to survey topics to be covered. Discussion questions and self tests are included for review at the end of each chapter.

In this 6th edition, some chapters have been split and expanded for better continuity and understanding. The complete book has been updated and partially rewritten for clarity. Color figures are included in Chapter 3 to help students visualize the appearance of certain microorganisms and methods of identification. Numerous Insight Boxes and Study Aids provide a more detailed look at particular aspects of topics being discussed in the chapters. Clinically-oriented Insight Boxes,

Study Aids, and Tables are identified with a caduceus symbol in the margin. Problems, questions, and self tests have been incorporated into each chapter to encourage students to review the material and to test themselves while learning the information. In this way, the student will have greater insight into the important facts that will be stressed on examinations. Several new appendices have been added, including a "mini-Parasitology course," clinical microbiology laboratory procedures, and useful information about weights and measures.

Although this book is intended primarily for nonscience majors, it is not an easy text—microbiology is not an easy topic. As the student will discover, the concise nature of this text has made each sentence significant. Thus, the reader will be intellectually challenged to learn each new concept as it is presented. It is our hope that students will enjoy their study of microbiology and be motivated to further explore this fascinating field, especially as it relates to their occupations.

We are deeply indebted to those colleagues (especially Elmer Koneman, MD, and Pat Hidy, Ed.D.), friends, family members, and students who provided illustrations used in the text or served as sources of advice and encouragement. We particularly wish to acknowledge Lawrence McGrew, sponsoring editor, and Holly Chapman, editorial assistant at Lippincott Williams & Wilkins, for their editorial assistance in the preparation of this manuscript, as well as Bob Magee for coordinating the production of this book.

Gwendolyn R. W. Burton, Ph.D.
Paul G. Engelkirk, Ph.D.

Contents

CHAPTER 1

Introduction to Microbiology 1

MICROBIOLOGY, THE SCIENCE 2
THE TOOLS OF MICROBIOLOGY 16
UNITS OF MEASUREMENT 22

CHAPTER 2

Cell Structure and Taxonomy 29

CELLS: EUCARYOTES AND PROCARYOTES 30
TAXONOMY 46

CHAPTER 3

Diversity of Microorganisms 53

CATEGORIES OF MICROORGANISMS 54
BACTERIA 54
RUDIMENTARY FORMS OF BACTERIA 65
PHOTOSYNTHETIC BACTERIA 68
ARCHAEBACTERIA 68
ALGAE 69
PROTOZOA 70
FUNGI 75
LICHENS 82
SLIME MOLDS 82
ACELLULAR INFECTIOUS AGENTS 83

CHAPTER 4

Basic Chemistry Concepts   98

BASIC CHEMISTRY   99

CHAPTER 5

Biochemistry: The Chemistry of Life   112

ORGANIC CHEMISTRY   113
BIOCHEMISTRY   115

CHAPTER 6

Microbial Physiology and Genetics   137

NUTRITION   139
ENZYMES, METABOLISM, AND ENERGY   141
MICROBIAL GROWTH   152
BACTERIAL GENETICS   159
GENETIC ENGINEERING   166
GENE THERAPY   166

CHAPTER 7

Controlling the Growth of Microorganisms   174

DEFINITION OF TERMS   176
FACTORS INFLUENCING MICROBIAL GROWTH   178
ANTIMICROBIAL METHODS   182
CHEMOTHERAPY   192

CHAPTER 8

Microbial Ecology   211

INTERACTIONS BETWEEN HUMANS AND MICROBES   212
MICROBES IN AGRICULTURE   220
BIOTECHNOLOGY   222
BIOREMEDIATION   223
BIOLOGICAL WARFARE AGENTS   223

CHAPTER 9

Microbial Pathogenicity and Epidemiology *229*

MICROBIAL PATHOGENICITY   230

EPIDEMIOLOGY AND DISEASE TRANSMISSION   243

RESERVOIRS OF INFECTION   249

MODES OF DISEASE TRANSMISSION   254

CONTROL OF EPIDEMIC DISEASES   259

CHAPTER 10

Preventing the Spread of
Communicable Diseases *266*

PREVENTION OF HOSPITAL-ACQUIRED INFECTIONS   267

SPECIMEN COLLECTING, PROCESSING, AND TESTING   283

ENVIRONMENTAL DISEASE CONTROL MEASURES   297

CHAPTER 11

Human Defenses Against Infectious Diseases *309*

NONSPECIFIC MECHANISMS OF DEFENSE   310

IMMUNE RESPONSE TO DISEASE: THIRD LINE OF DEFENSE   323

CHAPTER 12

Major Infectious Diseases of Humans *360*

SKIN INFECTIONS   362

EYE INFECTIONS   374

INFECTIOUS DISEASES OF THE MOUTH   378

EAR INFECTIONS   381

INFECTIOUS DISEASES OF THE RESPIRATORY SYSTEM   382

INFECTIOUS DISEASES OF THE GASTROINTESTINAL (GI) TRACT   395

INFECTIOUS DISEASES OF THE GENITOURINARY (GU) TRACT   412

INFECTIOUS DISEASES OF THE CIRCULATORY SYSTEM   421

INFECTIOUS DISEASES OF THE NERVOUS SYSTEM   437

APPENDIX A
Taxonomic Categories of Selected Medically Important Bacteria   449

APPENDIX B
Compendium of Important Bacterial Pathogens of Humans   450

APPENDIX C
Parasitology   453

APPENDIX D
Microbiology Laboratory Procedures   467

APPENDIX E
Useful Conversions   473

APPENDIX F
Suggested Reading   475

Glossary   477
Lippincott Williams & Wilkins Credits   495
Index   I-i

MICROBIOLOGY
for the Health Sciences

CHAPTER 1

Introduction to Microbiology

MICROBIOLOGY, THE SCIENCE
The Scope of Microbiology
  General Microbiology
  Medical Microbiology
  Veterinary Microbiology
  Agricultural Microbiology

Sanitary Microbiology
Industrial Microbiology
Microbial Physiology and
  Genetics
Environmental Microbiology
Milestones of Microbiology

THE TOOLS OF MICROBIOLOGY
Microscopes
  Light Microscopes
  Electron Microscopes
UNITS OF MEASUREMENT

OBJECTIVES

After studying this chapter, you should be able to:

- *Define microbiology*
- *List some important functions of microbes in the environment*
- *Explain the relevance of microbiology to the health professions*
- *List some areas of microbiological study*
- *Outline some contributions of Leeuwenhoek, Pasteur, and Koch to microbiology*
- *Explain the biological theory of fermentation*

- *Explain the germ theory of disease*
- *Learn Koch's postulates and cite some circumstances in which they may not apply*
- *Describe differences between light microscopes and electron microscopes and the applications of both*
- *List the metric units used in microscopic measurements and indicate their relative sizes*

NEW TERMS

Abiogenesis
Acquired immunodeficiency
  syndrome (AIDS)
Algae (sing. *alga*)
Angstrom
Antiseptic surgery
Aseptic techniques
Bacteria (sing. *bacterium*)
Bacteriologist

Bacteriology
Biogenesis
Compound microscope
Contagious
Electron microscope
Fungi (sing. *fungus*)
Immunology
Indigenous microflora
Infectious disease

Koch's postulates
Light microscope
Microbiologist
Microbiology
Micrometer
Microorganisms (microbes)
Microscope
Mycologist
Mycology

Nanometer	Prion	Tyndallization
Nonpathogen	Protozoa (sing. *protozoan*)	Ubiquitous
Opportunistic pathogens	Protozoologist	Variolation
Pasteurization	Protozoology	Viroid
Pathogen	Resolving power (resolution)	Virologist
Petri dish	Saprophyte	Virology
Phycologist	Simple microscope	Viruses (sing. *virus*)
Phycology	Sterile techniques	Zoonoses (sing. *zoonosis*)

MICROBIOLOGY, THE SCIENCE

The word biology comes from *bios,* referring to living organisms, and *logy,* meaning "the study of." Thus, biology is the study of living organisms. *Microbiology* is a branch of biology. *Micro* means very small—anything so small that it must be viewed with a *microscope* (an optical instrument used to observe very small objects). Microbiology is, therefore, the study of very small living organisms—organisms called *microorganisms* or *microbes.* These microorganisms include *bacteria, algae, protozoa, fungi,* and *viruses* (Fig. 1-1). Because many scientists do not consider viruses to be living organisms, the terms "infectious agents" or "infectious particles" are often used in reference to viruses.

Because microorganisms are so small, you may not be aware of the effect they have on your daily life. Occasionally, you may become conscious of their effect on your body when, for example, a cut or burn becomes infected or when you have a sore throat. Have you ever been very sick after a picnic and wondered which of the foods you ate contained harmful germs? Most of us are aware of "germs" or disease-causing microorganisms (technically called *pathogens*) only when we are personally affected by them, when we hear about them on television or read about them in newspapers (Table 1-1). Actually, only a small percentage of microbes (about 3% of known microbes) is capable of causing disease (pathogenic) (Fig. 1-2). The others (the *nonpathogens*) are either beneficial or have no effect upon us. Microbes that live on and in our bodies (*e.g.,* on the skin, in the mouth, and in the intestine) are collectively known as our *indigenous microflora* (or *indigenous microbiota*). Some of them can cause disease should they accidentally invade the "wrong place" at the "wrong time," such as when the host's resistance is low and growth conditions are right. Such microbes are referred to as *opportunistic pathogens* (or *opportunists*). Opportunistic pathogens ordinarily cause us no harm, but because they are potentially pathogenic, they can cause disease should they gain entrance to wounds, the bloodstream, or organs such as the urinary bladder. In a sense, they are microorganisms awaiting the opportunity to be pathogens.

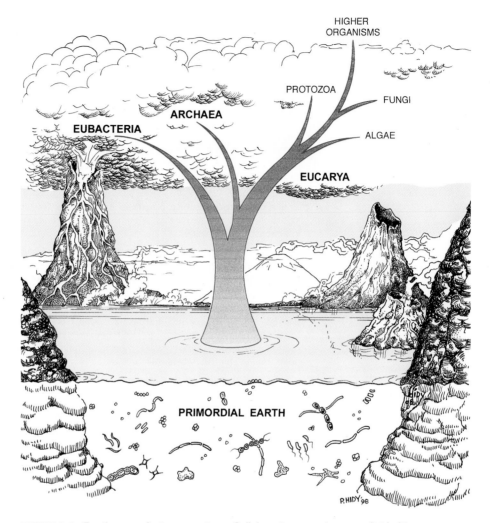

FIGURE 1-1. Family tree of microorganisms. Cellular microorganisms are divided into *eucaryotes* (organisms having a true nucleus, such as algae, fungi and protozoa) and *procaryotes* (organisms lacking a true nucleus, such as Archaea and Eubacteria). Viruses are not considered to be cells (they are said to be acellular) and are, therefore, not included on this family tree. (The various categories of microorganisms are described in Chapter 3.)

Diseases caused by microorganisms are called *infectious diseases.* As a group, infectious diseases are the leading cause of death in the world and the third leading cause of death in the United States (after heart disease and cancer). Each year, infectious diseases cause between 150,000 and 200,000 deaths in the United States. In 1997, infectious diseases claimed 17 million lives worldwide; the majority of deaths occurred in developing countries.

Microorganisms are said to be *ubiquitous,* meaning that they are virtually everywhere. Many of these microorganisms contribute to our welfare. The indigenous

TABLE 1-1. Pathogens	
Category	**Examples of Diseases They Cause**
Algae	a rare cause of infections; intoxications (resulting from ingestion of toxins)
Bacteria	anthrax, botulism, cholera, diarrhea, diphtheria, ear and eye infections, food poisoning, gas gangrene, gonorrhea, hemolytic uremic syndrome (HUS), Legionnaires' disease, leprosy, Lyme disease, meningitis, plague, pneumonia, Rocky mountain spotted fever, scarlet fever, staph infections, strep throat, syphilis, tetanus, tuberculosis, tularemia, typhoid fever, typhus, urethritis, urinary tract infections, whooping cough
Fungi	allergies, cryptococcosis, histoplasmosis, intoxications (resulting from ingestion of toxins), meningitis, pneumonia, thrush, tinea (ringworm) infections, yeast vaginitis
Protozoa	African sleeping sickness, amebic dysentery, babesiosis, Chagas' disease, cryptosporidiosis, diarrhea, giardiasis, malaria, meningoencephalitis, pneumonia, toxoplasmosis, trichomoniasis
Viruses	acquired immunodeficiency syndrome (AIDS), certain types of cancer, chickenpox, cold sores (fever blisters), common cold, dengue, diarrhea, encephalitis, genital herpes infections, German measles, hantavirus pulmonary syndrome (HPS), hepatitis, infectious mononucleosis, influenza, measles, meningitis, mumps, pneumonia, polio, rabies, shingles, warts, yellow fever

microflora actually inhibit the growth of pathogens in those areas of the body where they live by occupying the space, using the food supply, and secreting materials (waste products, toxins, antibiotics, etc.) that may prevent or reduce the growth of pathogens. Other microorganisms are used in the production of foods, alcoholic beverages, and many other products (Table 1-2).

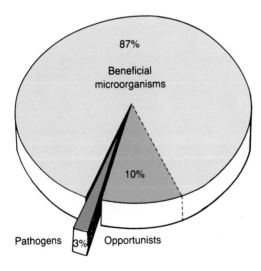

FIGURE 1-2. Pathogens comprise about 3% of all known microorganisms. Another 10% are opportunistic pathogens that may cause disease if given the opportunity (see text for explanation).

TABLE 1-2. Products Requiring Microbial Participation in the Manufacturing Process

Category	Examples
Foods	Bread, butter, buttermilk, chocolate, coffee, cottage cheese, cured hams, green olives, pickles, sauerkraut, sour cream, sourdough bread, soy sauce, various cheeses, yogurt
Alcoholic beverages	Beer, sherry, whiskey, wine
Chemicals	Acetone, ethanol, butanol

Many bacteria and fungi are *saprophytes*, which aid in fertilization by returning inorganic nutrients to the soil. Saprophytes break down dead and dying organic materials (plants and animals) into nitrates, phosphates, carbon dioxide, water, and other chemicals necessary for plant growth (Fig. 1-3). These saprophytes also destroy paper, feces, and other biodegradable substances, although they cannot break down most plastics or glass. Nitrogen-fixing bacteria, which live in the root nodules of certain plants called legumes (*e.g.,* peas, peanuts, alfalfa, clover), are able to return nitrogen from the air to the soil in the form of ammonia for use by other plants (Fig. 1-4). The involvement of various types of bacteria in the nitrogen cycle is described in Chapter 8. Knowledge of these microbes is important to farmers who practice crop rotation to replenish their fields and to gardeners who keep compost pits as a source of natural fertilizer. In both cases, dead organic material is broken down into inorganic nutrients (nitrates and phosphates) by microorganisms.

The purification of waste water is partially accomplished by bacteria in the holding tanks of sewage disposal plants, in which feces, garbage, and other organic

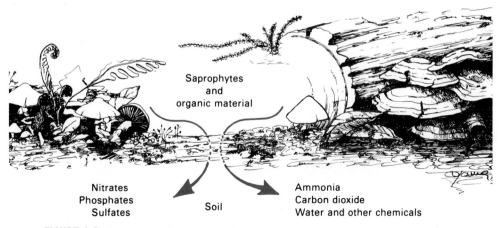

Saprophytes
and
organic material

Nitrates
Phosphates
Sulfates

Soil

Ammonia
Carbon dioxide
Water and other chemicals

FIGURE 1-3. Saprophytes break down dead and decaying organic material into inorganic nutrients in the soil.

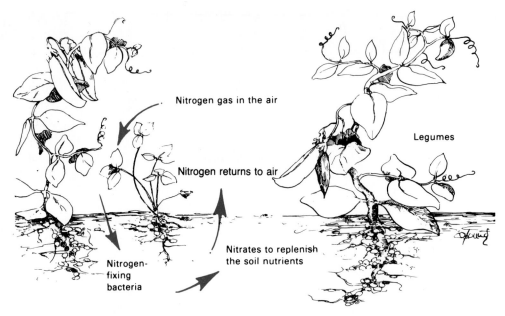

Nitrogen gas in the air

Nitrogen returns to air

Legumes

Nitrogen-fixing bacteria

Nitrates to replenish the soil nutrients

FIGURE 1-4. Nitrogen fixation. Nitrogen-fixing bacteria that live on or near the roots of legumes convert free nitrogen from the air into ammonia. Nitrifying bacteria then convert the ammonia into nitrites and nitrates, which are nutrients used by plants.

materials are collected and reduced to harmless waste. Some microorganisms, such as the iron- and sulfur-utilizing bacteria even break down metals and minerals. The beneficial activities of microbes affect every part of our environment—in the land, water, and air.

Those who work in the health professions must be particularly aware of pathogens, their sources, and how they may be transmitted from one person to another. Physicians' assistants, nurses, dental assistants, laboratory technologists, respiratory therapists, orderlies, nurses' aides, and all others associated with patient care must take precautions to prevent the spread of pathogens. Harmful microorganisms may be transferred from health workers to patients; from patient to patient; from contaminated mechanical devices, instruments, and syringes to patients; from contaminated bedding, clothes, dishes, and food to patients; and from patients to health workers and other susceptible persons.

The Scope of Microbiology

There are many fields of study within the science of microbiology. For example, one may specialize in the study of specific types of microorganisms. A *bacteriologist* concentrates on *bacteriology,* the study of the structure, functions, and activities of bacteria. *Phycology* is the study of the various types of algae by scientists called *phycologists.*

Those who specialize in the study of fungi, or *mycology*, are called *mycologists*. *Proto-zoologists* explore the area of *protozoology*, the study of protozoa and their activities. *Virology* encompasses the study of viruses and their effects on living cells of all types. *Virologists* and cell biologists may become genetic engineers who transfer genetic material (deoxyribonucleic acid or DNA) from one cell to another. Virologists may also study *prions* and *viroids*, acellular infectious agents that are even smaller than viruses.

Within the general field of microbiology there are many specialized areas in which a knowledge of specific types of microorganisms and of their applications is important. These areas include medical, veterinary, and agricultural microbiology as well as applications in sewage disposal, industrial production, space research, microbial ecology, biodegradation, and genetic engineering. People employed in these fields of microbiology are called *microbiologists;* such individuals may have a bachelor's, masters, or doctoral degree in microbiology.

GENERAL MICROBIOLOGY

The study and classification of microorganisms and how they function is known as general microbiology. It encompasses all areas of microbiology.

MEDICAL MICROBIOLOGY

The field of medical microbiology involves the study of pathogens, the diseases they cause, and the body's defenses against disease. This field is concerned with epidemiology, transmission of pathogens, disease-prevention measures, aseptic techniques, treatment of infectious diseases, immunology, and the production of vaccines to protect people and animals against infectious diseases. The complete or almost complete eradication of diseases like smallpox and polio, the safety of modern surgery, and the treatment of victims of *acquired immunodeficiency syndrome* (AIDS) are due to the many technological advances in this field. A branch of medical microbiology, called clinical microbiology, is concerned with the laboratory diagnosis of infectious diseases of humans.

VETERINARY MICROBIOLOGY

The spread and control of infectious diseases among animals is the concern of veterinary microbiologists. The production of food from livestock, the raising of other agriculturally important animals, the care of pets, and the transmission of diseases from animals to humans are areas of major importance in this field. Infectious diseases of humans acquired from animal sources are called *zoonoses* or zoonotic diseases. Zoonoses are discussed in Chapter 9.

AGRICULTURAL MICROBIOLOGY

Included in the field of agricultural microbiology are studies of the beneficial and harmful roles of microbes in soil formation and fertility; in carbon, nitrogen, phosphorus, and sulfur cycles; in diseases of plants; in the digestive processes of cows and other ruminants; and in the production of crops and foods. A food microbiologist is concerned with the production, processing, storage, cooking, and serving of food, as well as the prevention of food spoilage, food poisoning, and food toxicity. A dairy microbiologist oversees the grading, pasteurization, and processing of milk and cheeses to prevent contamination, spoilage, and transmission of diseases from environmental, animal, and human sources.

SANITARY MICROBIOLOGY

The field of sanitary microbiology includes the processing and disposal of garbage and sewage wastes, as well as the purification and processing of water supplies to ensure that no pathogens are carried to the consumer by drinking water. Sanitary microbiologists also inspect food processing installations and eating establishments to ensure that proper food handling procedures are being enforced.

INDUSTRIAL MICROBIOLOGY

Many businesses and industries depend on the proper growth and maintenance of certain microbes to produce beer, wine, alcohol, and organic materials such as enzymes, vitamins, and antibiotics. Industrial microbiologists monitor and maintain the essential microorganisms for these commercial enterprises. Applied microbiologists conduct research aimed at producing new products and more effective antibiotics. The scope of microbiology has broad, far-reaching effects on humans, pathogens, and the relationships of microorganisms to the environment.

MICROBIAL PHYSIOLOGY AND GENETICS

Research in microbial physiology has contributed greatly to a clearer understanding of the function of microorganisms, the structure of DNA, and the science of genetics (the study of heredity) in general. Genetic manipulation is much easier and faster with viruses and bacteria than is it with more complex cells; thus, everyday organisms such as the intestinal bacterium, *Escherichia coli,* are invaluable tools in this study.

ENVIRONMENTAL MICROBIOLOGY

The field of environmental microbiology, or microbial ecology, has become important because of increased concern about the environment. This field encompasses the areas of soil, air, water, sewage, food, and dairy microbiology, as well as the cycling of elements by microbial, environmental, and geochemical processes. In addition, the biodegradation of toxic chemicals by various microorganisms is being used as a new method for cleaning up hazardous materials found in soil and water.

Milestones of Microbiology

Scientists tell us that the earth was formed about 4.5 billion years ago and for the first billion years of earth's existence, there was no life on this planet. The first definitive evidence of life—fossils of primitive microorganisms found in ancient rock formations in Northwestern Australia—dates back to about 3.5 billion years ago. In our present form, humans (*Homo sapiens*) have existed for only the past 100,000 years or so. We know that human pathogens have existed for thousands of years because damage caused by them has been observed in the bones of mummies and early human fossils, indicating that bacterial diseases such as tuberculosis and syphilis and parasitic worm infections such as schistosomiasis and dracunculiasis (guinea worm infection) were present.

In the ancient civilizations of Egypt and China, people kept clean by washing with water in an effort to prevent disease. They knew that some diseases were easily transmitted from one person to another, and they learned to isolate the sick to prevent spread of these diseases, which we recognize as being *contagious*. The Egyptians were aware of the effectiveness of biological warfare. They often used the blood and bodies of their diseased dead to contaminate water supplies of their enemies and to spread diseases among them.

The Book of Leviticus in the Bible was probably the first recording of laws concerning public health. The Hebrew people were told to practice personal hygiene by washing and keeping clean. They were also instructed to bury their waste material away from their campsites, to isolate those who were sick, and to burn soiled dressings. They were prohibited from eating animals that had died of natural causes. The procedure for killing an animal was clearly described, and the edible parts were designated.

Most of the knowledge about public sanitation and transmission of disease was lost in Europe during the Middle Ages, when there was a general stagnation of culture and learning for almost 1,000 years. However, during the Renaissance, widespread epidemics of smallpox (a viral disease), syphilis, rabies (a viral disease), and other diseases prompted physicians and alchemists to search for explanations as to how contraction and transference of diseases occurred. Most people believed that diseases were caused by curses of the gods and, as a result, many bizarre treatments (bleeding, drilling holes in the head, attaching leeches) were used to drive away the devils or evil spirits and relieve the symptoms.

An Italian physician, Girolamo Fracastorius, having observed the syphilis epidemic of the 1500s, proposed in 1546 that the agents of communicable diseases were living germs that could be transmitted by direct contact with humans and animals and indirectly by objects. Because the agents of disease could not be observed and experimental evidence was lacking, proof for vague theories such as this was long delayed. Until the development of magnifying lenses and microscopes that could sufficiently magnify microorganisms to allow them to be visualized, the discovery of disease-causing agents was impossible. Actually, it is not known who built the first microscope, but *compound microscopes,* which use two or more magnifying lenses, were developed by Johannes Janssen (1590), Galileo Galilei (1609), and Robert Hooke (1660). When Antony van Leeuwenhoek first described bacteria in 1667, he used a small, *simple microscope* with one magnifying lens the size of a large pinhead and observed material that he placed on the point of a pin (see Insight

INSIGHT
Leeuwenhoek—Father of Bacteriology and Protozoology

The discovery of various bacteria and protozoa by Antony van Leeuwenhoek, using his small, simple microscopes over 300 years ago, has long been an area for study and discussion. How did he build the microscopes and use them so effectively to observe the "wee beasties," as he called them? Some believe he may have used polished clear glass beads of various sizes for lenses and various intensities of outdoor light to view the microbes. By changing the direction and intensity of the light source, he could have developed the darkfield capability to enable him to observe microbial movement. See Insight Figure 1-1 for an idea of what Leeuwenhoek's microscopes looked like and what he saw.

Perhaps it was curiosity about the reason for pepper's potent taste that led Leeuwenhoek to the discovery of bacteria. He steeped peppercorns for 3 weeks to soften them, examined the water, and observed the "incredibly small organisms" we now know as bacteria. Though this has been acknowledged as the first recorded observation of bacteria, definitive evidence for the discovery of bacteria was not provided until he wrote a letter to the Royal Society of London on September 17, 1686. In this letter, he described his regimen for keeping his teeth clean, detailing his examination of the white matter (plaque) that grew between his teeth. When he added this white matter to saliva (which he thought was free of microorganisms) and examined it microscopically, he described the "many very small living animals that moved very prettily."

Leeuwenhoek, determined to keep his microscopic methods to himself, shared his techniques with no one. Thus, to this day, we don't know how he was able to grind such marvelous little lenses or observe such tiny organisms.

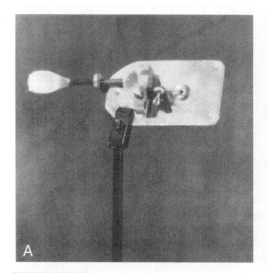

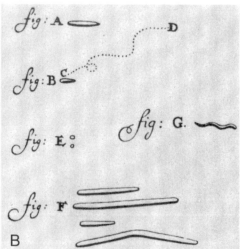

INSIGHT FIGURE 1–1. (A) Leeuwenhoek used a microscope with a single biconvex lens to view bacteria suspended in a drop of liquid placed on a moveable pin. (B) Although his microscope was capable of only 200- to 300-fold magnification, Leeuwenhoek was able to create these remarkable drawings of different bacteria types. (Volk WA, et al.: Essentials of Medical Microbiology, 4th ed. Philadelphia, JB Lippincott, 1990)

Box). Leeuwenhoek, the "Father of Microbiology," is so called because he was the first person to see live bacteria and protozoa. As a result of his microscopic observations of pepper water, tooth scrapings, gutter water, semen, blood, urine, and feces, in his letters to the Royal Society of London, he described the three general shapes of bacteria as well as protozoa, sperm, and blood cells. His letters convinced scientists of the late 17th century of the existence of microorganisms, which Leeuwenhoek referred to as "animalcules." He did not speculate on the origin of these microbes nor did he associate them with the cause of disease. Such relationships were not established until the work of Louis Pasteur and Robert Koch in the late 19th century. Leeuwenhoek's fine art of grinding a single lens that would magnify an object to 300 times its size was lost at his death because he had not taught this skill to anyone.

Detailed descriptions of microorganisms did not occur until the development of better microscopes in the 19th and 20th centuries. Modern microscopy is discussed at the end of this chapter.

Although Leeuwenhoek was probably not concerned about the origin of microorganisms, many other scientists were searching for an explanation of the appearance of living creatures in decaying meat, stagnating ponds, fermenting grain, and infected wounds. On the basis of observation, many of the "scientists" of that time believed that life could develop spontaneously from decomposing, nonliving material, a theory referred to as spontaneous generation or *abiogenesis*. For more than two centuries, from 1650 to 1850, this theory was debated and tested. Following the work of many others, Louis Pasteur (Fig. 1-5) and John Tyndall finally disproved the theory of spontaneous generation and proved that life must arise from preexisting life, referred to as the theory of *biogenesis,* first proposed by Rudolf Virchow in 1858.

Experiments devised by Pasteur and other scientists in the mid-1800s resulted in several major advances in the field of microbiology: (1) the concept that life must arise from preexisting life; (2) the techniques of sterilization and pasteurization; (3) an understanding of the biological process of fermentation; (4) the development of the germ theory of disease; and (5) the development of vaccines from inactivated (killed) anthrax bacteria and attenuated (weakened) rabies viruses.

The sterilization techniques used by Pasteur showed that boiled broth remains sterile until it is contaminated by particles in the air. While repeating Pasteur's experiment, Tyndall found bacterial forms, called endospores (described in Chap. 2), that were not destroyed by the first boiling and that germinated into reproducing (vegetative) bacteria. This discovery led to the development of the fractional sterilization process, often called *tyndallization,* in which endospores of bacteria are destroyed by boiling and cooling three times, allowing the spores to germinate between boilings.

When Pasteur was investigating reasons for the spoilage of beer and wine, he developed the "biological theory of fermentation," which states that a specific microbe produces a specific change in the substance on which it grows or a specific microorganism produces a specific fermentation product. Just as yeasts ferment sugar in grape juice to produce ethyl alcohol (ethanol) in wine, some contaminating bacteria, such as *Acetobacter,* may change the alcohol into acetic acid (vinegar); this, of course, ruins the taste of the wine. To eliminate harmful contaminating

FIGURE 1-5. Pasteur in his laboratory. A 1925 wood engraving by Timothy Cole. (Zigrosser C: *Medicine and the Artist* [Ars Medica]. New York, Dover Publications, Inc., 1970. By permission of the Philadelphia Museum of Art.)

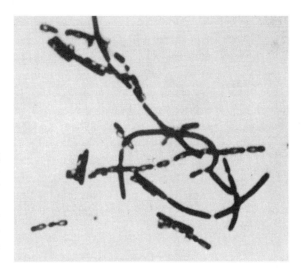

FIGURE 1-6. *Bacillus anthracis,* from a 48-hour culture, showing endospores. This bacterium causes anthrax and was used by Pasteur and Koch in many experiments on disease causation (original magnification ×1200). (Burrows W: Textbook of Microbiology. Philadelphia, WB Saunders, 1963)

bacteria from beer and wines, Pasteur heated them to 50° to 60°C (see footnote). This process, now called *pasteurization,* has been adapted to destroy pathogens in milk by heating it to 63°C for 30 minutes or to 72°C for 15 seconds.

By extending the biological theory of fermentation to animals and humans, Pasteur (together with others, such as Robert Koch) developed the "germ theory of disease," which states that each specific infectious disease is caused by a specific microorganism (Fig. 1-6). After isolating the causative pathogens of chicken cholera and rabies, Pasteur prepared vaccines against these diseases. He used attenuated or weakened pathogens, which were no longer pathogenic (capable of causing disease), but which made the injected animals immune to the disease. Pasteur used rabies virus from the brain and spinal cord of rabid dogs, for example, and discovered that by transferring the virus from rabbit to rabbit many times, the virus became so weakened that it no longer caused rabies in rabbits or dogs. Pasteur used this attenuated virus to immunize animals and people against rabies.

The beginnings of *immunology* (the study of the immune system) can be traced to ancient China, where a type of vaccination against smallpox was practiced, in which healthy people inhaled a powder made from the scabs of healing pustules of smallpox. Although many Chinese developed the disease as a result of this custom, others became immune. This method is called *variolation,* because the inactivated smallpox virus (variola) was used. In the late 1700s, Edward Jenner used cowpox virus (vaccinia) to immunize people against smallpox after he observed that milkmaids who caught cowpox, a mild disease transmitted to them from cows, were protected against smallpox, a far more serious disease. The words "vaccine" and "vaccination" come from the word vaccinia.

("C" is an abbreviation for Celsius. Although Celsius is also referred to as centigrade, Celsius is preferred. Appendix E contains formulas for converting Celsius to Fahrenheit and vice versa.)

Although Pasteur used the technique of isolating and then growing a microorganism in a pure culture in nutrient media in the laboratory, Robert Koch, a German physician, is usually given credit for most pure-culture research methods in microbiology. He and his assistant, Julius Petri, developed the *petri dish,* which is still in use today for microbial growth on solid media. At the suggestion of an associate's wife, Frau Hesse, they used agar (an extract from a red marine alga used at that time to make jelly) to solidify the growth medium so that distinct colonies of bacteria could be observed. Each bacterial colony contains millions of bacteria.

In addition, Robert Koch, in 1876, established an experimental procedure to prove the germ theory of disease, which states that "a specific infectious disease is caused by a specific pathogen." This scientific procedure is known as *Koch's postulates* (Fig. 1-7).

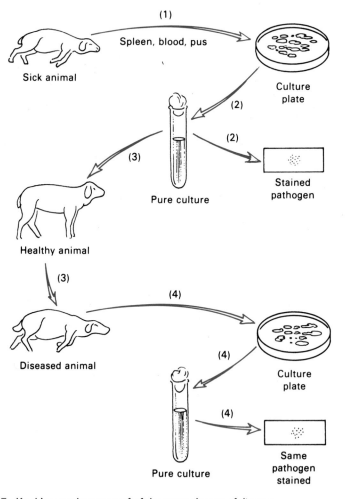

FIGURE 1-7. Koch's postulates: proof of the germ theory of disease.

KOCH'S POSTULATES:

1. The causative agent must be present in every case of the disease and must not be present in healthy animals.
2. The pathogen must be isolated from the diseased host animal and must be grown in pure culture.
3. The same disease must be produced when microbes from the pure culture are inoculated into healthy susceptible animals.
4. The same pathogen must be recoverable once again from this artificially infected host animal, and it must be able to be grown again in pure culture.

Koch's postulates not only proved the germ theory of disease, but also gave a tremendous boost to the development of microbiology by stressing laboratory culture and identification of microorganisms. However, some circumstances exist in which these postulates may not be easy to apply, as outlined below:

EXCEPTIONS TO KOCH'S POSTULATES:

- Many healthy people carry pathogens, but do not exhibit symptoms of the disease. These "carriers" may transmit the pathogens to others who then may become diseased. This is usually the case with epidemics of certain hospital-acquired (nosocomial) infections, gonorrhea, typhoid fever, diphtheria, pneumonia, and AIDS.
- Some microbes are very difficult or impossible to grow *in vitro* (in the laboratory) in artificial media; these include viruses, rickettsias, chlamydias, and the bacteria that cause leprosy and syphilis. Thus, pure cultures of such pathogens are difficult to obtain. However, many of these fastidious pathogens (those having complex nutritional requirements) can be grown in cultures of living human or animal cells of various types, in embryonated chicken eggs, or in certain animals. The leprosy pathogen thrives in armadillos; the spirochetes of syphilis grow well in the testes of rabbits and chimpanzees; and human immunodeficiency virus (HIV), also known as "the AIDS virus," proliferates in human lymphocyte cultures.
- To induce a disease from a pure culture, the experimental animal must be susceptible to that pathogen. Many animals, such as rats, are very resistant to microbial infections. Many pathogens are species-specific, which means that they grow in only one species of animal. For example, the pathogen that causes cholera in humans does not cause hog cholera, and *vice versa*. Because human volunteers are difficult to find and ethical reasons limit their use, the researcher may only be able to observe the changes caused by the pathogen in human cells that can be grown in the laboratory.
- Certain diseases develop only when an opportunistic pathogen invades a weakened host. These secondary invaders or opportunists cause disease in a person who is ill or recovering from another disease. Examples are pneumonia and ear infections, which may follow influenza. If researchers were

looking for the influenza virus, they might be misled by isolating the bacteria that caused the pneumonia.

It is also important to remember that not all diseases are caused by microorganisms. Many diseases, such as rickets and scurvy, result from diet deficiencies. Some diseases are inherited or are caused by an abnormality in the chromosomes, as in sickle cell anemia. Others, such as diabetes, result from malfunction of a body organ or system; still others, such as cancer of the lungs and skin, are influenced by environmental factors. However, all infectious diseases are caused by microorganisms.

The period of rapid development of microbiological techniques in the late 1800s is known as the "Golden Age of Microbiology." Efficient surgical and sterilization techniques were developed. During this period, a physician named Ignaz Semmelweis showed that puerperal sepsis (an infection that follows childbirth) was caused by infectious agents present on the hands of doctors and midwives. In 1847, he demonstrated that washing and disinfecting the hands with a solution of chlorinated lime greatly reduced the number of these infections. However, his work was not widely accepted. It was nearly 20 years later (1865) when Joseph Lister showed that there were fewer complications from infections following surgery and childbirth if surgical instruments were boiled and if the hands of surgeons and the wound were disinfected with carbolic acid (phenol) (Table 1-3).

This technique of using sterilization and disinfectants to prevent microorganisms from entering a surgical wound became known as *antiseptic surgery*. The application of antiseptic principles to surgery paved the way for many advances in surgical techniques, including *sterile techniques* (techniques that exclude all microorganisms) and *aseptic techniques* (techniques that exclude pathogens), which are practiced throughout the modern world in operating rooms and microbiology laboratories.

Modern nursing techniques logically followed a better understanding of the concepts of disease causation, transmission, and sterilization. Florence Nightingale, an English nurse of the 19th century, developed modern principles of nursing, methods of training nurses, and procedures for organizing hospitals to reduce the spread of disease.

THE TOOLS OF MICROBIOLOGY

Microscopes

Because microorganisms cannot be seen without the aid of powerful magnifying lenses or a microscope, the expansion of the field of microbiology beyond the advances made during the 19th century became dependent on the development of better microscopes to properly observe these organisms. Although such extremely small infectious agents as rabies and smallpox viruses were known to exist, they could not be seen until the electron microscope was developed.

TABLE 1-3. Major Contributors to the Development of Microbiology As a Science Before 1900

Contributor	Contribution	Date(s)
Antony van Leeuwenhoek	First to observe live microorganisms, using a simple microscope	1685
Francisco Redi	Demonstrated that animals do not arise spontaneously from dead organic matter	1660
Abbe Spallanzani	One of the first to demonstrate that heated broth, in the absence of air, did not support spontaneous generation	1770
Schröder and von Dusch	Demonstrated that broth heated in the presence of filtered air did not support spontaneous generation	1854
John Tyndall	Demonstrated that open tubes of broth remained free of bacteria if air was free of dust. Developed tyndallization to destroy spores	1860
Louis Pasteur	Disproved the theory of spontaneous generation (1861). Contributed to the understanding of fermentation (1858). Developed technique for selective destruction of microorganisms (pasteurization) (1866). Study of bacterial contamination of wine (1866) and diseases of silkworms (1868). Attenuated vaccines for anthrax (1881) and chicken cholera. Immunization against rabies (1885)	1855–1890s
Joseph Lister	Contributed to concept of aseptic technique	1865–1870
Robert Koch	Developed postulates for proving the cause of infectious disease (1884) and pure culture concept. Observed anthrax bacilli (1876). Developed solid culture media (1882). Discovered organisms causing tuberculosis (1882)	1870s to 1890s
Paul Ehrlich	Formulated humoral theory of resistance (see Chap. 11) Developed new staining techniques. Developed first chemotherapeutic agent	1890s to 1900
Elie Metchnikoff	Formulated cellular theory of resistance	1890s
Emil von Behring	Developed method for producing immunity by using antitoxin against diphtheria	1890s

LIGHT MICROSCOPES

A single-lens magnifying glass usually magnifies the image of an object from about 3 to 20 times the object's actual size. The *light microscope* used in laboratories today is a compound brightfield microscope with two magnifying-lens systems and a visible light source (hence the terms "light" or "brightfield" microscope) that passes through the specimen and the lenses to the observer's eye (Fig. 1-8). The eyepiece contains an ocular lens. The second magnifying-lens system is in the objective, which is positioned near the object to be viewed.

The two-lens system of the typical compound microscope can magnify 40 to 1000 times. The magnification is usually preceded by an "×", such as "×1000," in which "×" means "times." The total magnification of a compound microscope is

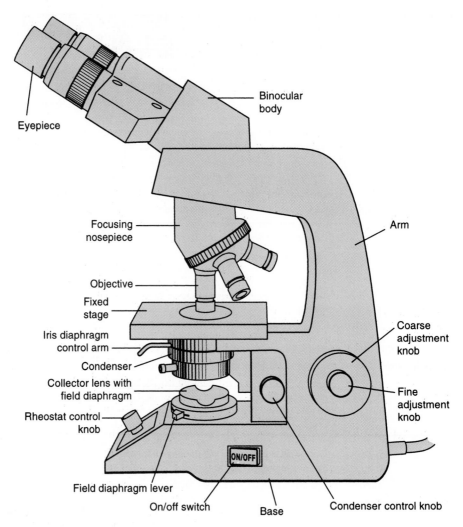

FIGURE 1-8. A modern light microscope.

obtained by multiplying the magnifying power of the ocular lens (usually ×10) by the magnifying power of the objective lens (usually ×4, ×10, ×40, or ×100). Thus, with the low-power (×10) objective in place, the total magnification is 10 multiplied by 10, or ×100. Usually this objective is used to locate the microorganism to be studied. With the high power or "high-dry" ×40 lens, the total magnification is 10 times 40, or ×400; this lens is used to study algae, protozoa, and other large microorganisms. With the oil-immersion (×100) objective, a total magnification of ×1000 is obtained, which is useful for observing the general characteristics of bacteria.

The oil-immersion objective must be used with a drop of immersion oil placed between the specimen and the objective lens; the oil reduces the scattering of light. For clear observation of the specimen, the light must be properly adjusted and focused. The condenser, located beneath the stage, focuses light onto the specimen, adjusts the amount of light, and shapes the cone of light entering the objective. Generally, the higher the magnification, the more light that is needed.

Magnification alone is of little value unless the enlarged image possesses increased detail and clarity. Image clarity is dependent upon the microscope's *resolving power* (or *resolution*), which is the ability of the lens to distinguish two adjacent points or objects between a particular distance. The resolving power is dependent upon the wavelength of the source of illumination and the numerical aperture (NA) of the objective lens. The greater the numerical aperture, the greater the resolving power. The resolving power of the unaided human eye is approximately 0.2 millimeters (mm). The resolving power of a compound microscope is approximately 0.2 micrometers (μm) when the oil immersion lens is used at a maximum numerical aperture (focus). This means that two bacteria could be distinguished as separate entities if they were separated by 0.2 μm or more; it also means that objects smaller than 0.2 μm cannot be seen using the compound light microscope. The resolving power of the compound microscope is approximately 1000 times better than the resolving power of the unaided human eye. In practical terms, this means that objects can be examined with the compound microscope that are up to 1000 times smaller than the smallest objects that can be seen with the unaided human eye.

Modified light microscopes are used for darkfield techniques, phase-contrast microscopy, and fluorescence microscopy. For darkfield microscopy, the regularly used condenser is replaced with a darkfield condenser, which provides a dark background (a "dark field") in place of the usual bright background ("bright field"). In darkfield microscopy, bacteria and other objects appear brightly illuminated against a dark background. This technique can be used to observe such very thin bacteria as the spirochete *Treponema pallidum*, which causes syphilis (Fig. 1-9). Primary syphilis (the first stage of syphilis) is usually diagnosed by examining patient specimens

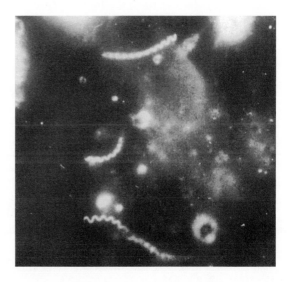

FIGURE 1-9. Spiral-shaped *Treponema pallidum,* the etiologic agent of syphilis, as seen by darkfield microscopy.

using darkfield microscopy. The phase-contrast microscope may be used to observe living microbes without staining because the light refracted by living cells is different from the surrounding medium, and, thus, they are more easily seen. The microscope used for fluorescence microscopy has an ultraviolet (UV) light source that illuminates the object, but does not pass into the objective of the microscope. When UV light strikes certain dyes and pigments, they emit a certain type of light; for example, different types of chlorophyll emit green, yellow, or orange light that can be seen in the microscope against a dark background. The UV light microscope is often used in immunology laboratories to show that antibodies stained with a fluorescent dye have combined with specific antigens on bacteria—a type of immunodiagnostic procedure (see Chapter 11).

ELECTRON MICROSCOPES

Electron microscopes use an electron beam as a source of illumination instead of visible light and they use magnets instead of lenses to focus the beam. In transmission electron microscopy, electrons pass through an extremely thin specimen, which is mounted in resin, and the image is seen on a fluorescent screen. The object may then be magnified up to approximately 1 million times—about 1000 times higher magnification than with light microscopes. Thus, even very tiny microbes (*e.g.,* viruses) may be observed. By examining thin sections of cells with the transmission electron microscope, microbes' internal structure can be studied. Another type of electron microscope—the scanning electron microscope—is very useful for observing surfaces and the three-dimensional structure of an object. Photographs taken using transmission and scanning electron microscopes are called transmission electron micrographs (TEMs) and scanning electron micrographs (SEMs). Figures 1-10, 1-11, and 1-12 show the differences in magnification and detail between electron micrographs and light photomicrographs. Table 1-4 lists the characteristics of various types of microscopes.

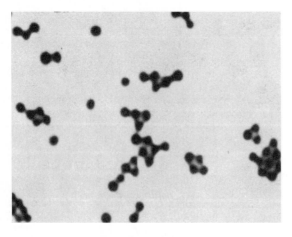

FIGURE 1-10. *Staphylococcus aureus,* as seen by light microscopy; original magnification ×1000. (Photograph courtesy of W. L. Wong)

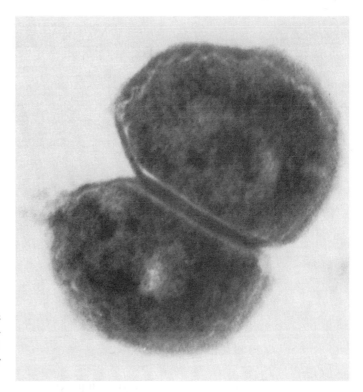

FIGURE 1-11. *Staphylococcus aureus,* as seen by transmission electron microscopy; original magnification, × 40,000. (Photograph courtesy of Ray Rupel)

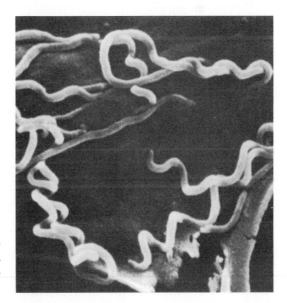

FIGURE 1-12. The three-dimensional qualities of scanning electron microscopy clearly reveal the corkscrew shape of cells of the syphilis-causing spirochete, *Treponema pallidum,* attached here to rabbit testicular cells grown in culture (original magnification ×8000).

TABLE 1-4. Characteristics of the Various Types of Microscopes

Type	Resolving Power	Useful Magnification	Characteristics
Brightfield	0.2000 μm	1,000	Used to observe morphology of microorganisms, such as bacteria, protozoa, fungi, and algae in living (unstained) and nonliving (stained) state. Cannot resolve organisms less than 0.2 μm, such as spirochetes and viruses.
Darkfield	0.2000 μm	1,000	Background is dark, and unstained organisms can be seen. Useful for examining spirochetes. Slightly more difficult to operate than brightfield.
Phase contrast	0.2000 μm	1,000	Can observe dense structures in living procaryotic and eucaryotic microorganisms.
Fluorescence	0.2000 μm	1,000	Fluorescent dye attached to organism. Primarily a diagnostic technique (immunofluorescence) to detect microorganisms in cells, tissue, and clinical specimens. Training required in specimen preparation and microscope operation.
Transmission electron microscope (TEM)	0.0002 μm (0.2 nm)	200,000	Specimen can be viewed on screen. Excellent resolution. Allows examination of cellular ultrastructure and viruses. Specimen is nonliving. Image is two dimensional.
Scanning electron microscope (SEM)	0.0200 μm (20 nm)	10,000	Specimen can be viewed on screen. Three-dimensional view of specimen. Useful in examining surface structure of cells and viruses. Specimen is nonliving. Resolution limited compared with TEM.

(ADAPTED FROM BOYD, 1988)

UNITS OF MEASURE

In microbiology, several common metric units are used to describe the size of microorganisms. The meter, the basic unit of the metric system, is equivalent to approximately 39 inches. The meter may be divided into 10 decimeters, 100 centimeters, 1000 millimeters, 1 million micrometers, or 1 billion nanometers. Relationships among these units are shown in Figure 1-13.

It should be noted that the old term "micron" (μ) has been replaced by the term *micrometer* (μm); the term "millimicron" (mμ) has been replaced by the term *nanometer* (nm); and an *angstrom* (Å) is 0.1 nanometer (0.1 nm), according to the International System adopted by scientists worldwide. On this scale, human red blood cells are about 7 μm in diameter. A typical spherical bacterium is approximately 1 μm in diameter. A typical rod-shaped bacterium is about 0.5 to 1 μm wide

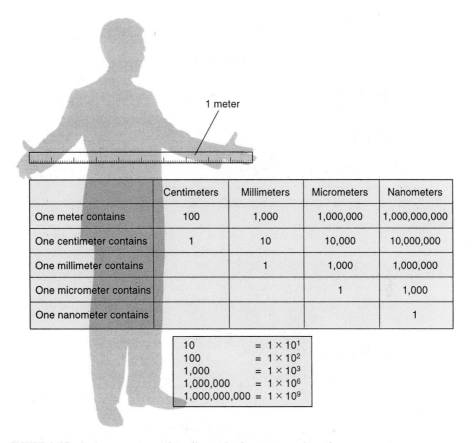

1 meter

	Centimeters	Millimeters	Micrometers	Nanometers
One meter contains	100	1,000	1,000,000	1,000,000,000
One centimeter contains	1	10	10,000	10,000,000
One millimeter contains		1	1,000	1,000,000
One micrometer contains			1	1,000
One nanometer contains				1

$$
\begin{aligned}
10 &= 1 \times 10^1 \\
100 &= 1 \times 10^2 \\
1{,}000 &= 1 \times 10^3 \\
1{,}000{,}000 &= 1 \times 10^6 \\
1{,}000{,}000{,}000 &= 1 \times 10^9
\end{aligned}
$$

FIGURE 1-13. Representations of metric units of measure and numbers.

× 1 to 3 μm long, although they can be smaller or may form very long filaments. Most viruses range in size from about 10 to 300 nm, although some (*e.g.*, Ebola virus) can be as large as 1000 nm (1 μm). Some very large protozoa reach a length of 2000 μm (2 mm).

The sizes of microorganisms are measured using an ocular micrometer, a tiny ruler within the eyepiece (ocular) of the light microscope. Before it can be used to measure objects, however, the ocular micrometer must first be calibrated using a microscope stage measuring device called a stage micrometer. Calibration must be performed for each of the objective lenses, to determine the distance between the marks on the ocular micrometer. The ocular micrometer can then be used to measure lengths and widths of microbes and other objects on the specimen slide.

Each type of microscope has its limits of visibility. The light microscope can be used for observation of cells larger than 0.2 μm, whereas the transmission electron microscope can discern objects as small as 0.2 nm (0.0002 μm) in diameter.

■ REVIEW OF KEY POINTS

- ■ Microorganisms, also called microbes, include bacteria, algae, protozoa, fungi, and viruses.
- ■ Microorganisms are ubiquitous (found virtually everywhere). Those that live on and in various parts of the human body are called our indigenous microflora.
- ■ Only a small percentage of microbes cause disease. Those that do are called pathogens and the diseases they cause are referred to as infectious diseases. Microorganisms that do not cause disease are called nonpathogens. Opportunistic pathogens do not cause disease under ordinary circumstances; however, they have the potential to cause disease should they gain access to the "wrong place" at the "wrong time."
- ■ Healthcare professionals must be particularly aware of pathogens, their sources, and how they may be transmitted from place to place, and person to person. Sterile, aseptic and antiseptic antimicrobial techniques are used everywhere in hospitals, operating rooms, and laboratories.
- ■ Persons having an interest in microbiology have many different areas in which to work, including medical, veterinary, and agricultural microbiology; microbiologists also play important roles in genetic engineering, antibiotic production, the food and beverage industries, space research, ecology, biodegradation, water treatment, and sewage disposal.
- ■ The development of simple and compound light microscopes enabled the discovery and visualization of microorganisms. Viruses and even smaller infectious agents can only be seen with electron microscopes.
- ■ Leeuwenhoek, Virchow, Pasteur, Tyndall, Koch, and many others contributed significantly to the development of the field of microbiology.
- ■ The metric system is used to describe the size of microorganisms. The sizes of bacteria and protozoa are expressed in micrometers (μm), whereas the sizes of viruses are expressed in nanometers (nm).

Problems and Questions

1. What does the study of microbiology include? What types of microorganisms?
2. Why is the study of microbiology important to people working in health occupations?
3. Why is Leeuwenhoek called the "Father of Microbiology"?
4. Why was the theory of spontaneous generation debated for 200 years?
5. What contributions did Pasteur make to microbiology?
6. What contributions to microbiology were made by Semmelweis, Lister, Jenner, and Nightingale?
7. What are Koch's postulates used to prove?
8. List Koch's postulates and the circumstances under which they might not be easily applied.
9. What types of microscopes are used to observe bacteria and viruses?
10. What units of measurement are used to describe the size of bacteria and viruses?

Self Test

After you have read Chapter 1, reviewed the chapter outline, examined the objectives, studied the new terms, and answered the problems and questions above, complete the following self test:

MATCHING EXERCISES

Complete each statement from the list of words provided within each section.

Historical Milestones of Microbiology

Leeuwenhoek	Nightingale	Petri
Tyndall	Hesse	Jenner
Lister	Janssen	Koch
Fracastorius	Pasteur	Semmelweis

1. Modern nursing techniques that reduce the spread of disease were developed by _Nightingale_.
2. The man who helped prove the germ theory of disease using laboratory procedures and a list of postulates was _Koch_.
3. The chemist who stated the biological theory of fermentation and the germ theory of disease and helped to disprove the theory of spontaneous generation was _Pasteur_
4. The physician who demonstrated the effectiveness of hand washing in reducing infections following childbirth was _Semmelweis_.
5. The "Father of Microbiology," who first described living microorganisms, was _Leeuwenhoek_.
6. The physician who described the transmission of diseases in 1546 was _Fracastorius_.
7. An early compound microscope was built by _Janssen_.

8. The process, which bears his name, of boiling and cooling repeatedly to destroy spores was developed by _Tyndall_.
9. The person who developed vaccines for chicken cholera and rabies was _Pasteur_.
10. Antiseptic surgical technique was proved effective by _Lister_.
11. A small, flat glass dish for the growth of microorganisms was developed by _Petri_.
12. The wife of a researcher in Koch's laboratory who suggested the use of agar growth medium was _Hesse_.
13. The physician who demonstrated that cowpox virus could be used to vaccinate against smallpox was _Jenner_.
14. The technique for pure culture research methods is attributed to the laboratory group led by _Koch_.

Some Types of Microbes

pathogenic	saprophytes	nonpathogenic
opportunists	nitrogen-fixing	prions
indigenous microflora	microbes	

1. The microorganisms that usually live on or within a person are called the _indigenous microflora_.
2. Microbes that are usually harmless but that may cause disease when a person's resistance is low are called _opportunists_.
3. Soil microbes that return nitrogen from the air to the soil are _nitrogen-fixing microbes_.
4. Bacteria and fungi that break down decaying organic materials to return plant nutrients to soil are called _saprophytes_.
5. Microorganisms that cause diseases are known as _pathogenic_ microorganisms.
6. Legumes are plants, such as clover, alfalfa, and peanuts, that have microbes living on or near their roots. Some of these microbes replenish the soil because they are _nitrogen-fixing microbes_.

Theories and Terms

abiogenesis	germ theory of	pasteurization
sterile	disease	pathogen
technique	biological theory of	pure culture
biogenesis	fermentation	saprophytes
contaminant	medium	sterilization
fermentation	micrometer	vaccination

1. The procedure for killing all microorganisms is _sterilization_.
2. The process used to destroy harmful microbes in milk and beer is _pasteurization_.
3. The surgical technique in which all microorganisms are excluded from the surgical field is _sterile technique_.

4. The theory stating that a specific disease is caused by a specific pathogen is the _germ theory of disease_

5. The theory stating that a specific microorganism produces a specific change in the material on which it grows is the _biolog. theory of fermentation_

6. The theory explaining that life must arise from preexisting life is the theory of _biogenesis_.

7. The technique of isolating and growing one species of microorganism on a growth medium results in a _pure culture_.

8. For centuries it was believed that living animals could arise spontaneously from nonliving material. This is the theory of _abiogenesis_.

9. The artificial process by which people can be made immune to certain diseases is _vaccination_.

10. The unit of measurement equal to 1/1000 of a millimeter is a _micrometer_.

11. A _pathogen_ is a disease-causing microbe.

12. The breakdown of carbohydrates to produce alcohol is _fermentation_

13. A culture containing only one species of organisms is a _pure culture_.

14. The liquid or solid within or on which microorganisms are grown is a culture _medium_.

15. Organisms that live on dead or dying organic matter are _saprophytes_.

16. An unwanted organism in an otherwise pure culture is a _contaminant_.

17. The concept of a living organism originating from dead organic matter without prior existence of organisms of its own type is _abiogenesis_.

18. The _germ theory of disease_ is an explanation of the cause of disease based on the existence of pathogenic microorganisms.

TRUE OR FALSE (T OR F)

F 1. Most microorganisms are harmful to humans.

F 2. Most viruses can be seen with a light microscope.

_ 3. Pasteurization kills all organisms in milk.

F 4. Pasteur believed in abiogenesis.

F 5. All bacteria live on or in animals.

T 6. Sanitary microbiologists are concerned with the microorganisms found in sewage, garbage, water, and food.

T 7. Tyndallization is a process of repeated boiling and cooling to destroy spores.

F 8. Koch's postulates can be applied when studying animals with viral infections.

T 9. Pneumonia and ear infections are often caused by opportunistic bacteria that invade a weakened host.

F 10. Bacteria can best be described when using the low-power objective of a light microscope.

MULTIPLE CHOICE

1. To see viruses, one must use
   a. a light microscope.
   b. a phase-contrast microscope.
   c. an electron microscope.
   d. viruses are too small to be seen.

2. Magnification of ×1000 would be achieved by
   a. an ocular lens of ×10 and an objective lens of ×100.
   b. an ocular lens of ×15 and an objective lens of ×40.
   c. an ocular lens of ×100 and an objective lens of ×100.

3. Which of the following is an exception to Koch's postulates?
   a. The causative agent is present in every case of the disease.
   b. The causative agent may be present in healthy animals that are carriers.
   c. The pathogen inoculated into a healthy animal produces the disease.

4. The study of algae is called
   a. virology.
   b. phycology.
   c. mycology.
   d. algonology.

5. Pasteur is credited with all of the following *except*
   a. the development of most pure culture techniques.
   b. the germ theory of disease.
   c. pasteurization.
   d. the development of vaccines.

6. Attenuated viruses are used
   a. to grow pure cultures.
   b. for vaccines against viral diseases.
   c. to prove Koch's postulates.
   d. during tyndallization.

7. For routine diagnosis of primary syphilis, the spirochete *Treponema pallidum* is observed using
   a. a phase-contrast microscope.
   b. a fluorescence microscope.
   c. a darkfield microscope.
   d. a light ("brightfield") microscope.

8. Which of the following are types of light microscopes? (1) ultraviolet, (2) fluorescence, (3) electron, (4) darkfield, (5) phase-contrast.
   a. 1, 2, 4, and 5 only.
   b. 2, 3, and 4 only.
   c. 3, 4, and 5 only.
   d. 1 and 2 only.
   e. all of the above (1 through 5).

CHAPTER 2

Cell Structure and Taxonomy

CELLS: EUCARYOTES AND PROCARYOTES
Eucaryotic Cell Structure
  Cell Membrane
  Nucleus
  Cytoplasm
  Cell Wall
  Flagella and Cilia

Procaryotic Cell Structure
  Chromosome
  Cytoplasm
  Cytoplasmic Particles
  Cell Membrane
  Bacterial Cell Wall
  Capsules
  Flagella

Pili or Fimbriae
Spores (Endospores)
Differences Between Procaryotic and Eucaryotic Cells
TAXONOMY
Microbial Classification

OBJECTIVES

After studying this chapter, you should be able to:

- *State the cell theory*
- *Give a function for each part of a eucaryotic animal cell*
- *Cite a function for each part of a bacterial cell*
- *Explain the differences among plant, animal, and bacterial cells*
- *Distinguish between the 5-kingdom and 3-kingdom systems of classification*

NEW TERMS

Amphitrichous bacteria
Archaebacteria
Autolysis
Axial filament
Binary fission
Capsule
Cell
Cell membrane
Cell theory
Cell wall
Cellulose

Centrioles
Chitin
Chloroplast
Chromatin
Chromosome
Cilium (pl. *cilia*)
Conjugation
Cytology
Cytoplasm
Deoxyribonucleic acid (DNA)

Endoplasmic reticulum (ER)
Endospores
Eucaryotic cells
Fimbriae (sing. *fimbria*)
Flagella (sing. *flagellum*)
Flagellin
Gene
Gene product
Genus (pl. *genera*)
Glycocalyx
Golgi complex

Lophotrichous bacteria
Lysosomes
Lysozyme
Metabolism
Microtubules
Mitochondria (sing.
  mitochrondrion)
Mitosis
Monotrichous bacteria
Negative stain
Nuclear membrane
Nucleolus
Nucleoplasm
Nucleus (pl. *nuclei*)

Organelles
Peptidoglycan
Peritrichous bacteria
Phagocyte
Phagocytosis
Photosynthesis
Pili (sing. *pilus*)
Plasmid
Plastid
Polyribosomes
Procaryotic cells
Protoplasm
Ribonucleic acid (RNA)
Ribosomes

Rough endoplasmic
  reticulum (RER)
Selective permeability
Sex pilus
Slime layer
Smooth endoplasmic
  reticulum (SER)
Species (pl. *species*)
Specific epithet
Spirochetes
Sporulation
Taxa (sing., *taxon*)
Taxonomy
Teichoic acids

CELLS: EUCARYOTES AND PROCARYOTES

In 1665, while peering through his crude microscope, Robert Hooke observed the small empty chambers in the structure of cork. He named them *cells* because they reminded him of the bare rooms (called cells) in a monastery. More than a century later, when biologists had access to more advanced microscopes, they found that cells are not empty, but rather contain a sticky (viscous) fluid. They called this material *protoplasm,* meaning "the substance of life." Now we know that a cell is composed of many different substances and contains tiny structures and particles called *organelles* that have important functions. The living material of the cell is still occasionally called protoplasm. It consists of two parts: *cytoplasm,* which lies outside the nucleus; and *nucleoplasm,* which lies inside the nucleus. Various chemicals within the cytoplasm and nucleoplasm enable the cell to live and reproduce.

Around 1838–1839, a German botanist named Matthias Schleiden and a German physiologist named Theodor Schwann concluded that all plant and animal tissues were composed of cells. Then in 1858, the German pathologist Rudolf Virchow refined the *cell theory*—stating that "all cells arise from cells" and that the cell is the basic unit of life. In other words, life must arise from preexisting life. The cell theory does not address the issue of the origin of living cells, a complex topic about which much has been written.

In biology, the cell is defined as the fundamental living unit of any organism because, like the organism, the cell exhibits the basic characteristics of life. A cell obtains food from the environment to produce energy and nutrients for metabolism (Fig. 2-1). *Metabolism* is an inclusive term to describe all the chemical reactions that occur within a cell (see Chapter 6 for a detailed discussion of metabolism). Through

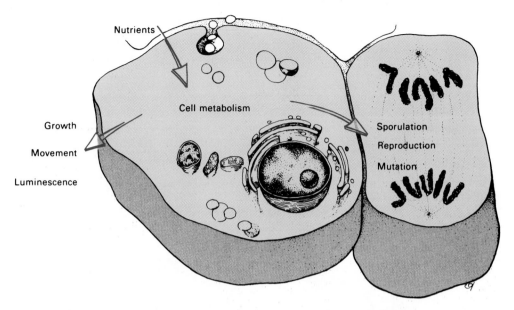

FIGURE 2-1. Cell metabolism. The metabolic reactions within cells enable them to use nutrients for growth, movement, luminescence, sporulation, reproduction, and mutation.

its metabolism, a cell can grow and reproduce. It can respond to such changes in its environment as light, heat, cold, and the presence of chemicals. It can mutate (change) as a result of accidental changes in its genetic material—the *DNA* (*deoxyribonucleic acid*) that makes up the genes of its *chromosomes*—and thus become better or less suited to its environment. As a result of these genetic changes, the mutant organism may be better adapted for survival and development into a new species of organism.

Considerable evidence exists to indicate that almost 4 billion years ago the first bit of life to appear on earth was a very primitive cell similar to the simple bacteria of today. Bacterial cells exhibit all the characteristics of life, although they do not have the complex system of membranes and organelles found in the more-advanced single-celled organisms. These less-complex cells, which include eubacteria and archaebacteria, are called *procaryotes* or procaryotic cells. The more complex cells, containing a true nucleus and many membrane-bound organelles, are called *eucaryotes* or eucaryotic cells. Eucaryotes include such organisms as green, brown, and red algae; protozoa; fungi; plants; and animals, including humans. Some microorganisms are procaryotic, some are eucaryotic, and some are not cells at all (see Fig. 2-2).

Because they are composed of only a few genes protected by a protein coat and sometimes contain a few enzymes, viruses appear to be the result of regressive or reverse evolution. Viruses depend on the energy and metabolic machinery of a host cell to reproduce. Therefore, because they are not truly viable cells, they are usually placed in a completely separate category and are not classified with the simple procaryotic cells. Viruses are said to be acellular.

Microorganisms

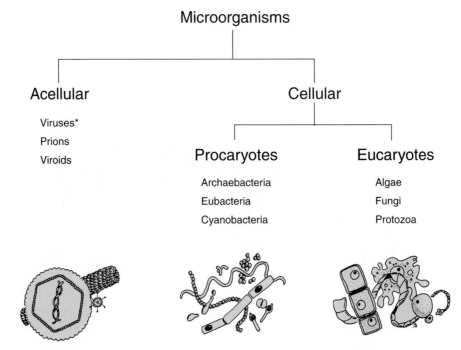

Acellular

Viruses*
Prions
Viroids

Cellular

Procaryotes

Archaebacteria
Eubacteria
Cyanobacteria

Eucaryotes

Algae
Fungi
Protozoa

FIGURE 2-2. Acellular and cellular microbes. Acellular microbes include viruses, prions, and viroids. Cellular microbes include the less complex procaryotes (bacteria) and the more complex eucaryotes (like algae, fungi, protozoa, and multicellular helminths).

For those in the health professions, it is important to understand the structure of different types of cells, not only to classify the microorganisms, but also to understand differences in their structure and metabolism. These factors must be known before we can determine or explain how the chemicals of modern chemotherapy can destroy pathogens, but not healthy human cells.

Cytology, the study of the structure and function of cells, has developed during the past 40 years with the aid of the electron microscope and sophisticated biochemical research. Many books have been written about the details of these tiny functional factories, but only a brief discussion of their structure and activities is presented here.

Eucaryotic Cell Structure

Eucaryotes (eu means true; *caryo* refers to a nut or nucleus) are so named because they have a true nucleus, in that their DNA is enclosed by a *nuclear membrane.* Most animal and plant cells are 10 to 30 μm in diameter, about 10 times larger than most procaryotic cells. Figure 2-3 illustrates a typical eucaryotic animal cell. This illustration is a composite of most of the structures that might be found in the various types of human body cells. The electron photomicrograph in Figure 2-4 is of an actual yeast cell. A discussion of the functional parts of eucaryotic cells can be better understood by keeping the illustrated structures in mind.

CELL MEMBRANE

The cell is enclosed and held intact by the *cell membrane,* which is also often called the plasma membrane, cytoplasmic membrane, or cellular membrane. Structurally, it is a mosaic composed of large molecules of proteins and phospholipids (certain types of fats). These large molecules regulate the passage of nutrients, waste products, and secretions across the cellular membrane. Because the cell membrane has the property of *selective permeability,* only certain substances may enter and leave the cell. The cell membrane is similar in structure and function to all the other membranes that are a part of the organelles of eucaryotic cells.

NUCLEUS

The organelle within the cell that unifies, controls, and integrates the functions of the entire cell is the *nucleus,* which is enclosed in the nuclear membrane. The nucleus

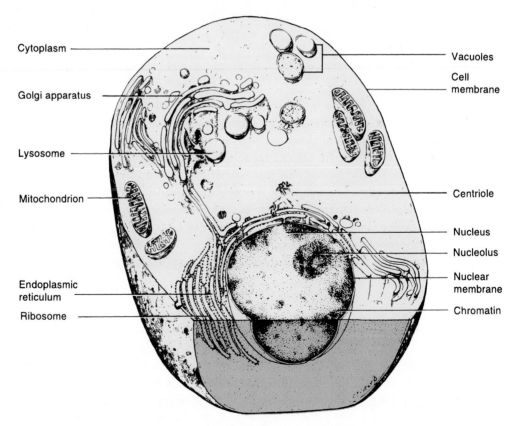

FIGURE 2-3. A typical eucaryotic animal cell.

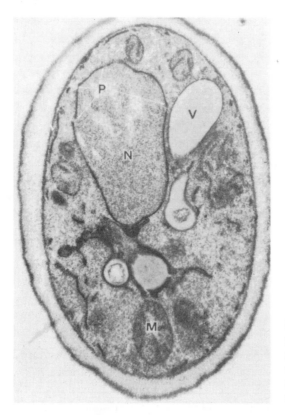

FIGURE 2-4. Cross section through a yeast cell showing the nucleus (*N*) with pores (*P*), mitochondrion (*M*), and vacuole (*V*). The cytoplasm is surrounded by the cell membrane. The thick, outer portion is the cell wall.

contains one or more chromosomes. The number and composition of chromosomes and the number of genes on each chromosome are characteristic of the particular species of organism. Different species have different numbers and sizes of chromosomes. Human diploid cells, for example, have 46 chromosomes (23 pairs), each consisting of thousands of genes. A *gene* is the unit that codes for, or determines, a particular trait or characteristic of an individual organism. Each gene contains the genetic code used by the cell to produce a *gene product*. Most gene products are proteins, but genes also code for the production of ribosomal *ribonucleic acid* (rRNA) and transfer ribonucleic acid (tRNA) molecules (discussed in Chapter 5). When genes are broken apart chemically, they are found to be coiled strands of DNA and proteins. DNA contains the genetic information for the production of essential proteins that enable the cell to function properly. To understand more about how the base coding of the DNA can chemically control the entire organism, refer to Chapter 5.

Close observation of the nucleus of nondividing cells reveals the *chromatin*, which are loosely wound strands of chromosomes that are suspended in the nucleoplasm. Nucleoplasm is the nutrient gelatinous matrix or base material of the nucleus. These chromatin strands condense into tightly coiled chromosomes just before the cell divides. Chromatin and chromosomes are composed of DNA and proteins called histones. In the very dense dark area of the nucleus, called the *nucleolus,* rRNA molecules are manufactured, which then move to the cytoplasmic portion of the cell. The nuclear membrane contains holes (nuclear pores) through which molecules may pass into or out of the nucleus.

CYTOPLASM

The part of the cell where most of the work is performed is the *cytoplasm,* which is controlled by information carried in the DNA of the nucleus. Cytoplasm is the cellular material (a type of protoplasm) outside the nucleus, enclosed by the cell membrane. It is composed of a semifluid, gelatinous, nutrient matrix (referred to as the cytosol); insoluble storage granules; and a variety of cytoplasmic organelles, including endoplasmic reticulum, ribosomes, Golgi complex, mitochondria, centrioles, microtubules, lysosomes, and other membrane-bound vacuoles. Each of these organelles has a highly specific function, and all of the functions are interrelated to maintain the cell and allow it to properly perform its activities.

The *endoplasmic reticulum* (ER) is a highly convoluted system of membranes that is interconnected and arranged to form a network of tubules connecting the outside of the cell to the nucleus. The ER transports nutrients to the nucleus and also provides some structural support for the cell. Much of the ER has a rough, granular appearance when observed by electron microscopy, and is designated as *rough endoplasmic reticulum* (RER). This rough appearance is due to the many *ribosomes* attached to the outer surface of the membranes. Eucaryotic ribosomes are 18 to 22 nm in diameter. They consist mainly of ribosomal RNA (rRNA) and protein and play an important part in the synthesis (manufacture) of essential proteins. Clusters of ribosomes (called polysomes) are sometimes observed held together by a molecule of messenger RNA (mRNA). Endoplasmic reticulum to which ribosomes are not attached is called *smooth endoplasmic reticulum* (SER).

Each eucaryotic ribosome is composed of two subunits—a large subunit (the 40S subunit) and a small subunit (the 60S subunit)—that are produced in the nucleolus. The subunits are then transported to the cytoplasm where they remain separate until such time as they join together with a messenger RNA (mRNA) molecule to initiate protein synthesis (Chapter 5). When united, the 40S and 60S subunits form an 80S ribosome. The "S" refers to Svedberg units and 40S, 60S, and 80S are sedimentation coefficients. A sedimentation coefficient expresses the rate at which a particle or molecule moves in a centrifugal field; it is determined by the size and shape of the particle or molecule.

The *Golgi complex,* also known as the Golgi apparatus or Golgi body, usually connects or communicates with the ER. This stack of flattened membranous vesicles completes the synthesis of secretory products and packages them into small, membrane-enclosed sacs called vesicles for storage within the cell or export outside the cell (exocytosis or secretion).

Lysosomes are small (about 1 μm diameter) vesicles that originate from the Golgi complex. They contain *lysozyme* and other digestive enzymes that break down foreign material taken into the cell by *phagocytosis* (the engulfing of large particles by

INSIGHT
The Origin of Mitochondria and Chloroplasts

Symbiosis is the living together or close association of two dissimilar organisms, usually two different species. In such a relationship, each party is referred to as a symbiont. Endosymbionts are organisms that live *inside* of other organisms, the latter of which are referred to as hosts.

Many scientists believe that the mitochondria and chloroplasts of eucaryotic cells were originally derived from bacterial endosymbionts—bacteria that once led a free-living, independent existence. The theory known as the serial endosymbiosis hypothesis proposes that, at some point in time—perhaps 1.5 billion years ago—certain bacteria were engulfed (phagocytized) by other procaryotic cells. At first, the engulfed bacteria continued to live an independent existence within the host cells. But, in time, an interdependence developed between the two organisms and the endosymbionts developed into the organelles known as mitochondria and chloroplasts.

Most of the evidence for the serial endosymbiosis theory is based upon similarities between these organelles and bacteria. Mitochondria possess a circular chromosome, a specific type of RNA, and ribosomes which are very much like those of bacteria and, similar to bacteria, mitochondria arise only from preexisting mitochondria. Chloroplasts are very much like photosynthetic bacteria. They contain DNA and ribosomes quite similar to those found in bacteria and they too arise independently of other organelles. This theory becomes even more plausible when one considers that many simple marine animals and protists existing today contain photosynthetic endosymbionts. Based upon 16S rRNA sequence data, the most likely candidates to have evolved into mitochondria and chloroplasts are purple bacteria and cyanobacteria, respectively (see text for information on 16S sequences).

Not all scientists agree with the serial endosymbiosis theory, however. Another theory—the autogenous hypothesis—states that mitochondria and chloroplasts, as well as other membranous structures found within eucaryotic cells, were derived from the cytoplasmic membrane. Undoubtedly, additional research will determine which of these hypotheses is correct.

amebas and certain types of white blood cells called *phagocytes*). These enzymes also aid in breaking down worn out parts of the cell and may destroy the entire cell by a process called *autolysis* if the cell is damaged or deteriorating.

The energy necessary for cellular function is provided by the formation of such high-energy phosphate molecules as adenosine triphosphate (ATP). The ATP molecules are the major energy carrying or energy storing molecules within cells. The *mitochondria* are the "power plants" or "energy factories" of the cell where most of the ATP molecules are formed by cellular respiration. During this process, energy is released from glucose molecules and other food nutrients to drive other cellular functions (see Chapter 6). The number of mitochondria in a cell varies greatly depending on the activities required of that cell. Mitochondria are about 1 μm in diameter. Some scientists believe that mitochondria and chloroplasts arose from bacteria living within eucaryotic cells (see Insight Box).

Two cylindrical organelles called *centrioles* lie perpendicular to each other near the nucleus. Centrioles are involved in the formation of spindle fibers for eucaryotic cell division. This process, which results in two daughter cells with the same number of chromosomes as the parent cell, is called *mitosis,* or "the dance of the chromosomes." Eucaryotic cilia and flagella also appear to arise from centriole material as their complex internal protein fibril configuration is very similar.

Other structures that may be present in the cell include microfilaments, microtubules, granules, and vacuoles containing food, secretory products, and pigments. *Plastids* are membrane-bound structures containing various photosynthetic pigments. They are the sites of photosynthesis. *Chloroplasts,* one type of plastid, contain a green, photosynthetic pigment called chlorophyll. Chloroplasts are always found in plant cells and algae. *Photosynthesis* is the process by which light energy is used to convert carbon dioxide and water into carbohydrates and oxygen (Chapter 6). Plant cells contain mitochondria and plastids.

CELL WALL

Some eucaryotic cells contain *cell walls,* external structures that provide rigidity, shape, and protection (Fig. 2-5). Eucaryotic cell walls, which are much simpler in structure than procaryotic (bacterial) cell walls, may contain cellulose, pectin, lignin, chitin, and some mineral salts (usually found in algae). Although the cell walls of plants and algae contain cellulose, those of fungi do not. The cell walls of fungi contain a polysaccharide called *chitin,* which is not found in the cell walls of other microorganisms.

FLAGELLA AND CILIA

Some eucaryotic cells (*e.g.,* spermatozoa and certain types of protozoa and algae) possess relatively long, thin structures called *flagella* (singular, *flagellum*). Such cells are said to be flagellated or motile; flagellated protozoa are called flagellates. The whipping motion of the flagella enables flagellated cells to "swim" through liquid environments. Thus, flagella are said to be organelles of locomotion (cell movement). Flagellated cells may possess one flagellum or two or more flagella. *Cilia* (singular, *cilium*) are also organelles of locomotion, but they tend to be shorter (hairlike), thinner, and more numerous than flagella. Cilia can be found on some species of protozoa (called ciliates) and on certain types of cells in our bodies (*e.g.,* the ciliated epithelial cells that line our respiratory tract). Unlike flagella, cilia tend to beat with a coordinated, rhythmic movement. Eucaryotic flagella and cilia, which

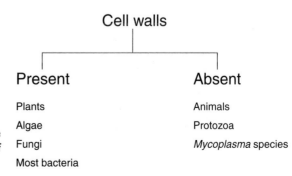

FIGURE 2-5. Presence or absence of cell walls in various types of cells.

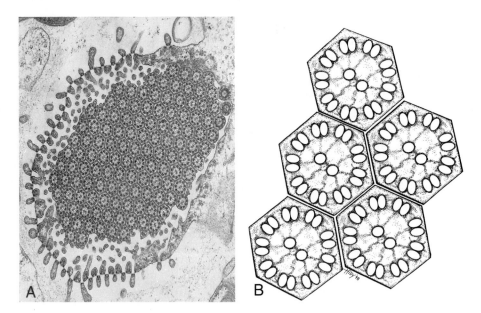

FIGURE 2-6. (*A*) Transmission electron micrograph showing the cross-section of a tapeworm flame cell (an excretory organ), containing numerous cilia. (*B*) Diagrammatic representation of cilia in cross-section, illustrating the 9 + 2 arrangement of microtubules.

contain an internal 9 + 2 arrangement of microtubules (Fig. 2-6) are structurally more complex than bacterial flagella. *Microtubules* are slender, tubular structures, composed primarily of a protein called tubulin.

Procaryotic Cell Structure

Procaryotic cells are about 10 times smaller than eucaryotic cells. A typical *E. coli* cell is about 2 to 3 μm long and 1 μm wide. Structurally, *procaryotes* are very simple cells when compared with the eucaryotic system of membranes, yet they are able to carry on the necessary processes of life. In these cells, division is by *binary fission*—the simple division of a cell into two parts following formation of a separating membrane and cell wall. All bacteria are procaryotes.

Refer to Figures 2-7, 2-8, and 2-9 as the structure of procaryotic cells is discussed, from internal to external structures. Within the cytoplasm, the chromosome, mesosomes, ribosomes, polyribosomes, and other cytoplasmic particles can be seen. Unlike eucaryotic cells, the cytoplasm of procaryotic cells is not filled with internal membranes. The cytoplasm is surrounded by a cell membrane, a cell wall (usually), and sometimes a capsule or slime layer. These latter three structures make up the bacterial cell envelope. Depending on the particular genus and species of bacterium, flagella or pili (description follows) or both may be observed outside the cell envelope, and a spore may sometimes be seen within the cell.

CHROMOSOME

The procaryotic chromosome is not surrounded by a nuclear membrane, does not have a definite shape, and has fewer proteins associated with it than are associated with eucaryotic chromosomes. It usually consists of a single, long, supercoiled, circular DNA molecule, and serves as the control center of the bacterial cell, containing the genetic information needed for producing several thousand enzymes and other proteins. The procaryotic chromosome is capable of duplicating itself, guiding cell division, and directing cellular activities. The thin and tightly folded chromosome of *Escherichia coli* is about 1.5-mm long and only 2-nm wide. Because a typical *E. coli* cell is about 1 μm × 0.5 μm, its chromosome is approximately 1500

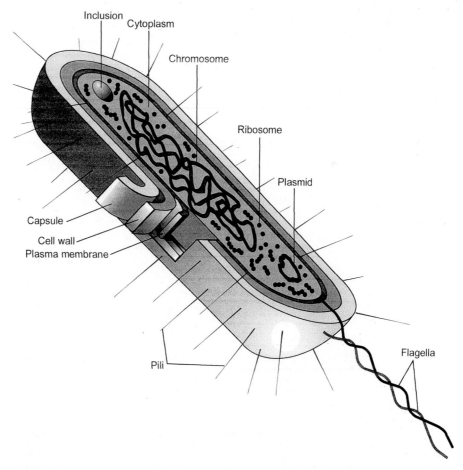

FIGURE 2-7. A typical procaryotic (bacterial) cell.

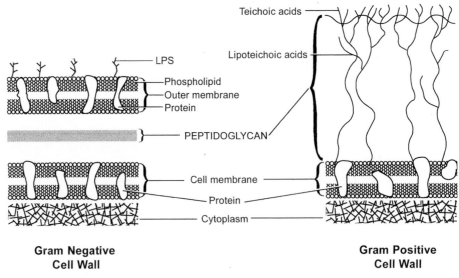

FIGURE 2-8. Differences between Gram-negative and Gram-positive cell walls. The relatively thin Gram-negative cell wall contains a thin layer of peptidoglycan, an outer membrane, and lipopolysaccharide (LPS). The thicker Gram-positive cell wall contains a thick layer of peptidoglycan and teichoic and lipoteichoic acids.

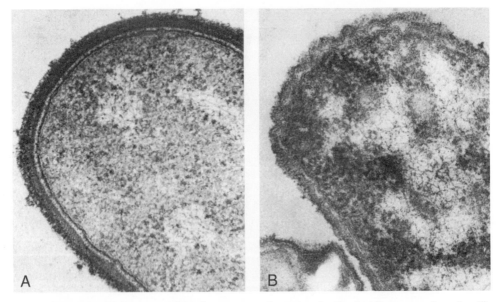

FIGURE 2-9. (*A*) A portion of the Gram-positive bacterium *Bacillus fastidiosus;* note the cell wall's thick peptidoglycan layer beneath which can be seen the cytoplasmic membrane. (*B*) The Gram-negative bacterium *Enterobacter aerogenes;* both the cytoplasmic membrane and the outer membrane are visible along some sections of the cell wall.

times longer than the cell itself—quite a packaging feat! Bacterial chromosomes contain from 3000 to 10,000 genes, depending upon the species. The DNA-occupied space within a bacterial cell is sometimes referred to as the bacterial nucleoid. Small, circular molecules of DNA that are not part of the chromosome (extrachromosomal DNA), called *plasmids,* may also be present in the cytoplasm of procaryotic cells. Plasmids are discussed in Chapter 6.

CYTOPLASM

The semiliquid cytoplasm, which surrounds the chromosome, is contained within the plasma membrane. Cytoplasm consists of water, enzymes, oxygen (in some cases), waste products, essential nutrients, proteins, carbohydrates, and lipids—a complex mixture of all the materials required by the cell for its metabolic functions.

CYTOPLASMIC PARTICLES

Within the bacterial cytoplasm, many tiny particles have been observed. Most of these are ribosomes, often occurring in clusters called *polyribosomes* (*poly* meaning many). Procaryotic ribosomes are smaller than eucaryotic ribosomes, but their function is the same—they are the sites of protein synthesis. A 70S procaryotic ribosome is composed of a 30S subunit and a 50S subunit. It has been estimated that there are about 15,000 ribosomes in the cytoplasm of an *E. coli* cell.

Cytoplasmic granules occur in certain species of bacteria. These may be stained, by use of a suitable stain, and then identified microscopically. The granules may consist of starch, lipids, sulfur, iron, or other stored substances.

CELL MEMBRANE

Enclosing the cytoplasm is the cytoplasmic, plasma, or cell membrane. This membrane is similar in structure to the eucaryotic cell membrane. Chemically, the plasma membrane consists of proteins and phospholipids, which are discussed further in Chapter 5. Being selectively permeable, the membrane controls which substances may enter or leave the cell. It is flexible and so thin that it cannot be seen with a light microscope. However, it is frequently observed in electron micrographs of bacteria.

Many enzymes are attached to the cell membrane and a variety of metabolic reactions take place there. Mesosomes are inward foldings of these membranes, and some scientists believe that this is where cellular respiration takes place in bacteria. This process is similar to that occurring in the mitochondria of eucaryotic cells, in which food nutrients are broken down to produce energy in the form of ATP molecules to be used in the cell's metabolic activities. On the other hand, some scientists think that mesosomes are nothing more than artifacts created during the processing of bacterial cells for electron microscopy.

In cyanobacteria and other photosynthetic bacteria (bacteria that use light as an energy source), some internal membranes, which are derived from the cell membrane, contain chlorophyll and other pigments that serve to trap light energy for photosynthesis. However, bacteria do not have complex internal membrane systems similar to the endoplasmic reticulum and Golgi complex of eucaryotic cells.

BACTERIAL CELL WALL

The rigid exterior cell wall that defines the shape of bacterial cells is chemically complex. Thus, they are quite different from the relatively simple structure of eucaryotic cell walls, although serving the same functions—primarily protection. The main constituent of most bacterial cell walls is a complex macromolecular polymer known as *peptidoglycan* (or murein), consisting of many polysaccharide chains linked together by small peptide (protein) chains. Peptidoglycan is only found in bacteria. The thickness of the cell wall and its exact composition vary with the species of bacteria. The cell walls of certain bacteria, called "Gram-positive bacteria" (to be explained later), have a thick layer of peptidoglycan combined with *teichoic acid*. The cell walls of "Gram-negative bacteria" (also explained later) have a much thinner layer of peptidoglycan, but this layer is covered with a complex layer of lipid macromolecules, usually referred to as the outer membrane, as shown in Figures 2-8 and 2-9. These macromolecules are discussed in Chapter 5. The cell walls of archaebacteria (described later) do not contain peptidoglycan, and bacteria in the genus *Mycoplasma* do not possess cell walls.

CAPSULES

Some bacteria have a layer of material (glycocalyx) outside the cell wall. *Glycocalyx* is a thick layer of slimy, gelatinous material produced by the plasma membrane and secreted outside of the cell wall. This layer of glycocalyx is called a *slime layer* if it is not highly organized and not firmly attached to the cell wall. It is called a *capsule* if it is highly organized and firmly attached. Capsules usually consist of complex sugars, or polysaccharides, which may combine with lipids and proteins, depending on the bacterial species. Knowledge of the chemical composition of capsules is useful in differentiating between different types of bacteria within a particular species; for example, different strains of *Haemophilus influenzae,* a cause of meningitis and ear infections in children, are identified by their capsular types. A vaccine, called Hib vaccine, is available for protection against disease caused by *H. influenzae* capsular type b.

Capsules can be detected using a *negative stain,* whereby the bacterial cell and background become stained, but the capsule remains unstained. Thus, the capsule appears as an unstained halo around the bacterial cell. Antigen-antibody tests (described in Chapter 11) may be used to identify specific strains of bacteria possessing unique capsular molecules called antigens.

Encapsulated bacteria usually produce colonies on nutrient agar that are smooth, mucoid, and glistening, and referred to as S-colonies. Nonencapsulated bacteria tend to grow as dry, rough colonies, called R-colonies. Bacterial capsules and slime layers may serve any of several functions. Slime layers enable certain bacteria to glide or slide along solid surfaces. Some capsules enable the bacterial cells to attach to mucous membranes and tooth surfaces so that they are not flushed away by body secretions. Frequently, capsules serve an antiphagocytic function, protecting the encapsulated bacteria from being phagocytized (ingested) by phagocytic white blood cells thus, allowing the bacteria to survive longer in the body.

FLAGELLA

Many bacteria have flagella, threadlike protein appendages with a whiplike motion that enable the bacteria to move. Flagellated bacteria are said to be motile, whereas nonflagellated bacteria are usually nonmotile. Bacterial flagella are about 10- to 20-nm thick.

The number and arrangement of flagella possessed by a certain type of bacterium are characteristics of the species and can, thus, be used for classification and identification purposes (Fig. 2-10). Bacteria possessing flagella over their entire surface (perimeter) are said to be *peritrichous* bacteria (Fig. 2-11). Those with a tuft of flagella at one or both ends are described as being *lophotrichous;* those having one flagellum at each end are *amphitrichous;* and those with a single polar flagellum are *monotrichous* bacteria.

Bacterial flagella consist of three, four, or more threads of protein (called *flagellin*) twisted like a rope, unlike eucaryotic flagella and cilia, which have a complex arrangement of internal microtubules running the length of the membrane-bound flagellum. The flagella of bacteria arise from a basal body in the cell membrane and project outward through the cell wall and capsule, as shown in Figure 2-7.

Some *spirochetes* (spiral-shaped bacteria) have two flagella-like fibrils called *axial filaments,* one attached to each end of the bacterium. These axial filaments

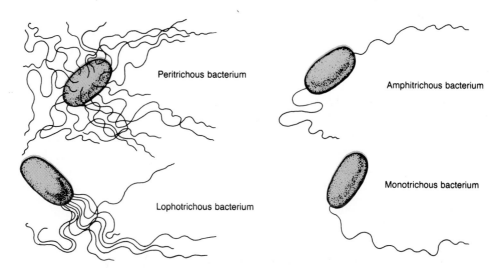

Peritrichous bacterium

Amphitrichous bacterium

Lophotrichous bacterium

Monotrichous bacterium

FIGURE 2-10. The four basic types of flagellar arrangement on bacteria: peritrichous = flagella all over surface; lophotrichous = a tuft of flagella at one or both ends; amphitrichous = one flagellum at each end; and monotrichous = one flagellum.

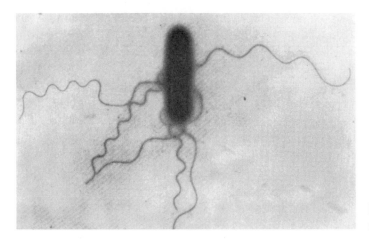

FIGURE 2-11. A peritrichous *Salmonella* cell.

extend toward each other, wrap around the organism between the layers of the cell wall, and overlap in the midsection. As a result, spirochetes can move in a spiral, helical, or inch-worm manner.

PILI OR FIMBRIAE

Pili (singular, *pilus*) or *fimbriae* (singular, *fimbria*) are hair-like structures most often observed on Gram-negative bacteria. They are much thinner than flagella, have a rigid structure, and are not associated with motility. These tiny appendages arise from the cytoplasm and extend through the plasma membrane, cell wall, and capsule (if present). There are two types of pili: one type enables bacteria to adhere or attach to surfaces; the other type (called a *sex pilus*) enables transfer of genetic material from one bacterial cell to another (discussed in Chapter 6).

The pili that enable bacteria to anchor themselves to surfaces (*e.g.*, tissues within the human body) are usually quite numerous (Fig. 2-12). In some species of bacteria, piliated strains (those possessing pili) are able to cause such diseases as urethritis and cystitis, whereas nonpiliated strains (those not possessing pili) are unable to cause these diseases.

A bacterial cell possessing a sex pilus (called a donor cell) is able to attach to another bacterial cell (called a recipient cell) by means of the sex pilus. Genetic material (usually in the form of a plasmid) is then transferred through the hollow sex pilus from the donor cell to the recipient cell—a process known as *conjugation* (described more fully in Chapter 6).

SPORES (ENDOSPORES)

A few genera of bacteria (*e.g.*, *Bacillus* and *Clostridium*) are capable of forming thick-walled spores as a means of survival when their moisture or nutrient supply is low. Bacterial spores are referred to as *endospores* and the process by which they are formed is called *sporulation*. During sporulation, the cell's genetic material becomes enclosed in several protein coats that are resistant to heat, cold, drying, and most chemicals. Spores have been shown to survive for many years in soil or dust, and some spores are quite resistant to disinfectants and boiling. When the dried spore lands on a moist, nutrient surface, it germinates, and a new vegetative bacterial cell (one capable of growing and dividing) emerges. Germination of a spore may be compared with germination of a seed. However, in bacteria, spore formation is related to the survival of the bacterial cell, not to reproduction. Although fungal spores are a means of reproduction, bacterial spores are a means of survival. Only one spore is produced in a bacterial cell and it germinates into only one vegetative bacterium (Fig. 2-13).

Differences Between Procaryotic and Eucaryotic Cells

Eucaryotic cells are divided into plant and animal types. Animal cells do not have a cell wall, whereas plant cells have a simple cell wall, usually containing cellulose. *Cellulose,* a type of polysaccharide, is a rigid polymer of glucose; polymers and polysaccharides are described in Chapter 5. Procaryotic cells have complex cell walls consisting of proteins, lipids, and polysaccharides. Eucaryotic cells contain membrane-bound organelles, such as the endoplasmic reticulum, whereas procaryotic cells

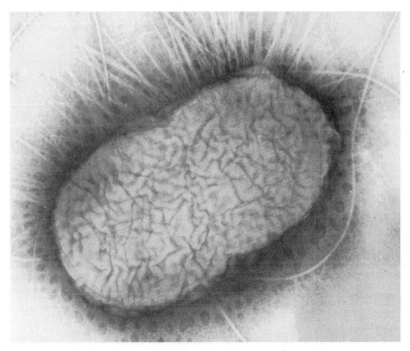

FIGURE 2-12. *Proteus vulgaris* cells possessing numerous short, straight pili and several longer, curved flagella.

FIGURE 2-13. A bacillus with a well-defined endospore.

have only a few (if any) cytoplasmic membranes (*e.g.,* mesosomes and photosynthetic membranes), which arise from the plasma membrane. Eucaryotic ribosomes (involved in protein synthesis) are larger and more dense (referred to as 80S ribosomes) than those found in procaryotes (70S ribosomes). The fact that 70S ribosomes are found in the mitochondria and chloroplasts of eucaryotes may indicate that these structures were derived from parasitic procaryotes during their evolutionary development. Other differences are listed in Table 2-1.

TAXONOMY

Taxonomy (the science of classification of living organisms) consists of three separate, but interrelated areas: classification, nomenclature, and identification. Classification is the arrangement of organisms into taxonomic groups (known as *taxa*) on the basis of similarities or relationships. Taxa include kingdoms, divisions, orders, classes, families, genera and species. Closely related organisms (*i.e.,* organisms having similar characteristics) are placed into the same taxon. Nomenclature is the assignment of names to the various taxa according to international rules. Identification is the process of determining that a new isolate belongs to one of the established, named taxa. When attempting to identify an organism that has been isolated from a patient specimen, laboratory technicians/technologists are very much like detectives. They gather "clues" (characteristics/properties/traits) about the organism until they have sufficient clues to identify (speciate) the organism. The "clues" that have been gathered will match the characteristics of an established species.

Microbial Classification

Since Aristotle's time, naturalists have attempted to name and classify plants, animals, and microorganisms in a meaningful way, based on their appearance and behavior. Thus, the science of *taxonomy* was established, based on the binomial system developed in the 18th century by the Swedish scientist, Carolus Linnaeus. In the binomial system, each organism is given two names (*e.g., Homo sapiens* for humans). The first name is the *genus* (plural, *genera*), and the second is the *specific epithet*. The first and second names together are referred to as the *species.*

Because written reference is often made to genera and species, biologists throughout the world have adopted a standard method of expressing these names that identify a specific organism. To express the genus, capitalize the first letter of the word and underline or italicize the whole word—for example, *Escherichia.* To express the species, capitalize the first letter of the genus name (the specific epithet is not capitalized) and then underline or italicize the entire species name—for example, *Escherichia coli.* Frequently, the genus is designated by a single letter abbreviation; in the example just given, *E. coli* indicates the species. In an essay or article about *Escherichia coli, Escherichia* would be spelled out the first time the organism is mentioned; thereafter, the abbreviated form, *E. coli,* could be used. The abbreviation

TABLE 2-1. Comparison Between Eucaryotic and Procaryotic Cells

	EUCARYOTIC CELLS		PROCARYOTIC CELLS
	Plant	Animal	
Biological distribution	All plants, fungi, and algae	All animals and protozoa	All bacteria
Nuclear membrane	Present	Present	Absent
Membranous structures other than cell membranes	Generally present	Generally present	Generally absent except mesosomes and photosynthetic membranes
Microtubules and centrioles	Present	Present	Absent
Cytoplasmic ribosomes (density)	80S	80S	70S
Chromosomes	Composed of DNA and proteins	Composed of DNA and proteins	Composed of DNA alone
Flagella or cilia	When present, have a complex structure similar to centrioles	When present, have a complex structure similar to centrioles	Flagella, when present, have a simple twisted protein structure; no cilia
Cell wall	When present, of simple chemical constitution, usually cellulose	Absent	Of complex chemical constitution, containing peptidoglycan
Active cytoplasmic movements	Present	Present	Absent (not observed)
Photosynthesis (chlorophyll)	Present	Absent	Present in cyanobacteria and some other bacteria

"sp." is used to designate a single species, whereas the abbreviation "spp." is used to designate more than one species. In addition to the proper scientific names for bacteria, acceptable terms like staphylococci (for *Staphylococcus* spp.), streptococci (for *Streptococcus* spp.), clostridia (for *Clostridium* spp.), pseudomonads (for *Pseudomonas* spp.), mycoplasmas (for *Mycoplasma* spp.), rickettsias (for *Rickettsia* spp.), and chlamydias (for *Chlamydia* spp.) are commonly used. Nicknames and slang terms frequently used within hospitals are GC and gonococci (for *Neisseria gonorrhoeae*), meningococci (for *Neisseria meningitidis*), pneumococci (for *Streptococcus pneumoniae*), staph (for staphylococci or staphylococcal), and strep (for streptococci or streptococcal). It is common to hear healthcare workers using terms like meningococcal meningitis, pneumococcal pneumonia, and staph infection.

Each organism is categorized into larger groups based on their similarities and differences. According to the presently popular Whittaker 5-Kingdom System of classification, all organisms are placed into five kingdoms: bacteria are in the Kingdom Procaryotae (or Monera), algae and protozoa are in the Kingdom Protista, fungi are in the Kingdom Fungi, plants are in the Kingdom Plantae, and animals are in the Kingdom Animalia. Viruses are not included because they are not living cells; they are acellular. Note that four of the five kingdoms consist of eucaryotic organisms. Each kingdom consists of divisions or phyla which, in turn, are divided into classes, orders, families, genera, and species (see Table 2-2). In some cases, species are subdivided into subspecies, their names consisting of a genus, a specific epithet, and a subspecific epithet. It should be noted that not all scientists agree with Whittaker's 5-Kingdom System of classification and that other taxonomic

TABLE 2-2. Comparison of Human and Bacterial Classification

	Human	*Escherichia coli*— a typical, medically-important, Gram-negative bacterium	*Staphylococcus aureus*— a typical, medically-important, Gram-positive bacterium
Kingdom	*Animalia*	*Procaryotae*	*Procaryotae*
Phylum (Division)	*Chordata*	*Gracilicutes*	*Firmicutes*
Class	*Mammalia*	*Scotobacteria*	*Firmibacteria*
Order	*Primates*	*Eubacteriales*	*Micrococcales*
Family	*Hominidae*	*Enterobacteriaceae*	*Micrococcaceae*
Genus	*Homo*	*Escherichia*	*Staphylococcus*
Species (note that a species has two names; the first name is the genus; the second is the specific epithet)	*Homo sapiens*	*Escherichia coli*	*Staphylococcus aureus*

classification schemes exist. Protozoa, for example, are sometimes placed into a subkingdom of the Animal Kingdom.

In 1978, Carl R. Woese devised a 3-Kingdom System of classification, which is gaining in popularity among some scientists. In this 3-Kingdom System, there are two kingdoms of bacteria (the kingdom Archaea or *Archaebacteria,* meaning "ancient" bacteria; and the kingdom Eubacteria, meaning "true" bacteria) and one kingdom (Eucarya) which includes all eucaryotic organisms. The 3-Kingdom System is based upon differences in the structure of certain ribosomal RNA (rRNA) molecules among organisms in the three kingdoms.

How do scientists determine how closely related one organism is to another? The most widely used technique for gauging diversity or relatedness involves the gene that codes for the so-called 16S rRNA molecule of ribosomes. Procaryotic ribosomes contain a single 16S rRNA molecule in the small (30S) subunit. In eucaryotes, the 16S rRNA molecule is part of mitochondria and chloroplast ribosomes. The gene that codes for the 16S rRNA molecule consists of a sequence of about 1500 DNA base pairs. If the 16S sequence of one organism is quite similar to the 16S sequence of another organism, then the organisms are closely related. The less similar the 16S sequences, the less related are the organisms. For example, the 16S sequence of a human is much more similar to that of a chimpanzee than to the 16S sequence of a fungus.

Perhaps taxonomists will someday combine the 3-Kingdom System and the 5-Kingdom System, producing a 6-Kingdom System consisting of the following six kingdoms: Archaea, Eubacteria, Protista, Fungi, Plantae, and Animalia.

■ REVIEW OF KEY POINTS

- The cell is the fundamental unit of any living organism; it exhibits the basic characteristics of life. All living organisms are composed of one or more cells.
- Complex eucaryotic cells contain membrane-bound organelles and a true nucleus, containing DNA. Procaryotic cells (*i.e.,* bacteria) exhibit all the characteristics of life but do not have a true nucleus or a complex system of membranes and organelles.
- Some eucaryotic cells have cell walls to provide rigidity, shape, and protection; these simple cell walls may contain cellulose, pectin, lignin, chitin, or mineral salts. Procaryotic bacterial cell walls are more complex, containing peptidoglycan and/or lipopolysaccharides.
- Mitochondria are the organelles that provide energy for eucaryotic cells. Procaryotic bacterial cells produce energy on their cell membranes.
- External to the cell wall, some bacteria have capsules or slime layers. Capsules serve an antiphagocytic function and have been used in the production of certain vaccines. Determining whether a bacterium possesses a capsule or not is of value when attempting to identify the organism.

- Many bacteria have flagella that enable motility and some produce spores for survival. Determining whether a bacterium possesses flagella or not is of value when attempting to identify the organism, as are the number and location of the flagella. Likewise, the presence or absence of spores is of value when identifying bacteria.
- In the binomial system of nomenclature, the first name is the genus, the second name is the specific epithet, and the two names together represent the species.
- Taxonomic classification of organisms separates them into kingdoms, divisions, orders, classes, families, genera and species, based on their characteristics, properties, and traits.
- In the 5-Kingdom System of classification, microorganisms are found in the first three kingdoms—Procaryotae (bacteria), Protista (algae and protozoa), and Fungi. In the 3-Kingdom System, microorganisms are found in all three kingdoms—Archaea, Eubacteria, and Eucarya.
- The most widely used technique for determining how closely one organism is related to another involves the gene that codes for the 16S rRNA molecule of ribosomes. The more similar the 16S sequences, the more closely related are the organisms. The less similar the 16S sequences, the less related are the organisms.

Problems and Questions

1. Which microbes are eucaryotic cells?
2. Which microorganisms are procaryotes?
3. What are the functions of mitochondria and chloroplasts?
4. Where does protein synthesis occur in eucaryotic cells? in procaryotic cells?
5. Which microorganisms do not have a nuclear membrane?
6. List five differences between eucaryotic animal cells and procaryotic bacterial cells.
7. What is the difference between bacterial and protozoal flagella?
8. Describe peritrichous, amphitrichous, lophotrichous, and monotrichous flagellation in bacteria.

Self Test

After you have read Chapter 2, reviewed the chapter outline, examined the objectives, studied the new terms, and answered the problems and questions above, complete the following self test.

MATCHING EXERCISES

Complete each statement from the list of words provided with each section.

Descriptive Terms

amphitrichous	eucaryotes	chitin
peptidoglycan	spirochete	monotrichous
procaryotes	cellulose	peritrichous
axial filaments	lophotrichous	

1. Cyanobacteria and bacteria are _____.
2. Bacteria with one flagellum at each end are _____.
3. A substance found in the cell walls of fungi that is not found in the cell walls of other types of microorganisms is _____.
4. Some spirochetes have two flagella-like fibrils attached at each end, called _____.
5. Fungi, protozoa, algae, plants, and animal cells are _____.
6. Bacteria that have a tuft of flagella at one or both ends are _____.
7. A substance found in the cell walls of bacteria that is not found in the cell walls of other types of microorganisms is _____.
8. A substance found in the cell walls of algae that is not found in the cell walls of other types of microorganisms is _____.
9. Cells that have no true nuclear membrane are _____.
10. Plant cells have cellulose-containing cell walls, and animal cells have no cell wall, but both are _____.
11. *Salmonella* organisms are motile because they have flagella covering their entire surfaces; they are _____.

12. Cells that have one "naked" chromosome composed of DNA are _____.

13. Some bacteria that live in the intestine have only one flagellum; they are _____.

14. Bacteria are classified as _____.

15. Cells that have membrane-bound organelles, such as mitochondria, endoplasmic reticulum, and Golgi bodies are _____.

TRUE OR FALSE (T OR F)

__   1. Mitochondria are the powerhouses that generate energy for eucaryotic cells.

__   2. The chromosomes of procaryotic cells consist of histone proteins and DNA.

__   3. Small, circular molecules of extrachromosomal DNA are called plastids.

__   4. Lysosomes contain lysozyme and other enzymes to aid in the digestion of dead or dying cells and bacteria.

__   5. The chromosome of cyanobacteria is surrounded by a well-defined nuclear membrane.

__   6. Bacterial flagella are structurally like protozoan flagella and cilia.

__   7. Plastids are the sites of protein synthesis.

__   8. In bacteria, ribosomes are found attached to the endoplasmic reticulum.

__   9. All bacteria have a cell wall.

__ 10. Ribosomes are found in both procaryotic and eucaryotic cells.

MULTIPLE CHOICE

1. The semipermeable structure controlling the transport of materials between the cell and its external environment is the
   a. cell wall
   b. protoplast
   c. cytoplasm
   d. plasma membrane

2. Centrioles are
   a. cylindrical organelles
   b. involved in cell division
   c. found in eucaryotes
   d. structurally similar to cilia and flagella
   e. all of the above

3. Sporulation in bacteria is
   a. a means of reproduction
   b. degeneration of organelles
   c. a means of survival
   d. development of a cell wall

4. In eucaryotic cells, _____ are the sites of photosynthesis.
   a. ribosomes
   b. golgi bodies
   c. plasmids
   d. plastids
   e. mitochondria

5. In binomial nomenclature, the second name is the
   a. genus
   b. species
   c. specific epithet
   d. kingdom

CHAPTER 3

Diversity of Microorganisms

CATEGORIES OF
MICROORGANISMS
BACTERIA
Characteristics
 Cell Morphology
 Staining
 Motility
 Colony Morphology
 Atmospheric Requirements
 Nutritional Requirements
 Biochemical and Metabolic
  Activities
 Pathogenicity
 Amino Acid Sequencing of Proteins
 Genetic Composition
RUDIMENTARY FORMS OF
BACTERIA

Rickettsias and
 Chlamydias
Mycoplasmas
PHOTOSYNTHETIC
BACTERIA
ARCHAEBACTERIA
ALGAE
PROTOZOA
Classification
FUNGI
Characteristics
 Reproduction
Classification of True Fungi
 Mushrooms
 Molds
 Yeasts

Dimorphism
Fungal Diseases
 Superficial and Cutaneous
  Mycoses
 Subcutaneous and Systemic
  Mycoses
LICHENS
SLIME MOLDS
ACELLULAR INFECTIOUS
AGENTS
Viruses
 Human Immunodeficiency
  Virus
 Bacteriophages
 Viruses and Genetic Changes
Viroids and Prions

OBJECTIVES

After studying this chapter, you should be able to:

- List the characteristics used to classify bacteria
- State the differences among rickettsias, chlamydias, and mycoplasmas
- Name several important bacterial diseases
- List the classes of protozoa and characteristics for classifying them
- List five pathogenic protozoa
- State some important characteristics of fungi
- List five diseases caused by fungi

- Discuss the important characteristics of procaryotic and eucaryotic algae
- Discuss the important characteristics that make algae different from protozoa and fungi
- Describe the characteristics used to classify viruses
- Compare some of the differences between viruses and bacteria
- List several important viral diseases

NEW TERMS

Acid-fast stain
Aerotolerant anaerobe
Ameba (pl. *amebae*)
Anaerobe
Bacillus (pl. *bacilli*)
Bacteriophage
Capnophile
Capsid
Capsomeres
Chitin
Ciliates (sing. *ciliate*)
Ciliophora
Coccobacilli
Coccus (pl. *cocci*)
Conidium (pl. *conidia*)
Cytostome
Dimorphism
Diplobacilli
Diplococci
Facultative anaerobe
Fastidious microbes

Flagellate
Gram stain
Hyphae (sing. *hypha*)
Inclusion bodies
L-forms
Lichen
Lysogenic bacteria
Lysogenic conversion
Lysogenic cycle
Lysogeny
Lytic cycle
Mastigophora
Microaerophiles
Mycelium (pl. *mycelia*)
Mycosis (pl. *mycoses*)
Mycotoxins
Obligate aerobe
Obligate anaerobe
Obligate intracellular
   pathogen
Octad

Pellicle
Pleomorphism
Prion
Prophage
Pseudopodium (pl.
   pseudopodia)
Sarcodina
Sarcomastigophora
Simple stain
Slime mold
Sporozoea
Staphylococci
Streptobacilli
Streptococci
Temperate
   bacteriophage
Tetrad
Vector
Virion
Viroid
Virulent bacteriophage

CATEGORIES OF MICROORGANISMS

Microorganisms can be divided into those that are truly cellular (bacteria, algae, protozoa, and fungi) and those that are acellular (viruses, viroids, and prions) [Refer back to Figure 2-2]. The cellular microorganisms can be subdivided into those that are procaryotic (bacteria) and those that are eucaryotic (algae, protozoa, and fungi). For a variety of reasons, acellular microorganisms are not considered to be living organisms by many scientists. Thus, viruses, viroids, and prions are often referred to as infectious agents or infectious particles rather than microorganisms.

BACTERIA

Characteristics

The bacteriologist's most important reference (sometimes referred to as the bacteriologist's "bible") is a four-volume set of books entitled *Bergey's Manual of Systematic Bacteriology,* which was published in 1984. An outline of these volumes can be found in Appendix A. Although about 3000 different bacteria are described in this reference, some authorities estimate that these represent only from less than 1% to a few percent of the total number of bacteria in nature. Using computers, microbiologists have established numerical taxonomy systems that not only help to identify bacteria by their characteristics, but also can help to establish how closely related these organisms are by comparing the composition of their genetic material and other cell substances.

Many characteristics of bacteria are examined to provide data for identification and classification. These characteristics include cell morphology (shape), staining reactions, motility, colony morphology, atmospheric requirements, nutritional requirements, biochemical and metabolic activities, pathogenicity, amino acid sequencing of proteins, and genetic composition.

CELL MORPHOLOGY

With the light microscope, the size, shape, and morphological arrangement of various bacteria are easily observed. Bacteria vary widely in size, usually ranging from

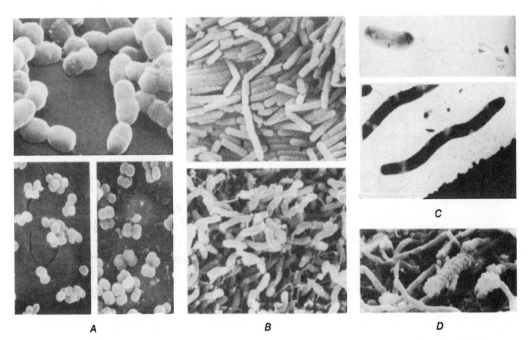

FIGURE 3-1. Forms of bacteria. (A) Cocci. *Top: Streptococcus mutans,* demonstrating pairs and short chains (original magnification ×9400). *Bottom left:* Single cells and small clusters of *Staphylococcus epidermidis* (original magnification ×3000). *Bottom right:* Pairs, tetrads, and regular clusters of *Micrococcus luteus* (original magnification ×3000). (B) Bacilli. *Top:* Single cells and short chains of *Bacillus cereus* (original magnification ×1700). *Bottom:* Flagellated bacilli (unnamed) associated with periodontitis (original magnification ×3700). (C) *Top:* A cell of *Vibrio cholerae;* note curved cell and single flagellum (original magnification ×8470). *Bottom:* The spirillum *Aquaspirillum bengal;* note polar tufts of flagella (original magnification ×2870). (D) Variety of organisms (cocci and bacilli) in dental plaque after 3 days without brushing (original magnification ×1360).

TABLE 3-1. Arrangements of Cocci			
Arrangement	**Description**	**Example**	**Disease**
Diplococci	Cocci in pairs	*Neisseria gonorrhoeae*	Gonorrhea
Streptococci	Cocci in chains	*Streptococcus pyogenes*	Strep throat
Staphylococci	Cocci in clusters	*Staphylococcus aureus*	Boils
Tetrads	Four cocci in a packet	*Micrococcus luteus*	Rarely pathogenic
Octads	Eight cocci in a packet	*Sarcina ventriculi*	Rarely pathogenic

spheres measuring about 0.2 μm in diameter to 10.0 μm long spiral-shaped bacteria to even longer filamentous bacteria. Some unusually large bacteria have also been discovered (see Insight Box). There are three basic shapes (as shown in Figure 3-1): (1) round or spherical bacteria—the *cocci* (singular, *coccus*); (2) rectangular or rod-shaped bacteria—the *bacilli* (singular, *bacillus*); and (3) curved and spiral-shaped bacteria (sometimes referred to as spirilla). Cocci are observed in various arrangements (*e.g.,* pairs or *diplococci,* chains or *streptococci,* clusters or *staphylococci,* packets of four or *tetrads,* packets of eight or *octads*), depending on the particular species and the manner in which the cells divide (Table 3-1). A typical coccus is approximately 1 μm in diameter.

INSIGHT
The World's Largest Bacterium?

The typical size of a spherical bacterium or coccus (*e.g.,* a *Staphylococcus aureus* cell) is 1 μm in diameter. A typical rod-shaped bacterium or bacillus (*e.g.,* an *Escherichia coli* cell) is about 0.5 to 1.0 μm wide × 1.0 to 3.0 μm long, although some bacilli are long thin filaments —up to about 12 μm in length or even longer—but still only about 1-μm wide. Thus, most bacteria are microscopic, requiring a microscope to be seen.

An enormous bacillus has been isolated from the intestines of the reef surgeonfish. Called *Epulopiscium fishelsonii,* this bacterium is over 600 μm long (over 0.6 mm in length). Thus, *Epulopiscium* cells, which are about five times longer than eucaryotic *Paramecium* cells, can be seen with the unaided human eye. The volume of an *Epulopiscium* cell is about a million times greater than the volume of a typical bacterial cell.

Although classified as a bacterium, this organism does not reproduce by binary fission as most other bacteria do. *Epulopiscium* cells produce intracellular daughter cells which are then released through a slit in the wall of the parent cell. Genetic studies have shown that *Epulopiscium* is most closely related to *Clostridium* species, which are sporeformers. In some ways, the method of reproduction in *Epulopiscium* is similar to the sporulation process.

Sporeforming bacteria called metabacteria, found in the intestines of herbivorous rodents, are also closely related to *Epulopiscium,* but they reach lengths of only 20 to 30 μm. Although shorter than *Epulopiscium,* metabacteria are much longer than most bacteria.

Bacilli (often referred to as rods) may be short or long, thick or thin, pointed or with curved or blunt ends. They may occur singly, in pairs (*diplobacilli*), in chains (*streptobacilli*), in long filaments, or branched. Some rods are very short, resembling elongated cocci; they are called *coccobacilli*. *Listeria monocytogenes*, a common cause of neonatal meningitis, is a coccobacillus. Some bacilli stack up next to each other, side by side in a palisade arrangement, which is characteristic of *Corynebacterium diphtheriae* (the cause of diphtheria) and organisms that resemble it in appearance (diphtheroids). A typical bacillus (like *E. coli*) is approximately 0.5 to 1.0 μm in width × 1.0 to 3.0 μm in length.

Curved and spiral-shaped bacilli are placed into a third morphological grouping. For example, *Vibrio* species, such as *Vibrio cholerae* (the cause of cholera), are curved (comma-shaped) bacilli. Spiral-shaped bacteria usually occur singly, but some species may form chains. Different species vary in size, length, rigidity, and the number and amplitude of their coils. Some have rigid cell walls that maintain the shape of a helix or coil. Spirochetes, such as *Treponema pallidum*, the causative agent of syphilis, have a flexible cell wall enabling them to move readily through tissues (refer back to Fig. 1-9). Its characteristic motility—spinning around its long axis—makes *T. pallidum* easy to recognize in clinical specimens obtained from patients with syphilis. Although less tightly coiled than *T. pallidum*, the etiologic agents of Lyme disease and relapsing fever—*Borrelia* species—are also spirochetes (Fig. 3-2).

Some bacteria may lose their characteristic shape because adverse growth conditions prevent the production of normal cell walls. Such cell wall-deficient bacteria are called *L-forms*. Some L-forms revert to their original shape when placed in

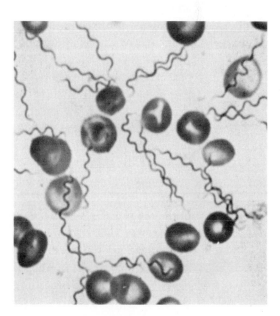

FIGURE 3-2. *Borrelia hermsii,* a cause of relapsing fever, in a Giemsa-stained blood smear.

favorable growth conditions, whereas others do not. Bacteria in the genus *Mycoplasma* do not have cell walls; thus, microscopically they appear in various shapes (known as *pleomorphism*). Mycoplasmas are tiny, have no rigid cell shape, reproduce slowly, are relatively fragile, are very susceptible to changes in osmotic pressure, and are resistant to antibiotics that inhibit cell wall synthesis.

STUDY AID   Bacteria having "coccus" in their names are cocci (spherical cells). Examples include genera such as *Enterococcus*, *Peptostreptococcus*, *Staphylococcus*, and *Streptococcus*. However, not all cocci have "coccus" in their names (*e.g.*, *Neisseria* spp.). Bacteria having "bacillus" in their names are bacilli (rod-shaped or rectangular cells). Examples include such genera as *Actinobacillus*, *Bacillus*, *Lactobacillus*, and *Streptobacillus*. However, not all bacilli have "bacillus" in their names (*e.g.*, *E. coli*).

STAINING

Because unstained bacteria are colorless, transparent, and difficult to see, various staining methods have been devised to examine bacteria. In preparation for staining, the bacteria are smeared onto a glass microscope slide, air-dried, and then "fixed." Fixation serves three purposes: (1) it kills the organisms, (2) it preserves their morphology (appearance), and (3) it anchors the smear to the slide. Specific stains and techniques are used to observe bacterial morphology (*e.g.*, size, shape, morphological arrangement, type of cell wall, nuclear material, capsules, flagella, endospores, fat globules, and various types of granules).

A *simple stain* is sufficient to determine bacterial shape and morphological arrangement (*e.g.*, pairs, chains, clusters). For this method, as shown in Figure 3-3, a dye (such as methylene blue) is applied to the fixed smear, rinsed, dried, and examined using the oil immersion lens of the microscope.

In 1884, Dr. Hans Gram developed a very important staining technique that bears his name—the *Gram stain*. This staining procedure, which differentiates between "Gram-positive" and "Gram-negative" bacteria, has four steps: (1) cover the fixed smear with crystal violet (a purple dye) for 30 seconds; (2) rinse gently with water and cover the smear with Gram's iodine solution for 30 seconds; (3) wash off the iodine with water and decolorize with ethanol; then (4) counterstain with safranin (a bright red dye) for 1 minute, rinse, dry, and examine using the oil immersion objective. The color of the bacteria at the end of the Gram-staining procedure (either blue-to-purple or pink-to-red) depends on the chemical composition of the cell wall (Table 3-2). Gram-positive bacteria retain the blue-to-purple color of the crystal violet. In Gram-negative cells, the crystal violet is removed during the decolorization step and the cells are subsequently stained pink-to-red by the safranin. Some strains of bacteria are neither consistently blue-to-purple nor pink-to-red following this procedure; they are referred to as Gram-variable bacteria. Examples of such bacteria are members of the genus *Mycobacterium*, such as *M.*

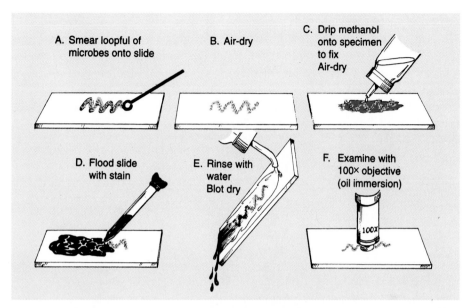

FIGURE 3-3. Simple bacterial staining technique. (*A*) With a flamed loop, smear a loopful of bacteria suspended in broth or water onto a slide. (*B*) Allow slide to air-dry. (*C*) Fix the smear with absolute (100%) methanol. (*D*) Flood the slide with the stain. (*E*) Rinse with water. (*F*) Blot dry with bibulous paper or paper towel. (*G*) Examine the slide with the ×100 microscope objective, using a drop of immersion oil directly on the smear.

tuberculosis and *M. leprae*. Appendix D contains additional information about the Gram staining procedure.

Mycobacterium species are more often identified using the *acid-fast stain*. In this method, carbol fuchsin (a bright red dye) is driven into the bacterial cell wall with heat so that the decolorizing agent (a mixture of acid and alcohol) does not remove the red color from the mycobacteria; because they are not decolorized by the acid-alcohol mixture, they are said to be acid-fast. Most other bacteria are decolorized by the acid-alcohol treatment, and are said to be non-acid-fast. The acid-fast stain is

TABLE 3-2. Differences Between Gram-Positive and Gram-Negative Bacteria		
	Gram-Positive Bacteria	**Gram-Negative Bacteria**
Color at the end of the Gram staining procedure	Blue-to-purple	Pink-to-red
Peptidoglycan in cell walls	Thick layer	Thin layer
Teichoic acids in cell walls	Present	Absent
Lipopolysaccharide in cell walls	Absent	Present

TABLE 3-3. Characteristics of Some Important Pathogenic Bacteria

Bacterium	Diseases	Type	Gram-Stain Reaction[a]
Bacillus anthracis	Anthrax	Spore-forming rod	+
Bordetella pertussis	Whooping cough	Rod	−
Brucella abortus and *B. melitensis*	Brucellosis, undulant fever	Rod	−
Chlamydia trachomatis	Lymphogranuloma venereum, trachoma	Coccoid	−
Clostridium botulinum	Botulism (food poisoning)	Spore-forming rod	+
Clostridium perfringens	Gas gangrene, wound infections	Spore-forming rod	+
Clostridium tetani	Tetanus (lockjaw)	Spore-forming rod	+
Corynebacterium diphtheriae	Diphtheria	Rod	+
Escherichia coli	Urinary tract infections	Rod	−
Francisella tularensis	Tularemia	Rod	−
Haemophilus ducreyi	Chancroid	Rod	−
Haemophilus influenzae	Meningitis, pneumonia	Rod	−
Klebsiella pneumoniae	Pneumonia	Rod	−
Mycobacterium leprae	Leprosy	Rod	+/−
Mycobacterium tuberculosis	Tuberculosis	Rod	+/−
Mycoplasma pneumoniae	Atypical pneumonia	Pleomorphic	−
Neisseria gonorrhoeae	Gonorrhea	Diplococcus	−
Neisseria meningitidis	Nasopharyngitis, meningitis	Diplococcus	−
Proteus vulgaris and *P. morgani*	Gastroenteritis, urinary tract infections	Rod	−
Pseudomonas aeruginosa	Respiratory, urogenital, and wound infections	Rod	−
Rickettsia rickettsii	Rocky Mountain spotted fever	Rod	−
Salmonella typhi	Typhoid fever	Rod	−
Salmonella species	Gastroenteritis	Rod	−
Shigella species	Shigellosis (bacillary dysentery)	Rod	−
Staphylococcus aureus	Boils, carbuncles, pneumonia, septicemia	Cocci in clusters	+
Streptococcus pyogenes	Strep throat, scarlet fever, rheumatic fever, septicemia	Cocci in chains	+
Streptococcus pneumoniae	Pneumonia, meningitis	Diplococcus	+
Treponema pallidum	Syphilis	Spirochete	−
Vibrio cholerae	Cholera	Curved rod	−
Yersinia pestis	Plague	Rod	−

[a]+ = gram-positive; − = gram-negative; +/− = gram-variable

TABLE 3-4. Types of Bacterial Staining Procedures		
Category	**Example(s)**	**Purpose**
Simple stains	Staining with methylene blue	Merely to stain the cells so that size, shape, and morphological arrangement can be determined
Structural stains	Capsule stains	To determine if the organism is encapsulated
	Flagella stains	To determine if the organism possesses flagella and, if so, their number and location on the cell
	Endospore stains	To determine if the organism is a sporeformer
Differential stains	Gram stain	To differentiate between gram-positive and gram-negative bacteria
	Acid-fast stain	To differentiate between acid-fast and non-acid-fast bacteria

especially useful in the tuberculosis laboratory ("TB lab") where the acid-fast mycobacteria are readily seen as red organisms against a blue or green background in a sputum specimen from a tuberculosis patient (see Color Figures 12 and 13 in this chapter). The Gram and acid-fast staining procedures are referred to as differential staining procedures because they enable microbiologists to differentiate one group of bacteria from another; *i.e.,* Gram-positive bacteria from Gram-negative bacteria, and acid-fast bacteria from non-acid-fast bacteria.

The procedures for staining bacteria to observe capsules, spores, and flagella are referred to as structural staining procedures; they are described in most microbiology laboratory manuals. Refer to Table 3-3 and the Color Figures for the staining characteristics of certain pathogens. Table 3-4 summarizes the various types of bacterial staining procedures.

MOTILITY

The ability of an organism to move by itself is called motility. Bacterial motility is usually associated with the presence of flagella or axial filaments, although some bacteria exhibit gliding motility on secreted slime on solid agar. Most spiral-shaped bacteria and about one half of the bacilli are motile by means of flagella, but cocci are generally nonmotile. A flagella stain can be used to demonstrate the presence, number, and location of flagella on bacterial cells. Various terms (*e.g.,* monotrichous, amphitrichous, lophotrichous, peritrichous) are used to describe the number and location of flagella on bacterial cells (if necessary, refer back to Chapter 2). Motility can be demonstrated by stabbing organisms into a tube of semisolid medium or by using the hanging-drop technique.

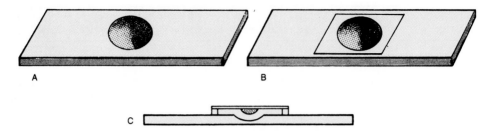

FIGURE 3-4. Hanging drop preparation for study of living bacteria. (*A*) Depression slide. (*B*) Depression slide with coverglass over the depression area. (*C*) Side view of hanging drop preparation, showing the drop of culture hanging from the center of the coverglass above the depression.

In semisolid medium, nonmotile organisms will grow only along the stab line, but motile organisms will spread away from the stab line. In the hanging drop method (Fig. 3-4), motile bacteria suspended in a hanging drop of liquid can be seen darting about in every direction within the drop, when examined microscopically.

COLONY MORPHOLOGY

A bacterial colony is a mound or pile of bacteria on an agar surface; it contains millions of organisms. The colony morphology (appearance of the colonies) of bacteria varies from one species to another. Colony morphology includes the size, color, overall shape, elevation, and consistency of the colony. As with cell morphology and staining characteristics, colony features serve as important "clues" in the identification (speciation) of bacteria. Size of colonies is determined by the organism's rate of growth (generation time), and is an important characteristic of a particular bacterial species. Colony morphology also includes the results of enzymatic activity on various types of culture media, such as those shown in Color Figures 14, 15, and 16 in this chapter.

ATMOSPHERIC REQUIREMENTS

In the microbiology laboratory, it is useful to classify bacteria on the basis of their relationship to oxygen (O_2) and carbon dioxide (CO_2). Regarding oxygen, a bacterial isolate can be classified into one of five major groups: obligate aerobes, microaerophilic aerobes (microaerophiles), facultative anaerobes, aerotolerant anaerobes, and obligate anaerobes. In a liquid medium, the region in which the organism grows depends on the oxygen needs of that particular species.

To grow and multiply, *obligate aerobes* require an atmosphere that contains molecular oxygen in concentrations comparable to that found in room air (*i.e.,* 20 to 21% O_2). Mycobacteria and certain fungi are examples of microorganisms that are obligate aerobes. *Microaerophiles* (microaerophilic aerobes) also require oxygen for multiplication, but in concentrations lower than that found in room air. *Neisseria gonorrhoeae* (the cause of gonorrhea) and *Campylobacter* species (causes of bacterial diarrhea) are examples of microaerophilic bacteria that prefer an atmosphere containing about 5% oxygen.

Anaerobes can be defined as organisms that do not require oxygen for life and reproduction. However, they vary in their sensitivity to oxygen. The terms obligate anaerobe, aerotolerant anaerobe, and facultative anaerobe are used to describe the organism's relationship with molecular oxygen. An *obligate anaerobe* is an anaerobe that only grows in an anaerobic environment. It will not grow in a microaerophilic environment (about 5% O_2), in a CO_2 incubator (about 15% O_2), or in air.

An *aerotolerant anaerobe* does not require oxygen, grows better in the absence of oxygen, but can survive in atmospheres containing molecular oxygen (such as air and a CO_2 incubator). *Facultative anaerobes* are capable of surviving in either the presence or absence of oxygen, anywhere from 0% O_2 to 20–21% O_2. Many of the bacteria routinely isolated from clinical specimens are facultative anaerobes (*e.g.,* members of the family *Enterobacteriaceae,* streptococci, staphylococci).

Room air contains less than 1% CO_2. Some bacteria, referred to as *capnophiles* (capnophilic organisms), grow better in the presence of increased concentrations of CO_2. Some anaerobes (*e.g., Bacteroides* and *Fusobacterium* species) are capnophiles,

INSIGHT
The Discovery of Anaerobes

Scientists once believed that life in the absence of oxygen was impossible, but we now know differently. We know that there are organisms—obligate anaerobes—that can live in the total absence of oxygen.

Three scientists deserve credit for the discovery of anaerobes—Leeuwenhoek, Spallanzani, and Pasteur. They each made scientific observations that contributed to our knowledge and understanding of anaerobes.

In 1680, Antony van Leeuwenhoek performed an experiment using pepper and sealed glass tubes. He wrote that "animalcules developed although the contained air must have been in minimal quantity."

Lazzaro Spallanzani, an Italian scientist, performed similar experiments in the latter half of the 18th century. He drew the air from microbe-containing glass tubes, fully expecting the microbes to die—but some did not. He wrote in a letter to a friend, "The nature of some of these animalcules is astonishing! They are able to exercise in a vacuum the functions they use in free air. . . . How wonderful this is! For we have always believed there is no living being that can live without the advantages air offers it."

It was Louis Pasteur who actually introduced the terms "aerobe" and "anaerobe." In an 1861 paper, he wrote "these infusorial animals are able to live and multiply indefinitely in the complete absence of air or free oxygen. . . . These infusoria can not only live in the absence of air, but air actually kills them. . . . I believe this is . . . the first example of an animal living in the absence of free oxygen." In 1877, Pasteur discovered a pathogenic anaerobe, the bacterium that today is known as *Clostridium septicum.* (To learn more about these scientists and their discoveries, you should read *Microbe Hunters* by Paul De Kruif [1926] and *Milestones in Microbiology* by Thomas Brock [1961]).

as are some aerobes (*e.g.,* certain *Neisseria, Campylobacter,* and *Haemophilus* species). Capnophilic aerobes will grow in a candle extinction jar, but not in room air. A candle extinction jar (or candle jar, as it is sometimes called) generates a final atmosphere containing 12 to 17% O_2 and 3 to 5% CO_2. In the hospital bacteriology laboratory, CO_2 incubators are routinely calibrated to contain between 5% and 10% CO_2.

NUTRITIONAL REQUIREMENTS

All bacteria need some form of the elements carbon, hydrogen, oxygen, sulfur, phosphorus, and nitrogen for growth. Special elements, such as potassium, calcium, iron, manganese, magnesium, cobalt, copper, zinc, and uranium, are needed by certain bacteria. Some have specific vitamin requirements; others need organic substances secreted by other living microorganisms during their growth. Organisms with particularly demanding nutritional requirements are said to be *fastidious.* Special enriched media must be used to grow fastidious organisms in the laboratory. The nutritional needs of a particular organism are usually characteristic for that species of bacteria, and of value in identifying that organism. Nutritional requirements are discussed in Chapter 6.

BIOCHEMICAL AND METABOLIC ACTIVITIES

As bacteria grow, they produce many waste products and secretions, some of which are enzymes that enable them to invade their host and cause disease. The pathogenic strains of many bacteria, such as staphylococci and streptococci, can be tentatively identified by the enzymes they secrete. Also, in particular environments, some bacteria are characterized by the production of certain gases, such as carbon dioxide, hydrogen sulfide, oxygen, or methane. To aid in the identification (speciation) of certain types of bacteria in the laboratory, bacteria are inoculated into various substrates (*e.g.,* carbohydrates and amino acids) to determine if they possess the enzymes necessary to break down those substrates. Learning whether or not a particular organism is able to break down a certain substrate serves as a "clue" to the identity of that organism. Different types of culture media are also used in the laboratory to learn information about an organism's metabolic activities (see Color Figures 15, 16).

PATHOGENICITY

The disease-producing abilities of pathogens are important to understand. Many pathogens are able to cause disease because they have capsules or endotoxins (part of the cell wall of Gram-negative bacteria), or because they secrete exotoxins and exoenzymes that damage cells and tissues (described in Chapter 9). Frequently, pathogenicity (the ability to cause disease) is tested by injecting the organism into mice or cell cultures. Some common pathogenic bacteria are listed in Table 3-3.

AMINO ACID SEQUENCING OF PROTEINS

Some proteins found within a bacterium are unique to that species. Thus, by comparing the amino acid sequence of certain bacterial proteins, the species and how closely related it is to other bacteria can be determined.

Color Plates

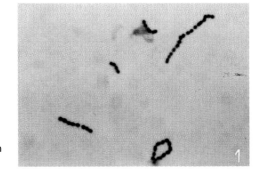

COLOR FIGURE 1. Chains of Gram-positive streptococci in a Gram-stained smear prepared from a broth culture.

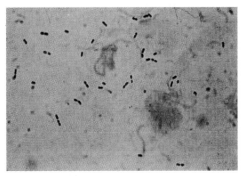

COLOR FIGURE 2. Gram-positive *Streptococcus pneumoniae* in a Gram-stained smear of a blood culture. Note the pairs of cocci (known as diplococci).

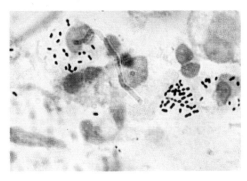

COLOR FIGURE 3. Gram-positive *Streptococcus pneumoniae* in a Gram-stained smear of a purulent (pus-containing) sputum. Note the diplococci. Several pink-staining polymorphonuclear leukocytes (PMNs) can also be seen.

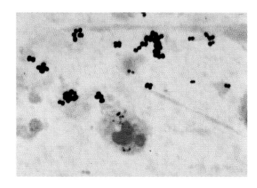

COLOR FIGURE 4. Gram-positive staphylococci in a Gram-stained smear of a purulent exudate. Note the grapelike clusters of cocci. A pink-staining PMN can also be seen.

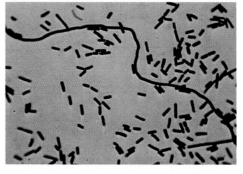

COLOR FIGURE 5. Gram-positive bacilli (*Clostridium perfringens*) in a Gram-stained smear prepared from a broth culture. Individual bacilli and chains of bacilli (streptobacilli) can be seen.

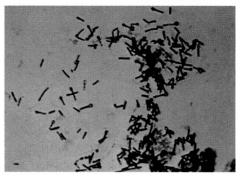

COLOR FIGURE 6. Gram-positive bacilli (*Clostridium tetani*) in a Gram-stained smear prepared from a broth culture. Terminal spores can be seen on some of the cells.

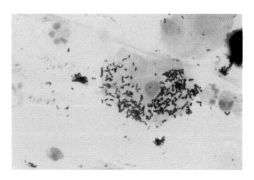

COLOR FIGURE 7. Many Gram-positive bacteria can be seen on the surface of a pink-stained epithelial cell in this Gram-stained sputum specimen. Several smaller pink-staining PMNs can also be seen.

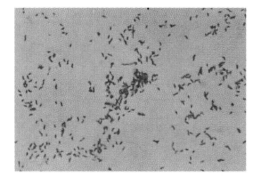

COLOR FIGURE 8. Gram-negative bacilli in a Gram-stained smear prepared from a bacterial colony. Individual bacilli and a few short chains of bacilli can be seen.

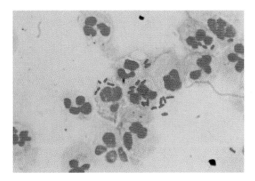

COLOR FIGURE 9. Many Gram-negative bacilli and many pink-staining PMNs can be seen in this Gram-stained urine sediment from a patient with cystitis.

COLOR FIGURE 10. Loosely coiled, Gram-negative spirochetes. *Borrelia burgdorferi,* the etiologic agent of Lyme disease.

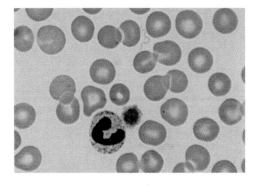

COLOR FIGURE 11. A Wright's-stained peripheral blood smear from a patient with a *Borrelia* infection. A loosely coiled spirochete can be seen between red blood cells near the top center of the photomicrograph. Two blue-stained white blood cells—a neutrophil and a small lymphocyte—can be seen near the bottom center of the photomicrograph.

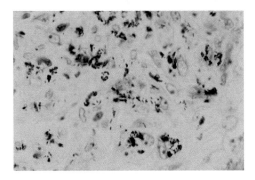

COLOR FIGURE 12. Many red acid-fast mycobacteria can be seen in this acid-fast stained liver biopsy specimen.

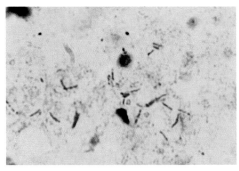

COLOR FIGURE 13. Many red acid-fast bacilli (*Mycobacterium tuberculosis*) can be seen in the acid-fast stained concentrate from a digested sputum specimen.

COLOR FIGURE 14. Colonies of a beta-hemolytic *Streptococcus* species on a blood agar plate. The clear zones (beta hemolysis) around the colonies are caused by enzymes (hemolysins) that lyse the red blood cells in the agar.

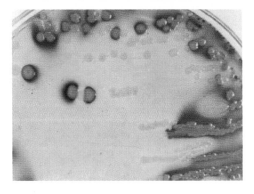

COLOR FIGURE 15. Bacterial colonies on MacConkey agar, which is a selective and differential medium. It is selective for Gram-negative bacteria, meaning that only Gram-negative bacteria will grow on this medium. Colonies of lactose-fermenters (pink colonies) and nonlactose-fermenters (clear colonies) can be seen.

COLOR FIGURE 16. Mannitol salt agar, a selective and differential medium, is used to screen for *Staphylococcus aureus*. Any bacteria capable of growing in a 7.5% sodium chloride concentration will grow on this medium, but *S. aureus* will turn the medium yellow due to its ability to ferment the mannitol in the medium.

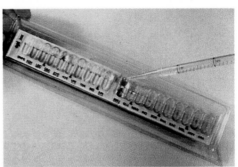

COLOR FIGURE 17. Shown here is a minisystem (API-20E) used to identify members of the family *Enterobacteriaceae*. Each of the 20 chambers contains a different substrate. If the organism is capable of breaking down a particular substrate, a change in pH will occur; this will cause the pH indicator to change color. Thus, a color change indicates a positive test result. No color change indicates a negative test result.

COLOR FIGURE 18. Shown here is a minisystem (Enterotube II) used to identify members of the family *Enterobacteriaceae*. As with the API 20-E strip, color changes represent positive test results. By totaling the numerical values of the positive tests, a 5-digit code number is generated. The identity of the organism is then determined by looking the number up in a code book.

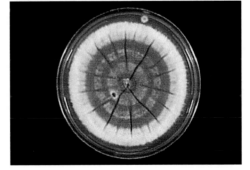

COLOR FIGURE 19. A colony (mycelium) of the mold *Aspergillus fumigatus,* a common cause of pulmonary infections in immunosuppressed patients.

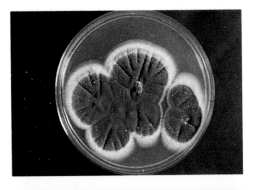

COLOR FIGURE 20. Colonies (mycelia) of a *Penicillium* species. Although penicillin is derived from *Penicillium*, this mold can also cause human infections in immunosuppressed patients.

COLOR FIGURE 21.   Colonies of the yeast, *Candida albicans*, on a blood agar plate. The footlike extensions from the margins of the colonies are typical of this species.

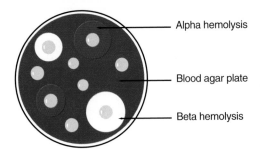

Alpha hemolysis

Blood agar plate

Beta hemolysis

COLOR FIGURE 22.   Diagram illustrating the three types of hemolysis: alpha hemolysis (green zone around colonies); beta hemolysis (clear zone around colonies); gamma hemolysis (no change in red blood cells that surround the colonies; *i.e.*, neither alpha nor beta hemolysis).

GENETIC COMPOSITION

The composition of the genetic material (DNA) is unique to each species. Thus, by determining the base composition and by comparing the cytosine/guanine ratio with the total amount of bases, a numerical ratio can be calculated (see Chapter 5). Also, by identifying or hybridizing a sequence of bases in portions of DNA or RNA, a researcher can determine the degree of relationship between two different bacteria and, perhaps, the particular species or strain of bacteria.

RUDIMENTARY FORMS OF BACTERIA

Rickettsias, chlamydias, and mycoplasmas are Gram-negative bacteria, but they do not possess all of the attributes of typical bacterial cells. Because they are so small and difficult to isolate, they were formerly thought to be viruses.

Rickettsias and Chlamydias

Rickettsias and chlamydias are coccoid, rod-shaped, or pleomorphic (irregular) Gram-negative bacteria with a bacterial-type cell wall. Unlike viruses, they contain *both* DNA and RNA. Most known forms are *obligate intracellular pathogens* that cause diseases in humans and other animals. As the name implies, an obligate intracellular pathogen is a pathogen that *must* live within a host cell. To grow such organisms in the laboratory, they must be inoculated into embryonated chicken eggs, laboratory animals, or cell cultures.

Because they appear to have leaky cell membranes, most rickettsias must live inside another cell to retain all necessary cellular substances. Diseases caused by *Rickettsia* species are transmitted by arthropod *vectors* (carriers). A closely related organism, *Coxiella burnetii* (the cause of Q fever) may also be airborne or foodborne.

Arthropods such as lice, fleas, and ticks are vectors of rickettsial diseases (Table 3-5). They transmit pathogens from one host to another by their bites or waste products. Diseases caused by arthropod-borne rickettsias include typhus, and typhus-like diseases (*e.g.*, Rocky Mountain spotted fever). A closely related organism called *Bartonella quintana* (formerly *Rochalimaea quintana*) is associated with cat scratch disease, bacteremia, and endocarditis. Other closely related organisms called *Ehrlichia* species cause human tick-borne diseases such as human monocytic ehrlichiosis (HME) and human granulocytic ehrlichiosis (HGE). *Ehrlichia* spp. are intraleukocytic pathogens; they live within certain types of white blood cells.

Chlamydias are probably the most primitive of all bacteria because they lack the enzymes required to perform many essential metabolic activities, particularly production of ATP. Sometimes called "energy parasites," chlamydias are obligate intracellular pathogens that are transferred by direct contact between hosts, not by arthropods.

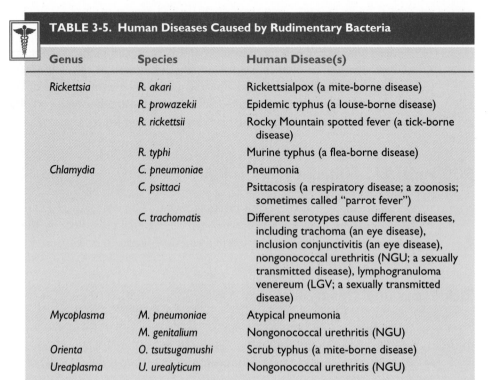

TABLE 3-5. Human Diseases Caused by Rudimentary Bacteria

Genus	Species	Human Disease(s)
Rickettsia	R. akari	Rickettsialpox (a mite-borne disease)
	R. prowazekii	Epidemic typhus (a louse-borne disease)
	R. rickettsii	Rocky Mountain spotted fever (a tick-borne disease)
	R. typhi	Murine typhus (a flea-borne disease)
Chlamydia	C. pneumoniae	Pneumonia
	C. psittaci	Psittacosis (a respiratory disease; a zoonosis; sometimes called "parrot fever")
	C. trachomatis	Different serotypes cause different diseases, including trachoma (an eye disease), inclusion conjunctivitis (an eye disease), nongonococcal urethritis (NGU; a sexually transmitted disease), lymphogranuloma venereum (LGV; a sexually transmitted disease)
Mycoplasma	M. pneumoniae	Atypical pneumonia
	M. genitalium	Nongonococcal urethritis (NGU)
Orienta	O. tsutsugamushi	Scrub typhus (a mite-borne disease)
Ureaplasma	U. urealyticum	Nongonococcal urethritis (NGU)

Chlamydias have two forms in their life cycle (Fig. 3-5). The infectious form, called the *elementary body,* attaches to the host cell. After it is engulfed by the host cell, the elementary body reorganizes into the larger, less infectious form, called the *reticulate body.* This form finally divides to produce many small infectious elementary bodies that are released to spread and to infect surrounding host cells or other individuals. Chlamydias are easily transmitted during sexual contact and cause infections of the urethra (urethritis), bladder (cystitis), fallopian tubes (salpingitis), prostate (prostatitis), or other complications. They cause many diseases in vertebrate animals (cats, dogs, sheep, cattle, birds, humans), including those listed in Table 3-5.

Mycoplasmas

Mycoplasmas are the smallest of the cellular microbes (Fig. 3-6). Because they lack cell walls, they assume many shapes, from coccoid to filamentous. Sometimes they are confused with the L-forms of bacteria, described earlier; however, even in the most favorable growth media, mycoplasmas are not able to produce cell walls, which is not true for L-forms. Mycoplasmas were formerly called pleuropneumonia-like organisms (PPLO), first isolated from cattle with lung infections. These

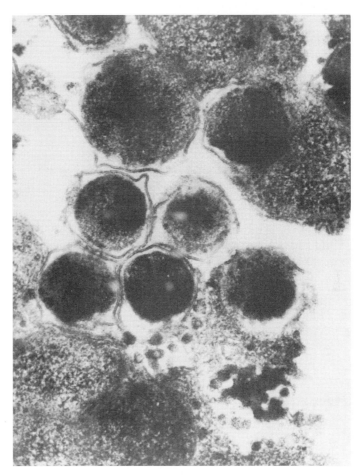

FIGURE 3-5. Elementary bodies of chlamydia. (Courtesy of S. Koester)

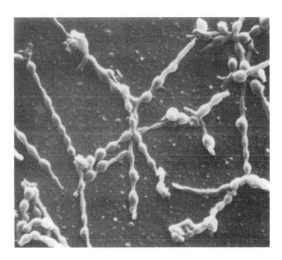

FIGURE 3-6. Scanning electron micrograph of *Mycoplasma pneumoniae*; note the coccoid structures within the filaments (original magnification ×10,000).

Gram-negative bacteria may be free-living or parasitic and are pathogenic to many animals and some plants. In humans, pathogenic mycoplasmas cause primary atypical pneumonia and genitourinary infections. Because they have no cell wall, they are resistant to treatment with penicillin and other antibiotics that work by inhibiting cell wall synthesis. Mycoplasmas can be cultured on artificial media in the laboratory, where they produce tiny colonies that resemble fried eggs in appearance. In the body, some species can grow intracellularly. Diseases caused by mycoplasmas and a closely related organism (*Ureaplasma ureolyticum*) are shown in Table 3-5.

PHOTOSYNTHETIC BACTERIA

Photosynthetic bacteria include purple bacteria, green bacteria, and cyanobacteria (erroneously referred to in the past as blue-green algae). Although all three groups use light as an energy source, they do not all carry out photosynthesis in the same way. For example, purple and green bacteria (which, in some cases, are not actually those colors) do not produce oxygen, whereas cyanobacteria do. Many scientists believe that cyanobacteria were the first organisms capable of carrying out oxygenic photosynthesis and, thus, played a major part in the oxygenation of the atmosphere. Fossil records reveal that cyanobacteria were already in existence 3.3 to 3.5 billion years ago.

When appropriate conditions exist, cyanobacteria in pond or lake water will overgrow, creating a waterbloom—a "pond scum" that resembles a thick layer of bluish-green (turquoise) oil paint. The conditions include mild conditions or no wind, a balmy water temperature (15 to 30° C), a water pH of 6 to 9, and an abundance of the nutrients nitrogen and phosphorous in the water. Many cyanobacteria are able to convert nitrogen gas (N_2) from the air into ammonium ions (NH_4+) in the soil; this process is known as *nitrogen fixation* (Chapter 8).

Some cyanobacteria produce toxins (poisons), such as neurotoxins (which affect the central nervous system), hepatotoxins (which affect the liver), and cytotoxins (which affect various cells). These toxins are harmful to birds, domestic animals, and wild animals that consume the cyanobacteria, as well as the minute animals (zooplankton) that live in the pond or lake water. In the Midwestern U.S., thousands of migrating ducks and geese have died after consuming cyanobacterial toxins. Thus far, no human deaths have been attributed to these toxins. There is concern, however, that certain cyanobacterial toxins may contribute to the development of cancer.

ARCHAEBACTERIA

Bacteria described in the previous sections are all members of the kingdom Eubacteria or "true" bacteria. Bacteria in the kingdom Archaea are known as archaebacteria (sometimes spelled archaeobacteria). "Archae" means "ancient," and this name was originally assigned when it was thought that these bacteria evolved earlier than eubacteria. Now there is considerable debate as to whether eubacteria or archaebacteria came first. Archaebacteria are more closely related to eucaryotes than they are to eubacteria; some possess genes otherwise found only in eucaryotes. Many scientists believe that eubacteria and archaebacteria diverged from a common ancestor relatively soon after life began on this planet. Later, the eucaryotes split off from the archaebacteria (Fig. 1-1).

Many, but not all, archaebacteria are "extremophiles," in the sense that they live in extreme environments, such as extremely acidic, extremely hot, and extremely salty environments. Some live at the bottom of the ocean near thermal vents, where, in addition to heat and salinity, there is extreme pressure. Other archaebacteria, called methanogens, produce methane, which is a flammable gas. Although archaebacteria possess cell walls, their cell walls contain no peptidoglycan. All eubacterial cell walls contain peptidoglycan. The 16S sequences of archaebacteria are quite different than the 16S sequences of eubacteria. The 16S sequence data suggests that archaebacteria are more closely related to eucaryotes than they are to eubacteria. You will recall that differences in rRNA structure form the basis of the 3-kingdom classification system.

ALGAE

Algae are eucaryotic, photosynthetic organisms which, together with protozoa, are classified in the second kingdom—Protista. Algae range from tiny, unicellular, microscopic organisms (*e.g.,* diatoms, dinoflagellates, and desmids) to large, multicellular, plantlike seaweeds (*e.g.,* kelp); see Table 3-6. They may be arranged in colonies or strands, and are found in fresh and salt water and soil and on trees, plants, and rocks. Algae produce their energy by photosynthesis, using energy from the sun, carbon dioxide, water, and inorganic nutrients from the soil to build cellular material. However, a few species use organic nutrients, and others survive with very little sunlight. Algal cell walls contain cellulose, a polysaccharide not found in the cell walls of any other microorganisms. Depending upon the types of photosynthetic pigments they possess, algae are classified as green, brown, or red algae. One genus of algae (*Prototheca*) is a rare cause of human infections, but several other genera secrete substances (*phycotoxins*) that are poisonous to humans, fish, and other animals (Table 3-7). One alga, called *Pfiesteria piscicida,* has killed billions of fish along the eastern seaboard in recent years. Its toxins also cause human disease.

Algae are easy to find. They include large seaweeds and the kelp found along ocean shores, the green scum floating on ponds, and the slippery material on wet rocks. There are also many microscopic forms in pond water that differ from the colorless, motile protozoa in that they are photosynthetic, pigmented, and may move very slowly. Some algae (*e.g., Euglena* and *Volvox*) have characteristics (*e.g.,* primitive mouth, flagella) that cause them to be classified as protozoa by some taxonomists.

Algae are an important source of food, iodine and other minerals, fertilizers, emulsifiers for pudding, and stabilizers for ice cream and salad dressings; they are also

TABLE 3-6. Characteristics of Algae

Phylum (and Common Name)	Structural Arrangement	Predominant Color	Photosynthetic Pigments*	Habitat
Bacillariophyta (diatoms)	Unicellular	Olive brown	Chlorophyll c, carotenoids, xanthophylls	Fresh water and sea water
Chlorophyta (green algae)	Unicellular and multicellular	Green	Chlorophyll b, carotenoids	Fresh water (predominantly) and sea water
Chrysophyta (golden algae)	Unicellular	Golden olive	Chlorophyll c, carotenoids, xanthophylls	Fresh water
Dinoflagellata (dinoflagellates)	Unicellular	Brown	Chlorophyll c, carotenoids, xanthophylls	Fresh water and sea water
Euglenophyta (*Euglena* spp. and closely related organisms)	Unicellular	Green	Chlorophyll b, carotenoids, xanthophylls	Fresh water
Phaeophyta (brown algae)	Multicellular seaweeds	Olive brown	Chlorophyll c, carotenoids, xanthophylls	Sea water; most commonly, cold environments
Rhodophyta (red algae)	Multicellular seaweeds	Red to black	Chlorophyll d (in some), carotenoids, phycobilins	Sea water (predominantly) and fresh water; most commonly, tropical environments

*In addition to chlorophyll a, which is possessed by all algae. Carotenoids are yellow-orange, chlorophylls are greenish, phycobilins are red or blue, and xanthophylls are brownish.

used as a gelling agent for jams and nutrient media for bacterial growth. The agar that is used to produce solid laboratory culture media is derived from a red marine alga. Damage to water systems is frequently caused by algae clogging filters and pipes where many nutrients are present. Some typical algae are shown in Figure 3-7.

PROTOZOA

Protozoa are eucaryotic, usually single-celled, animal-like microorganisms, ranging in length from 3 to 2000 μm. Most are free-living organisms found in soil and water (Fig. 3-8). They have no chlorophyll and, therefore, cannot make their own foods by photosynthesis. Some ingest whole algae, yeasts, bacteria, and other smaller protozoans as their source of nutrients (Fig. 3-9); others live on dead, decaying organic matter. Parasitic protozoa break down and absorb nutrients from the body of the host in which they live. Many parasitic protozoa are pathogens, such as those that cause malaria, giardiasis, and amebic dysentery (see Chapter 12 and Appendix C). Other parasitic protozoa co-exist with the host animal in a type of symbiotic (living together) relationship, wherein both organisms benefit. A typical example of such a symbiotic relationship is the termite and its intestinal protozoa. The protozoa digest the wood eaten by the termite, enabling both organisms to absorb the nutrients necessary for life. Without the parasitic protozoa, the termite would be unable to digest the wood that it eats and would starve to death.

TABLE 3-7. Human Diseases Caused by Phycotoxins

Disease	Cause
Amnesic shellfish poisoning (one of the most serious illnesses associated with red tide toxins; causes gastrointestinal and neurological symptoms; can be fatal)	Ingestion of shellfish (mussels) containing the toxins of *Nitzchia pungens*, a diatom
Ciguatera fish poisoning (one of the most frequently reported non-bacterial illnesses associated with eating fish in the United States and its territories, especially Southern Florida, Puerto Rico, and Hawaii; causes gastrointestinal, neurological, and cardiovascular symptoms; can cause paralysis; can be fatal)	Ingestion of fish (usually tropical fish) containing the toxins of dinoflagellates such as *Gambierdiscus toxicus*, *Prorocentrum mexicanum*, *Ostreopsis lenticularis*, *Coolia monotis*, *Thecadinium* sp., and *Amphidinium carterae*
Diarrhetic shellfish poisoning (usually a mild gastrointestinal disorder)	Ingestion of shellfish (mussels, oysters, scallops) containing the toxins of dinoflagellates in the genus *Dinophysis*
Neurotoxic shellfish poisoning (causes gastrointestinal and neurological symptoms; not fatal)	Ingestion of shellfish (oysters, clams) containing the toxins of the dinoflagellate *Pytchodiscus brevis*; wave action can produce aerosols which, when inhaled, can produce respiratory asthma-like symptoms
Paralytic shellfish poisoning (causes neurological symptoms; can cause death)	Ingestion of shellfish (mussels, clams, cockles, scallops) containing toxins of *Alexandrium* spp., *Gymnodinium catenatum*, *Pyrodinium bahamense*, and other dinoflagellates; primarily occurs in the Pacific Northwest and Alaska
Neurological and gastrointestinal symptoms and skin sores	Ingestion of, or contact with, toxins of the dinoflagellate *Pfiesteria piscicida*

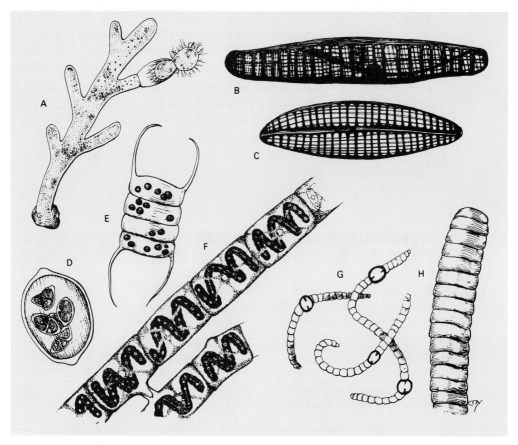

FIGURE 3-7. Typical algae. (A) *Vaucheria*. (B) Diatom. (C) *Navicula*. (D) *Oocystis*. (E) *Scenedesmus*. (F) *Spirogyra*. (G) *Nostoc*. (H) *Oscillatoria*.

Classification

Protozoa are divided into groups (phyla, subphyla, or classes) according to their method of locomotion (Table 3-8). *Amebas* or amebae, which are in the subphylum *Sarcodina* in the phylum *Sarcomastigophora*, move by means of *pseudopodia* (false feet). An ameba first extends a pseudopodium in the direction the ameba wishes to move and then the rest of the cell slowly flows into it; this process is called *ameboid movement*. An ameba ingests food by surrounding the food particle with pseudopodia, which then fuse together; this process is known as *phagocytosis*. The ingested food, surrounded by a membrane, is referred to as a food vacuole (or phagosome). Digestive enzymes then digest or break down the food into nutrients. Some of the white blood cells in our bodies ingest and digest materials in the same manner as amebae. Phagocytosis is further discussed in Chapter 11. When fluids are ingested in a similar manner, the process is known as *pinocytosis*. One important

Representative Pond Protozoa

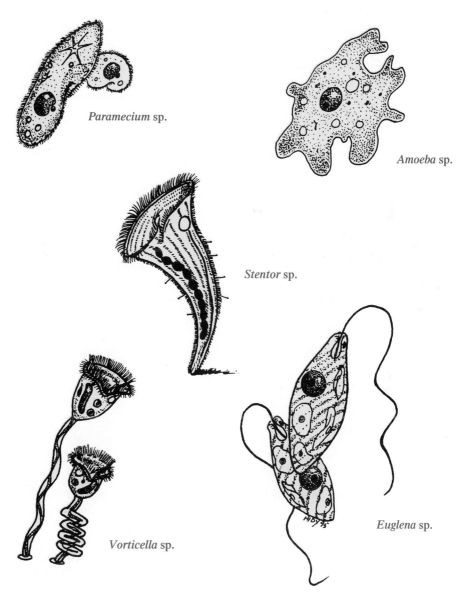

Paramecium sp.

Amoeba sp.

Stentor sp.

Vorticella sp.

Euglena sp.

FIGURE 3-8. Pond protozoa. (Illustrations from *Introduction to Microbiology and the Clinical Micro-biology Laboratory* by Paul G. Engelkirk, Ph.D., 1995; reproduced courtesy of Hunter Textbooks, Inc., Winston-Salem, NC)

pathogenic ameba is *Entamoeba histolytica,* which causes amebic dysentery (amebiasis) and extraintestinal (outside of the intestine) amebic abscesses.

The *flagellates* (subphylum *Mastigophora* in the phylum *Sarcomastigophora*) move by means of whiplike flagella. Some species are pathogenic. For example, *Trypanosoma brucei* subspecies *gambiense* is transmitted by the tsetse fly and causes African sleeping sickness in humans; *Trichomonas vaginalis* causes persistent sexually transmitted infections (trichomoniasis) of the male and female genital tracts; and *Giardia lamblia* causes a persistent diarrheal disease (giardiasis).

Ciliates (phylum *Ciliophora* or *Ciliata*) move about as a result of large numbers of hair-like cilia on their surfaces. They are the most complex of all the protozoa. A pathogenic ciliate, *Balantidium coli,* causes dysentery in underdeveloped countries. It is usually transmitted to humans via drinking water that has been contaminated by swine feces. *B. coli* is the only ciliated protozoan that causes disease in humans. Protozoa do not have cell walls but some, including some flagellates and some ciliates, possess a thickened cell membrane (called a *pellicle*) which serves the

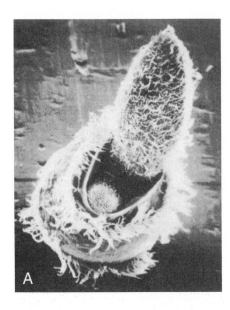

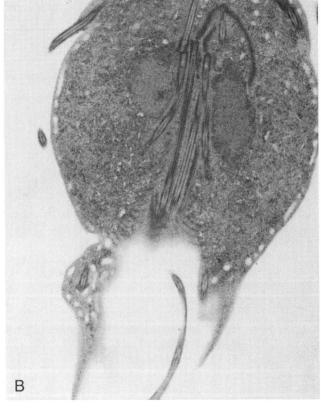

FIGURE 3-9. Protozoa. (A) *Didinium nasutum* with partially ingested prey. (B) TEM longitudinal section of *Giardia lamblia.* (From S. Koester and P. Engelkirk)

TABLE 3-8. Characteristics of Major Protozoa

Phylum	Means of Movement	SELECTED DIFFERENTIATING PROPERTIES (METHOD OF REPRODUCTION)		Representatives
		Asexual	Sexual	
Ciliophora	Cilia	Transverse fission	Conjugation	*Balantidium coli, Paramecium, Stentor, Tetrahymena, Vorticella*
Sarcodina	Pseudopodia (false feet)	Binary fission	When present, involves flagellated sex cells	*Amoeba, Naegleria, Entamoeba histolytica*
Mastigophora	Flagella	Binary fission	None	*Chlamydomonas, Giardia lamblia, Trichomonas, Trypanosoma*
Sporozoea	Generally nonmotile except for certain sex cells	Multiple fission	Involves flagellated sex cells	*Plasmodium, Toxoplasma gondii, Cryptosporidium*

same purpose as a cell wall. Some flagellates and some ciliates ingest food through a primitive mouth or opening, called a *cytostome.*

Nonmotile protozoa are classified as sporozoa (phylum *Sporozoea*). The most important pathogens are the *Plasmodium* species that cause malaria in many areas throughout the world. One of these, *Plasmodium vivax,* causes a relatively mild form of malaria in some parts of the United States. These pathogens are carried and transmitted by the female *Anopheles* mosquito, which becomes infected when it takes a blood meal from a person with malaria. Another sporozoan, not previously recognized as a serious pathogen, *Cryptosporidium parvum,* causes severe diarrheal disease (cryptosporidiosis) in immunosuppressed patients, especially those with acquired immunodeficiency syndrome (AIDS) (see Chapter 12). A 1993 epidemic in Milwaukee, Wisconsin, caused by *Cryptosporidium* oocysts in drinking water, resulted in over 400,000 cases of cryptosporidiosis, including some deaths. Pathogenic protozoa are described in Appendix C.

FUNGI

Fungi are found almost everywhere on earth, living on organic matter in water and soil and living upon and within animals and plants. Some are harmful whereas others are beneficial. Fungi also live on many unlikely materials, causing deterioration of leather and plastics and spoilage of jams, pickles, and many other foods. Beneficial fungi are important in the production of cheeses, yogurt, beer and wine, and other foods as well as certain drugs and antibiotics.

Characteristics

Fungi are eucaryotic organisms that include mushrooms, molds, and yeasts. As saprophytes, their main source of food is dead and decaying organic matter. Fungi are the "garbage disposers" of nature, the "vultures" of the microbial world. By secreting digestive enzymes into dead plant and animal matter, they decompose this material into absorbable nutrients for themselves and other living organisms; thus they are the original "recyclers." Imagine living in a world without saprophytes, stumbling through endless piles of decaying debris! Some fungi also live as parasites, on and in living animals and plants.

Fungi are often incorrectly referred to as plants. One way they differ from plants and algae is that they are not photosynthetic; they have no chlorophyll or other photosynthetic pigments. Whereas the cell walls of algal and plant cells contain cellulose, fungal cell walls contain a substance called *chitin*. Although many fungi are unicellular (*e.g.,* yeasts), others grow as filaments called *hyphae* (singular: *hypha*), which intertwine to form a mass called a *mycelium* (plural, *mycelia*) or thallus; thus, they are different from saprophytic bacteria. Remember that bacteria are procaryotic, whereas fungi are eucaryotic.

REPRODUCTION

Depending upon the particular species, fungal cells can reproduce by budding, hyphal extension, or the formation of spores (Fig. 3-10). Sexual reproduction in many fungi involves the nuclear fusion of two gametes and their subsequent division into many sexual spores. Some species of fungi produce both asexual and sexual spores or more than one type of asexual spore. Asexual spores are also called *conidia* (singular, *conidium*). Sexual spores have a variety of names (*e.g.,* ascospores, basidiospores, zygospores), depending upon the manner in which they are formed. Fungi are classified in accordance with the type of sexual spore that they produce or the type of structure upon which spores are produced. Figure 3-11 illustrates some typical fungi. Fungal spores are very resistant structures that are carried great distances by wind. They are resistant to heat, cold, acids, bases, and other chemicals. Many people are allergic to fungal spores.

Classification of True Fungi

Mycologists, scientists who study fungi, have separated the so-called "true fungi" into five classes: Basidiomycetes, Oomycetes, Zygomycetes, Ascomycetes, and

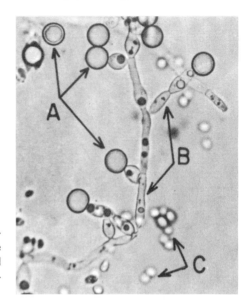

FIGURE 3-10. A culture of *C. albicans* showing (*A*) chlamydospores. (*B*) pseudohyphae (elongated yeast cells, linked end to end), and (*C*) budding yeast cells (blastospores) (original magnification ×450).

Deuteromycetes. These classes are based on mode of reproduction and types of mycelia, spores, and gametes that are produced. The characteristics of each of these classes are shown in Table 3-9.

MUSHROOMS

Mushrooms (Basidiomycetes) are a class of true fungi that consist of a network of filaments or strands (mycelium) that grow in the soil or in a rotting log, and the fruiting body (mushroom) that forms and releases spores. The spore, much like the seed of a plant, germinates into a new organism. Many mushrooms are delicious to eat, but many that resemble edible fungi are extremely toxic and may cause permanent brain damage or death. Toxins produced by fungi are called *mycotoxins.*

MOLDS

Molds are the fungi often seen in water and soil and on food. They grow in the form of filaments or hyphae that make up the mycelium of the mold. Reproduction is by spore formation, either sexually or asexually, on the reproductive hyphae. Various species of molds are found in each of the classes of fungi except Basidiomycetes (mushrooms). An interesting mold in class Oomycetes is *Phytophtera,* the potato blight mold that caused a famine in Ireland in the mid-19th century. The black bread mold, *Rhizopus,* is a zygomycete, which can cause respiratory disease in immunosuppressed patients. Both of these genera are primitive molds with aseptate hyphae, meaning that the hyphae are not divided into individual cells by cross-walls (septa). Although there may be many nuclei in

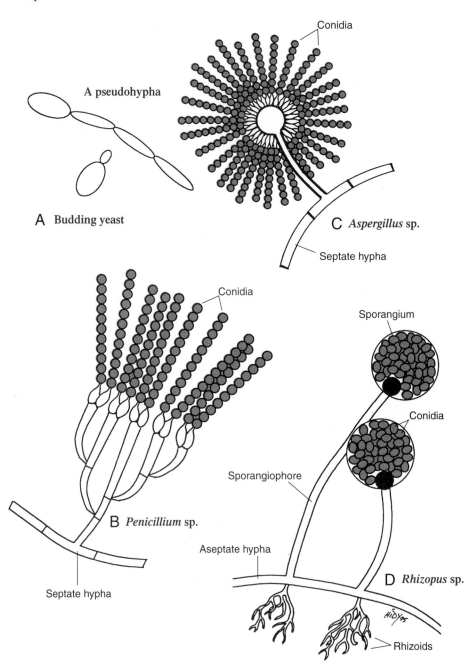

FIGURE 3-11. Microscopic appearance of various fungi. (A) Yeast cells. A pseudohypha is a chain of elongated buds. (B) *Penicillium* mold resembles a camel-hair paint brush or candelabra. (C) *Aspergillus* mold resembles dandelion seeds. (D) *Rhizopus* is a bread mold with rootlike structures called rhizoids. (Illustrations from *Introduction to Microbiology and the Clinical Microbiology Laboratory* by Paul G. Engelkirk, Ph.D., 1995; reproduced courtesy of Hunter Textbooks, Inc., Winston-Salem, NC)

TABLE 3-9. Selected Characteristics of the Major Classes of Fungi

Class	Type of Mycelium	SITE OF SPORE FORMATION		Representative Groups
		Asexual	Sexual	
Ascomycetes	Septate	Tips of hyphae	Within sacs	Common antibiotic-producing fungi, *Penicillium,* yeasts
Basidiomycetes	Septate	Tips of hyphae	Surface of basidium	Mushrooms, rusts
Deuteromycetes (fungi imperfecti)	Septate	Tips of hyphae	None present	Most human pathogenic molds and yeasts
Oomycetes	Aseptate	In sacs	Within a unicellular female sex organ (oogonium)	Some aquatic forms, mildew, plant blights, fish infections
Zygomycetes	Usually aseptate	In sacs	In mycelium	Bread mold (*Rhizopus nigricans*), aquatic species

aseptate hyphae, they are not separated by cell walls. Hyphae containing septa are referred to as septate hyphae. Among the Ascomycetes and Deuteromycetes classes are found many antibiotic-producing molds, such as *Penicillium* and *Cephalosporium.*

Molds have great commercial importance. They are the main source of antibiotics. Many new antibiotics are developed by growing soil cultures and isolating any molds that inhibit the growth of bacteria. The antibiotic penicillin was accidentally discovered when a *Penicillium notatum* mold contaminated an agar plate containing a *Staphylococcus,* and inhibited growth of the bacteria. Today, to increase their spectrum of activity, antibiotics can be chemically altered in pharmaceutical company laboratories, as has been done with the various penicillins. Some molds are also used to produce large quantities of enzymes (such as amylase, which converts starch to glucose) and of citric acid and other organic acids that are used commercially. The flavor of cheeses such as bleu, Roquefort, camembert, and limburger, are the result of molds that grow on them.

Molds also can be harmful. Like yeasts, they cause a variety of infectious diseases—called *mycoses* (discussed in a following section). The rusts and molds of crop plants, grains, corn, and potatoes not only destroy crops, but some are also toxic. The aflatoxin from an *Aspergillus* mold on peanuts and cottonseed and the ergot (a type of smut) on rye and wheat are extremely toxic to humans and farm animals, causing aflatoxicosis and ergotism, respectively; both diseases occur in the United States. Aflatoxins have also been shown to be carcinogenic (cancer causing). Although diseases due to mycotoxins were originally thought to be caused only by ingestion of mycotoxin-contaminated foods, recent evidence suggests that mycotoxins can also enter the body via the airborne route.

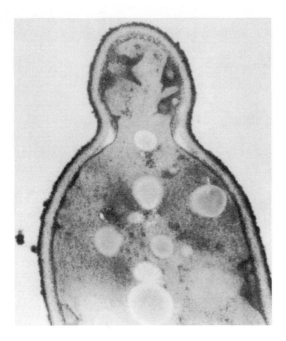

FIGURE 3-12. Longitudinal section of a budding yeast cell (original magnification ×15,500).

YEASTS

Yeasts are microscopic, eucaryotic, single-celled Ascomycetes or Deuteromycetes that lack mycelia. They usually reproduce by budding (Fig. 3-12), but occasionally do so by spore formation. Yeasts are found in soil and water and on the skins of many fruits and vegetables. People produced wine, beer, and alcoholic beverages for centuries before Pasteur discovered that naturally occurring yeasts on the skin of grapes and other fruits and grains were responsible for these fermentation processes. The common yeast *Saccharomyces cerevisiae* ferments sugar to alcohol under anaerobic conditions (without oxygen). In aerobic conditions (with oxygen), this yeast breaks down simple sugars to carbon dioxide and water; for this reason, it has long been used to leaven light bread. Yeasts are also a good source of nutrients for humans because they produce many vitamins and proteins. Some yeasts are human pathogens. *Candida albicans* is the yeast most frequently isolated from human clinical specimens.

Dimorphism

A few fungi, usually pathogens, can live either as molds or as yeasts depending on growth conditions. The phenomenon is called *dimorphism* and the organisms are referred to as dimorphic fungi. When they are isolated from living tissues at body temperature (37°C), they appear in a unicellular parasitic form as yeasts. If they are grown at room temperature (25°C) or isolated from the soil or dust, they grow in the saprophytic form as a mold with hyphae and spores. *Histoplasma,* which causes histoplasmosis; *Sporothrix,* which causes sporotrichosis; *Coccidioides,* which causes coccidioidomycosis; and *Blastomyces,* which causes blastomycosis, are examples of dimorphic fungi.

Fungal Diseases

Considering the large number of fungal species, very few are pathogenic for humans, and most of those are found in the class Deuteromycetes (see Table 3-9). Diseases caused by fungi are called *mycoses* (singular, *mycosis*) and are categorized as superficial, cutaneous, subcutaneous, or systemic mycoses. In some cases the infection may progress through all of these stages. Representative mycoses are listed in Table 3-10.

SUPERFICIAL AND CUTANEOUS MYCOSES

Superficial and cutaneous fungal infections are caused by *dermatophytes* (fungi living on or within the skin) or other fungi that live on skin, hair, fingernails, toenails, or mucous membranes at body openings. Dermatophytes cause forms of "ringworm" or tinea infections like athlete's foot (tinea pedis) and lesions of the nails

TABLE 3-10. Selected Fungal Diseases of Humans

Category	Genus/species	Diseases
Yeasts	*Candida albicans*	Thrush; yeast vaginitis; nail infections; systemic infection
	Cryptococcus neoformans	Cryptococcosis (lung infection; meningitis; etc.)
Molds	*Aspergillus* spp.	Aspergillosis (lung infection; systemic infection)
	Mucor and *Rhizopus* spp. and other species of bread molds	Mucormycosis or zygomycosis (lung infection; systemic infection)
	Various dermatophytes	Tinea ("ringworm") infections
Dimorphic fungi	*Blastomyces dermatitidis*	Blastomycosis (primarily a disease of lungs and skin)
	Coccidioides immitis	Coccidioidomycosis (lung infection; systemic infection)
	Histoplasma capsulatum	Histoplasmosis (lung infection; systemic infection)
	Sporothrix schenckii	Sporotrichosis (a skin disease)
Other	*Pneumocystis carinii*	*Pneumocystis carinii* pneumonia (PCP)

(tinea unguium), scalp (tinea capitis), face and neck (tinea barbae), trunk (tinea corporis), groin area (tinea cruris), and other locations.

Candida albicans is an opportunistic yeast that lives harmlessly on the skin and mucous membranes of the mouth, intestine, and reproductive tract. However, when the chemical balance (homeostasis) is upset and the number of indigenous bacteria is reduced, this yeast flourishes to cause infections of the mouth (oral thrush), skin, and vagina (yeast vaginitis). This type of local infection may become a focal site from which the organisms invade the bloodstream to become a generalized or systemic infection in many internal areas.

SUBCUTANEOUS AND SYSTEMIC MYCOSES

Spores of some pathogenic fungi may be inhaled with dust from contaminated soil and dried bird and bat feces (guano), or they may enter through wounds of the hands and feet. If the spores are inhaled into the lungs, they may germinate there to cause a respiratory infection similar to tuberculosis. This type of deep-seated pulmonary infection is caused by species of *Coccidioides* (coccidioidomycosis), *Histoplasma* (histoplasmosis), *Blastomyces* (blastomycosis), and *Cryptococcus* (cryptococcosis). In each case, the pathogens may invade further to cause systemic infections, especially in immunosuppressed individuals.

Skin tests for most of these infections are available, but the diseases are difficult to cure. Mycotic diseases are most effectively treated with antifungal agents like nystatin, amphotericin B, or 5-fluorocytosine. Because these chemotherapeutic agents may be toxic to humans, they are prescribed with due consideration and caution.

LICHENS

Nearly everyone has seen lichens. They appear as colored, often circular patches on tree trunks and rocks. A *lichen* is actually a combination of two organisms—an alga (or a cyanobacterium) and a fungus—living in such a close relationship that they appear to be one organism. Close relationships of this type are referred to as symbiotic relationships; a lichen represents a particular type of symbiotic relationship known as mutualism (discussed further in Chapter 8). There are about 20,000 different species of lichens. Lichens may be brown, black, orange, various shades of green, and other colors, depending upon the specific combination of alga and fungus.

SLIME MOLDS

Slime molds, which are found in soil and on rotting logs, have both fungal and protozoal characteristics and very interesting life cycles. Some slime molds (known as cellular slime molds) start out in life as independent amebae, ingesting bacteria and fungi by phagocytosis. When they run out of food, they fuse together to form a motile, multicellular form known as a slug, which is only about 0.5 mm long. The slug then becomes a fruiting body, consisting of a stalk and a spore cap. Spores produced within the spore cap become disseminated and from each spore emerges an ameba. Cellular slime molds represent cell differentiation at the lowest level and scientists are studying them in an attempt to determine how some of the cells in the slug "know" to become part of the stalk, how others "know" to become part of the spore cap, and still others "know" to differentiate into spores within the spore cap. Other slime molds, known as plasmodial (or acellular) slime molds, also produce stalks and spores, but their life cycles differ somewhat from those of cellular slime molds. In the life cycle of a plasmodial slime mold, haploid cells fuse to become diploid cells, which develop into very large masses of motile, multinucleated protoplasm, each such mass being known as a plasmodium. Slime molds are classified as protists.

ACELLULAR INFECTIOUS AGENTS

Viruses

Complete virus particles, called *virions*, are so small and simple in structure that they do not fit the living cell classification. Most viruses range in size from 10 to 300 nm in diameter, although some—like Ebola virus—can be as large as 1 μm. The smallest virus is about the size of the large hemoglobin molecule of a red blood cell. No type of organism is safe from viral infections; viruses infect humans, other animals, plants, algae, fungi, protozoa, and bacterial cells (Table 3-11). Some viruses—called *oncogenic viruses* or *oncoviruses*—cause specific types of cancer, including human cancers such as Burkitt's lymphoma, nasopharyngeal carcinoma, and certain types of leukemia. In addition, all human warts are caused by viruses.

Virions are said to have five specific properties that distinguish them from living cells: (1) they possess *either* DNA or RNA, never both; (2) their replication (multiplication) is directed by the viral nucleic acid within a host cell; (3) they do not divide by binary fission or mitosis; (4) they lack the genes and enzymes necessary for energy production; and (5) they depend on the ribosomes, enzymes, and nutrients of the infected cells for protein production.

A typical virus particle consists of a genome (genetic material) of DNA or RNA surrounded by a *capsid* (protein coat), which is composed of many small protein units called *capsomeres* (Fig. 3-13). Some virions (called enveloped viruses) have a protective envelope composed of fats and polysaccharides. Bacterial viruses may also have a sheath and tail fibers. There are no ribosomes for protein synthesis or sites of energy production; hence, the virus must take over a functioning cell to produce new virus particles.

Viruses are classified by the following characteristics: (1) type of genetic material (either DNA or RNA); (2) shape of the capsid; (3) number of capsomeres; (4) size of capsid; (5) presence or absence of an envelope; (6) host that it infects (animal,

TABLE 3-11. Relative Sizes and Shapes of Some Viruses

Viruses	Nucleic Acid Type	Shape	Size Range (nm)
Animal Viruses			
Vaccinia	DNA	Complex	200 by 300
Mumps	RNA	Helical	150–250
Herpes simplex	DNA	Polyhedral	100–150
Influenza	RNA	Helical	80–120
Retroviruses	RNA	Helical	100–120
Adenoviruses	DNA	Polyhedral	60–90
Retroviruses	RNA	Polyhedral	60–80
Papovaviruses	DNA	Polyhedral	40–60
Polioviruses	RNA	Polyhedral	28
Plant Viruses			
Turnip yellow mosaic	RNA	Polyhedral	28
Wound tumor	RNA	Polyhedral	55–60
Alfalfa mosaic	RNA	Polyhedral	18 by 36–40
Tobacco mosaic	RNA	Helical	18 by 300
Bacteriophages			
T2	DNA	Complex	65 by 210
λ	DNA	Complex	54 by 194
φχ-174	DNA	Complex	25

plant, or microorganism); (7) the type of disease produced; (8) target cell; and (9) immunological properties.

The genetic material of most viruses is either double-stranded DNA or single-stranded RNA, but a few have single-stranded DNA or double-stranded RNA. Viral genomes are usually circular molecules, but some are linear (with two ends). Capsids of viruses have various shapes and symmetry. They may be polyhedral (many sided), helical (coiled tubes), bullet shaped, spherical, or a complex combination of these shapes. Polyhedral capsids have 20 sides or facets, geometrically referred to as icosahedrons. Each facet consists of several capsomeres; thus, the size of the virus is determined by the size of each facet and the number of capsomeres in each. Frequently, the envelope around the capsid makes the virus appear spherical or irregular in shape in electron micrographs. The envelope is usually acquired by certain animal viruses from the host nuclear or cellular membrane as the new virus particle leaves the cell. Apparently, viruses are able to alter these membranes and add protein fibers, spikes, and knobs that enable the virus to recognize the next host cell to be invaded. A list of some viruses, their characteristics, and diseases they cause is presented in Table 3-12. Sizes of viruses are depicted in Figure 3-14.

The virion usually infects a cell by injecting its DNA or RNA into the cell (Fig. 3-15) or by being phagocytized by the cell (Fig. 3-16). The viral genetic material may become incorporated into the host cell's DNA, where it can remain latent or inactive, and be transferred to each daughter cell when the host cell divides. This is known as *lysogeny* or the *lysogenic cycle* of viral infection. Later, the viral genetic material may be induced by heat (fever), ultraviolet (UV) light, or certain chemical agents to take over the cell and make viruses.

When the genetic material from the virus takes over the metabolic "machinery" of the cell to produce new viruses, it has entered the *lytic cycle* of virus infection. It breaks up the cellular DNA and produces viral DNA or RNA. It then synthesizes viral protein capsids and assembles many new virus particles until the cell bursts (lyses) and releases new viruses into the area to infect neighboring cells. The five steps of the lytic cycle are shown in Table 3-13.

A good example of the lysogenic (latent) and lytic cycles is the ordinary cold sore or fever blister which is caused by herpes simplex virus. Infected persons harbor the latent or temperate virus in a nerve ganglion in the lysogenic cycle. A fever, stress, or excessive sunlight can trigger the viral genetic material to take over the cells and produce more viruses (the lytic cycle); in the process, a cold sore develops.

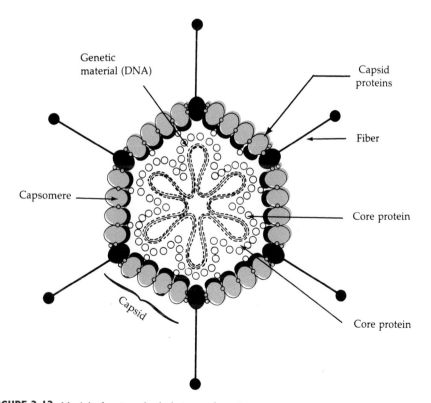

FIGURE 3-13. Model of an icosahedral virus: adenovirus.

TABLE 3-12. Selected Important Groups of Viruses and Viral Diseases

Virus Type	Viral Characteristics	Virus	Disease
Poxviruses	Large, brick shape with envelope, d.s. DNA	Variola Vaccinia	Smallpox Cowpox
Polyoma-Papilloma	d.s. DNA, polyhedral	Papillomavirus Polyomavirus	Warts Some tumors, some cancer
Herpesviruses	Polyhedral with envelope, d.s. DNA	Herpes simplex I Herpes simplex II Herpes zoster Varicella	Cold sores or fever blisters Genital herpes Shingles Chickenpox
Adenoviruses	d.s. DNA, icosahedral, no envelope		Respiratory infections, pneumonia, conjunctivitis, some tumors
Picornaviruses (the name means small RNA viruses)	s.s. RNA, tiny icosahedral, with envelope	Rhinovirus Poliovirus Hepatitis types A and B Coxsackievirus	Colds Poliomyelitis Hepatitis Respiratory infections, meningitis
Reoviruses	d.s. RNA, icosahedral with envelope	Enterovirus	Intestinal infections
Myxoviruses	RNA, helical with envelope	Orthomyxoviruses types A & B Myxovirus parotidis Paramyxovirus Rhabdovirus	Influenza Mumps Measles (rubeola) Rabies
Arbovirus	Arthropod-borne RNA, cubic	Mosquito-borne type B Mosquito-borne types A and B Tick-borne, corona-virus	Yellow fever Encephalitis (many types) Colorado tick fever
Retrovirus	d.s. RNA, helical with envelope	RNA tumor virus HTLV virus HIV (Human immunodeficiency virus)	Tumors Leukemia AIDS (Acquired immunodeficiency syndrome)

d.s. = double-stranded
s.s. = single-stranded

Latent viral infections are usually limited by the body's defenses—phagocytes and antiviral proteins called interferons that are produced by virus-infected cells. Shingles, a painful nerve disease, is another example of a latent viral infection. When the body's immune defenses become weakened by old age or disease, the latent chickenpox virus resurfaces to cause shingles.

Antibiotics are not effective against viral infections because these drugs work by inhibiting certain metabolic activities within cellular pathogens. However, for certain patients with colds and influenza, antibiotics are given to prevent secondary bacterial infections that might occur after the viral infection. Several new chemicals, called antiviral agents, have been developed that interfere with virus-specific enzymes and virus production by either disrupting critical phases in viral cycles or inhibiting the synthesis of viral DNA or RNA. Sometimes, agents that affect the synthesis of

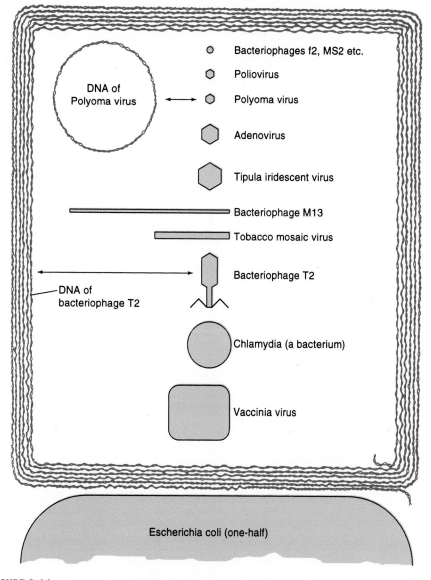

FIGURE 3-14. Comparative sizes of virions, their nucleic acids, and bacteria.

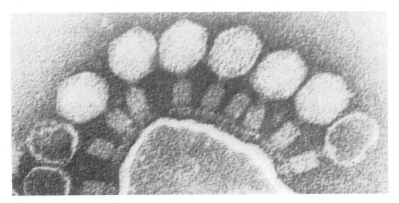

FIGURE 3-15. A partially lysed cell of *Vibrio cholerae* with attached virions of phage CP-T1. Note the empty capsids, full capsids, contracted tail sheaths, base plates and spikes (original magnification ×257,000). (Courtesy of R. W. Taylor and J. E. Ogg, Colorado State University, Fort Collins, Colorado)

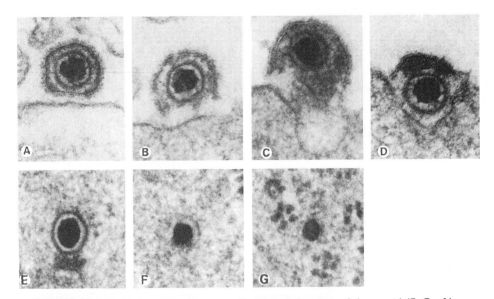

FIGURE 3-16. Adsorption (A), penetration (B–D), and digestion of the capsid (E–G) of herpes simplex on HeLa cells, as deduced from electron micrographs of infected cell sections. Penetration involves local digestion of the viral and cellular membranes (B,C), resulting in fusion of the two membranes and release of the nucleocapsid into the cytoplasmic matrix (D). The naked nucleocapsid is intact in (E), is partially digested in (F), and has disappeared in (G), leaving a core containing DNA and protein. (Morgan C, et al.: J Virol 2:507, 1968)

TABLE 3-13. The 5 Steps of the Lytic Cycle

Step	Name of Step	What Occurs During This Step
1	Attachment	The virus attaches to the host cell; it can only attach to cells possessing the correct receptors
2	Penetration	The viral nucleic acid or, in some cases, the entire virus, enters the cell; if the entire virus enters, uncoating must occur to release the viral nucleic acid from the capsid
3	Biosynthesis	Genetic information contained in the viral nucleic acid directs the production of viral proteins and nucleic acids
4	Assembly	The viral proteins and nucleic acids are assembled into complete viral particles (virions)
5	Release	The newly formed virions are released either by lysis or by budding; in most cases, the host cell is destroyed in the process

new virus particles may also cause damage to host cells. Refer to Chapter 7 for a discussion of chemotherapy.

Many tumors and cancers are apparently caused by viruses that change the genetic composition of cells and cause uncontrolled growth of abnormal cells under certain environmental conditions. Cancer-causing viruses are known as oncogenic viruses. Some chemotherapeutic drugs used to treat cancer are chemicals that interfere with DNA and RNA synthesis in rapidly dividing cells, thus inhibiting or destroying tumor cells and also some human cells (*e.g.*, hair, blood cells, sperm).

Remnants or collections of viruses, called *inclusion bodies,* are often seen in infected cells and are used as a diagnostic tool to identify certain diseases. Inclusion bodies may be found in the cytoplasm (cytoplasmic inclusion bodies) or within the nucleus (intranuclear inclusion bodies), depending on the particular disease. In rabies, the cytoplasmic inclusion bodies in nerve cells are called Negri bodies. The inclusion bodies of AIDS and the Guarnieri bodies of smallpox are also cytoplasmic. Herpes and poliomyelitis viruses cause intranuclear inclusion bodies. In each case, inclusion bodies may be merely aggregates or collections of viruses. Some important human viral diseases include AIDS, chickenpox, cold sores, the common cold, Ebola virus infections, genital herpes infections, German measles, hantavirus pulmonary syndrome, infectious mononucleosis, influenza, measles, mumps, poliomyelitis, rabies, and viral encephalitis. In addition, all human warts are caused by viruses.

Where did viruses come from? Two main theories have been proposed to explain the origin of viruses. One theory states that viruses existed prior to cells, but this seems unlikely because viruses require cells for their replication. The other theory states that cells came first and that viruses represent ancient derivatives of degenerate cells or cell fragments. The question of whether or not viruses are alive depends upon one's definition of life and, thus, is not an easy question to answer.

However, most scientists agree that viruses lack most of the basic features of cells and, thus, they consider viruses to be nonliving entities.

HUMAN IMMUNODEFICIENCY VIRUS

Human immunodeficiency virus (HIV), the cause of acquired immunodeficiency syndrome (AIDS), is an enveloped, double-stranded RNA virus (Fig. 3-17). It is in a member of a genus of viruses called Lentiviruses, in a family of viruses called *Retroviridae* (retroviruses). HIV is able to attach to and invade cells bearing receptors that the virus recognizes. The most important of these receptors is designated CD_4, and cells possessing that receptor are called CD_4+ cells. The most important of the CD_4+ cells is the helper T cell (discussed in Chapter 11); HIV infections destroy these important cells of the immune system. Macrophages also possess CD_4 receptors and can, thus, be invaded by HIV. In addition, HIV is able to invade certain cells that do not possess CD_4 receptors.

BACTERIOPHAGES

Viruses that infect bacteria are called *bacteriophages* or simply phages. Most bacteriophages are species- and strain-specific, meaning that they only infect a particular

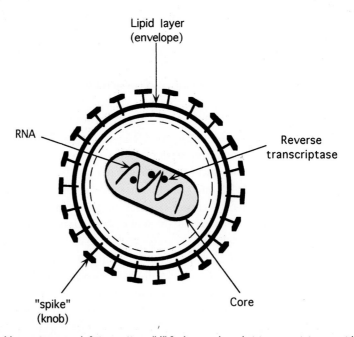

FIGURE 3-17. Human immunodeficiency virus (HIV). An enveloped virus, containing two identical RNA strands. Each of its 72 surface knobs contains a glycoprotein (designated gp 120) capable of binding to a CD_4 receptor on the surface of certain host cells (e.g., T-helper cells). The "stalk" that supports the knob is a transmembrane glycoprotein (designated gp 41), which may also play a role in attachment to host cells. Reverse transcriptase is an RNA-dependent DNA polymerase.

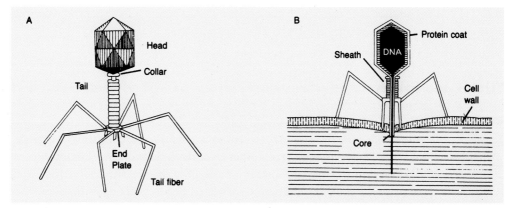

FIGURE 3-18. (*A*) The bacteriophage T4 is an assembly of protein components. The head is a protein membrane with 20 facets, filled with DNA. It is attached to a tail consisting of a hollow core surrounded by a sheath and based on a spiked end-plate to which six fibers are attached. (*B*) The sheath contracts, driving the core through the cell wall, and viral DNA enters the cell.

species or strain of bacteria. Those that infect *Escherichia coli* are called coliphages. Refer to Figures 3-15 and 3-18 to observe the complexity of phages. The end plate, tail fibers, and spikes enable the phages to attach to bacterial cells. The phage DNA is then injected into the cell, much like injecting water from a syringe (Fig. 3-18).

Phages that lyse host bacteria following the production of new bacteriophages are referred to as *virulent bacteriophages.* However, many DNA phages are *temperate bacteriophages;* their viral genome does not take over and destroy the host cell. Temperate phage DNA is injected into the bacterium, integrates into the bacterial chromosome, but causes no damage to the cell. Then, each time the host cell divides, the viral genome also replicates and is passed to each daughter cell. Thus, each daughter cell is infected with the phage DNA. This relationship between the phage and its host cell is called *lysogeny.* The infected bacteria are referred to as *lysogenic bacteria* and the latent viral genome is called a *prophage.* In some cases, certain environmental conditions (*e.g.,* heat, UV light, certain chemicals) can induce the prophage to produce new complete phages and lyse the bacterial cell.

VIRUSES AND GENETIC CHANGES

Much genetic research has been accomplished using bacteria and bacteriophages. In the laboratory, some bacterial species and strains are identified by the specific bacteriophages that will infect them. When bacteriophage DNA is injected into a bacterial cell, the cell has acquired new genetic information (new genes), and can now produce whatever products (gene products) are coded for by the viral genes. This process is called *lysogenic conversion.* An example of such a conversion is the diphtheria bacterium (*Corynebacterium diphtheriae*), which is pathogenic only when a certain phage gene is present in the cell. Evidently, the gene (called the tox gene) that enables diphtheria bacteria to produce diphtheria toxin is actually a viral gene that is injected with the phage DNA. Genetic changes produced in bacteria by bacteriophages are discussed in greater detail in Chapter 6.

Viroids and Prions

Although viruses are very small, nonliving infectious agents, viroids and prions are even smaller and less complex infectious agents. *Viroids* consist of short, naked fragments of RNA (about 300 to 400 nucleotides in length) that can interfere with the metabolism of plant cells and stunt the growth of plants, sometimes killing the plants in the process. They are transmitted between plants in the same manner as viruses are, usually by insects or damaged plant surfaces. Plant diseases thought or known to be caused by viroids include potato spindle tuber (producing small, cracked, spindle-shaped potatoes), citrus exocortis (stunting of citrus trees), and diseases of chrysanthemums, coconut palms, and tomatoes. Thus far, no animal diseases have been discovered that are caused by viroids.

Prions (pronounced "pree-ons") are small infectious proteins that apparently cause fatal neurological diseases in animals, such as scrapie (pronounced "scrape-ee") in sheep and goats; bovine spongiform encephalopathy ("mad cow disease"); and kuru, Creutzfeldt-Jakob (C-J) disease, Gerstmann-Sträussler-Scheinker (GSS) disease, and fatal familial insomnia in humans. Similar diseases in mink, mule deer, elk, and cats may also be caused by prions.

Kuru is a disease that was once common among natives in Papua New Guinea, where women and children ate prion-infected human brains as part of a traditional burial custom (ritualistic cannibalism). Kuru, C-J disease, and GSS disease involve loss of coordination and dementia. Dementia, a general mental deterioration, is characterized by disorientation and impaired memory, judgement and intellect. In fatal familial insomnia, insomnia and dementia follow difficulty sleeping. All of these diseases are fatal spongiform encephalopathies, in which the brain becomes riddled with holes (sponge-like).

Scientists are currently investigating a possible link between "mad cow disease" and a form of C-J disease (called variant CJD or vCJD) in humans. As of 1998, a total of 26 cases of vCJD had been diagnosed in the United Kingdom, possibly the result of eating prion-infected beef. The cattle may have acquired the disease via ingestion of cattle feed that contained ground-up parts of prion-infected sheep. The 1997 Nobel Prize for Physiology or Medicine was awarded to Stanley B. Prusiner, the scientist who coined the term prion, and studied the role of these proteinaceous infectious particles in disease. Of all microorganisms, prions are believed to be the most resistant to disinfectants. The mechanism by which prions cause disease remains a mystery, although it is thought that prions convert normal protein molecules into dangerous ones by causing the normal molecules to change their shape. Many scientists remain unconvinced that proteins alone can cause disease.

■ REVIEW OF KEY POINTS

- Microbes can be divided into those that are cellular (bacteria, algae, protozoa, and fungi) and acellular infectious particles (viruses, virions, and prions).
- Characteristics used for identification and classification of bacteria include cell morphology, staining reactions, motility, colony morphology, atmospheric requirements, nutritional needs, biochemical and metabolic activities, pathogenicity, amino acid sequencing of proteins, and genetic composition.
- The three basic shapes of bacteria are cocci, bacilli, and curved or spiral-shaped. Cocci occur in pairs (diplococci), chains (streptococci), clusters (staphylococci), or packets of four (tetrads) or eight (octads). Bacilli occur singly, in pairs (diplobacilli), in chains (streptobacilli), or they may be branched or filamentous. Very short bacilli are called coccobacilli. Curved bacteria may occur singly, or in pairs or chains. Spiral-shaped bacteria usually occur singly.
- Fixation of a bacterial smear, in preparation for staining, serves to kill the organisms, preserve their morphology, and anchors the smear to the slide.
- Most motile bacteria possess whiplike structures called flagella. Various terms (*e.g.*, monotrichous, amphitrichous, lophotrichous, peritrichous) are used to describe the number and location of flagella on the bacterial cell.
- Bacterial colony morphology includes size, color, overall shape, elevation, and consistency.
- On the basis of its oxygen requirements, a bacterial isolate can be classified as an obligate aerobe, microaerophile, facultative anaerobe, aerotolerant anaerobe, or obligate anaerobe. Bacteria requiring increased concentrations of carbon dioxide are called capnophiles.
- All bacteria need some form of the elements, carbon, hydrogen, oxygen, sulfur, phosphorus, and nitrogen. In addition, certain bacteria require potassium, calcium, iron, manganese, magnesium, cobalt, copper, zinc, and uranium. Fastidious (nutritionally demanding) microbes also need specific vitamins, amino acids, and other organic compounds.
- Pathogenic bacteria may produce capsules, endotoxins, exotoxins and exoenzymes that enable them to cause disease.
- Rickettsias, chlamydias, and mycoplasmas are rudimentary gram-negative bacteria. Mycoplasmas differ from other bacteria because they have no cell walls. Rickettsias and chlamydias are unique because they are obligate intracellular pathogens.
- Archaebacteria are more closely related to eucaryotic organisms than to eubacteria. They differ from eubacteria in several ways: the type of rRNA they possess, their cell walls contain no peptidoglycan, they live in extreme environments, and some (called methanogens) produce methane.
- Algae are eucaryotic, photosynthetic organisms that range in size from tiny, unicellular, microscopic cells to large, multicellular, plantlike seaweeds. They are an

important source of food, iodine and other minerals, fertilizers, emulsifiers, stabilizers, and gelling agents. Some produce toxins, but infections due to algae are extremely rare.

■ Protozoa are eucaryotic, usually single-celled and non-photosynthetic, animal-like microbes; many are parasitic and some are pathogens. They are classified as amebae, flagellates, ciliates, or sporozoa, based primarily on their mode of locomotion.

■ Fungi are eucaryotic, non-photosynthetic organisms, that include mushrooms, toadstools, bracket fungi, puffballs, molds, and yeasts. Many fungi are saprophytic decomposers in nature, and many others are parasitic on animals or plants. The human diseases caused by fungi are classified as superficial, cutaneous, subcutaneous, and systemic mycoses.

■ Mature virus particles, called virions, may be distinguished from living cells because they possess either DNA or RNA, but not both; most consist merely of nucleic acid and structural proteins; they must invade host cells in order to replicate; they lack the enzymes necessary for the production of energy, proteins, and nucleic acid.

■ Viruses are classified by type of nucleic acid, shape of the capsid, size of capsid, number of capsomeres, presence or absence of an envelope, host(s) and host cell(s) that it infects, type of disease caused, and immunological properties.

■ Bacteriophages are viruses that infect certain bacteria. They may destroy the host cell or change the cell genetically.

■ Viroids are fragments of RNA that interfere with the metabolism of plant cells. Prions are infectious proteins that cause certain diseases in animals.

Problems and Questions

1. Describe a symbiotic relationship of a parasitic protozoan.
2. What is the role of fungi in recycling nutrients?
3. What are the differences between protozoa and algae?
4. List five properties of viruses that distinguish them from living cells.
5. Describe the lytic and lysogenic cycles of viruses.

Self Test

After you have read Chapter 3, reviewed the chapter outline, examined the objectives, studied the new terms, and answered the problems and questions above, complete the following self test.

MATCHING EXERCISES

Complete each statement from the list of words provided with each section.

Descriptive Terms

diplococci	staphylococci	diplobacilli
spirochete	cocci	streptococci
bacilli	streptobacilli	coccobacilli

1. Rod-shaped bacteria are called _____.
2. *Treponema pallidum* is an example of a _____.
3. Cocci that are found in pairs are called _____.
4. The spherical or round bacteria are called _____.
5. The term used to describe grape-like clusters of cocci is _____.
6. Bacilli that appear in pairs are called _____.
7. Very short bacilli are called _____.
8. Bacilli that form chains are called _____.
9. Cocci that form chains are called _____.
10. A corkscrew-shaped bacterium is called a _____.

Characteristics of Microorganisms

algae	protozoa	mycoplasmas
viruses	rickettsias	prions
chlamydias	fungi	

1. Organisms as diverse as yeasts and mushrooms are examples of
   _____.
2. All _____ are photosynthetic.
3. Single-celled (usually), eucaryotic microbes that are classified by their means of locomotion are _____.
4. _____ are obligate intracellular pathogens that must obtain their energy (ATP) from host cells.
5. _____ are bacteria that lack a cell wall.

6. The cell walls of _____ contain chitin.
7. Bacteria that must be transmitted from person to person by arthropod vectors such as fleas, lice, and ticks are _____.
8. Acellular microorganisms, most consisting only of nucleic acid and a protein coat, are called _____.
9. The cell walls of _____ contain cellulose.
10. _____ consist only of protein and are thought to cause "mad cow disease."

Diseases and Microorganisms

Match the following diseases with the type of microorganisms that cause them, using the following words:

virus	protozoa	algae
mycobacteria	staphylococci	mycoplasmas
streptococci	rickettsias	spirochete
chlamydias	fungi	curved rods

1. Strep throat _____
2. Typhus _____
3. Cold sores _____
4. Boils and wound infections _____
5. Atypical pneumonia and sinusitis _____
6. Cholera _____
7. Rabies _____
8. Syphilis _____
9. Smallpox _____
10. Tuberculosis _____
11. Poliomyelitis _____
12. Lymphogranuloma venereum _____
13. Influenza _____

TRUE OR FALSE (T OR F)

___ 1. All organisms classified as microorganisms are capable of a free-living existence.
___ 2. Viruses contain both DNA and RNA.
___ 3. An icosahedron is a 20-sided figure.
___ 4. Bacteriophages are viruses that infect bacteria.
___ 5. Fungi may reproduce sexually or asexually.
___ 6. Pathogenic rickettsias differ from viruses in that they require living cells for reproduction.
___ 7. Molds differ from bacteria in that they are multicellular and reproduce by spores.
___ 8. During the cyst stage, protozoa resemble bacterial spores because they have thick walls, are dormant, and resist drying.

__ 9. Bacteria reproduce by budding, and yeasts reproduce by binary fission.
__ 10. Organisms in the genus *Vibrio* are curved bacilli.
__ 11. PPLOs, like the L-forms, have a rigid cell wall.
__ 12. Diseases caused by *Rickettsia* species are transmitted by arthropod vectors.

MULTIPLE CHOICE

1. Procaryotic cells reproduce by
   a. gamete production
   b. budding
   c. mitosis
   d. binary fission

2. The group of bacteria that lack rigid cell walls and take on irregular shapes is
   a. rickettsias
   b. chlamydias
   c. *Clostridium* spp.
   d. mycoplasmas

3. Lysogenic bacteria may be induced to enter a lytic cycle by exposure to
   a. ultraviolet light
   b. sunlight
   c. heat
   d. certain chemicals
   e. all of the above

4. Gram-positive bacteria stain
   a. blue—to—purple
   b. red
   c. yellow
   d. green

5. Rickettsias are the causative agents of
   a. vaginitis
   b. Rocky Mountain spotted fever
   c. trichomoniasis
   d. typhoid fever

6. Infections of the male and female genital tracts may be caused by *Trichomonas vaginalis,* which is a
   a. protozoan
   b. virus
   c. yeast
   d. fungus

7. During lysogeny, the virus
   a. causes lysis of the host cell
   b. is latent in the host cell
   c. has not yet infected the cell
   d. induces the production of more viruses

CHAPTER 4

Basic Chemistry Concepts

BASIC CHEMISTRY
Atoms, Molecules, Elements,
  Compounds
Chemical Bonding

**Importance of Water in Living
  Cells and Systems**
Solutions
Acids

Bases
Salts
pH

OBJECTIVES

After studying this chapter, you should be able to:

- *Differentiate among atoms, ions, isotopes, elements, molecules, and compounds*
- *Calculate the atomic weight and atomic number of an element when told the number of protons and neutrons in the nucleus of an atom of that element*

- *Describe an acid-base reaction*
- *Discuss the importance of water in biochemical reactions*
- *Distinguish between organic and inorganic compounds*

NEW TERMS

Acid
Anion
Atom
Atomic number
Atomic weight
Base
Biochemistry
Cation
Covalent bond
Compound

Dehydration synthesis
  reactions
Electrolyte
Element
Glycosidic bonds
Hydrogen bonding
Hydrolysis reactions
Inorganic chemistry
Ion
Ionic bond

Isotopes
Molecule
Organic chemistry
Peptide bonds
Polarity
Salt
Solute
Solution
Solvent

BASIC CHEMISTRY

Initially, it is not always obvious to students why it is necessary to learn chemistry in a microbiology course. The reason why chemistry is an important part of a microbiology course is the answer to the question, "What exactly is a microorganism?" A microbe can be thought of as a "bag" of chemicals which interact with each other in a variety of ways. Even the "bag" itself is composed of chemicals. Everything a microorganism is and does has to do with chemistry. The various ways microorganisms function and survive in their environment depend on their chemical makeup.

To understand microbial cells and how they work, one must have a basic knowledge of the chemistry of atoms, molecules, and macromolecules (large, complex molecules). As explained in Chapter 2, even the simplest procaryotic cells consist of such very large molecules (macromolecules) as DNA, RNA, proteins, lipids, and polysaccharides, as well as many combinations of these macromolecules that combine to make up structures like capsules, cell walls, cell membranes, cytoplasm, and flagella. These macromolecules can be broken down into smaller units, or molecules, of nucleotides, amino acids, glycerol, fatty acids, and monosaccharides (simple sugars). Each of these molecules, in turn, may be broken down into inorganic molecules of water, carbon dioxide, ammonia, sulfides, and phosphates, and finally into the atoms of carbon, hydrogen, oxygen, nitrogen, sulfur, and phosphorus. Basic inorganic chemistry is introduced in this chapter; organic chemistry and biochemistry are discussed in Chapter 5. *Organic chemistry* is the study of compounds that contain carbon; *biochemistry* is the chemistry of living cells; *inorganic chemistry* involves all other chemical reactions.

Only when all of these units are in place and working together properly can the cell function as a well-managed industrial plant. As in industry, the cell must have the appropriate structures and parts, the regulatory molecules (enzymes) to control its activities, the fuel (nutrients or light) to provide energy, and raw materials (nutrients) for manufacturing end products.

Atoms, Molecules, Elements, Compounds

All substances, whether gas, liquid, or solid, have certain fundamental characteristics in common. If one could break down any substance into its smallest elemental units, it would be composed of *atoms*. Although it is difficult to observe atomic structure, atoms are known to consist of a mass of positively charged protons, noncharged (or neutral) neutrons, tiny negatively charged electrons, and other even smaller subparticles.

An atom has equal numbers of electrons and protons; thus it has a net charge of zero (neither positive nor negative). When an atom loses or gains electrons it becomes a positively or negatively charged *ion* (*e.g.,* Na^+, Cl^-). A positively charged ion is called a *cation* and a negatively charged ion is called an *anion*. The protons and neutrons are found in a central nucleus, and the electrons travel around the nucleus, like negatively charged satellites attracted to a positively charged planet. Figure 4-1 illustrates an oxygen atom. When a multitude of atoms with identical numbers of electrons and protons and the same chemical properties exist together, the substance is known as an *element*. About 30 elements are essential to living organisms, but four of them—hydrogen, oxygen, nitrogen, and carbon—make up over 99% of the mass of most cells. Sulfur (S) and phosphorus (P) are also important elements found in living organisms.

The atoms of each element have their own special weight (or mass) called their *atomic weight*, which is the total mass of the number of protons and neutrons, each of which is assigned a value of 1. Oxygen has eight protons and eight neutrons; thus, it has an atomic weight of 16. Electrons are negligible in weight. *Isotopes* are atoms of the same element, with identical chemical characteristics, but with different atomic weights. They differ in weight because they have various numbers of neutrons within the atoms. For example, carbon-12 has six protons and six neutrons for an atomic weight of 12, whereas carbon-14 has six protons and eight neutrons for an atomic weight of 14; both are isotopes of the element carbon. Elements are also identified by a number that indicates the number of positive charges, or protons, in the nucleus. This is called the *atomic number*. Thus, the atomic number of carbon is 6 and the atomic number of oxygen is 8. Atomic weights and atomic numbers can be found in a periodic table of the elements in any chemistry book; a sample is shown in Figure 4-2. The atomic weights and numbers of the first 20 elements are shown in Table 4-1.

Two or more atoms of the same or different elements that combine into a single stable unit form a *molecule;* examples include ammonia (NH_3), carbon dioxide

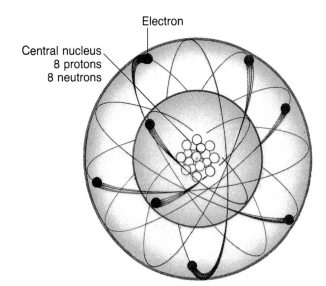

Electron

Central nucleus
8 protons
8 neutrons

FIGURE 4-1. Representation of an oxygen atom. Eight protons and eight neutrons are tightly bound in the central nucleus, around which the eight electrons revolve.

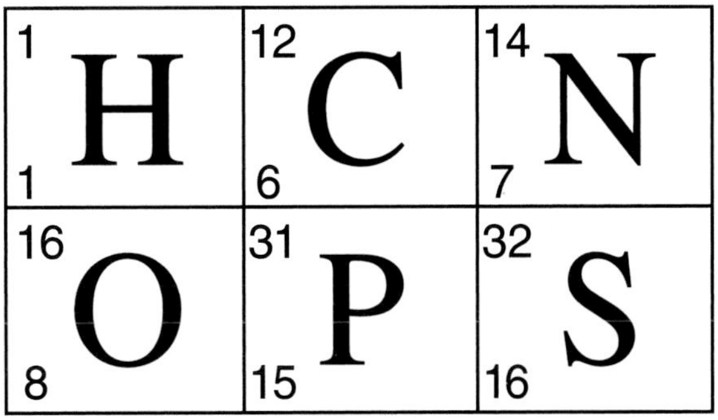

FIGURE 4-2. Selected elements as they would appear in a periodic table of the elements. The atomic weight appears in the upper left corner and the atomic number appears in the lower left corner of each box. By studying the information in each box, you should be able to determine the name of the element and the number of protons, neutrons, and electrons possessed by atoms of that element. (See text for details.)

| 1 H 1 | 12 C 6 | 14 N 7 |
| 16 O 8 | 31 P 15 | 32 S 16 |

(CO_2), oxygen (O_2), sulfuric acid (H_2SO_4), and water (H_2O). When two or more elements exist together in a substance that can be seen or weighed, the substance is called a *compound; e.g.,* common table salt (NaCl). Although there are only about 110 known elements, the number of known compounds is already in the millions, and new ones are being created and discovered each year. The component parts of

TABLE 4-1. The First 20 Elements

Element	Symbol	Atomic Number	Approximate Atomic Weight
Hydrogen	H	1	1
Helium	He	2	4
Lithium	Li	3	6.9
Beryllium	Be	4	9
Boron	B	5	10.8
Carbon	C	6	12
Nitrogen	N	7	14
Oxygen	O	8	16
Fluorine	F	9	19
Neon	Ne	10	20.1
Sodium	Na	11	23
Magnesium	Mg	12	24.3
Aluminum	Al	13	27
Silicon	Si	14	28.1
Phosphorus	P	15	31
Sulfur	S	16	32.1
Chlorine	Cl	17	35.5
Argon	Ar	18	40
Potassium	K	19	39.1
Calcium	Ca	20	40.1

living cells and systems are composed of many macromolecules that interact to produce and maintain life. You will learn about many of these macromolecules, such as polysaccharides, structural proteins, regulatory enzymes, and DNA, although many of the most complex are still not fully understood.

Chemical Bonding

In general, atoms form molecules by gaining or losing electrons (forming *ionic bonds*) or by sharing electrons (forming *covalent bonds*). Ionic bonds hold molecules together in compounds such as NaCl and many others that ionize in water. When sodium chloride dissolves in water, the sodium atom (Na) loses a negative electron and assumes a positive charge, thus becoming a positive ion (Na^+, a cation) in the water. The chlorine atom (Cl) accepts one electron from the donor sodium atom to become a chloride ion (Cl^-, an anion). The ionic bonds between sodium and chlorine atoms are easily broken; however, they re-form in water because the opposite charges are attracted to each other. They may temporarily form NaCl molecules, and then dissociate into ions again (Fig. 4-3).

Other molecules are held together by covalent bonds. A *covalent bond* is formed when atoms share one or more pairs of electrons. In Figure 4-4, the methane (CH_4) molecule has four hydrogen atoms, each of which shares two electrons with the carbon atom. A covalent bond is a much stronger type of bond than an ionic bond; thus, the molecule is much more stable. Carbon compounds are especially good examples of covalent bonding. All of organic chemistry is based on stable, covalently bonded carbon molecules and how they react with other molecules. Biochemistry, the chemistry of living cells and systems, is a study of large macromolecules composed of covalently bonded carbon, hydrogen, oxygen, nitrogen, phosphorus, and sulfur atoms. Loosely bound electrons on the surface of macromolecules enable them to react and bind with other molecules in living organisms. *Peptide bonds,* which hold amino acids together in a protein, and *glycosidic bonds,* which bind the sugar groups of polysaccharides together, are examples of covalent bonds.

Types of Chemical Bonds
Ionic Bonds
Covalent Bonds
    Peptide Bonds
    Glycosidic Bonds
Hydrogen Bonds
Polar Bonds

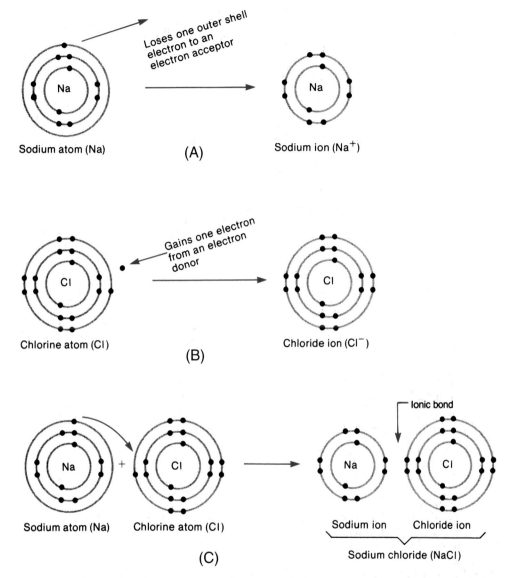

FIGURE 4-3. Ionic bond formation. (*A*) A sodium atom loses an electron to become a positively charged sodium ion (Na$^+$). (*B*) A chlorine atom gains an electron to become a negatively charged chloride ion (Cl$^-$). (*C*) The positive sodium ion is attracted to the negative chloride ion to form a sodium chloride molecule, which is held together by a weak ionic bond.

Hydrogen bonding occurs when a hydrogen atom that is covalently bonded to an oxygen or nitrogen atom of one molecule or chemical group is electrostatically attracted to an oxygen or nitrogen atom on a different molecule or group. These relatively weak hydrogen bonds are found in DNA, proteins, and holding water molecules together in water.

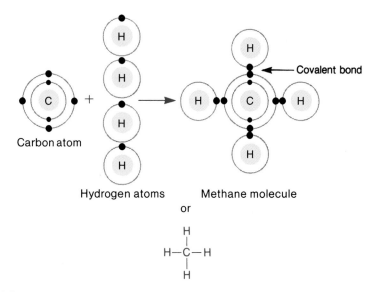

Carbon atom

Hydrogen atoms   Methane molecule

or

$$H-\overset{\displaystyle H}{\underset{\displaystyle H}{C}}-H$$

FIGURE 4-4. Covalent bond formation. A methane molecule with four hydrogen atoms sharing electrons with a carbon atom, forming four covalent bonds.

Importance of Water in Living Cells and Systems

Water is the most abundant molecule in living cells and is essential for the functioning of living cells. Typically, water accounts for 70% to 80% of a cell's total mass. The human body and an *E. coli* cell each contain about 70% water.

A water molecule (H_2O), consisting of two hydrogen atoms on one side of an oxygen atom, has *polarity;* that is, a water molecule has positive and negative regions (Fig. 4-5). The polarity of water molecules makes water an ideal solvent, or suspending medium; its positively and negatively charged regions interact with other polar molecules and ions. In water, the water molecules are held together by hydrogen bonds; a hydrogen atom of one water molecule interacts weakly with the oxygen atom of an adjacent water molecule. The polarity of a water molecule accounts for the following four major characteristics that make it an essential part of living cells:

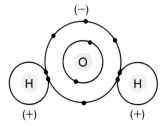

FIGURE 4-5. A water molecule, showing two hydrogen atoms covalently bonded to an oxygen atom. The arrangement gives the molecule a more positive charge near the hydrogen protons and a negative charge near the oxygen electrons, thus providing an attraction between water molecules (polar bonding).

1. Water is an excellent *solvent* (a liquid in which molecules of one or more other chemicals are dissolved). A dissolved substance is known as a *solute* and the combination of solvent plus solute is called a *solution*. Nutrients and waste materials move into and out of a cell, crossing the cell membrane, because they are dissolved in water.
2. Its polarity makes the molecules adhere to each other, producing surface tension and capillary action; hence, water can move within and among cells and tissues.
3. It also serves as an excellent buffer against heat changes because it can absorb and hold large amounts of heat; thus, the cell is protected from temperature fluctuations.
4. Water enters into *hydrolysis reactions* (breakdown or digestion of large macromolecules like cellulose by breaking apart the molecular units) and *dehydration synthesis reactions* (in which macromolecules are synthesized by removal of water molecules). The latter are also known as condensation reactions.

SOLUTIONS

A solution is a homogeneous mixture of one or more substances (solutes) dispersed molecularly (dissolved) in a sufficient quantity of dissolving medium (solvent). The solute may be gas, liquid, or solid. The solvent is usually liquid, but may be solid. The most common solutions are those in which a compound is dissolved in water. Other solvents, such as alcohol, are more commonly used with organic compounds. Two types of compounds—*acids* and *bases*—have characteristics that set them apart from all others. One interesting feature of these two groups is that any acid will react with any base, almost without exception. This certain reactivity is not true for other classes of compounds. Also, it is relatively easy to measure the acidity or alkalinity of solutions of acids and bases.

ACIDS

The sour tastes of lemons, grapefruit, and vinegar are familiar sensations. Acids can be recognized by this sour taste. However, this is *not* a safe method of identification because some acids are poisonous or very destructive to human tissues. Acids are identified by more reliable, safer techniques using chemical indicators or dyes that exhibit a distinct color in an acid solution. Some common acids are shown in Table 4-2.

All acids behave in a similar way because they all share a common feature: the ability to produce hydrogen ions (H^+) when in solution. An acid molecule has at least one hydrogen atom. As an illustration, consider hydrogen chloride or hydrochloric acid, which has the formula HCl. Hydrogen chloride in pure form is a gas, but when bubbled into water, it releases a hydrogen ion (H^+), by donating an electron to form the chloride ion (Cl^-). The result is a cation (the positively charged hydrogen ion) and an anion (the negatively charged chloride ion), as illustrated in the reaction equation shown in Fig. 4-6.

TABLE 4-2. A Few Common Acids and Their Formulas

Acid	Formula	Location or Use
Hydrochloric acid	HCl	Acid in stomach
Sulfuric acid	H_2SO_4	Industrial mineral acid and battery acid
Boric acid	H_3BO_3	Eye wash
Nitric acid	HNO_3	Industrial oxidizing acid
Acetic acid	$HC_2H_3O_2$	Acid in vinegar
Formic acid	$HCHO_2$	Acid in some insect venom
Carbonic acid	H_2CO_3	Acid in carbonated drinks

Hydrochloric acid is an *electrolyte,* defined as a substance with charged particles that can conduct electricity. All acids produce hydrogen ions, which are the particles that contribute to the sour taste of acidic foods and also affect the color of indicators. Indicator dyes are acidic or basic compounds that have distinctive colors when in acidic or basic solutions.

BASES

If you have had the misfortune to taste soap, you know the bitter taste of bases. However, due to their potentially toxic nature, this is not a wise method to use for identification of a base. When dissolved in water, all bases release hydroxyl ions (OH^-) and a cation, and become electrolytes. Table 4-3 lists some common bases. Notice the hydroxyl ion in their formulas.

As mentioned before, all acids will react with all bases. When they react, they form one common product , water (H_2O). To illustrate this, three reactions among different acids and bases are shown in Fig. 4-7.

Acid-base reactions can be summarized by a general statement: *An acid plus a base produces a salt and water.*

SALTS

Salts are another general class of electrolytic compounds. Chemical formulas are often written in a manner to assist in determining whether a substance is an acid, a base, or a salt. For example, acids are written with the "H" at the left, as in HCl

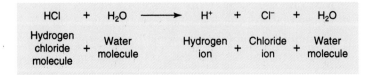

FIGURE 4-6. The ionization of hydrogen chloride in water.

TABLE 4-3. Some Common Bases

Base	Formula	Location or Use
Sodium hydroxide	NaOH	Caustic soda
Potassium hydroxide	KOH	Caustic potash
Ammonium hydroxide	NH_4OH	Household cleaners
Calcium hydroxide	$Ca(OH)_2$	Limewater
Magnesium hydroxide	$Mg(OH)_2$	Antacid drugs

and H_2SO_4 (sulfuric acid). Bases have an "OH" at the right, such as NaOH (sodium hydroxide) and $Mg(OH)_2$ (magnesium hydroxide). The formulas for salts have neither an H at the left nor an OH at the right. Sodium chloride (NaCl) and sodium bicarbonate ($NaHCO_3$) are examples.

It is important for all living creatures, including microorganisms, to have and maintain the appropriate acid, base, and salt balance to metabolize and function properly. When a nutrient medium is prepared for growing microbial cells in the laboratory, the correct acid-base-salt balance is as important as the availability of the necessary nutrients.

pH

pH is a measure of acidity and relates to the hydrogen ion (H^+) concentration. The pH scale used to indicate acidity or alkalinity ranges from 0 to 14, with 7 as the neutral point (Fig. 4-8). Solutions with a pH less than 7 are said to be acidic, while those having a pH greater than 7 are said to be alkaline or basic. Pure water has a pH of 7. If an acid is added to the water, the pH *decreases* to between 0 and 7. Adding a base to water *increases* the pH between 7 and 14. Table 4-4 lists some common substances and the pH for each.

The pH of a substance can be measured with an instrument called a pH meter, with indicator solutions, or with chemically treated papers (indicator papers) that turn a definite color within certain pH ranges. In the human body, the pH of various fluids (blood, lymph) must remain within very narrow ranges to keep the delicately balanced metabolic processes working properly. As with microbial cells, if the pH varies too far from optimum, metabolism is disrupted and death may result.

$$HCl + NaOH \longrightarrow NaCl + H_2O$$
$$H_2SO_4 + Mg(OH)_2 \longrightarrow MgSO_4 + 2\ H_2O$$
$$H_2CO_3 + NaOH \longrightarrow NaHCO_3 + H_2O$$

FIGURE 4-7. General equation: acid + base → salt + water

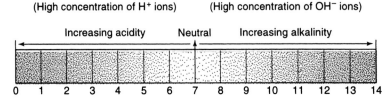

FIGURE 4-8. The pH scale is derived by using a complex mathematical formula to determine the free hydrogen ion (H^+) concentration.

■ REVIEW OF KEY POINTS

■ Atoms, molecules, and macromolecules, including nucleic acids, proteins, lipids, and carbohydrates, are integral parts of every cell, and are necessary for proper metabolism and function of cells and organisms.

■ An atom has a neutral charge because the number of electrons in orbit around the nucleus equals the number of protons in the nucleus. When an atom loses an electron, it becomes a positively charged ion, called a cation. When an atom gains an electron, it becomes a negatively charged ion, called an anion.

■ An element consists of a multitude of atoms with identical numbers of electrons and protons and the same chemical properties. Isotopes are atoms of the same element, with identical chemical characteristics, but with different atomic weights.

■ The atomic weight of an element equals the sum of the protons and neutrons in an atom of that element. The atomic number of an element is equal to the number of protons in an atom of that element.

■ A molecule is a single stable unit of two or more atoms of the same or different elements, such as carbon dioxide, oxygen, and glucose; they are held together by ionic, covalent, hydrogen and other types of bonds.

TABLE 4-4. The pH of Some Common Substances	
Substance	**pH**
Household ammonia cleaner	11.9
Blood (human)	7.35–7.45
Water	7
Milk	6.9
Black coffee	5
Orange juice	2.9
Gastric juice (stomach acid)	1.5

- Water is an essential part of living cells because it serves as a solvent for nutrients. The polarity of water molecules causes the molecules to adhere to each other. Water serves as a buffer against temperature changes, and it is involved in hydrolysis and dehydration synthesis reactions.
- An acid is defined as a substance that produces hydrogen ions (H^+) when added to distilled water. A base is a substance that produces hydroxyl ions (OH^-) when added to water. An acid plus a base produces a salt and water.
- pH is a measure of acidity or alkalinity, and relates to the hydrogen ion concentration of a solution. A pH of 7.0 is considered to be neutral. A pH greater than 7.0 is considered to be alkaline or basic, whereas a pH less than 7.0 is considered to be acidic. A neutral pH is preferred by most cells.

Problems and Questions

1. Describe the relationships between electrolytes, solvents, solutes, and solutions.
2. What are cations, anions, and isotopes?
3. Differentiate between acids and bases, and cite some examples of each.
4. If a nutrient growth medium is too acidic or too basic, might a bacterial culture die? Why?

Self Test

After you have read Chapter 4, reviewed the chapter outline, examined the objectives, studied the new terms, and answered the problems and questions above, complete the following self test.

MATCHING EXERCISE

Complete each statement from the list of words provided with each section.

salts	covalent	dehydration synthesis
neutrons	ionic	hydrogen
protons	hydrolysis	ions
electrons	atoms	

1. Matter is composed of fundamental units called _____.
2. Atoms that have gained or lost electrons are called _____.
3. The various isotopes of an element differ in the number of _____.
4. Molecules will ionize when added to water if the atoms of that molecule are held together by _____ bonds.
5. The atomic number of an element is equal to the number of _____ in the nucleus of an atom of that element.
6. In a neutral atom, the number of _____ orbiting around the nucleus is equal to the number of _____ in the nucleus.
7. When water participates in the breakdown of a large molecule, it is known as a/an _____ reaction.
8. If a water molecule is removed during the synthesis of a molecule, it is known as a/an _____ or dehydrolysis reaction.
9. The bonds holding amino acid molecules together in a protein molecule are examples of _____ bonds.
10. The two strands of a double-stranded DNA molecule are held together by _____ bonds.

MULTIPLE CHOICE

1. Which of the following are examples of covalent bonds?
   a. The bonds that hold water molecules together
   b. The bonds that hold the two strands of a double-stranded DNA molecule together
   c. Peptide bonds and glycosidic bonds
   d. The bonds that hold Na and Cl together in molecules of table salt

2. If the atoms of an element all contain one extra neutron, this form of the element is known as a/an
   a. cation
   b. compound
   c. anion
   d. molecule
   e. isotope

3. The atomic number of an element is equal to
   a. the number of electrons orbiting the nucleus
   b. the number of protons in the nucleus
   c. the number of neutrons in the nucleus
   d. the sum of protons and neutrons in the nucleus

4. A chloride ion carries a net charge of −1 because it has
   a. an extra electron
   b. an extra proton
   c. an extra neutron
   d. lost a proton

CHAPTER 5

Biochemistry: The Chemistry of Life

ORGANIC CHEMISTRY
Carbon Bonds
Cyclic Compounds
BIOCHEMISTRY
Carbohydrates
  Monosaccharides
  Disaccharides
  Polysaccharides

Lipids
  Waxes
  Compound Lipids
  Derived Lipids
Proteins
  Amino Acid Structure
  Protein Structure
  Enzymes

Nucleic Acids
  Function
  Structure
  DNA Structure
  DNA Replication
  Protein Synthesis

OBJECTIVES

After studying this chapter, you should be able to:

- *Describe four main types of biochemical molecules*
- *List the characteristics of monosaccharides, disaccharides, and polysaccharides, and cite examples of each*
- *Discuss the structure of carbohydrates, fats, proteins, and nucleic acids and their breakdown products*

- *Describe the role of enzymes in metabolism*
- *Discuss how DNA directs cellular activities*
- *Cite important differences between the structures of DNA and RNA*
- *Define what is meant by "the central dogma"*
- *Describe the processes of DNA replication, transcription, and translation*

NEW TERMS

Amino acids	Dipeptide	Glycogen
Anticodon	Disaccharide	Hydrocarbon
Carbohydrates	DNA polymerase	Lipids
Catalyst	DNA replication	Messenger RNA (mRNA)
Cistron	Enzyme	Monosaccharides
Codon	Fatty acid	Mutation
Coenzyme	Genetic code	Nucleic acids
Cofactor	Glucose	Nucleotides

Organic compounds	Purine	Substrate
Polymer	Pyrimidine	Transcription
Polypeptide	Ribosomal RNA (rRNA)	Transfer RNA (tRNA)
Polysaccharide	RNA polymerase	Translation
Proteins	Starch	Tripeptide

Everything that a microorganism is and does involves biochemistry. Biochemicals make up the structure of a microorganism and a multitude of biochemical reactions take place within the microorganism. And what is true for microbes is also true for every other living organism. The three characteristics that distinguish living organisms from inanimate objects—(1) their complex and highly organized structure, (2) their ability to extract, transform, and use energy from their environment, and (3) their capacity for precise self-replication and self-assembly—all result from the nature, function, and interaction of biomolecules. Because biochemistry is a branch of organic chemistry, a brief introduction to organic chemistry is presented first.

ORGANIC CHEMISTRY

Organic compounds are compounds that contain carbon and *organic chemistry* is that branch of the science of chemistry that specializes in the study of organic compounds. The term "organic" is somewhat misleading, as it implies that all of these compounds are produced by or are in some way related to living organisms. This is not true. Although some organic compounds are associated with living organisms, many are not. A typical *Escherichia coli* cell contains more than 6000 different kinds of organic compounds, including about 3000 different proteins and approximately the same number of different molecules of nucleic acid. Proteins make up about 15% of the total weight of an *E. coli* cell, whereas nucleic acids, polysaccharides, and lipids make up about 7%, 3%, and 2%, respectively.

Organic chemistry is a broad and important branch of chemistry, involving the chemistry of fossil fuels (petroleum and coal), dyes, drugs, paper, ink, paints, plastics, gasoline, rubber tires, food, and clothing. The number of compounds that contain carbon far exceeds the number of compounds that do not contain carbon. Some carbon-containing compounds are very large and complex, some containing thousands of atoms.

Carbon Bonds

In our current understanding of life, carbon is the primary requisite for all living systems. The element carbon exists in three forms: diamond, graphite, and carbon or carbon black. These three forms have dramatically different physical properties, and it is difficult to believe that they are truly the same element. In addition to these unique physical differences, carbon atoms have a valence of 4, meaning that a carbon atom can bond to four other atoms. For convenience, the carbon atom is illustrated in this text

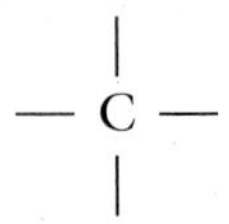

$$-\overset{\displaystyle |}{\underset{\displaystyle |}{C}}-$$

The uniqueness of carbon lies in the ability of its atoms to bond to each other to form a multitude of compounds. The variety of carbon compounds increases still more when atoms of other elements also attach in different ways to the carbon atom.

There are three ways in which carbon atoms can bond to each other: *single bonds, double bonds,* and *triple bonds.* In the illustrations below, each line between the carbon atoms represents a pair of shared electrons (a covalent bond). In a carbon-carbon single bond, the two carbon atoms share one pair of electrons; in a carbon-carbon double bond, two pairs of electrons; and in a carbon-carbon triple bond, three pairs of electrons. Covalent bonds are typical of the compounds of carbon and are the bonds of primary importance in organic chemistry.

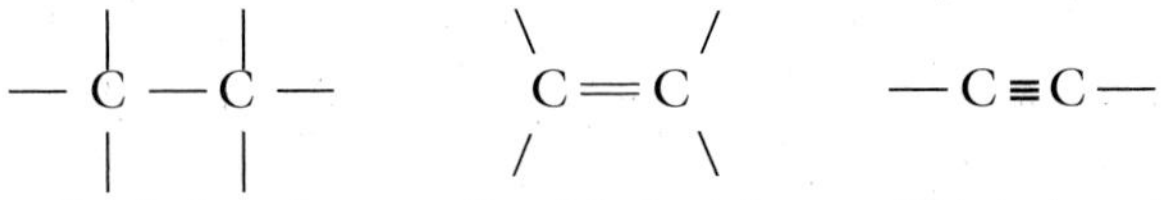

Single bond	Double bond	Triple bond

When atoms of other elements attach to additional available bonds of the carbon atoms, stable compounds are formed. If only hydrogen atoms are bonded to the available bonds, for example, compounds called *hydrocarbons* are formed. Just a few of the many hydrocarbon compounds are shown in Figure 5-1.

When more than two carbons are linked together, longer molecules are formed. A series of many carbon atoms bonded together is logically called a *chain.* The long-chain hydrocarbons are usually liquid or solid, whereas the short-chain hydrocarbons, such as the ones in Figure 5-1, are gases.

Cyclic Compounds

Carbon atoms may link to carbon atoms to close the chain, forming *rings* or cyclic compounds. An example is benzene, which has six carbons and six hydrogens, as shown in Figure 5-2. Although benzene contains six carbon atoms, other ring

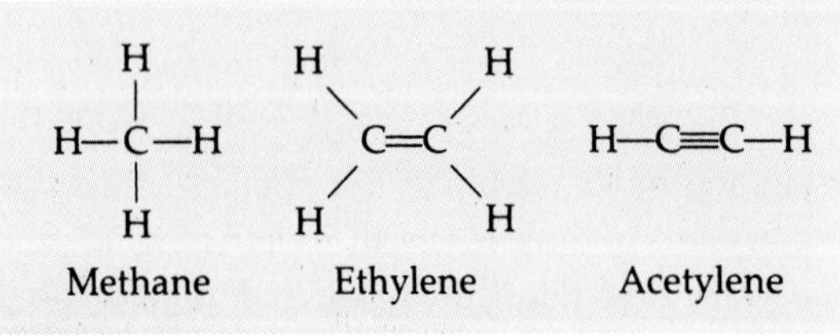

Methane	Ethylene	Acetylene

FIGURE 5-1. Simple hydrocarbons.

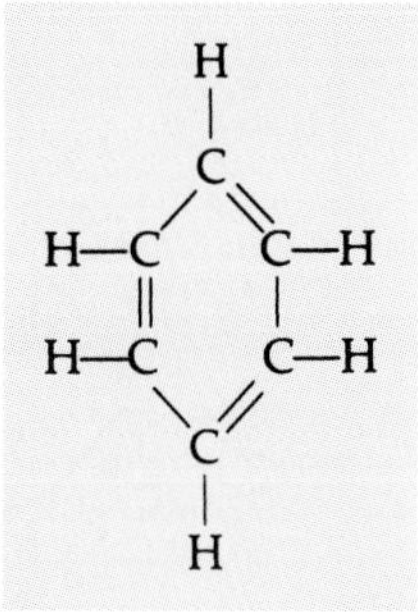

FIGURE 5-2. The benzene ring.

structures contain fewer or more carbon atoms, and some compounds contain fused rings.

BIOCHEMISTRY

Biochemistry is the study of biology at the molecular level. Not only is biochemistry a branch of biology, but it is also a branch of organic chemistry. Biochemistry involves the study of biomolecules present within living organisms. These biomolecules are usually large (called macromolecules) and include carbohydrates, lipids, proteins, and nucleic acids. Vitamins, enzymes, hormones, and energy-carrying molecules, such as adenosine triphosphate (ATP), are also included among these biomolecules.

Humans obtain their nutrients from the foods they eat. The carbohydrates, fats, nucleic acids, and proteins contained in these foods are digested, and their components are absorbed into the blood and carried to every cell in the body. In these cells, components from the food biochemicals are absorbed and rearranged. In this way, the proper compounds necessary for cell structure and function are synthesized. Microorganisms absorb their essential nutrients into the cell by various means (described in Chapter 6). These nutrients are then used in metabolic reactions as sources of energy and as building blocks for enzymes, structural macromolecules, and genetic materials.

Carbohydrates

Carbohydrates are biomolecules composed of carbon (C), hydrogen (H), and oxygen (O) in the ratio of 1:2:1, or simply CH_2O. Glucose, fructose, sucrose, lactose, maltose, starch, cellulose, and glycogen are all examples of carbohydrates.

MONOSACCHARIDES
The simplest carbohydrates are sugars and the smallest sugars are called *monosaccharides* (Gr. "mono" = "one"; "sakcharon" = "sugar"). The most important

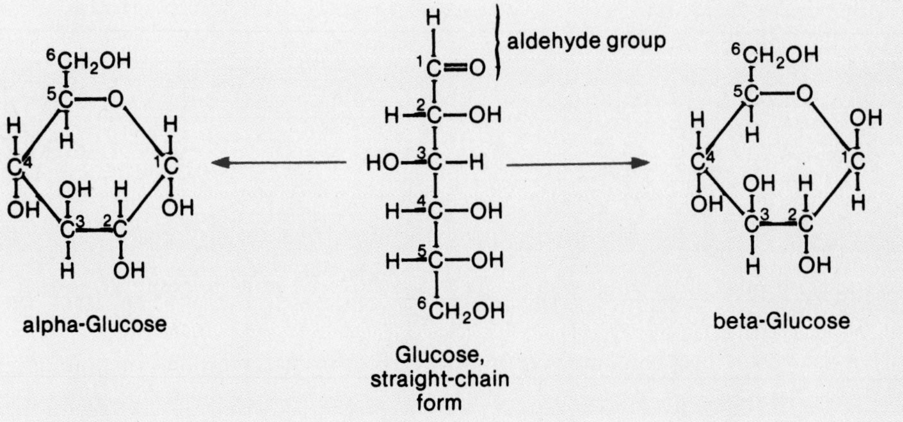

FIGURE 5-3. Glucose. The straight-chain and α and β forms may all exist in equilibrium in solution.

monosaccharide in nature is *glucose* ($C_6H_{12}O_6$), which may occur as a chain or in alpha or beta ring configurations, as shown in Figure 5-3. Monosaccharides may contain from three to seven carbon atoms. A three-carbon monosaccharide is called a triose; one containing four carbons is called a tetrose; five, a pentose; six, a hexose; and seven, a heptose. Ribose and deoxyribose are pentoses that are found in ribonucleic acid (RNA) and deoxyribonucleic acid (DNA), respectively. Glucose (also called dextrose) is a hexose.

The main source of energy for body cells, glucose, is found in most sweet fruits and in blood. The glucose carried in the blood to the cells is oxidized to produce the energy-carrying molecule ATP with its high-energy phosphate bonds. This ATP molecule is the main source of energy used to drive most metabolic reactions. Other monosaccharides are galactose and fructose, both of which are hexoses. Fructose (Fig. 5-4), the sweetest of the monosaccharides, is found in fruits and honey.

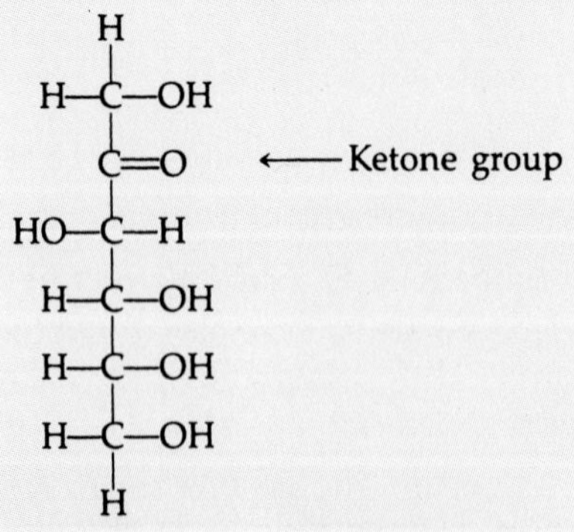

FIGURE 5-4. Fructose in straight-chain form. (Fructose may also exist in the ring form shown in Figure 5-5.)

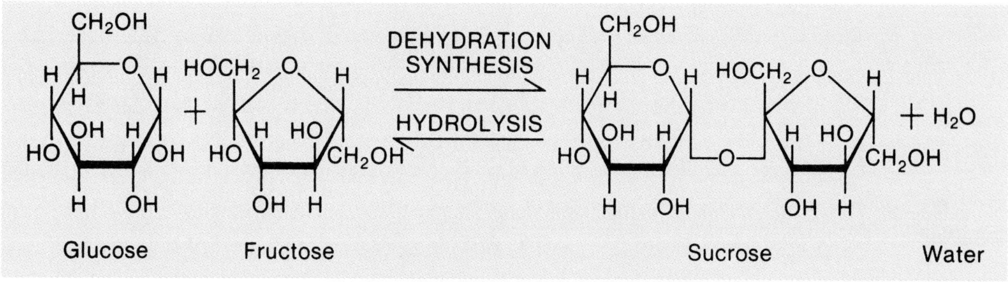

FIGURE 5-5. The dehydration synthesis and hydrolysis of sucrose.

DISACCHARIDES

Disaccharides are sugars that result from the combination of two monosaccharides. The synthesis of a disaccharide from two monosaccharides by removal of a water molecule is called a *dehydration synthesis reaction* (Fig. 5-5). Sucrose (table sugar) is a sweet disaccharide made from a glucose molecule and a fructose molecule. Sucrose comes from sugar cane, sugar beets, and maple sugar. Lactose (milk sugar) and maltose (malt sugar) are also disaccharides. Lactose is made from a molecule of glucose and a molecule of galactose. Maltose is made from two molecules of glucose. In a disaccharide, the monosaccharides are held together by glycosidic bonds. A glycosidic bond is a type of covalent bond.

Disaccharides react with water in a process called a *hydrolysis reaction,* which causes them to break down into two monosaccharides:

$$\text{disaccharide} + H_2O \rightarrow 2 \text{ monosaccharides}$$

$$\text{maltose} + H_2O \rightarrow \text{glucose} + \text{glucose}$$

$$\text{lactose} + H_2O \rightarrow \text{glucose} + \text{galactose}$$

$$\text{sucrose} + H_2O \rightarrow \text{glucose} + \text{fructose}$$

Carbohydrates composed of three monosaccharides are called trisaccharides; those containing four monosaccharides are called tetrasaccharides; and so on until one comes to polysaccharides.

POLYSACCHARIDES

The definition of a *polysaccharide* varies from one reference book to another; some stating that a polysaccharide consists of more than six monosaccharides; others stating more than eight; and still others stating more than ten. *Poly* means "many," and in reality, most polysaccharides contain many monosaccharides—up to hundreds or even thousands of monosaccharides. Thus, in this book, polysaccharides are defined as carbohydrate polymers containing many monosaccharides. Examples include cellulose, starch, and glycogen, which are made up of hundreds of repetitive glucose units, held together by different types of glycosidic bonds (also called glycosidic linkages). Polysaccharides are *polymers,* molecules consisting of many similar sub-units. Some of these molecules are so large that they are insoluble in water. In the presence of the proper enzymes or acids, polysaccharides may be hydrolyzed or broken down into disaccharides and then finally into monosaccharides (Fig. 5-6).

Polysaccharides serve two main functions. One is to store energy that can be used when the external food supply is low. The common storage molecule in animals is *glycogen,* which is found in the liver and in muscles. In plants, glucose is stored as *starch* and is found in potatoes and other vegetables and seeds. Some algae store starch, whereas bacteria contain glycogen granules as a reserve nutrient supply. The other function of polysaccharides is to provide a "tough" molecule for structural support and protection. Many bacteria secrete polysaccharide capsules for protection against drying and phagocytosis. Plant and algal cells have cellulose cell walls to provide support and shape, as well as protection against the environment.

Cellulose is insoluble in water and indigestible for humans and most animals. Some protozoa, fungi, and bacteria have enzymes that will break the β-glycosidic bonds linking the glucose units in cellulose. These microorganisms (saprophytes) are able to disintegrate dead plants in the soil and live in the digestive organs of herbivores (plant eaters). Protozoa in the gut of termites digest the cellulose in the wood that the termites eat. Starch and glycogen are easily digested by animals because they have the digestive enzyme that hydrolyzes the α-glycosidic bonds that link the glucose units into long, helical or branched polymers (Fig. 5-7). Fibers of cellulose extracted from certain plants are used to make paper, cotton, linen, and rope. These fibers are relatively rigid, strong, and insoluble because they consist of 100 to 200 parallel strands of cellulose.

When polysaccharides combine with other chemical groups (amines, lipids, and amino acids), extremely complex macromolecules are formed that serve specific purposes. Glucosamine and galactosamine (amine derivatives of glucose and

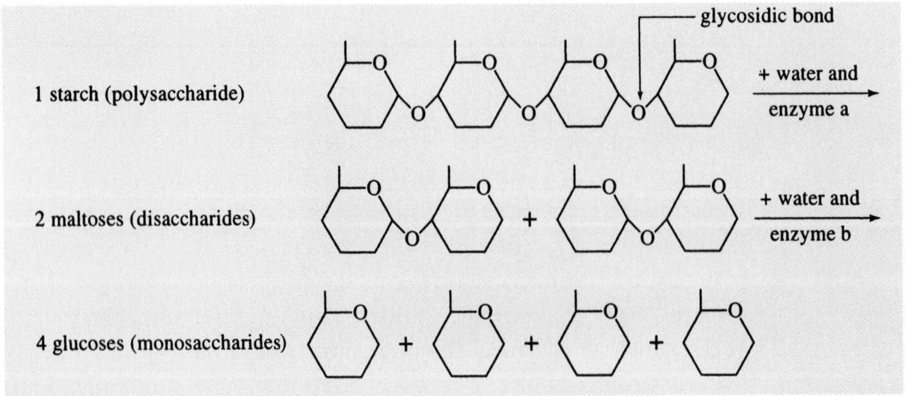

FIGURE 5-6. The hydrolysis of starch.

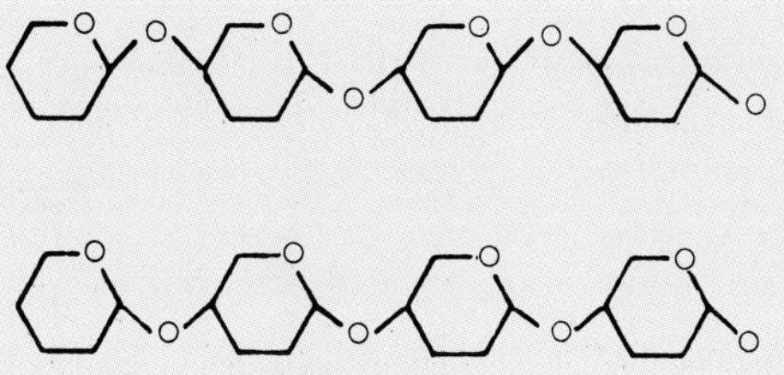

β (beta) linkage (alternating "up and down") in cellulose

α (alpha) linkage (no **alternation) in starch**

FIGURE 5-7. The difference between cellulose and starch.

galactose, respectively) are important constituents of the supporting polysaccharides in connective tissue fibers, cartilage, and chitin. Chitin is the main component of the hard outer covering of insects, spiders, and crabs, and is also found in the cell walls of fungi. The main portion of the rigid cell wall of bacteria consists of amino sugars and short polypeptide chains that combine to form the peptidoglycan layer.

Lipids

Lipids are a class of biomolecules consisting mainly of fats, oils, steroids, and waxes. Most lipids are insoluble in water but soluble in fat solvents such as ether, chloroform, and benzene. Lipids are essential constituents of almost all living cells. They may be classified into the following categories:

1. Simple lipids (contain C, H, O)
   a. Fats and oils (butter, vegetable oils)
   b. Waxes (beeswax, lanolin)
2. Compound lipids (contain C, H, O, N, P)
   a. Phospholipids (the main lipid constituents of cell membranes)
   b. Glycolipids (in nerve cells and cell membranes)
   c. Lipoproteins (in cell membranes)
3. Precursor and derived lipids (contain C, H, O)
   a. Fatty acids
   b. Glycerol
   a. Steroids (sex hormones, cholesterol, vitamin D)
   b. Lipid-soluble vitamins A, D, E, and K

Simple lipids (fats and oils) consist of one molecule of glycerol and three fatty acid molecules joined together by the removal of three molecules of water (dehydration synthesis), which produces a triglyceride fat or oil (Fig. 5-8).

Fatty acids are acids that have been derived from fats by hydrolysis. If the fatty

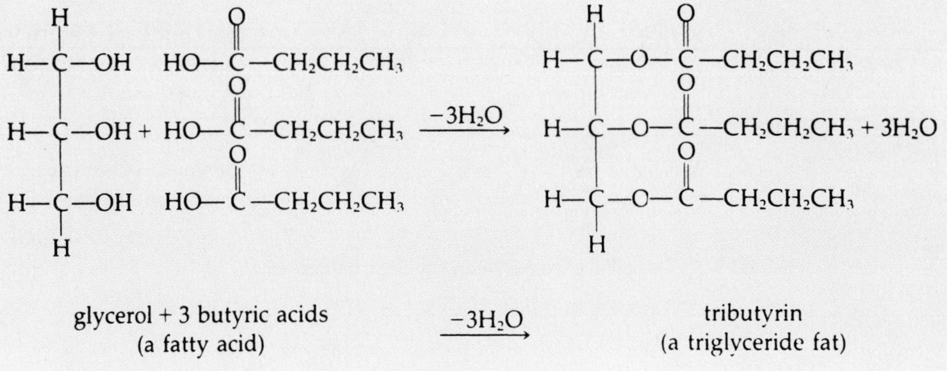

FIGURE 5-8. The synthesis of a fat.

acid contains no double bonds between carbons, it is called a saturated fatty acid because every carbon atom is saturated with hydrogen. When the fatty acid contains one or more double bonds between the carbons ($C=C$), it is said to be an unsaturated fatty acid. A monounsaturated fatty acid contains one double bond, whereas a polyunsaturated fatty acid contains two or more double bonds. Sources of unsaturated fats are peanut, olive, corn, soybean, and cottonseed oils. Fats are liquids or low-melting point solids at room temperature, depending on the relative composition of the fatty acids. Unsaturated fatty acids with short carbon chains are usually liquids. Many people are on low-saturated fat diets to help prevent lipid cholesterol deposits in their hearts and arteries.

A nutritious human diet should include at least the two essential fatty acids that the body cannot synthesize. These are linoleic acid ($C_{17}H_{31}COOH$), from fish liver oils and vegetable oils, and arachidonic acid ($C_{19}H_{31}COOH$) from egg yolk, liver, kidney tissues, and fish liver oils. However, different fatty acids are produced by and are necessary for the growth of bacteria and other microorganisms, depending on the species of the organism.

Saponification is the hydrolysis of a fat into glycerol (glycerin) and its three fatty acids. When NaOH or KOH (strong alkalis) are used to hydrolyze fats, the results are sodium or potassium salts of the fatty acids, which are soaps. Detergents, bile salts, and fat-digestive enzymes also saponify fats.

WAXES

Waxes are chemically different from fats because they are long-chain or complex alcohols, other than glycerol, with fatty acids attached. Waxes are semisolid substances that serve as a protective coating on the surfaces of leaves, stems, fruits, insects, and some bacteria. Waxes in the cell walls of *Mycobacterium tuberculosis,* the bacterium that causes tuberculosis, protect the organism from digestion by phagocytic white blood cells.

COMPOUND LIPIDS

Compound lipids include the phospholipids of cell membranes and the glycolipids of nerve and brain cells. Even when an organism is starving, the amount of these nonfat lipids in the cells does not vary.

DERIVED LIPIDS

Derived lipids are classified as lipids because they are also soluble in fat solvents. These complex molecules of four interlocking carbon rings are called steroids. Many steroid anti-inflammatory medications are produced by fungi and bacteria and are used in the treatment of arthritis and cancer patients. Other examples of steroids include cholesterol, male and female hormones, adrenal cortex hormones, and vitamin D. Vitamins A, E, and K are *not* steroids but are fat-soluble vitamins.

Cholesterol is a necessary body metabolite (Chapter 6). It is found in most eucaryotic membranes and in nervous tissues and aids in fatty acid adsorption. It is also essential for the production of sex hormones, vitamin D, and adrenal cortex hormones. However, an excess of cholesterol may be deposited in either the arteries, causing heart disease, or in the gallbladder, becoming gallstones.

Vitamin D helps maintain the calcium and phosphorus balance in the body. Good sources of vitamin D are fish oils and fortified foods. Vitamin A can be obtained from liver, eggs, butter, cheese, carrots, and green leafy vegetables; a lack of this vitamin can lead to night blindness or hardening of the mucous membranes.

Vitamin E (from plant oils, leafy vegetables, and eggs) reduces sterility in animals and the wasting effects of aging. Vitamin K, which is synthesized in the body, is necessary for normal blood clotting. Some colon bacteria (specifically *Escherichia coli*) also secrete vitamin K, which is absorbed into the blood.

Proteins

Proteins are among the most essential chemicals in all living cells, referred to by some scientists as "the substance of life." Some are the structural components of membranes, cells, and tissues, whereas others are enzymes and hormones that chemically control the metabolic balance within both the cell and the entire organism. All proteins are polymers of amino acids; however, they vary widely in the number of amino acids present and in the sequence of amino acids, as well as in their size, configuration, and functions. Proteins contain C, H, O, N, and sometimes S (sulfur).

AMINO ACID STRUCTURE

A total of 23 different *amino acids* have been found in proteins; 20 primary amino acids plus three secondary amino acids (derived from primary amino acids). Each is composed of carbon, hydrogen, oxygen, and nitrogen; three of the amino acids also have sulfur atoms in the molecule. Humans can synthesize certain types of the

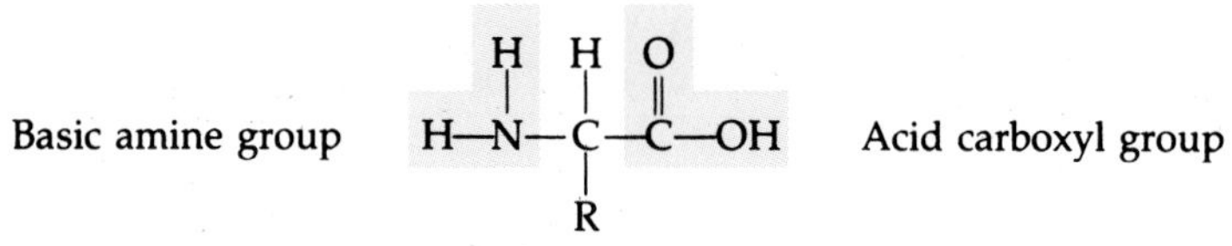

$$H-N-C-C-OH$$

Basic amine group   with H, H, O atoms; R group below C   Acid carboxyl group

FIGURE 5-9. The basic structure of amino acids.

amino acids, but not others. Those that cannot be synthesized (called *essential amino acids*) must be ingested as part of our diets. The term "essential amino acids" is somewhat misleading in view of the fact that *all* of the amino acids are necessary for protein synthesis. Because we cannot manufacture the "essential amino acids," it is *essential* that they be included in our diets.

The general formula for amino acids is shown in Figure 5-9. In this figure, the "R" group represents any of the 23 groups that may be substituted into that position to build the various amino acids. "H" in place of the "R" represents glycine, for example, and "CH_3" in that position results in the structural formula for alanine.

Amino Acids

Alanine (1°)　　　Glutamic acid (1°)　Isoleucine (1°, E)　Serine (1°)
Arginine (1°, E*)　Glutamine (1°)　　Leucine (1°, E)　　Threonine (1°, E)
Asparagine (1°)　　Glycine (1°)　　　Lysine (1°, E)　　Tryptophan (1°, E)
Aspartic acid (1°)　Histidine (1°, E*)　Methionine　　　Tyrosine (1°)
　　　　　　　　　　　　　　　　　　　(1°, E)
Cysteine (1°)　　　Hydroxylysine (2°)　Phenylalanine　　Valine (1°, E)
　　　　　　　　　　　　　　　　　　　(1°, E)
Cystine (2°)　　　Hydroxyproline　　Proline (1°)
　　　　　　　　　　(2°)

Key:　1° = a primary amino acid
　　　2° = a secondary amino acid
　　　E = an essential amino acid
　　　E* = additional essential amino acid in infants

The thousands of different proteins in the body are composed of a great variety of amino acids in various arrangements and amounts. The number of proteins that can be synthesized is virtually unlimited. Proteins are not limited by the number of different amino acids, just as the number of words in a written language is not limited by the number of letters in the alphabet. The actual number of proteins produced by an organism and the amino acid sequence of those proteins are determined by the particular genes present on the organism's chromosome(s).

PROTEIN STRUCTURE

When water is removed, by dehydration synthesis, amino acids become linked together by a peptide bond as shown in Figure 5-10. A *dipeptide* is formed by bonding two amino acids; three amino acids form a *tripeptide*. A chain (polymer) consisting of three or more amino acids is referred to as a *polypeptide*. Polypeptides are said to have primary protein structure—a linear sequence of amino acids in a chain (Fig. 5-11).

Most polypeptide chains naturally twist into helices or sheets as a result of the charged side chains protruding from the carbon-nitrogen backbone of the molecule. This helical or sheetlike configuration is referred to as secondary protein structure and is found in fibrous proteins. Fibrous proteins are long, threadlike molecules that are insoluble in water. They make up keratin (found in hair, nails, wool, horns, feathers), collagen (in tendons), myosin (in muscles), and the microtubules and microfilaments of cells.

Because a long coil can become entwined by folding back on itself, a polypeptide helix may become globular (Fig. 5-11). In some areas the helix is retained, but other areas curve randomly. This globular, tertiary protein structure is stabilized, not only by hydrogen bonding, but also by disulfide bond cross-links between two sulfur groups (S–S). This three-dimensional configuration is characteristic of enzymes, which work by fitting on and into specific molecules (see the next section). Other examples of globular proteins include many hormones (*e.g.,* insulin), albumin in eggs, and hemoglobin and fibrinogen in blood. Globular proteins are soluble in water.

When two or more polypeptide chains are bonded together by hydrogen and disulfide bonds, the resulting structure is referred to as quaternary protein structure (Fig. 5-11). Hemoglobin, for example, consists of four globular myoglobins. The size, shape, and configuration of a protein is specific for the function it must perform. If the amino acid sequence and, thus, the configuration of hemoglobin in red blood cells is not perfect, then the red blood cells may become distorted and assume a sickle shape (as in sickle cell anemia). In this state they are unable to carry oxygen, which is necessary for cellular metabolism. Myoglobin, the oxygen-binding protein found in skeletal muscles, was the first protein to have its primary, secondary, and tertiary structure defined.

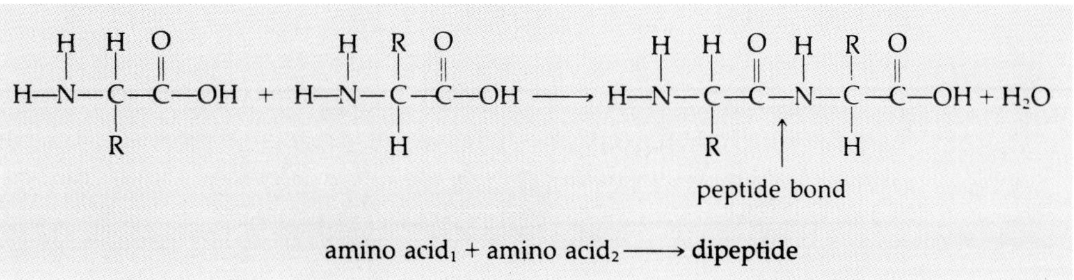

amino acid$_1$ + amino acid$_2$ ⟶ dipeptide

FIGURE 5-10. The formation of a dipeptide (R = any amino acid side-chain group).

Ser—Tyr—Ser—Met—Glu—His—Phe—Arg—Trp—Gly—Lys—Pro—Val—Gly—Lys

A

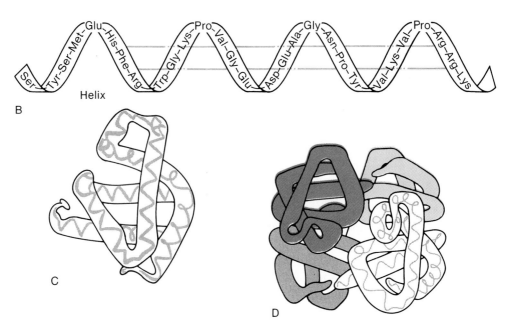

B   Helix

C

D

FIGURE 5-11. Basic protein structure. (A) Primary sequence of amino acids. (B) Secondary helix. (C) Tertiary globular structure. (D) Quaternary structure with four polypeptide chains.

ENZYMES

Enzymes are protein molecules produced by living cells as "instructed" by genes on the chromosomes. Enzymes are essential as *catalysts* in the biochemical reactions of metabolism; they increase the speed at which the reactions occur. A catalyst is defined as an agent that speeds up a chemical reaction without being consumed in the process. In some cases, a reaction will not occur at all in the absence of a catalyst. Almost every reaction in the cell requires the presence of a specific enzyme. Virtually all enzymes are proteins. Many consist solely of protein, whereas others (called apoenzymes) require a nonprotein *cofactor*, such as a metal ion (Ca^{2+}, Fe^{2+}, Mg^{2+}, Cu^{2+}) or an organic compound (called a *coenzyme*) such as vitamin C to function as an enzyme (a holoenzyme). Although enzymes influence the direction of the reaction and increase its rate of reaction, they do not provide the energy needed to activate the reaction.

apoenzyme + cofactor = holoenzyme (a functional enzyme)

Enzymes are usually named by adding the ending "-ase" to the word to indicate the compound or types of compounds on which an enzyme exerts its effect.

Proteases, carbohydrases, and lipases, for example, are enzymes specific for proteins, carbohydrates, and lipids, respectively. The specific molecule upon which an enzyme acts is referred to as that enzyme's *substrate*. Each enzyme has a particular substrate upon which it exerts its effect; thus, enzymes are said to be very specific. Although most enzymes end in "ase," some do not; lysozyme and hemolysins are examples.

Examples of Enzymes

Proteases	RNA polymerase
Peptidases	Catalase
Lipases	Coagulase
DNAse	Oxidase
RNAse	Hemolysins
DNA polymerase	Lysozyme

Poisons and toxins often cause damage to the body by interfering with the action of certain necessary enzymes. Cyanide poison, for example, binds to the iron and copper ions in the cytochrome systems of the mitochondria of eucaryotic cells. As a result, the cells cannot use oxygen to synthesize ATP, which is essential for energy production, and they soon die.

Proteins, including enzymes, may be denatured (structurally altered) by heat or certain chemicals. In a denatured protein, the bonds that hold the molecule in a tertiary structure are broken. With the bonds broken, the protein is no longer functional. Enzymes are discussed further in Chapter 6.

Nucleic Acids

The fourth major group of biomolecules in living cells are the *nucleic acids:* deoxyribonucleic acid (DNA) and ribonucleic acid (RNA). DNA is the macromolecule that makes up the major portion of chromosomes. Its important task is to contain the genetic information for each cell of an organism. The information from the DNA must be carried to the rest of the cell for the cell to function properly; this is accomplished by RNA. Thus, RNA is found not only in the nucleus, but in the cytoplasm as well.

FUNCTION

Nucleic acids have two very important functions. First, they determine precisely all the proteins that are synthesized, many of which control the metabolism of the organism. DNA's second role is to act as the genetic "molecular blueprint of life."

The genetic information must be passed from one generation to the next, from each parent cell to each daughter cell via the DNA.

STRUCTURE

In addition to the elements C, H, O, and N, DNA and RNA also contain P (phosphorus). The building blocks of these nucleic acid polymers are called *nucleotides*. These are more complex monomers (single molecular units that can be repeated to form a polymer) than amino acids, which are the building blocks of proteins. Nucleotides consist of three subunits: a nitrogen-containing (nitrogenous) base, a five-carbon sugar (pentose), and a phosphate group, joined together, as shown in Figure 5-12.

Nucleotides		
3 Parts to Every Nucleotide	**4 DNA Nucleotides (deoxyribonucleotides)**	**4 RNA Nucleotides (ribonucleotides)**
1. Nitrogenous base	1. Adenine (a purine)	1. Adenine (a purine)
	2. Guanine (a purine)	2. Guanine (a purine)
	3. Cytosine (a pyrimidine)	3. Cytosine (a pyrimidine)
	4. Thymine (a pyrimidine)	4. Uracil (a pyrimidine)
2. Pentose	Deoxyribose	Ribose
3. Phosphate group	Phosphate group	Phosphate group

As previously stated, there are two kinds of nucleic acids in cells: DNA and RNA. DNA contains deoxyribose as its pentose, whereas RNA contains ribose as its pentose. There are three types of RNA, which are named for the function they serve: *messenger RNA (mRNA)*, *ribosomal RNA (rRNA)*, and *transfer RNA (tRNA)*. The five most important nitrogenous bases are adenine (A), guanine (G), thymine (T), cytosine (C), and uracil (U). Thymine occurs only in DNA and uracil occurs only in RNA. The other three bases (A, G, C) occur in both DNA and RNA. Both A and G are *purines* (double-ring structures), whereas T, C, and U are *pyrimidines* (single-ringed structures) (see Fig. 5-13).

The nucleotides join together between their sugar and phosphate groups to form very long polymers, 100,000 or more monomers long, as shown in Figure 5-14.

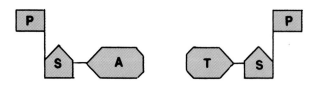

FIGURE 5-12. Two nucleotides, each consisting of a nitrogenous base (A or T), a five-carbon sugar (S), and a phosphate group (P).

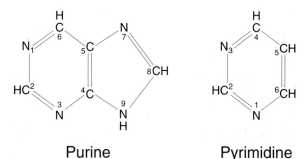

FIGURE 5-13. General structures of purines and pyrimidines.

Purine   Pyrimidine

DNA STRUCTURE

In 1953, James Watson and Francis Crick proposed a double-stranded, helical structure for DNA and the method by which a DNA molecule could copy (replicate) itself exactly, so that identical genetic information could be passed on to each daughter cell. Watson and Crick (along with Maurice Wilkins) received a Nobel Prize in Chemistry in 1962 for their contributions to our understanding of DNA.

STUDY AID Here's one way to remember the difference between purine and pyrimidines. Think of the double-ring structure of purine (adenine and guanine) as "pure and un-CUT." The single-ringed pyrimidines can be thought of as "CUT," where the "C" stands for cytosine, the "U" stands for uracil, and the "T" stands for thymine.

For a double-stranded DNA molecule to form, the nitrogenous bases on the two separate strands must bond together. It was found that A (a purine) always bonds with T (a pyrimidine) via two hydrogen bonds, and G (a purine) always bonds with C (a pyrimidine) via three hydrogen bonds because of the size and bonding attraction between the molecules. The bonding forces of the double-stranded polymer

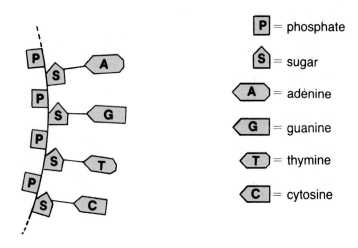

P = phosphate
S = sugar
A = adenine
G = guanine
T = thymine
C = cytosine

FIGURE 5-14. One small section of a nucleic acid polymer.

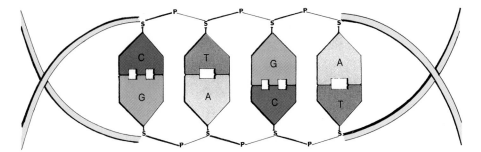

FIGURE 5-15. The DNA double helix.

cause it to assume the shape of a double α-helix, which is similar to a right-handed spiral staircase (Fig. 5-15).

DNA REPLICATION

When a cell is preparing to divide, all of the DNA molecules in the chromosomes of that cell must duplicate, thereby ensuring that the same genetic information is passed to both daughter cells. This process is called *DNA replication*. It occurs by separation of the DNA strands and the building of complementary strands by the addition of the correct nucleotides, as indicated in Figure 5-16. The point on the molecule where DNA replication starts is called the replication fork. The duplicated DNA of the chromosomes can then be separated during ordinary cell division so that the same number of chromosomes, the same genes, and the same amount of DNA as in the parent cell are found in each daughter cell (except during meiosis, the reduction division to produce egg and sperm cells). The most important enzyme required for DNA replication is *DNA polymerase* (or DNA-dependent DNA polymerase). Other enzymes are also required, including DNA helicase and DNA topoisomerase (which initiate the separation of the two strands of the DNA

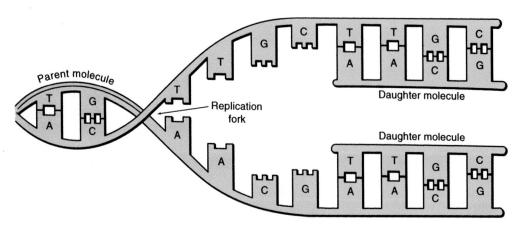

FIGURE 5-16. Replication of DNA before cell division.

molecule), primase (which synthesizes a short RNA primer), and DNA ligase (which connects fragments of newly synthesized DNA). There are subtle differences between DNA replication in procaryotes and eucaryotes.

STUDY AID Francis Crick provided this method of visualizing what happens during DNA replication. First, remember that DNA is a double-stranded molecule, much like a hand within a glove. When the hand is removed from the glove, a new glove is formed around the hand. Simultaneously, a new hand is formed within the glove. What you end up with are two gloved hands, each of which is identical to the original gloved hand.

PROTEIN SYNTHESIS

In a normally functioning living cell, the DNA of the chromosomes controls all of the cell's metabolic activities. It accomplishes this by directing the synthesis of the protein enzymes that regulate the chemical reactions occurring in metabolism.

The translation and interpretation of the information carried by the DNA molecule to produce the proper proteins is a complex procedure. It is the sequence of the four nitrogenous bases (A, T, C, and G) that carries the code for the amino acid sequences of the protein to be synthesized in the cytoplasm of the cell. A diagrammatic representation of this process is shown in Figure 5-17.

You should remember that each chromosome consists of many genes that carry the genetic information for the inherited traits of the organism. Each gene consists of one or more *cistrons,* which are the portions of the chromosomal DNA that code for *one* particular protein or enzyme. The terms "cistron" and "gene" are often used synonymously.

Major Differences Between DNA and RNA
1. DNA is double-stranded, whereas RNA is single-stranded.
2. DNA contains deoxyribose, whereas RNA contains ribose.
3. DNA contains thymine, whereas RNA contains uracil.

When a cell is stimulated (by need) to produce a particular protein, such as insulin, the DNA of the appropriate cistron is activated to unwind temporarily from its helical configuration. This unwinding exposes the bases, which then attract the bases of free nucleotides, and a messenger RNA (mRNA) molecule begins to be built on one strand, the activated strand, of the opened DNA. Thus, the DNA strand has served as a template, or pattern, and has coded for a complementary mirror image of its structure in the RNA. On the growing mRNA molecule, an A will be introduced opposite a T on the DNA molecule, a G opposite a C, a C opposite a G, and a U opposite an A. Remember that there is no T in RNA molecules. This procedure is called *transcription* because the message from the DNA is transcribed onto the mRNA. After the mRNA has been synthesized over the length of the cistron, it is released from the active DNA strand to carry the message to the cytoplasm and direct the synthesis of that particular protein, which is insulin in this case. The primary enzyme that is required for transcription to occur is called *RNA polymerase* (or DNA-dependent RNA polymerase). Located along the DNA template are various "traffic signals" that let the RNA polymerase know where to start and stop the transcription process. Each mRNA molecule carries the same genetic information that was contained in one gene on the DNA molecule. The information in the mRNA molecule will be used to synthesize one protein. It was Francis Crick who, in 1957, proposed what is referred to as "the central dogma":

$$DNA \rightarrow mRNA \rightarrow protein$$

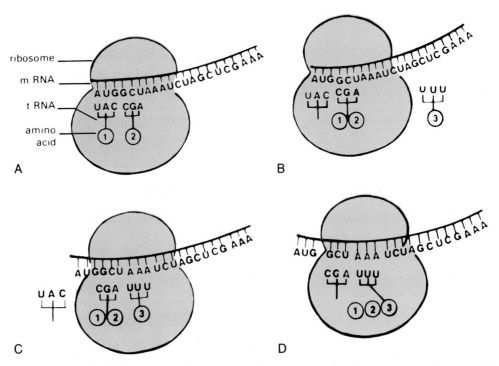

FIGURE 5-17. Outline of protein synthesis. (A) Messenger RNA bound to ribosome. Amino acids 1 and 2 are bonded to their transfer RNAs. The t-RNA-amino acid 1 is bound to the peptidyl binding site. The t-RNA-amino acid 2 is bound to the amino acid binding site. (B) Amino acid 1 is released from its t-RNA and is joined to amino acid 2 through a peptide bond. (C) The ribosome moves along the m-RNA, and t-RNA 1 leaves the peptidyl binding site. The t-RNA 2 with the attached dipeptide moves to the peptidyl site and t-RNA amino acid 3 binds to the amino acid binding site. (D) Amino acid 2 is bonded to amino acid 3 as in step B. The process repeats until the protein is completed.

> **The Central Dogma**
> **(the one gene—one protein hypothesis)**
> 1. The genetic information contained in one gene of a DNA molecule is used to make one molecule of mRNA by a process known as *transcription*.
> 2. The genetic information in that mRNA molecule is then used to make one protein by a process known as *translation*.

In eucaryotes, transcription occurs within the nucleus. The newly formed mRNA molecules then travel through the pores of the nuclear membrane and into the cytoplasm, where they take up positions on the protein "assembly line." Ribosomes, which contain ribosomal RNA (rRNA), attract the mRNA molecules. Ribosomes are usually attached to endoplasmic reticulum membranes in eucaryotic cells.

In procaryotes, transcription occurs in the cytoplasm. Ribosomes attach to the mRNA molecules as they are being transcribed at the DNA; thus, transcription and *translation* (protein synthesis) may occur simultaneously.

Where Various Processes Occur	Procaryotic Cells	Eucaryotic Cells
DNA replication	in the cytoplasm	in the nucleus
Transcription	in the cytoplasm	in the nucleus
Translation	in the cytoplasm	in the cytoplasm

The base sequence of the mRNA is read or interpreted in groups of three bases, called *codons*. The sequence of a codon's three bases is the code that determines which amino acid is inserted in that position in the protein being synthesized. Also located on the mRNA molecule are various codons that act as start and stop signals.

Amino acids must first be activated by attaching to an appropriate *transfer RNA (tRNA)* molecule, which then carries each amino acid from the cytoplasmic matrix to the site of protein assembly. The enzyme responsible for attaching amino acids to their corresponding tRNA molecules is amino acyl-tRNA synthetase.

The three-base sequence of the codon determines which tRNA brings its specific amino acid to the ribosome, because the tRNA has an *anticodon;* a three-base sequence which is complementary to, or attracted to, the codon of the mRNA. For example, the tRNA with the anticodon base sequence UUU carries the amino acid lysine to the mRNA codon AAA. Similarly, the mRNA codon CCG codes for the tRNA anticodon GGC, which carries the amino acid proline at the opposite end of the tRNA molecule. This information system is called the *genetic code*. The following chart illustrates the sequence of three bases in the DNA that codes for a particular codon in mRNA, which, in turn, attracts a particular anticodon on the tRNA carrying a specific amino acid:

DNA	mRNA (codon)	tRNA (anticodon)	Amino Acid
G	C	G	
G	C	G	Proline
C	G	C	

The process of translating the message carried by the mRNA, whereby particular tRNAs bring amino acids to be bound together in the proper sequence to make a specific protein (*e.g.*, insulin), is called *translation*. It should be noted that a eucaryotic cell is constantly producing mRNAs in its nucleus, which direct the synthesis of all the proteins, including metabolic enzymes necessary for the normal functions of that specific type of cell. Also, mRNA and tRNA are short-lived nucleic acids that may be reused many times, then destroyed and resynthesized in the nucleus. The rRNA is made in the dense portion of the nucleus called the nucleolus. The ribosomes last longer in the cell than the mRNA.

As tRNA molecules attach to mRNA while it is sliding over the ribosomes, they bring the correct activated amino acids into contact with each other so that peptide bonds are formed and a polypeptide is synthesized. Recent evidence suggests a role for rRNA in the formation of the peptide bonds. As the polypeptide grows and becomes a protein, it folds into the unique shape determined by the amino acid sequence. This characteristic shape allows the protein to perform its specific function. If one of the bases of a DNA cistron is incorrect or out of sequence (known as a *mutation*), then the amino acid sequence of the gene product will be incorrect and the altered protein configuration will not allow the protein to function properly. Some diabetics, for example, may not produce insulin properly because a mutation in one of their chromosomes caused a rearrangement of the bases in the cistron that codes for insulin. Such errors are the basis for most genetic and inherited diseases, such as phenylketonuria (PKU), sickle cell anemia, cerebral palsy, cystic fibrosis, cleft lip, clubfoot, extra fingers, albinism, and many other birth defects. Likewise, nonpathogenic microbes may mutate to become pathogens, and pathogens may lose the ability to cause disease by mutation.

The relatively new sciences of genetic engineering and gene therapy attempt to repair the genetic damage in some diseases. As yet, the morality of manipulation of human genes has not been resolved by society. However, many genetically engineered microbes are able to produce substances, such as human insulin, interferon, growth hormones, new pharmaceutical agents, and vaccines, that will have a substantial effect on the medical treatment of humans.

■ REVIEW OF KEY POINTS

- ■ Organic compounds contain carbon atoms which bond together by single, double or triple bonds, forming small single molecules, cyclic molecules or long chain molecules, which may combine with other atoms.

- ■ Biochemistry is a branch of biology and organic chemistry; it involves the study of biomolecules, the macromolecules of carbohydrates, fats, proteins and nucleic acids, and their metabolic reactions.

- ■ Microorganisms must absorb essential nutrients to be used in metabolic reactions to produce energy, and building blocks for enzymes, structural macromolecules, and genetic materials.

- ■ Carbohydrates such as monosaccharides, disaccharides, and polysaccharides are used for energy production and synthesis of other metabolic and structural macromolecules.

- ■ Lipids are essential constituents of most living cells. They are grouped as simple, compound, or derived lipids, determined by their complexity and function in the cell.

- ■ The thousands of different proteins in an organism are composed of various amino acids in varied arrangements and amounts. The actual number of proteins and their amino acid sequence is determined by the particular genes present on the organism's chromosome(s).

- ■ The nucleic acids are deoxyribonucleic acid (DNA; found as chromosomes) and ribonucleic acid (RNA; 3 types of which are found at the sites of protein synthesis in the cell).

- ■ The DNA of chromosomes controls all of the cell's metabolic activities by directing the synthesis of the protein enzymes that regulate the chemical reactions occurring in metabolism.

- ■ Protein synthesis is a very complex process, that follows the sequence DNA → mRNA → protein. The information (genetic code) in one gene of a DNA molecule is used to produce a messenger RNA (mRNA) molecule; this part of the process is known as transcription. The mRNA molecule attaches to a ribosome, where the information in the mRNA molecule is used to synthesize a protein; this part of the process is known as translation. Transfer RNA (tRNA) molecules activate amino acids and transfer them to the growing protein chain. Specific amino acids are added at the correct locations because 3-nucleotide sequences (anticodons) on the tRNA molecules recognize 3-nucleotide sequences (codons) on the mRNA molecule. The newly formed protein molecule (polypeptide chain) twists into secondary spirals, that can be used as fibrous structural cell proteins, or the spirals may fold back on themselves to become tertiary globular structures. Quaternary globular proteins, like hemoglobin, consist of more than one globular protein.

- ■ The size, shape, and configuration of a protein is specific for the function it must perform, as determined by the DNA genes on the chromosome.

Problems and Questions

1. Why is it possible for humans to digest starch but not cellulose and chitin? Which microorganisms can digest cellulose?
2. Is the peptidoglycan layer of a bacterial cell wall a polymer? If so, provide reasons to support this claim.
3. Are the fat-soluble vitamins lipids? Why? Where would lipids be found in bacterial cells?
4. What are the differences between structural proteins and enzymes in a cell? Describe the primary, secondary, tertiary, and quaternary structure of proteins. How does an enzyme function?
5. Draw a typical bacterial cell and indicate the types of macromolecules found in its various parts.
6. Explain the chemical characteristics of polymers, carbohydrates, monosaccharides, disaccharides, polysaccharides, lipids, proteins, enzymes, and nucleic acids.
7. Differentiate between DNA, mRNA, rRNA, and tRNA. Which four nitrogenous bases are found in DNA? In RNA?

Self Test

After you have read Chapter 5, reviewed the chapter outline, examined the objectives, studied the new terms, and answered the problems and questions above, complete the following self test.

MATCHING EXERCISE

Complete each statement from the list of words provided with each section.

primary	glycerol	DNA
monosaccharides	tertiary	proteins
RNA	substrate	nucleotides
secondary	disaccharides	
enzymes	polysaccharides	

1. A fat molecule is composed of three fatty acid molecules and one _____ molecule.
2. The order, or sequence, of amino acids in a protein molecule constitutes its _____ structure.
3. _____ is the genetic material of living systems, the subunits (building blocks) of which are called _____.
4. Glycogen, starch, and cellulose are examples of _____.
5. Most metabolic reactions are made possible through the action of _____, or biological catalysts.
6. The substance acted on by an enzyme is called a/an _____.
7. Enzymes are types of _____.
8. Glucose, ribose and deoxyribose are examples of _____.
9. Sucrose, lactose and maltose are examples of _____.
10. In translation, amino acids are activated by a type of _____.

Match the chemicals in Column I with an appropriate group in Column II. An answer from Column II may be used more than once.

Column I

___ 1. Cholesterol
___ 2. Triglyceride
___ 3. Hemoglobin
___ 4. Glucose
___ 5. DNA
___ 6. Sucrose
___ 7. Glycogen
___ 8. RNA
___ 9. Phospholipid
___ 10. Cellulose

Column II

___ a. Lipid
___ b. Carbohydrate
___ c. Nucleic acid
___ d. Protein

TRUE OR FALSE (T OR F)

___ 1. In general, DNA molecules are double-stranded.
___ 2. RNA molecules contain thymine but do not contain uracil.
___ 3. Transcription occurs within the nucleus of eucaryotic cells, whereas translation occurs on ribosomes.
___ 4. In DNA, the molecule guanine normally pairs with cytosine.
___ 5. Lipids are generally insoluble in organic solvents.
___ 6. Steroids are a form of lipid.
___ 7. The chief polysaccharide of animal cells is starch.
___ 8. RNA molecules are generally single-stranded.
___ 9. The type of chemical bond that links amino acids to one another in a polypeptide molecule is a glycosidic bond.
___ 10. Purines are double-ringed structures, whereas pyrimidines are single-ringed structures.

MULTIPLE CHOICE

1. Which of the following are the building blocks of proteins?
   a. monosaccharides
   b. amino acids
   c. nucleotides
   d. fatty acids

2. Glucose, sucrose and cellulose are examples of
   a. polypeptides
   b. polysaccharides
   c. carbohydrates
   d. monosaccharides
   e. disaccharides

3. Which of the following nitrogenous bases is not usually found in an RNA molecule?
   a. adenine
   b. guanine
   c. cytosine
   d. thymine
   e. uracil

4. Which of the following are purines?
   a. adenine and thymine
   b. guanine and cytosine
   c. cytosine and uracil
   d. adenine and guanine

5. Which one of the following is not found at the site of protein synthesis?
   a. DNA
   b. mRNA
   c. rRNA
   d. tRNA

6. Which of the following statements about DNA is not true?
   a. Within cells, DNA molecules are usually double-stranded.
   b. DNA molecules contain ribose.
   c. DNA contains thymine but not uracil.
   d. In a double-stranded DNA molecule, adenine on one strand will usually be hydrogen bonded to thymine on the complementary strand.

7. The amino acids in a polypeptide chain are connected by
   a. glycosidic bonds
   b. hydrogen bonds
   c. peptide bonds
   d. ionic bonds

8. Which of the following is not one of the 3 parts of a nucleotide?
   a. a pentose
   b. a nitrogenous base
   c. a disulfide bridge
   d. a phosphate group

9. A heptose contains how many carbon atoms?
   a. 3
   b. 4
   c. 5
   d. 6
   e. 7

10. Virtually all enzymes are
   a. carbohydrates
   b. nucleic acids
   c. substrates
   d. proteins
   e. lipids

11. The notation C=C means that the two carbon atoms are sharing _____ electrons.
   a. one
   b. two
   c. three
   d. four

Microbial Physiology and Genetics

NUTRITION
Nutritional Requirements
Nutritional Types

ENZYMES, METABOLISM, AND ENERGY
Enzymes
 Metabolic Enzymes
 Inhibition of Enzymes
Energy Metabolism
Energy Production (Catabolism)
 Aerobic Respiration of Glucose
 Anaerobic Fermentation

Aerobic Oxidation by Chemolithotrophs
Anaerobic Respiration by Chemotrophs
Metabolic Biosynthesis (Anabolism)
Energy Conversion
Energy Use
Photosynthesis
Chemosynthesis

MICROBIAL GROWTH
Culture Media
Population Counts

Population Growth
 Curve

BACTERIAL GENETICS
Changes in Bacterial Genetic Constitution
 Lysogenic Conversion
 Transduction
 Transformation
 Conjugation

GENETIC ENGINEERING

GENE THERAPY

OBJECTIVES

After studying this chapter, you should be able to:

- List the various nutritional types of bacteria
- Discuss how these nutritional types fit into the biosphere
- State the meaning of phototroph, autotroph, chemotroph, and heterotroph
- Define producers, consumers, and decomposers
- Describe and give an example of catabolism, anabolism, respiration, and photosynthesis

- List six uses for energy in a cell
- Draw and label a bacterial growth curve
- List the reasons bacteria die during the death phase
- Describe the bacterial chromosome
- List and describe five ways by which the genetic constitution of bacteria can be changed

NEW TERMS

Adenosine triphosphate (ATP)
Anabolism
Autotroph

Catabolism
Catalyze
Chemoautotroph
Chemoheterotroph

Chemolithotroph
Chemostat
Chemosynthesis
Chemotroph

Citric acid cycle
Coenzyme
Competence
Conjugation
Death phase
Dehydrogenation
Differential media
Ecosystem
Electron transport
   system
Endoenzyme
Enriched media
Episome
Exoenzyme
Fastidious bacterium
Fermentation
Gene therapy

Genetic engineering
Genetics
Genotype
Glycolysis
Growth curve
Heterotroph
In vitro
In vivo
Lag phase
Lithotroph
Logarithmic growth
   phase
Lysogenic conversion
Metabolism
Metabolite
Mutagen
Mutant

Mutation
Oxidation
Oxidation-reduction
   reactions
Phenotype
Photoautotroph
Photoheterotroph
Photolithotroph
Phototroph
Plasmid
Reduction
Selective media
Stationary phase
Transduction
Transformation
Viable plate count

Microorganisms, especially bacteria, are ideally suited for use in studies of the basic metabolic processes of life. They are inexpensive to maintain; they take up little space; and they reproduce quickly. More importantly, species of bacteria can be found that represent each of the nutritional types of organisms on earth. We can learn much about our own cells by studying the nutritional needs of bacteria, their metabolic pathways, and why they grow or die under certain conditions. Population growth cycles of bacteria illustrate the growth phases of any population of a species of organism, including humans.

Each tiny single-celled bacterium strives to produce more cells like itself and, as long as water and a nutrient supply are available, it often does so at a rate that is alarming. Under favorable conditions, in 24 hours, the offspring of a single *Escherichia coli* bacterium would outnumber the entire human population on the earth!

Bacteria are easy to find, grow, and maintain in the laboratory. Their morphology, nutritional needs, and some of their metabolic reactions are easily observable; thus, when these usual characteristics change in a pure culture, the resultant mutant (a genetically changed organism) can be quickly identified. Because some bacteria, molds, and viruses produce generation after generation so rapidly and easily, they have been used extensively in genetic studies. In fact, most of the genetic knowledge of today was and is being obtained from the study of these microorganisms.

NUTRITION

The study of bacterial nutrition and other phases of microbial physiology helps us understand the vital chemical processes that occur within every living cell, including those of the human body.

Nutritional Requirements

All living protoplasm contains six major chemical elements: carbon, hydrogen, oxygen, nitrogen, phosphorus, and sulfur. Other elements usually necessary in lesser amounts include sodium, potassium, chlorine, magnesium, calcium, iron, iodine, and some trace elements. Combinations of all of these elements make up the vital macromolecules of life, including carbohydrates, fats, proteins, and nucleic acids (DNA and RNA).

Each organism must have a source of energy and nutrient chemicals to build the necessary cellular materials of life. Those materials that organisms cannot synthesize, but are required in building the macromolecules of protoplasm, are termed essential nutritional requirements. These are the nutrients that must be continually supplied to every organism for it to live. Essential nutrients vary from species to species.

Nutritional Types

Because microorganisms have been evolving since the beginning of life on earth, there are microbes representing each of the various nutritional types. Various terms are used to indicate the type of energy source and type of carbon source. As you will see, the various terms can be used in combination (Table 6-1).

The terms phototroph and chemotroph are used to describe an organism's energy source. A *phototroph* uses light as an energy source. Organisms able to convert light energy into chemical energy are called *photosynthetic organisms* and the process by which they do so is called *photosynthesis. Chemotrophs* use either inorganic or organic chemicals as an energy source.

The terms autotroph, lithotroph, and heterotroph (or organotroph) are used to describe an organism's carbon source. *Autotrophs* use carbon dioxide (CO_2) as their carbon source, *lithotrophs* use inorganic compounds other than CO_2, and *heterotrophs* use organic compounds. Photosynthetic organisms such as plants, algae, and cyanobacteria are examples of autotrophs. All animals (including humans), protozoa, and fungi are examples of heterotrophs. Both saprophytic fungi, which live on dead and decaying organic matter, and parasitic fungi are heterotrophs. Most bacteria are heterotrophs.

TABLE 6-1. Terms Relating to Energy and Carbon Sources

Terms Relating to Energy Source	Terms Relating to Carbon Source		
	Autotrophs (use CO_2 as a carbon source)	**Lithotrophs** (use inorganic chemicals other than CO_2 as a carbon source)	**Heterotrophs** (also known as organotrophs; use organic chemicals as a carbon source)
Phototrophs (use light as an energy source)	**Photoautotrophs** (e.g., plants, algae, some bacteria, including cyanobacteria)	**Photolithotrophs** (e.g., some bacteria)	**Photoheterotrophs** (also known as photoorganotrophs; e.g., some bacteria)
Chemotrophs (use chemicals as an energy source)	**Chemoautotrophs** (e.g., some bacteria)	**Chemolithotrophs** (e.g., some bacteria)	**Chemoheterotrophs** (also known as chemoorganotrophs; e.g., protozoa, fungi, animals, most bacteria)

Terms can be combined to indicate both an organism's energy source and its carbon source. For example, *photoautotrophs* are organisms (such as plants, algae, cyanobacteria, purple and green sulfur bacteria) that use light as an energy source and CO_2 as a carbon source. *Photoheterotrophs* (or *photoorganotrophs*), like purple nonsulfur and green nonsulfur bacteria, use light as an energy source and organic compounds as a carbon source. *Chemoautotrophs* (such as nitrifying, hydrogen, iron, and sulfur bacteria) use chemicals as an energy source and CO_2 as a carbon source. *Chemolithotrophs* use chemicals as an energy source and inorganic compounds other than CO_2 as a carbon source. *Chemoheterotrophs* (or *chemoorganotrophs*) use chemicals as an energy source and organic compounds as a carbon source. All animals, protozoa, fungi, and most bacteria are chemoheterotrophs.

Ecology is the study of the interactions between organisms and the world around them. The term *ecosystem* refers to the interactions between living organisms and their nonliving environment. Interrelationships among the different nutritional types are of prime importance in the functioning of the ecosystem. Photolithotrophs (like algae and plants) are the producers of food and oxygen for the chemoheterotrophs (such as animals). Dead plants and animals would clutter the earth as debris if the chemoheterotrophic saprophytic decomposers (certain fungi and bacteria) did not break down the dead organic matter into inorganic compounds (carbon dioxide, nitrates, phosphates) of the soil and air so that they could be used and recycled by the photolithotrophs. Plants, algae, and photosynthetic bacteria are photoautotrophs. They contribute energy to the ecosystem by trapping energy from the sun and using it to build organic compounds (carbohydrates, fats, nucleic acids, and proteins) from inorganic materials in the soil, water, and air. In photosynthesis, oxygen also is released for respiration by animals.

ENZYMES, METABOLISM, AND ENERGY

Microorganisms are able to grow only if they obtain the proper raw materials for use as nutrients and for the manufacture of the enzymes necessary to promote metabolic reactions. These processes are similar to those in our own body cells. The term *metabolism* refers to all chemical reactions that occur within any cell. Metabolic reactions are enhanced and regulated by enzymes.

Enzymes

METABOLIC ENZYMES

Enzymes are biological *catalysts;* they *catalyze* biochemical reactions. That is to say, enzymes are biomolecules that accelerate the rate of biochemical reactions at certain temperatures without being used up in the process. These reactions might occur at the same temperature without enzymes, but at a much slower rate. In some

INSIGHT
The Oxygen Holocaust

In the beginning, all the world was anaerobic—there was no oxygen. Scientists tell us that the first organisms were anaerobic microorganisms that evolved some 3 to 4 billion years ago. Life on earth then remained anaerobic for hundreds of millions of years.

Then, about 2 billion years ago, the first worldwide pollution crisis occurred. "The oxygen holocaust" (as described by Margulis and Sagan) came about as the result of the evolution of the purple and green photosynthetic microbes. These organisms were able to make use of the hydrogen in water, by photosynthesis, leaving a waste product called oxygen. Yes, the oxygen that we humans consider so precious was originally a gaseous poison dumped into the atmosphere.

As oxygen gains electrons (becomes reduced), highly reactive, short-lived chemicals (called free radicals) are produced. These free radicals wreak havoc with the organic compounds that are the very basis of life. They destroy membranes and enzymes and are lethal to cells.

As stated by Lovelock, "the first appearance of oxygen in the air heralded an almost fatal catastrophe for early life." Many anaerobic microbes were immediately destroyed. The microbes able to survive were those that responded to the crisis by developing ways to detoxify and eventually exploit the dangerous pollutant. These were the organisms that developed the ability to produce enzymes—like catalase, peroxidase, and superoxide dismutase—that break down and neutralize the various toxic reduction products of oxygen.

Those organisms lacking such enzymes either died or were forced to retreat to ecological niches devoid of oxygen, such as soil and mud and deep within the bodies of animals. Those anaerobes that constitute part of our own indigenous microflora, for example, lead a rather pampered existence. We provide them with warmth and nutrients and a safe haven from their worst enemy—oxygen. (To learn more about this subject, refer to *Microcosmos* by Lynn Margulis and Dorion Sagan [1986] and *Gaia—a New Look at Life on Earth* by J.E. Lovelock [1979]).

cases, the reaction will not occur at all in the absence of the enzyme. A *substrate* is a compound on which an enzyme exerts its effect. The enzyme must fit the combining site of the substrate, as a key fits a lock (Fig. 6-1). Usually, an enzyme is specific; that is, it works on only one type of substrate, but occasionally it can attach to different substrates having similar combining sites.

Virtually all enzymes are protein molecules. The three-dimensional shape of the protein enables it to attach to one or more substrates to accelerate a particular biochemical reaction, without causing the enzyme to change in the process. The enzyme continues to move from substrate molecule to substrate molecule at a rate of several hundred each second, producing a supply of the end product for as long as this particular end product is needed by the cell. However, enzymes do not last indefinitely; they finally degenerate and lose their activity. Therefore, the cell must synthesize and replace these important proteins. Because there are thousands of metabolic reactions continually occurring in the cell, there must be thousands of enzymes available to control and direct the essential metabolic pathways. At any particular time, all of the required enzymes need not be present; this situation is controlled by genes on the chromosomes and the needs of the cell, which are determined by the internal and external environment. If no lactose was present in the organism's environment, for example, it would not need the enzyme required to break down lactose. Enzymes that remain within the cell are called *endoenzymes.* Enzymes produced within the cell, but then released from the cell to perform extracellular functions are called *exoenzymes.* The digestive enzymes within phagocytes

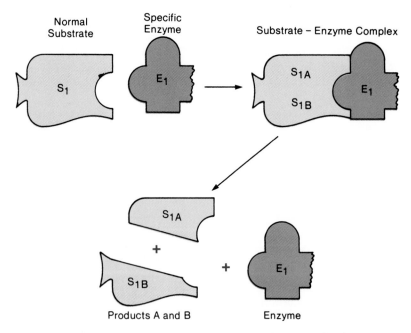

FIGURE 6-1. Action of a specific enzyme breaking down a substrate molecule.

are good examples of endoenzymes; they are used to digest materials that the phagocytes have ingested. Examples of exoenzymes are enzymes like cellulase and pectinase, which are secreted by saprophytic fungi to digest cellulose and pectin in the external environment. Large molecules outside the cell are broken down into smaller molecules, which can then be absorbed into the organism.

Some enzymes (called apoenzymes) exist in the cell in an inactive state; these require cofactors to activate them to perform their intended functions. These cofactors are usually mineral ions such as magnesium, calcium, or iron cations. Other apoenzymes function only in the presence of a *coenzyme* that acts as a carrier of small chemical groups (such as H_2) that are removed from the substrate. Coenzymes are small organic, vitamin-type molecules such as flavin-adenine dinucleotide (FAD) and nicotinamide-adenine dinucleotide (NAD). These coenzymes participate in the citric acid cycle, which is discussed later in this chapter. Coenzymes, like enzymes, do not have to be present in large amounts because they are recycled through many reactions. However, the lack of certain vitamins from which the coenzymes are synthesized will halt all reactions involving that particular coenzyme-enzyme complex.

Certain enzymes, called hydrolases, break down macromolecules by the addition of water in a process called hydrolysis. These hydrolytic processes enable saprophytes to break apart such complex materials as leather, wax, cork, wood, rubber, hair, some plastics, and even mechanical equipment. Some of the enzymes involved in the formation of large polymers like DNA and RNA are called polymerases. These polymerases are active each time the DNA of a cell is replicated and during the synthesis of RNA molecules (Chapter 5).

INHIBITION OF ENZYMES

Many factors affect the activity of enzymes. Any physical or chemical change may diminish or completely stop enzyme activity, because these protein molecules function properly only under optimal conditions. Optimal conditions for enzyme activity include a relatively limited range of pH and temperature, and the appropriate concentration of enzyme and substrate. Extremes in heat and acidity can denature (or alter) enzymes by breaking the bonds responsible for their three-dimensional shape, resulting in the loss of enzymatic activity. This explains why a particular bacterium grows best at a certain temperature and pH; these are the optimal conditions for the enzymes possessed by that bacterium. Optimal pH and temperature vary from one species to another.

Although mineral ions, calcium, magnesium, and iron, enhance the activity of enzymes by serving as cofactors, other heavy metal ions such as lead, zinc, mercury, and arsenic usually act as poisons to the cell. These toxic ions inhibit enzyme activity by replacing the cofactors, or sometimes replacing only hydrogen, at the combining site of the enzyme, thus inhibiting normal metabolic processes. Some disinfectants containing mineral ions are effective in inhibiting the growth of bacteria by this means.

Sometimes, a similar substrate can be used as an inhibitor to deliberately interfere with a particular metabolic pathway. It binds with the enzyme; thus the end product is not produced. A chemotherapeutic agent, such as a sulfonamide drug, for example, may bind with certain enzymes to prevent essential metabolites from being formed and thereby inhibits the growth of a pathogen.

The term *metabolism* refers to all of the chemical reactions occurring within in a cell, including the production of energy, intermediate products, and end products. A *metabolite* is any molecule that is a nutrient (energy source), intermediary product, or end product in a metabolic reaction. Within a cell, metabolic reactions proceed in many directions simultaneously, breaking down some materials and synthesizing (building) others. Most metabolic reactions fall into two categories: catabolism and anabolism. *Catabolism* is the metabolic degradation (breakdown) of organic compounds that results in the production of energy and smaller molecules. Catabolic reactions involve the breaking of chemical bonds. Any time chemical bonds are broken, energy is released. Breaking a disaccharide down into its two monosaccharides by hydrolysis is an example of a catabolic reaction.

Anabolism refers to those biosynthetic processes that use energy for the synthesis of protoplasmic materials needed for growth, maintenance, and other cellular functions. Anabolic reactions require energy because chemical bonds are being formed. It takes energy to create a chemical bond. Examples of anabolic reactions include creating a disaccharide from two monosaccharides by dehydration synthesis, the biosynthesis of polypeptides by linking amino acids molecules together, and the biosynthesis of nucleic acid molecules by linking nucleotides together.

Catabolism	Anabolism
All of the catabolic reactions in a cell	All of the anabolic reactions in a cell
Catabolic reactions release energy	Anabolic reactions require energy
Catabolic reactions involve the breaking of bonds; whenever chemical bonds are broken, energy is released	Anabolic reactions involve the creation of bonds; it takes energy to create chemical bonds
Larger molecules are broken down into smaller molecules (sometimes referred to as degradative reactions)	Smaller molecules are bonded together to create larger molecules (sometimes referred to as biosynthetic reactions)

The energy that is released during catabolic reactions is used to drive anabolic reactions. In this manner, the cell works much like a factory. It gathers and produces raw materials to be used in the production of macromolecules for building, maintenance, and repair. It must also have fuel or energy available to run this metabolic machinery. This energy may be trapped from the rays of the sun (as in photosynthesis), or it may be produced by certain catabolic reactions. Then the energy

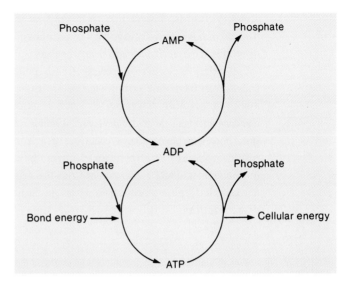

FIGURE 6-2. Conversion of bond energy to cellular energy.

is bound into high-energy bonds in special molecules, usually *adenosine triphosphate* (ATP) molecules. These high-energy molecules serve as the fuel, just as coal is used to fire the furnaces in the production of steel and other important alloys. Although ATP molecules are not the only high-energy compounds found within a cell, they are the most important ones. ATP molecules are the major energy-storing or energy-carrying molecules in a cell.

ATP molecules are found in all cells because they are used to transfer energy from energy-yielding molecules, like glucose, to an energy-requiring reaction. Thus, ATP is a temporary, intermediate molecule. If ATP is not used shortly after it is formed, it is soon hydrolyzed to adenosine diphosphate (ADP), a more stable molecule, and adenosine monophosphate (AMP) in catabolic reactions. ADP can also be used as an emergency energy source by the removal of another phosphate group to produce AMP (a catabolic reaction; Fig. 6-2). Both AMP and ADP bind with high-energy phosphate groups to produce ATP when the energy is removed from energy-yielding reactions.

In addition to the energy required for metabolic pathways, energy must also be available to the organism for growth, reproduction, sporulation, and movement. Some organisms even use energy for bioluminescence, such as the plankton that glow in the darkness of the ocean. Much of the energy of any system is lost in the form of heat.

Energy Metabolism

Chemical reactions are essentially energy transformation processes during which the energy that is stored in chemical bonds is transferred to other newly formed chemical bonds. The cellular mechanisms that release small amounts of energy as the cell needs it usually involve a sequence of catabolic and anabolic reactions, many of which are *oxidation-reduction reactions.*

Oxidation-reduction reactions are paired reactions, in which electrons are transferred from one compound to another. Whenever an atom, ion, or molecule loses one or more electrons (e^-) in a reaction, the process is called *oxidation,* and the molecule is said to be *oxidized.* The electrons lost do not float about at random but, because they are very reactive, attach immediately to another molecule. The resulting gain of one or more electrons by a molecule is called *reduction* and the molecule is said to be *reduced.* Within the cell, an oxidation reaction is always paired (or coupled) with a reduction reaction; thus the term oxidation-reduction or "redox" reaction.

STUDY AID   In the illustration below, an electron has been transferred from compound "A" to compound "B." Two reactions have occurred. Compound "A" has lost an electron (an oxidation reaction) and compound "B" has gained an electron (a reduction reaction). Oxidation is the loss of an electron. Reduction is the gain of an electron. Compound "A" has been oxidized and compound "B" has been reduced. The term "reduction" relates to the fact that an electron has a negative charge. When "B" receives an electron, its electrical charge is reduced.

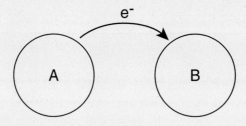

Many biological oxidations are referred to as *dehydrogenations* because hydrogen ions (H^+) and electrons are removed. Concurrently, those hydrogen ions must be picked up in a reduction reaction. Many good illustrations are found in the metabolism of glucose to form pyruvic acid and the concurrent synthesis of ATP and water (see the discussion of the citric acid cycle that follows).

Energy Production (Catabolism)

AEROBIC RESPIRATION OF GLUCOSE
The complete catabolism of glucose by the process known as aerobic respiration (or cellular respiration) takes place in three phases: (1) glycolysis, (2) the citric acid cycle, and (3) the electron-transport system. The first phase is anaerobic, whereas the last two require aerobic conditions (Fig. 6-3).

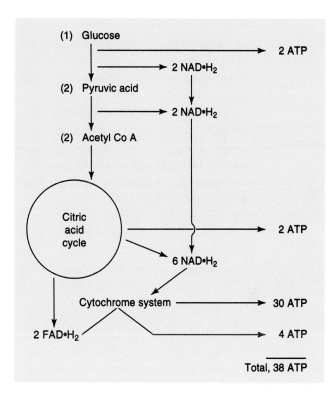

FIGURE 6-3. Summary tabulation of high-energy molecules produced by aerobic cellular respiration of one molecule of glucose.

Glycolysis. Glycolysis, also known as the glycolytic pathway or Embden-Meyerhof-Parnas pathway, is a 10-step biochemical pathway, involving 10 separate biochemical reactions, each of which requires a specific enzyme. In glycolysis, glucose is ultimately broken down into pyruvic acid. Glycolysis can take place in the presence or absence of oxygen, as oxygen does not participate in this phase of aerobic respiration. Somewhere in the metabolism of almost all cells, glycolysis, or the degradation of glucose, takes place to produce small amounts of ATP. Heterotrophs can degrade starch and glycogen to provide glucose for these glycolytic reactions. Other sugars, such as fructose, can also be used in these reactions. Autotrophs synthesize glucose during photosynthesis so that they can then derive energy from the glucose to drive other metabolic reactions. The amount of energy (ATP) derived from a glucose molecule depends on how much oxygen is available to bond with the hydrogen atoms released during the aerobic phase of aerobic respiration. It should be noted that aerobes and facultative anaerobes are much more efficient in energy production than are obligate anaerobes because they have oxygen available to aid in the production of many more ATP molecules.

Citric Acid Cycle and Electron Transport System. In the aerobic phases of aerobic respiration, oxygen is used as the final hydrogen acceptor following a long series of molecular reactions controlled by specific enzymes. Aerobic microorganisms and facultative anaerobes use pyruvic acid to produce about 18 times more energy than obligate anaerobes can produce by the fermentation of glucose and the other sugars. Obligate anaerobes do not have the appropriate enzymes and coenzymes to catalyze this metabolic pathway.

The *citric acid cycle* is also known as the tricarboxylic acid (TCA) cycle and is often called the Krebs cycle after the scientist who defined this phase of aerobic respiration. The pyruvic acid that was produced during glycolysis is first converted to acetyl-CoA which enters the TCA cycle—a series of 10 reactions, each of which is controlled by a different enzyme. In the first step in the TCA cycle, acetyl-CoA combines with oxaloacetate to produce citric acid (a tricarboxylic acid). The end products of the TCA cycle are carbon dioxide, reducing equivalents (hydrogen atoms and electrons), and regeneration of an oxaloacetate molecule. The reducing equivalents enter the *electron transport system* (also called the electron transport chain or respiratory chain), where cytochromes aid in the oxidative phosphorylation of ADP to ATP, and hydrogen bonds with oxygen to form water (cellular water).

The complete process of aerobic respiration of one molecule of glucose yields 36 or 38 ATP molecules; 36 in procaryotic cells and 38 in eucaryotic cells. In eucaryotic cells, two ATP molecules are gained from the glycolysis phase and two from the citric acid cycle; the other 34 result from oxidative phosphorylation in the electron transport system. Thus, most of the energy produced during aerobic respiration is produced via the electron transport chain. The chemical equation representing this highly efficient catabolic reaction is

$$C_6H_{12}O_6 + 6\,O_2 + 38\,ADP + 38\,\textcircled{P} \rightarrow 6\,H_2O + 6\,CO_2 + 38\,ATP$$

where $\textcircled{P}$ indicates activated phosphate groups.

In eucaryotic cells, glycolysis occurs in the cytoplasm, whereas the citric acid cycle and electron transport system occur within mitochondria. In procaryotic cells, all of these reactions occur in the cytoplasm. It is important to keep in mind, however, that neither the citric acid cycle nor the electron transport system occur in anaerobic bacteria.

Although the metabolic pathway and amount of energy that can be produced from aerobic respiration of glucose has been shown as an illustration, one must be aware that there are many variations to this pathway, depending on the individual organism and its available nutrient and energy resources. Some bacteria degrade glucose to pyruvic acid by other metabolic pathways. Also, glycerol, fatty acids from lipids, and amino acids from protein digestion may be fed into the citric acid cycle to produce energy for the cell when necessary; that is, when there are insufficient carbohydrates available (Fig. 6-4).

ANAEROBIC FERMENTATION
When the hydrogen atoms that are released from the breakdown of sugars bind to organic molecules instead of oxygen, the glycolytic process is referred to as

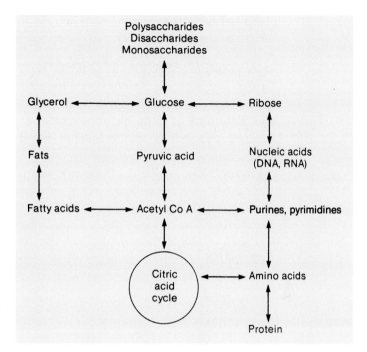

FIGURE 6-4. Other nutrients entering and exiting from the metabolism of glucose.

fermentation. Pyruvic acid usually accepts the hydrogen atoms to produce lactic acid or ethanol (ethyl alcohol), but other end products may be formed. The specific end products depend on the species of organism and on the sugar used as the source of carbohydrate. In human muscle cells, the lack of oxygen during extreme exertion results in pyruvic acid being converted to lactic acid. The presence of lactic acid in muscle tissue is the cause of the soreness that develops in exhausted muscles. Some bacteria (*e.g.*, *Lactobacillus* and *Streptococcus*) also produce lactic acid during the fermentation process. These organisms are found in the mouth, where the presence of lactic acid can promote tooth decay, and their presence in milk causes the normal souring of milk into curd and whey. The yeast, *Saccharomyces*, can ferment grain and fruit sugars into ethanol, the alcohol found in beer, wines, and liquors. Acetic acid bacteria (*Acetobacter*) are able to oxidize ethanol to acetic acid and spoil beer and wine by changing them to vinegar. These and other various end products of fermentation have many industrial applications.

AEROBIC OXIDATION BY CHEMOLITHOTROPHS

Chemolithotrophs are able to perform respiration reactions by oxidizing hydrogen (H_2) to water (H_2O), carbon (C) and carbon monoxide (CO) to carbon dioxide (CO_2), ammonia (NH_3) and nitrite ions (NO_2^-) to nitrate ions (NO_3^-), hydrogen sulfide (H_2S) and free sulfur (S) to sulfates (SO_4^{2-}), and iron to iron oxides. As one can imagine, these microorganisms (usually soil and water bacteria) are very important in recycling the elements and some compounds into forms more usable by plants and other microorganisms. They also must be acknowledged as major factors in the destruction of iron parts of machinery, as they promote rust.

ANAEROBIC RESPIRATION BY CHEMOTROPHS

Oxidation-reduction reactions occur, biologically, in certain anaerobes in the absence of oxygen. The electron donor is usually an organic compound such as glucose, but it may be inorganic iron or sulfur compounds (as used by *Thiobacillus*). Inorganic compounds (nitrates, sulfates, and carbonates) serve as the final electron acceptors in place of oxygen. Some facultative anaerobes convert nitrates (NO_3^-) to nitrites (NO_2^-) and free atmospheric nitrogen gas (N_2). Others reduce carbonates (CO_3^{2-}) to carbon dioxide (CO_2) and methane (CH_4). Some obligate anaerobes convert sulfate (SO_4^{2-}) to free sulfur (S) or to hydrogen sulfide gas (H_2S). These bacteria are found in soil, sea water, marine mud, fresh water, acid mine waters, sewage, and sulfur springs. Obviously, they are unable to produce as much ATP from their energy sources as heterotrophic organisms because their nutrients do not contain as much bound chemical energy.

Metabolic Biosynthesis (Anabolism)

ENERGY CONVERSION

In general, chemoheterotrophs produce energy from organic compounds by fermentation, anaerobic respiration, and aerobic respiration, as previously discussed. However, all of the phototrophs (algae, cyanobacteria, other photosynthetic bacteria, and plants) must derive their energy from light, usually the sun, by photosynthesis. This process provides energy to the greatest mass of organisms, not only all of the photosynthetic organisms but also the heterotrophs as well, because heterotrophs consume phototrophs.

ENERGY USE

The biosynthesis of organic compounds requiring the use of energy is called anabolism, or an anabolic reaction. In living cells this biosynthetic metabolism may be one of two types: *photosynthesis* by chemoautotrophs, photolithotrophs or photoheterotrophs; or *chemosynthesis* by chemoautotrophs, chemolithotrophs or chemoheterotrophs.

In photosynthesis, light energy is converted to chemical bond energy to be used to synthesize organic biochemicals. Those phototrophic organisms using inorganic raw materials (H_2O, H_2S, S) for biosynthesis are the photolithotrophs, whereas those using CO_2 are the photoautotrophs. Those phototrophs using small organic molecules, such as acids and alcohols, to build carbohydrates, fats, proteins, nucleic acids, and other important biochemicals are the photoheterotrophs.

PHOTOSYNTHESIS

The goal of photosynthetic processes is to trap the radiant energy of light and convert it into chemical bond energy in ATP and carbohydrates, particularly glucose, which can then be converted into more molecules of ATP via the respiratory pathways. The general overall photosynthesis reaction is

$$6\ CO_2 + 12\ H_2O \overset{light}{\underset{ATP}{\rightarrow}} C_6H_{12}O_6 + 6\ O_2 + 6\ H_2O + ADP + \textcircled{P}$$

Notice that this reaction is almost the reverse of the aerobic respiration reaction; it is nature's way of balancing substrates in the environment. Bacteria that produce oxygen by photosynthesis are called oxygenic photosynthetic bacteria.

Photosynthesis can take place in the absence of oxygen (anaerobically) by reactions that do not produce oxygen. Purple sulfur bacteria and green sulfur bacteria (obligate anaerobic photoautotrophs) are referred to as anoxygenic photosynthetic bacteria because their photosynthetic processes do not produce oxygen. These bacteria do not use H_2O to reduce CO_2, but instead use sulfur, sulfur compounds (*e.g.*, H_2S gas), or hydrogen gas to reduce CO_2. The overall reaction of anaerobic bacterial photosynthesis then becomes

$$6\ CO_2 + 12\ H_2S \overset{light}{\rightarrow} C_6H_{12}O_6 + 6\ H_2O + 12\ S$$

or

$$6\ CO_2 + 12\ H_2 \overset{light}{\rightarrow} C_6H_{12}O_6 + 6\ H_2O$$

The bacterial photosynthetic pigments use shorter wavelengths of light, which penetrate deep within a pond or into mud where it appears to be dark.

In the absence of light, some photolithotrophic organisms may survive anaerobically by the fermentation process alone. Other phototrophic bacteria also have a limited ability to use simple organic molecules in photosynthetic reactions; thus, they become photoheterotrophic organisms under certain conditions. A few species of cyanobacteria have also been found to exist as facultative phototrophs and facultative autotrophs, meaning that in certain environments they become photoheterotrophs. In other words, they have backup metabolic systems.

CHEMOSYNTHESIS

The chemosynthetic process involves a chemical source of energy and raw materials to synthesize the necessary metabolites and macromolecules for growth and function of the organisms. These chemotrophic organisms may be either autotrophs or heterotrophs.

The chemoautotrophs are the same chemolithotrophic bacteria, previously discussed, that obtain energy by aerobic oxidation of inorganic compounds or by anaerobic respiration of inorganic substances. These are the only organisms that do not depend on the radiant energy from the sun. They are considered among the most primitive bacteria and are frequently found near thermal vents deep within the ocean.

The chemoheterotrophs have been defined as those organisms that derive both their energy and nutrients from organic materials. Most bacteria, as well as all protozoa, fungi, and animals belong to this group. Although they vary greatly in the details of their metabolism, they all use carbohydrates, lipids, and proteins to synthesize their own carbohydrates, lipids, proteins, nucleic acids, and high-energy molecules such as ATP. These metabolic pathways may be carried on with or without oxygen by aerobic or anaerobic respiration, fermentation, and other biodegradation and biosynthetic reactions.

MICROBIAL GROWTH

Bacterial growth refers to an increase in the number of organisms rather than in their size. When each bacterial cell reaches its optimal size, it divides by binary fission ("bi" means "two") into two daughter cells; *i.e.,* each bacterium simply splits into two similar cells. These in turn divide, and as a result, a viable, healthy colony of cells is maintained as long as the nutrient supply, water, and space allow. This process continues until the waste products from cells build up to a toxic level or until the nutrients are depleted. The actual division of staphylococci by binary fission is shown in the electron micrograph in Figure 6-5.

The growth of microorganisms in the body (*in vivo*), in nature, or in the laboratory (*in vitro*) is greatly influenced by temperature, pH, moisture content, available nutrients, and the characteristics of other organisms present. Therefore, the number of bacteria in nature fluctuates unpredictably because these factors vary with the seasons, rainfall, temperature, and time of day.

In the laboratory, however, a pure culture of a single species of bacteria can usually be grown if the appropriate growth medium and environmental conditions are provided. The temperature, pH, and amount of oxygen are quite easily controlled to provide optimal conditions for growth. Then the appropriate nutrients must be provided in the growth medium. Some bacteria are so *fastidious* (having complex nutritional requirements) that they will not grow outside of living cells; thus, they must be cultured in living animals, embryonated chicken eggs, or cell cultures. Examples of organisms unable to grow on artificial media are viruses, rickettsias, chlamydias, *Treponema pallidum* (the etiologic agent of syphilis), and *Mycobacterium leprae* (the etiologic agent of leprosy).

Culture Media

Basically, there are two types of media for culturing bacteria: (1) a chemically defined synthetic medium and (2) a rich, natural, complex medium containing digested extracts from animal organs, meats, fish, yeasts, and plants providing the necessary nutrients, vitamins, and minerals. These media can be used in liquid (broth) form, which is available in tubes (and, thus, referred to as tubed media), or they may be solidified by the addition of agar and poured into tubes or petri dishes, where they solidify, so that the bacteria can be grown within or on the surface of

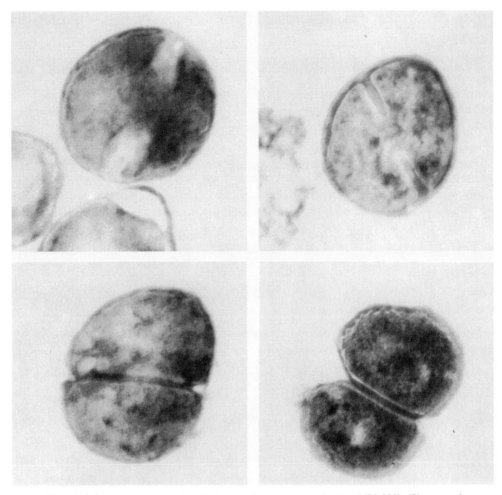

FIGURE 6-5. Binary fission of staphylococci (original magnification X30,000). (Photograph courtesy of Ray Rupel)

the agar. Agar, a complex polysaccharide obtained from a red marine alga, is used as a solidifying agent, much like gelatin is used in cooking.

An *enriched medium* is a broth or solid medium containing a rich supply of special nutrients that promotes the growth of fastidious organisms. It is usually prepared by adding extra nutrients to a basic medium called nutrient agar. Blood agar (nutrient agar plus 5% sheep red blood cells) and chocolate agar (nutrient agar plus powdered hemoglobin) are examples of solid enriched media that are used routinely in the clinical bacteriology laboratory. Blood agar is bright red, whereas chocolate agar is brown. Although both of these media contain hemoglobin, chocolate agar is considered to be more enriched than blood agar.

A *selective medium* has added inhibitors that discourage the growth of certain organisms without inhibiting growth of the organism being sought. For example, MacConkey agar inhibits growth of Gram-positive bacteria and is, thus, selective for Gram-negative bacteria. Phenylethyl alcohol (PEA) agar and colistin-nalidixic acid (CNA) agar are selective for Gram-positive bacteria. Thayer-Martin agar (a chocolate agar containing extra nutrients plus antimicrobial agents) is selective for *Neisseria gonorrhoeae*. Only salt-tolerant (haloduric) bacteria can grow on mannitol salt agar (MSA). Sabouraud dextrose agar is selective for fungi; its low pH (5.6) is inhibitory to most bacteria.

A *differential medium* permits the differentiation of organisms that grow on the medium. MacConkey agar is frequently used to differentiate between various Gram-negative bacilli that are isolated from fecal specimens. Gram-negative bacteria able to ferment lactose (an ingredient of MacConkey agar) produce pink colonies, whereas those unable to ferment lactose produce colorless colonies. Thus, MacConkey agar differentiates between lactose fermenting (LF) and nonlactose fermenting (NLF) Gram-negative bacteria. Mannitol salt agar is used to screen for *Staphylococcus aureus;* not only will *S. aureus* grow on MSA, but it turns the originally pink medium to yellow due to its ability to ferment mannitol. In a sense, blood agar is also a differential medium because it is used to determine the type of hemolysis (alteration or destruction of red blood cells) that the bacterial isolate produces. (See Color Figures 14, 15, 16.)

The various categories of media (enriched, selective, differential) are not mutually exclusive. For example, as just seen, blood agar is both enriched and differential. MacConkey agar and MSA are selective and differential. Because they start as blood agar to which selective inhibitory substances are added, PEA and CNA are enriched and selective. Thayer-Martin agar is highly enriched and highly selective.

Population Counts

Once the desired species of bacteria has been separated from other organisms present in a specimen, it can be grown as a pure culture under the best possible conditions. The changes in a bacterial population over an extended period follow a definite predictable pattern that can be shown by plotting the population growth curve on a graph (Fig. 6-7; described later).

Often, microbiologists need to know the rate of bacterial growth and how many bacteria are present at any given time. This information is particularly important in determining the degree of bacterial contamination in drinking water, milk, and other foods. The number can be determined by counting the total number of bacterial cells, living and dead, in 1 mL of solution, or by counting only the viable (living) bacteria present. A total count is easier and faster, but it differs from a viable cell count because it includes both living and dead cells; however, if the 1-mL sample is taken when the microorganisms are growing and dividing rapidly in the growth phase, few dead cells are found. In many hospitals, electronic cell counters are incorporated into the instruments used in blood, urine, and spinal fluid analyses. Many research laboratories use spectrophotometers, which determine the number of organisms present by measuring the turbidity (cloudiness) of the solution. Turbidity varies as a result of the number of organisms that are present. Turbidity increases (*i.e.,* the solution becomes more cloudy) as the number of organisms increases. Chemical analyses of nitrogen or carbon content also can be used to determine the number of bacteria present.

The *viable plate count* is usually the most accurate method for determining the number of living bacteria in a milliliter of liquid, which may be milk, water, diluted food, or broth. In this procedure, as shown in Figure 6-6, 1 mL of solution is diluted to 100 mL three times in sequence, and samples are taken from each dilution. Then 0.1-mL and/or 1-mL samples are inoculated onto nutrient agar. The number of colonies observed growing on the nutrient agar plates the following day indicates the number of viable bacteria present at that particular dilution. This number multiplied by the dilution factor indicates the number of living bacteria in the original culture at the time the sample was taken. For example, if 220 colonies were counted on an agar plate grown from a 1-mL sample from the second dilution bottle (1:10,000 dilution), there were $220 \times 10,000 = 2,200,000$ bacteria in 1 mL of the original material at the time the dilutions were made and cultured. Practically, it is easier to culture only 0.1 mL of each dilution, while increasing the dilution factor by 10, as shown in Figure 6-6. For the count to be statistically significant and most accurately representative of the number of living microorganisms in the solution, the number used in the calculations should be taken from the agar plate that has between 30 and 300 colonies.

A similar technique has been developed to count viruses. One mL of the diluted viruses is inoculated onto a "lawn" of bacteria or a layer of culture cells. Each cell lysed by a virus causes a clear zone (plaque) in the culture of cells following incubation. Thus, the number of plaques represents the number of viruses in 1 mL of the diluted solution. This number must then be multiplied by the dilution factor.

In the clinical microbiology laboratory, a viable cell count is part of a urine culture. A method of approximating the number and type of bacteria in urine involves inoculating the surface of an agar medium with a known volume (either 0.01 mL or 0.001 mL) of urine, using a calibrated inoculating loop. Following incubation, the colonies are counted and this number is then multiplied by the dilution factor (either 100 or 1000) to obtain the number of colony-forming units (CFU) per mL of urine. A CFU count that is greater than or equal to 100,000 (10^5) CFU/mL is indicative of a urinary tract infection (UTI), although high colony counts may also be due to contamination of the urine specimen with indigenous microflora during specimen collection or failure to refrigerate the specimen between collection and transport to the laboratory.

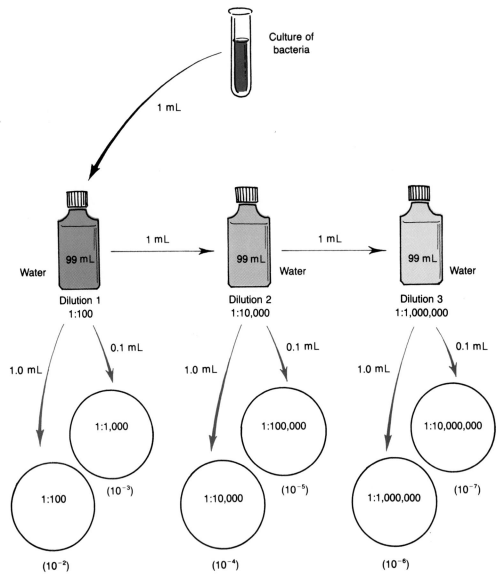

FIGURE 6-6. The viable plate count technique. One milliliter (mL) of the original culture is successively diluted and cultured on agar plates.

Population Growth Curve

The population *growth curve* for any particular species of bacteria may be determined by growing the organism in pure culture (a culture containing only one species of organism) at a constant temperature. The graph in Figure 6-7 is constructed by plotting the *logarithm* (refer to your math book) of the number of bacteria (on the vertical or Y axis) against the incubation time (on the horizontal or X axis).

The first stage of the growth curve is the *lag phase* (see *A* in Fig. 6-7), during which the bacteria absorb nutrients, synthesize enzymes, and prepare for reproduction. In the *logarithmic growth phase* (or log phase or exponential growth phase; *B* in Fig. 6-7), the bacteria multiply so rapidly that the population number doubles with each generation time. The generation time, which is the time that elapses between the formation of a new bacterium and its division into two daughter cells, varies with the species of bacteria. The growth rate is the greatest during the logarithmic growth phase. In the laboratory, under ideal growth conditions, *E. coli, Vibrio cholerae* (which causes cholera), *Staphylococcus,* and *Streptococcus* all have a generation time of about 20 minutes, whereas *Pseudomonas,* from the soil, may divide every 10 minutes, and *Mycobacterium tuberculosis* may divide only every 18 to 24 hours. The logarithmic growth phase is always brief unless the rapidly dividing culture is maintained by constant addition of nutrients and frequent removal of waste products and excess microorganisms.

Many industrial and research procedures depend on the maintenance of an essential species of microorganism. These are continuously cultured in a controlled environment called a *chemostat* (Fig. 6-8), which regulates the supply of nutrients and the removal of waste products and excess microorganisms. Chemostats are used in industries in which yeast is grown to produce beer and wine, where fungi and bacteria are cultivated to produce antibiotics, where *E. coli* cells are grown for genetic research, and in any other process needing a constant source of microorganisms.

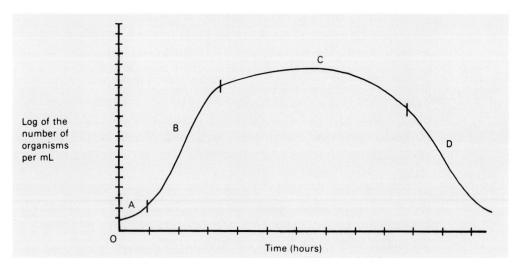

FIGURE 6-7. Population growth curve of living organisms. The logarithm of the number of bacteria per milliliter of medium is plotted against time. (*A*) Lag phase. (*B*) Logarithmic growth phase. (*C*) Stationary phase. (*D*) Death phase.

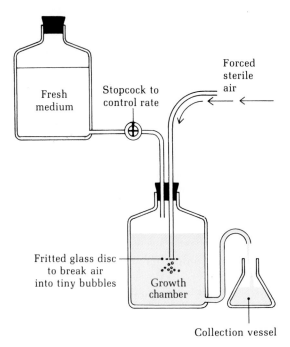

Fresh medium

Stopcock to control rate

Forced sterile air

Fritted glass disc to break air into tiny bubbles

Growth chamber

Collection vessel

FIGURE 6-8. Chemostat used for continuous cultures. Rate of growth can be controlled either by controlling the rate at which new medium enters the growth chamber or by limiting a required growth factor in the medium.

As the nutrients and oxygen in the culture tube are used up and waste products from the metabolizing bacteria build up and change the pH of the culture medium, the rate of division slows, such that the number of bacteria dividing equals the number dying. The result is the *stationary phase* (*C* in Fig. 6-7). It is during this phase that the culture is at its greatest population density.

As overcrowding occurs, the toxic waste products increase and the nutrient supply decreases. The microorganisms then die at a rapid rate; this is the *death phase* or decline phase (*D* in Fig. 6-7). The culture may die completely or a few microorganisms may continue to survive for months. If the bacterial species is a spore-former, it will form spores to survive beyond this phase. When cells are observed in old cultures of bacteria in the death phase, some of them look different from healthy organisms seen in the growth phase. As a result of unfavorable conditions, morphological changes in the cells may appear. Some cells undergo involution and assume a variety of shapes, becoming long, filamentous rods, or branching or globular forms that are difficult to identify. Some develop without a cell wall and are referred to as protoplasts, spheroplasts, or L-phase variants (L-forms). When these involuted forms are inoculated into a fresh nutrient medium, they usually revert to the original shape of the healthy bacteria.

A population growth curve may be plotted for all organisms, including humans. At present, the human population is in the logarithmic growth phase with a generation time of 35 years. No population of organisms is known to continue in this phase forever without careful control of the food supply, numbers of individuals, and proper disposal of waste products. Eventually, the forces of nature (food shortages, epidemics, wars, toxic waste products) will probably cause the world population to stabilize in the stationary growth phase and, hopefully, will not proceed beyond to the terminal death phase.

BACTERIAL GENETICS

Genetics, the study of heredity, involves many topics (*e.g.,* DNA, genes, the genetic code, chromosomes, DNA replication, transcription, translation), some of which have already been addressed in this book. An organism's *genotype* is its complete collection of genes, whereas an organism's *phenotype* is all of the organism's physical traits, attributes, or characteristics. Phenotypic characteristics of bacteria include the presence or absence of certain enzymes and such structures as capsules, flagella, and pili. An organism's phenotype is dictated by that organism's genotype. For example, an organism cannot produce a particular enzyme unless it possesses the gene that codes for that enzyme. It cannot produce flagella unless it possesses the genes for flagella production. Phenotype is the manifestation of genotype.

Most bacteria possess one chromosome, which usually consists of one long, continuous, double-stranded DNA molecule, with no protein on the outside as is found in eucaryotic chromosomes. The chromosome is a circular strand of genes, all linked together. Genes are the fundamental units of heredity that carry the information needed for the special characteristics of each different species of bacteria. Genes direct all functions of the cell, providing it with its own particular traits and individuality. Because there is only one chromosome that replicates just before cell division, identical traits of a species are passed from the parent bacterium to the daughter cells after binary fission has occurred. DNA replication must precede binary fission, to ensure that each daughter cell has exactly the same genetic composition as the parent cell.

The DNA of any gene on the chromosome is subject to accidental alteration, which changes the trait controlled by that gene. If the change in the gene alters or deletes (eliminates) a trait in such a way that the cell does not die or become incapable of division, the altered trait is transmitted to the daughter cells of each succeeding generation. A change in the characteristics of a cell caused by a change in the DNA molecule (genetic alteration) that is transmissible to the offspring is called a *mutation.* A lethal mutation is one that causes the cell to die because an essential functional gene is missing. It may perhaps be a gene that codes for an essential enzyme. Spontaneous mutations usually occur about once in every 10 million cell divisions. The mutation rate can be increased by exposing cells to physical or chemical agents that affect the DNA molecule. These agents are called *mutagens.* In the research laboratory, x-rays, ultraviolet light, and radioactive substances, as well as certain chemical agents, are used to induce more frequent mutations (Fig. 6-9). The organism containing the mutation is called a *mutant.* Bacterial mutants

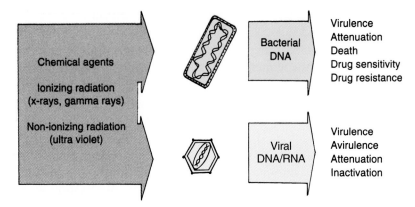

FIGURE 6-9. Agents that may cause mutagenic changes.

are used in genetic and medical research and in the development of vaccines. The types of mutagenic changes frequently observed in bacteria involve colony characteristics, cell shape, biochemical activities, nutritional needs, antigenic sites, virulence, pathogenicity, and drug resistance. Nonpathogenic live virus vaccines, such as the Sabin vaccine for polio, are examples of laboratory-induced mutations of pathogenic microorganisms.

Ways in Which Bacteria Acquire New Genetic Information
Mutations (involve changes in the base sequences of genes)
Lysogenic conversion (involves bacteriophages)
Transduction (involves bacteriophages)
Transformation (involves the uptake of "naked" DNA)
Conjugation (involves the transfer of genetic information from one
　cell to another through a hollow sex pilus)

Changes in Bacterial Genetic Constitution

There are at least four additional ways that the genetic composition of bacteria can be changed: lysogenic conversion, transduction, transformation, and conjugation. These types of gene transfers result in extra genetic material in the recipient cell. If this extra bit of DNA remains in the cytoplasm of the cell, it is called a *plasmid* (Fig. 6-10). Because they are not part of the chromosome, plasmids are referred to as extrachromosomal DNA. Some plasmids contain many genes, others only a few, but the cell is changed by the addition of these genetic components. Plasmids can replicate themselves simultaneously with chromosomal DNA replication or at various other times. A plasmid that can exist either autonomously or integrated into the chromosome is referred to as an *episome*. When a gene product (usually a protein) has been produced, the gene that codes for that particular gene product is said to have been expressed. Some plasmid genes can be expressed as extrachromosomal genes, but others must integrate into the chromosome before the genes become functional.

LYSOGENIC CONVERSION

Lysogenic conversion occurs when a temperate bacteriophage (bacterial virus) infects a bacterium, changing it to a *lysogenic bacterium* (*i.e.*, a bacterium having the potential to be lysed by viral gene products) (Fig. 6-11). The bacteriophage injects its DNA into the cytoplasm of the bacterium, and then the phage DNA incorporates into (becomes part of) the bacterial chromosome. When the phage DNA is integrated into the host cell's chromosome, the phage is referred to as a *prophage* (see the discussion of bacteria and bacteriophages in Chapter 3). The number of genes in the bacterium is increased by the number of genes injected by the phage, and the lysogenic bacterium is able to produce any gene products that are coded for by the prophage genes. Because the bacterium has been converted from a cell that could not produce those gene products to one that can, the phenomenon is referred to as lysogenic conversion.

A clinical example of lysogenic conversion involves the disease, diphtheria. The etiologic agent of diphtheria is a bacterium named *Corynebacterium diphtheriae.*, Because it is actually a viral gene (called the tox gene) that codes for the toxin, only cells of *C. diphtheriae* that contain prophage can produce the toxin that causes diphtheria. Strains of *C. diphtheriae* capable of producing diphtheria toxin are called toxigenic

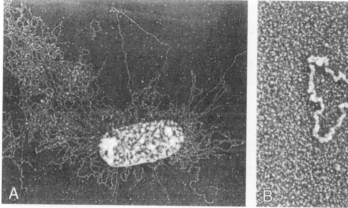
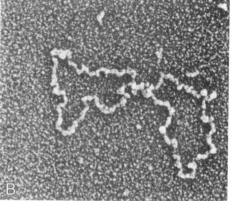

FIGURE 6-10. (A) Disrupted cell of *Escherichia coli*; the DNA has spilled out and a plasmid can be seen slightly to the left of top center. (B) Enlargement of a plasmid (about 1 μm from side to side).

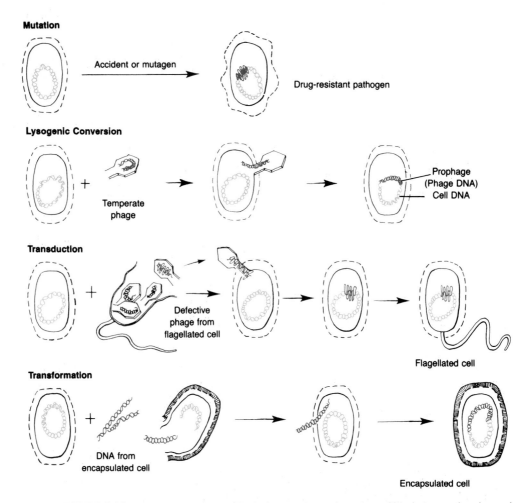

FIGURE 6-11. Four of the ways in which the genetic constitution of bacteria may be changed: mutation, lysogenic conversion, transduction, and transformation. See text for details.

strains and those unable to produce the toxin are called nontoxigenic strains. The phage that infects *C. diphtheriae* is known as a corynebacteriophage.

TRANSDUCTION

Transduction, like lysogenic conversion, involves bacteriophages. Transduction means "to carry across." Some bacterial genetic material may be "carried across" from one bacterial cell to another by a bacterial virus. This phenomenon may occur following infection of a bacterial cell by a temperate bacteriophage. The viral DNA combines with the bacterial chromosome, becoming a prophage. If a stimulating chemical, heat, or ultraviolet light activates the prophage, it begins to produce new viruses by the production of phage DNA and proteins. As the chromosome disintegrates, small pieces of bacterial DNA may remain attached to the maturing phage DNA. During the assembly of the virus particles, one or more bacterial genes may be incorporated into some of the mature bacteriophages. When all the phages are freed by cell lysis, they proceed to infect other cells, injecting bacterial genetic material as well as viral genes. Thus, bacterial genes that are attached to the phage DNA are carried to new cells by the virus.

There are two types of transduction: specialized and generalized. The explanation in the previous paragraph describes *specialized transduction* in which the infecting phage integrates into the bacterial chromosome or a plasmid. As the virus genome breaks away to replicate and produce more viruses, it carries a few identifiable bacterial genes with it to the newly infected cell. In this way, genetic capabilities involving the fermentation of certain sugars, antibiotic resistance, and other phenotypic characteristics can be transduced to other bacteria. This process has been shown in the laboratory (*in vitro*) to occur among species of *Bacillus, Pseudomonas, Haemophilus, Salmonella,* and *Escherichia,* and it is assumed to occur in nature.

In *generalized transduction,* the bacteriophage is a virulent lytic phage that does not incorporate into the bacterial genome or plasmid. Rather, it picks up fragments of bacterial DNA during the assembly of new virus particles and carries these genes to other cells that the new viruses infect. This generalized transduction has been observed in species of *Streptococcus, Staphylococcus,* and *Salmonella,* and in *Vibrio cholerae.*

Only small segments of DNA are transferred from cell to cell by transduction compared with the amount that can be transferred by transformation and conjugation.

TRANSFORMATION

In the *transformation* process, a recipient bacterial cell is genetically transformed following the uptake of DNA fragments from another strain of bacteria with at least one different observable characteristic (Fig. 6-11). It was transformation experiments that proved that DNA is, indeed, the genetic material (see Historical Note on the next page). When a DNA extract from encapsulated, pathogenic *Streptococcus pneumoniae* type 1 was added to a growing culture of nonencapsulated, nonpathogenic *S. pneumoniae* type 2, the resulting culture showed the presence of live, encapsulated type 2 because some of the type 1 DNA was incorporated into the dividing type 2 DNA. Thus, the nonencapsulated streptococci must have taken up (absorbed) some of the type 2 DNA and were transformed by the genes coding for capsules. Although this type of genetic recombination is not widespread, it has been demonstrated in several genera including *Bacillus, Escherichia, Haemophilus, Pseudomonas,* and *Neisseria.* Transformations have even been shown to occur between two different species (*e.g., Staphylococcus* and *Streptococcus*).

Pieces of DNA molecules from a donor cell can only penetrate the cell wall and cell membrane of certain bacteria. The ability to absorb "naked" DNA into the cell is referred to as *competence,* and bacteria capable of taking up "naked" DNA molecules are said to be *competent.* Recipient bacteria usually become competent during the late logarithmic growth phase when the cell secretes a protein competence factor that increases its permeability to DNA.

Some competent bacterial cells have incorporated DNA fragments from certain animal viruses (*e.g.,* cowpox), retaining the latent virus genes for long periods. This knowledge may have some importance in the study of viruses that remain latent in humans for many years before they finally cause disease, as may be the case in Parkinson's disease. These human virus genes may hide in the bacteria of the indigenous microflora until they are released to cause disease.

> **Historical Note:** Transformation was first demonstrated in 1928 by the British physician Frederick Griffith and his colleagues, performing experiments with *Streptococcus pneumoniae* and mice. Although the experiments demonstrated that bacteria could take up genetic material from the external environment and, thus, be transformed, it was not known at that time what molecule actually contained the genetic information. It was not until 1944 that Oswald Avery, Colin MacLeod and Maclyn McCarthy, who also experimented with S. *pneumoniae,* first demonstrated that DNA was the molecule that contained genetic information. Experiments conducted in 1952 by Alfred Hershey and Martha Chase, using *E. coli* and bacteriophages, confirmed that DNA carried the genetic code.

CONJUGATION

Conjugation involves a specialized type of pilus called a sex pilus. A bacterial cell (called the donor cell) possessing a sex pilus attaches by means of the sex pilus to another bacterial cell (called the recipient cell). Some genetic material (usually in the form of a plasmid) is then transferred through the hollow sex pilus from the donor cell to the recipient cell (Figs. 6-12 and 6-13). Although conjugation has nothing to do with reproduction, the process is occasionally referred to as "bacterial mating" and the terms "male" and "female" cells are sometimes used in reference to the donor and recipient cells, respectively. This type of genetic recombination occurs mostly among species of enteric, Gram-negative bacilli, but has been reported within species of *Pseudomonas* and *Streptococcus* as well. In electron micrographs, microbiologists have observed that sex pili are larger than other pili. Although many different genes may be transferred by conjugation, the ones most frequently noted include those coding for antibiotic resistance, colicin (a protein that kills certain enteric bacteria), and fertility factors (F^+, $Hfr+$), where F = fertility and Hfr = high frequency of recombination.

Bacteria possessing F^+ or HFr^+ genes have the ability to produce sex pili and become donor cells. If the fertility factor is on a plasmid, it is an F^+ gene; whereas, if it is incorporated within the chromosome, it is referred to as the HFr^+ gene. A complete copy of the F plasmid (containing the F^+ gene) usually moves to the

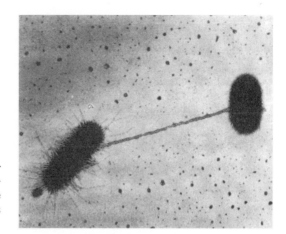

FIGURE 6-12. Conjugation in *Escherichia coli*. The donor cell (with numerous short pili) is connected to the recipient bacterium by a sex pilus (original magnification × 3,000).

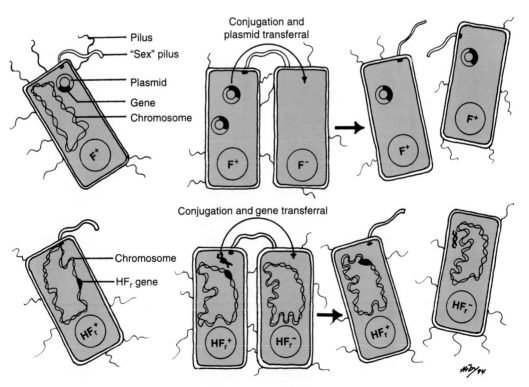

FIGURE 6-13. Bacterial conjugation. See text for details.

recipient (F^-) cell; thus, the recipient usually becomes F^+. However, the recipient cell usually receives only a portion of the chromosome from an HFr^+ cell, not including the HFr^+ gene; thus, the recipient remains HFr^- in that circumstance, does not produce a sex pilus, and cannot become a donor cell.

Transduction, transformation, and conjugation are excellent tools for mapping bacterial chromosomes and for studying bacterial and viral genetics. Although all of these methods are frequently used in the laboratory, it is believed that they also occur in natural environments under certain circumstances.

GENETIC ENGINEERING

An array of techniques has been developed to transfer eucaryotic genes, particularly human genes, into other easily cultured cells to facilitate the large-scale production of important gene products (proteins). This process is known as *genetic engineering* or recombinant DNA technology. Plasmids are frequently used as vectors or vehicles for inserting genes into cells. Bacteria, yeasts, human leukocytes, macrophages, and fibroblasts have been used as manufacturing plants for proteins such as human growth hormone (somatotropin), somatostatin (which inhibits the release of somatotropin), plasminogen activating factor, insulin, and interferon. Somatostatin and insulin were first produced by recombinant DNA technology in 1978.

Many industrial and medical benefits may be derived from genetic engineering research. In agriculture, there is a potential for incorporating nitrogen-fixing capabilities into more of the soil microorganisms, for making plants that are resistant to insects and bacterial and fungal diseases, and for increasing the size and nutritional value of foods. Genetically-engineered microorganisms can also be used to clean up the environment; *e.g.*, to get rid of toxic wastes. Consider this hypothetical example: A soil bacterium contains a gene that enables the organism to break oil down into harmless by-products, but, because the organism cannot survive in salt water, it cannot be used to clean up oil spills at sea. Remove the gene from the soil bacterium and, using a plasmid vector, insert it into a marine bacterium. Now the marine bacterium has the ability to break down oil and can, thus, be used to clean up oil spills at sea.

In medicine, there is potential for making engineered antibodies, antibiotics, and drugs; for synthesizing important enzymes and hormones for treatment of inherited diseases; and for making vaccines. Such vaccines would contain only part of the pathogen (for instance, the capsid proteins of a virus) to which the person would form protective antibodies (see Insight Box).

GENE THERAPY

Gene therapy of human diseases involves the insertion of a normal gene into cells to correct a specific genetic or acquired disorder that is being caused by a defective

INSIGHT
Genetically Engineered Bacteria and Yeasts

The term "genetic engineering" refers to the manufacture and manipulation of genetic material *in vitro* (in the laboratory). Genetic engineering has been possible only since the late 1960s, when a scientist named Paul Berg demonstrated that fragments of human or animal DNA can be attached to bacterial DNA. Such a hybrid DNA molecule is referred to as recombinant DNA. When a molecule of recombinant DNA is inserted into a bacterial cell, the bacterium is able to produce the gene product, usually a protein. Thus, microorganisms (primarily bacteria) can be genetically engineered to produce substances (gene products) that they would not normally manufacture. Paul Berg won a Nobel Prize in 1980 for his pioneering genetic engineering experiments.

Molecules of self-replicating, extrachromosomal DNA, called plasmids, are frequently used in genetic engineering and are referred to as vectors. A particular gene of interest is first inserted into the vector DNA, forming a molecule of recombinant DNA. The recombinant DNA is then inserted into or taken up by a bacterial cell. The cell is next allowed to multiply, creating many genetically identical bacteria (clones), each of which is capable of producing the gene product. From the clone culture, a genetic engineer may then remove ("harvest") the gene product.

The gram-negative bacillus, *Escherichia coli*, has often been used because it can be easily grown in the laboratory, has a relatively short generation time (about 20 minutes under ideal conditions), and its genetics are well-understood by researchers. A gram-positive bacterium, *Bacillus subtilis*, a yeast, *Saccharomyces cerevisiae*, and cultured plant and mammalian cells have also been used by genetic engineers to produce desired gene products.

An example of a product produced by genetic engineering is insulin, a hormone produced in *E. coli* cells and used to treat diabetic patients. Human growth hormone (somatotropin), somatostatin (a hormone used to limit growth), and interferon are also produced by genetically-engineered *E. coli*. The hepatitis B vaccine that is administered to health care workers is produced by a genetically-engineered yeast, called *Saccharomyces cerevisiae*. Bovine growth hormone (BGH) and porcine growth hormone (PGH) are produced in genetically-engineered *E. coli*.

New uses for recombinant DNA and genetic engineering are being discovered every day, causing profound changes in medicine, agriculture, and other areas of science.

gene. The first gene therapy trials were conducted in the United States in 1990. Viral delivery is currently the most common method for inserting genes into cells, where specific viruses are selected to target the DNA of specific cells. For example, a virus capable of infecting liver cells would be used to insert a therapeutic gene or genes into the DNA of liver cells. Viruses currently being used or considered for use as vectors include retroviruses, adenoviruses and possibly herpesviruses. It is likely that genes will someday be regularly prescribed as "drugs" in the treatment of certain diseases (*e.g.,* autoimmune diseases, cancer, cystic fibrosis, heart disease, hemoglobin defects, hemophilia, immune deficiencies, liver and lung diseases, and muscular dystrophy). In the future, synthetic vectors, rather than viruses, may be used to insert genes into cells.

■ REVIEW OF KEY POINTS

- ■ By studying bacterial nutrition and microbial physiology, scientists learn about the vital chemical and metabolic processes that occur in all living cells.

- ■ All living organisms require sources of energy and carbon so that they can build the necessary cellular materials of life. In addition, organisms must be provided with materials that they are unable to synthesize, but are required for survival; these essential nutritional requirements vary from species to species.

- ■ The energy source for certain organisms (called phototrophs) is light and for other organisms (called chemotrophs) is organic or inorganic chemicals.

- ■ An organism's carbon source may be CO_2 (autotrophs), inorganic compounds other than CO_2 (lithotrophs), or organic compounds (heterotrophs or organotrophs). All animals (including humans), protozoa, and fungi are heterotrophs, as are most bacteria.

- ■ Interrelationships among the different nutritional types are of prime importance in the functioning of the ecosystem. Photolithotrophs (plants, algae, and certain bacteria) are the producers of food and oxygen for the chemoheterotrophs (animals). Dead plants and animals are recycled by the chemoheterotrophic saprophytic decomposers (certain fungi and bacteria) into nutrients for photolithotrophs (certain bacteria) and chemolithotrophs (certain bacteria).

- ■ Metabolism refers to all the chemical reactions (catabolic and anabolic) that occur within any cell, including the production of energy and the synthesis of new molecules; such reactions are regulated by enzymes.

- ■ Enzymes are biological molecules (proteins) that serve as catalysts to control the rate of metabolic reactions. The enzymes produced by any particular cell are governed by the genotype of that cell and the presence or absence of a particular enzyme is part of the phenotype of that cell. All of the enzymes that a cell is capable of producing need not be present in the cell at a given time; they are produced to meet the metabolic needs of the cell as determined by the internal and external environment.

- ■ An enzyme operates at peak efficiency within a particular pH and temperature range and when there exists an appropriate concentration of the substrate for that enzyme. If the environment is too acidic, basic, hot, cold, or contains too much or too little substrate, the enzyme will not operate at peak efficiency and the reaction will not proceed at its maximum rate.

- ■ Adenosine triphosphate (ATP) is the principal energy-storing or energy-carrying molecule in the cell. Should a cell require energy, one of the high-energy bonds in an ATP molecule can be broken, producing energy, an ADP molecule, and a free phosphate. The energy can then be used for growth, reproduction, active transport of substances across membranes, sporulation, movement, anabolic reactions, and other energy-requiring activities.

- ■ Aerobes and facultative anaerobes are able to produce more energy than anaerobes because they can catabolize molecules via aerobic pathways.

- Phototrophic organisms (algae, plants, and photosynthetic bacteria) derive their energy from the sun by photosynthesis. Chemosynthetic organisms use a chemical source of energy and raw materials to synthesize metabolites and macromolecules for growth and function of the organisms.

- Bacterial growth refers to an increase in the number of organisms, rather than an increase in their size. Bacterial cell division continues for as long as nutrient supply, water, and space allow and/or until a lethal concentration of toxic waste products accumulates. Temperature, pH, atmosphere, and an appropriate nutrient-containing growth medium must be controlled for optimum growth of the cells *in vitro*.

- In bacteria, generation time is the amount of time it takes for one cell (the parent cell) to become two daughter cells by binary fission; generation time varies from one species to another. Generation time can be calculated by plotting a population growth curve, which consists of four phases: lag phase, logarithmic growth phase, stationary phase, and death phase.

- Cells are healthiest and the growth rate is greatest (shortest generation time) during the logarithmic growth phase. This phase may be perpetuated in a chemostat by maintaining optimum growth conditions (*i.e.*, by adding nutrients and removing toxic waste products and excess microorganisms).

- As with humans and other organisms, the genetics of microbes involves DNA, genes, the genetic code, chromosomes, DNA replication, transcription, and translation.

- The base sequence of any gene on a chromosome may be altered accidentally in many ways, resulting in a mutation. Mutations are expressed, not only in the cell in which the mutation occurred, but in subsequent generations as well. The altered genetic code will result in an altered protein, which could affect any of a number of different phenotypic characteristics (*e.g.*, changes in colony characteristics, cell shape, biochemical activities, nutritional needs, antigenic sites, virulence, pathogenicity, drug resistance). Mutant bacteria are used in genetic and medical research and the production of vaccines.

- In addition to mutations, genetic changes in a bacterial cell may be induced by lysogenic conversion, transduction, transformation, and conjugation, all of which occur in nature, as well as in the laboratory.

- Lysogenic conversion and transduction involve bacteriophages. Transformation involves the uptake of "naked" DNA from the environment. Conjugation involves the transfer of genetic material (often a plasmid) from a donor cell to a recipient cell through a hollow sex pilus.

- The field of genetic engineering involves the introduction of new genes into cells. When a cell receives a new gene, it can produce the gene product that is coded for by that gene. Genetically engineered bacteria are used to produce products such as insulin, interferon, human growth hormone, and materials for use as vaccines. Gene therapy involves the use of viruses and plasmids to introduce normal genes into cells that contain abnormal genes.

Problems and Questions

1. List five factors that influence the growth of microorganisms in nature.
2. What four factors are kept constant to produce a population growth curve in the laboratory?
3. List three factors that contribute to the death of a pure culture of bacteria in a tube of nutrient broth.
4. What are the six major elements found in living cells? List seven other elements that are also necessary for the metabolic functions of the cell.
5. Compare (a) autotrophs, lithotrophs, and heterotrophs; (b) phototrophs and chemotrophs; (c) chemolithotrophs and chemoheterotrophs; (d) photolithotrophs and photoheterotrophs.
6. Why are saprophytic decomposers necessary for ecological balance?
7. Describe catabolism, anabolism, respiration, fermentation, and photosynthesis.
8. Describe five ways by which the genetic constitution of bacteria may be altered.
9. What is a plasmid? an episome?
10. Describe the process of genetic engineering and cite an example.

Self Test

After you have read Chapter 6, reviewed the chapter outline, examined the objectives, studied the new terms, and answered the problems and questions above, complete the following self test.

MATCHING EXERCISES

Complete each statement from the list of words provided with each section.

Nutritional Types

heterotrophs	autotrophs	chemotrophs
phototrophs	chemoheterotrophs	photoheterotrophs
chemolithotrophs	photolithotrophs	

1. Organisms that use light as a source of energy are _____.
2. Organisms that get their energy from a chemical source are _____.
3. Organisms that use an organic carbon source are _____.
4. Organisms that use an inorganic source of carbon are _____.
5. Organisms that use a chemical source of energy and an organic source of carbon are _____.
6. Organisms that use a chemical source of energy and an inorganic source of carbon are _____.
7. Those organisms that use light as a source of energy and use an organic source of carbon are _____.
8. Organisms that use light as a source of energy and use an inorganic source of carbon are _____.
9. Which three terms describe algae? _____

10. Which three terms describe plants? _____
11. Which three terms describe animals? _____
12. Which three terms describe fungi and protozoa? _____
13. What is another term for organotrophs? _____

Metabolic Reactions

catabolism   fermentation   aerobic respiration
photosynthesis   anabolism

1. The metabolic process by which plants and algae use light, carbon dioxide, and water to build carbohydrates is called _____.
2. The process in which simple molecules are used to build complex ones is _____.
3. The process in which complex macromolecules are broken down into simple molecules is _____.
4. A chemical reaction in which oxygen participates with carbohydrates to yield energy and carbon dioxide is _____.
5. An anaerobic reaction that yields energy is _____.
6. The breakdown and recycling of red blood cells in the liver is an example of _____.
7. The synthesis of enzymes within a cell is an example of _____.
8. The production of alcoholic beverages from grain is an example of _____.

Growth Curve

lag phase   death phase   stationary phase
logarithmic growth phase

1. The organisms absorb nutrients, synthesize enzymes, and prepare to reproduce in the _____.
2. More organisms are dying than are reproducing in the _____.
3. The organisms are all alive and reproducing rapidly in the _____.
4. The number of living bacteria remain about the same in the _____.
5. The healthiest stage of growth is the _____.
6. A chemostat keeps the organisms in the _____.
7. An industry that harvests products from microorganisms maintains the microbes in the _____.
8. Sporulation of certain genera of bacteria occurs during the _____.

Bacterial Genetics

mutation   transformation   prophage
mutagens   transduction
conjugation   lysogenic conversion

1. When bacterial cells are genetically changed following the absorption and incorporation of DNA from the surrounding medium, the process is called _____.

2. A spontaneous change in the nucleotide arrangement of the DNA molecule within a living cell is called _____.

3. The process by which a nontoxigenic strain of *Corynebacterium diphtheriae* is changed into a toxigenic strain is called _____.

4. When a bacteriophage carries a bit of bacterial DNA from one bacterial cell to another, the process is called _____.

5. When the DNA of a temperate bacteriophage becomes incorporated (integrated) into a bacterial chromosome, the bacteriophage is known as a _____.

6. When two bacteria are joined by a pilus bridge and genetic material passes through the pilus from one bacterial cell to the other, the process is called _____.

7. When ultraviolet light, x-rays, and some chemicals are used to increase the rate of mutation, these agents are called _____.

TRUE OR FALSE (T OR F)

___ 1. Inorganic ions are basic nutrients required by living cells.

___ 2. A mutant is an organism that has survived mutation.

___ 3. Carbon, hydrogen, oxygen, and phosphorus are among the most necessary elements in living protoplasm.

___ 4. The macromolecules of living cells include carbohydrates, fats, proteins, and nucleic acids.

___ 5. Saprophytic fungi can get energy from the sun by photosynthesis.

___ 6. Autotrophs use carbon dioxide as a carbon source.

___ 7. Only a few groups of bacteria are chemoheterotrophs.

___ 8. The process of photosynthesis releases oxygen into the air for use by animals.

___ 9. Enzymes control metabolism in all living cells.

___ 10. All enzymes are proteins.

___ 11. All photosynthetic organisms must contain some form of photosynthetic pigment.

MULTIPLE CHOICE

1. A characteristic that humans, fungi, and saprophytic bacteria have in common is that they
   a. can be facultative anaerobes
   b. obtain carbon atoms from organic compounds
   c. use carbon dioxide as a carbon source
   d. obtain their energy from light

2. The largest number of ATP molecules are produced during which of the following phases of aerobic respiration?
   a. the citric acid cycle
   b. the electron transport chain
   c. glycolysis
   d. the Kreb cycle
   e. fermentation

3. In conjugation, bacteria acquire new _____ DNA.
   a. bacterial
   b. viral
   c. human
   d. fungal

4. In transduction, bacteria acquire new _____ DNA.
   a. bacterial
   b. viral
   c. human
   d. fungal

5. The process whereby "naked" DNA is absorbed into a bacterial cell is known as
   a. conjugation
   b. transformation
   c. transduction
   d. transcription
   e. translation

6. In lysogenic conversion, bacteria acquire _____ genes.
   a. bacterial
   b. viral
   c. human
   d. fungal

7. Saprophytic fungi are able to digest organic molecules outside of the organism by means of
   a. coenzymes
   b. endoenzymes
   c. holoenzymes
   d. exoenzymes
   e. apoenzymes

8. The process by which a nontoxigenic *Corynebacterium diphtheriae* cell is changed into a toxigenic cell is called
   a. conjugation
   b. transformation
   c. prestidigitation
   d. lysogenic conversion
   e. transduction

9. Which of the following does/do not occur in anaerobes?
   a. glycolysis
   b. fermentation reactions
   c. catabolic reactions
   d. electron transport system
   e. anabolic reactions

10. Proteins that must link up with a cofactor in order to function as an enzyme are called
   a. coenzymes
   b. endoenzymes
   c. holoenzymes
   d. exoenzymes
   e. apoenzymes

CHAPTER 7

Controlling the Growth of Microorganisms

DEFINITION OF TERMS
Sterilization
Disinfection
Microbicidal Agents
Microbistatic Agents
Asepsis
Sterile Technique

FACTORS INFLUENCING MICROBIAL GROWTH
Temperature
Moisture
Osmotic Pressure
pH

Barometric Pressure
Gases

ANTIMICROBIAL METHODS
Physical Antimicrobial
  Methods
  Heat
  Cold
  Drying
  Radiation
  Ultrasonic Waves
  Filtration
Chemical Antimicrobial
  Methods

Antisepsis
  How Antimicrobial Chemicals
    Work

CHEMOTHERAPY
Major Discoveries
Characteristics of Antimicrobial
  Agents
How Antimicrobial Agents
  Work
How Bacteria Become Resistant
  to Antimicrobial Agents
Side Effects of Chemotherapeutic
  Agents

OBJECTIVES

After studying this chapter, you should be able to:

- *List three reasons why microbial growth must be controlled*
- *Define sterilization, disinfection, bactericidal agents, and bacteriostatic agents*
- *Differentiate between sterilization, pasteurization, and lyophilization*
- *Describe aseptic, antiseptic, and sterile techniques*
- *List the factors that influence the growth of microbial life*
- *Describe the following types of microorganisms: psychrophilic, mesophilic, thermophilic,*

halophilic, haloduric, alkaliphilic, acidophilic, and barophilic
- *List several factors that influence the effectiveness of antimicrobial methods*
- *List common physical antimicrobial methods*
- *List common chemical antimicrobial compounds*
- *Describe the mode of action of sulfonamide drugs on bacteria*
- *Describe the action of penicillin on bacteria*
- *List four reasons why antibiotics should be used with caution*

NEW TERMS

Acidophile
Algicidal agent
Alkaliphile
Antibiotic
Antimicrobial agent
Antisepsis
Antiseptic
Antiseptic technique
Asepsis
Aseptic technique
Asymptomatic disease
Autoclave
Bactericidal agent
Bacteriostatic agent
Barophile
Biocidal agent
Chemotherapeutic agent
Chemotherapy
Contamination

Crenated
Crenation
Desiccation
Disinfectant
Disinfection
Fungicidal agent
Germicidal agent
Haloduric
Halophilic
Hemolysis
Hypertonic solution
Hypotonic solution
Infection
Isotonic solution
Lyophilization
Mesophile
Microbicidal agent
Microbistatic agent
Osmosis

Osmotic pressure
Plasmolysis
Plasmoptysis
Psychroduric organisms
Psychrophile
Psychrotroph
Sanitization
Sepsis
Sporicidal agent
Sterile technique
Sterilization
Subclinical disease
Superinfection
Thermal death point (TDP)
Thermal death time (TDT)
Thermoduric organisms
Thermophile
Tuberculocidal agent
Virucidal agent

The factors or agents that influence the growth of microorganisms are subject to continual study. On the basis of these studies, researchers learn how beneficial microbes can be encouraged to grow while growth of pathogenic microbes can be controlled and inhibited. Control of certain microorganisms is important (1) to prevent and control infectious diseases in humans, animals, and plants; (2) to preserve food; (3) to prevent contaminating microbes from interfering with certain industrial processes; and (4) to prevent contamination of pure culture research. Preventing the spread of infectious diseases and controlling infections require many different procedures. The source of infection can be controlled by (1) destroying or inhibiting disease-causing microbes; (2) blocking the sources, routes, and vectors of transmission of disease agents; and (3) protecting an infected person from the consequences of disease by building up the body's defenses and administering appropriate chemotherapeutic drugs.

An infectious disease is any disease caused by the invasion and multiplication of pathogenic microorganisms in the body. The word *infection* is often used as a synonym for infectious disease, as in "The patient has an ear infection," meaning that the patient has an infectious disease of the ear. But microbiologists use the word *infection* to mean colonization by a pathogen. If a pathogen enters a patient's body and starts multiplying there, the patient is infected with that particular pathogen.

The pathogen may or may not be causing disease and, if the pathogen is causing disease, the infected person may or may not be exhibiting signs and symptoms of disease. When disease is occurring, but the patient is not experiencing any symptoms, the disease is said to be *asymptomatic* or *subclinical.*

The word *contamination* also has various uses in microbiology. Fecally-contaminated drinking water may contain pathogens. Restaurant food can become contaminated with pathogens that are present on the hands of food handlers. If clinical specimens are not collected properly, they become contaminated with members of the patient's indigenous flora. If care is not taken, agar cultures can become contaminated with airborne fungal spores during the inoculation process. In the hospital setting, many materials (*e.g.,* eating utensils, wound dressings, bed linens) become contaminated with pathogens from patients' secretions and excretions.

Long before people were aware of the existence of microorganisms, they tried to prevent the spoilage of food and wines, the transmission of diseases, and the infection of wounds. They developed many primitive procedures and "cures" that were often more harmful than helpful to the recipient. Today we know that the avoidance, inhibition, and destruction of potentially pathogenic microorganisms is imperative to control diseases. It is essential that those who work in the health fields appreciate the importance of controlling microbes in patients' rooms, operating rooms, treatment rooms, and emergency rooms. Healthcare workers must be aware of proper aseptic procedures to be followed for dressing wounds, giving injections and respiratory treatments, and assisting physicians, dentists, and all other personnel who have the responsibility for patient care. Healthcare professionals must be able to properly handle contaminated linen and wound dressings, bedside equipment, and laboratory specimens to avoid infecting themselves, the patients for whom they are providing care, other patients, and visitors.

DEFINITION OF TERMS

Before discussing the various methods used to destroy or inhibit microbes, a number of terms should be understood as they apply to microbiology.

Sterilization

The complete destruction of all living organisms, including cells, viable spores, and viruses, is called *sterilization.* When something is *sterile,* it is devoid of microbial life. Sterilization of objects can be accomplished by heat, autoclaving (steam under pressure), gas (ethylene oxide), various chemicals (such as formaldehyde), and certain types of radiation (*e.g.,* ultraviolet light and gamma rays). These procedures are discussed later in this chapter.

Disinfection

Disinfection is the destruction or removal of pathogens from nonliving objects by physical or chemical methods. The heating process developed by Pasteur to disinfect beer and wines is called *pasteurization*. It is still used to eliminate pathogenic microorganisms from milk and beer. It should be remembered that pasteurization is not a sterilization procedure, because not all the microbes are destroyed. Chemical agents are also used to eliminate pathogens. Chemicals used to disinfect inanimate objects, such as bedside equipment and operating rooms, are called *disinfectants*. Disinfectants are strong chemical substances that cannot be used on living tissue. *Antiseptics* are solutions used to disinfect skin and other living tissues. *Sanitization* is the reduction of microbial populations to levels considered safe by public health standards, such as those applied to restaurants.

Microbicidal Agents

The suffix "-cide" or "-cidal" refers to "killing," as in the words homicide and suicide. General terms like *germicidal agents* (*germicides*), *biocidal agents* (*biocides*), and *microbicidal agents* (*microbicides*) are disinfectants that kill microbes; such agents might be used in sanitization procedures. *Bactericidal agents* (*bactericides*) are disinfectants that kill bacteria specifically, but not necessarily bacterial endospores. Because spore coats are thick and resistant to the effects of many disinfectants, *sporicidal agents* are required to kill bacterial endospores. *Fungicidal agents* (*fungicides*) kill fungi, including fungal spores. *Algicidal agents* (*algicides*) are used to kill algae in swimming pools and hot tubs. *Viricidal* (or *virucidal*) *agents* destroy viruses. *Pseudomonicidal agents* kill *Pseudomonas* species and *tuberculocidal agents* kill *Mycobacterium tuberculosis*.

The mechanism by which various biocides kill cells varies from one disinfectant to another. Some disinfectants target and destroy cell walls, whereas others attack cell membranes. Others destroy enzymes or structural proteins or nucleic acids. Various factors affect the effectiveness of a disinfectant and must be taken into consideration whenever a disinfectant is used. These factors include the concentration of the disinfectant, the amount of time the disinfectant remains in contact with the organisms, temperature, pH, and the presence of organic materials (such as blood, pus, etc.).

Microbistatic Agents

A *microbistatic agent* is a drug or chemical that inhibits growth and reproduction of microorganisms, whereas a *bacteriostatic agent* is one that specifically inhibits the metabolism and reproduction of bacteria. Some of the drugs used to treat bacterial diseases are bacteriostatic, whereas others are bactericidal (discussed later in this chapter). Freeze-drying (lyophilization) and rapid freezing (using liquid nitrogen) are microbistatic techniques that are used to preserve microbes for future use or study.

Asepsis

Sepsis refers to the growth of pathogens on living tissues, whereas *asepsis* means the absence of pathogens on living tissues. *Aseptic technique* is designed to eliminate and exclude all pathogens by sterilization of equipment, disinfection of the environment, and cleansing of body tissues with antiseptics. *Antiseptic technique,* developed by Lister in 1867, is a type of aseptic technique. Lister used dilute carbolic acid (phenol) to cleanse surgical wounds and equipment and a carbolic acid aerosol to prevent harmful microorganisms from entering the surgical field or contaminating the patient. *Antisepsis* is the prevention of infection.

Sterile Technique

When it is necessary to prevent *all* microorganisms from gaining entrance into a laboratory or onto a surgical field, *sterile technique* is followed.

FACTORS INFLUENCING MICROBIAL GROWTH

There are many environmental factors that enhance or inhibit the growth of microorganisms, including temperature, moisture, osmotic pressure, pH, barometric pressure, gases, radiation, chemicals, and the presence of neighboring microbes. Many concepts involving these factors may be applied to our everyday lives and to laboratory and hospital environments.

Temperature

For every microorganism, there is an optimal temperature at which the organism grows best, a minimal temperature below which it ceases to grow, and a maximal temperature above which it is destroyed. These temperature ranges differ greatly among organisms. In general, rate of growth and metabolism are slower at low temperatures and faster at higher temperatures. The effect of temperature changes varies from one species to another.

Microorganisms that grow best at high temperatures are called *thermophiles.* These heat-loving microbes may be found in hot springs, compost pits, silage, and near hydrothermal vents at the bottom of the ocean. Thermophilic cyanobacteria,

TABLE 7-1. Categories of Bacteria on the Basis of Temperature Tolerance			
Group	TEMPERATURE RANGE (°C)		
	Minimum	Optimum	Maximum
Psychrophiles	−20 to 5	0 to 20	19 to 35
Mesophiles	10 to 15	20 to 40	35 to 47
Thermophiles	40 to 45	55 to 75	60 to 90

other types of bacteria, and algae cause much of the color observed in the near-boiling hot springs found in Yellowstone National Park. Because thermophiles thrive at high temperatures, boiling is not an effective means of killing them. Some organisms other than thermophiles, such as endospores and some viruses, can survive or endure boiling; they are referred to as *thermoduric* organisms. Organisms that favor temperatures above 80°C are referred to as hyperthermophiles. The highest temperature at which bacteria have been found living is 105 to 113° C.

Microbes that thrive at moderate temperatures are called *mesophiles*. This group includes most of the species that grow on plants and animals and in warm soil and water. Most pathogens and members of the indigenous microflora are mesophilic, because they grow best at normal body temperature (37°C). Thus, most pathogens are easily destroyed by boiling. Exceptions are the endospores produced by spore-forming bacteria, mycobacteria having resistant cell walls, and microbes encased in a protective coating of organic material, such as mucus, vomitus, pus, or feces.

Psychrophiles prefer cold temperatures. They thrive in cold ocean water. At high altitudes, algae can be seen living on snow. Ironically, the optimal growth temperature of some microbes (called *psychrotrophs*) is refrigerator temperature (4°C); perhaps you encountered some of these organisms the last time you cleaned out your refrigerator. Microorganisms that prefer warmer temperatures, but can survive or endure very cold temperatures and can be preserved in the frozen state are known as *psychroduric* organisms. Fecal material left by early Arctic explorers contained psychroduric *Escherichia coli* that survived the Arctic temperatures. Refer to Table 7-1 for the temperature ranges of psychrophilic, mesophilic, and thermophilic bacteria.

Moisture

Living organisms require water to carry out their normal metabolic processes. However, some microorganisms (*e.g.*, bacterial endospores and protozoan cysts) can survive the complete drying process (*desiccation*). Such organisms are in a dormant or resting state after they have been dried; then, as soon as they are placed in a moist nutrient environment, they grow and reproduce normally.

Another method of inhibiting growth of microbes is by a combination of dehydration (or drying) and freezing, a process called *lyophilization*. Lyophilized materials are frozen in a vacuum; the container is then sealed to maintain the inactive state. This freeze-drying method is widely used in industry to preserve foods, antibiotics, antisera, microorganisms, and other biological materials. It should be remembered that lyophilization cannot be used to sterilize or kill microorganisms, but rather, it is used to prevent them from reproducing.

Osmotic Pressure

Osmotic pressure is that pressure exerted on the cell membrane by solutions inside and outside the cell (Fig. 7-1). When cells are suspended in a solution, the ideal situation is that the osmotic pressure inside the cell equals the pressure of the solution outside the cell. Substances dissolved in liquids are referred to as solutes. When the concentration of solutes in the environment outside of a cell is greater than the concentration of solutes inside the cell, the solution in which the cell is suspended is said to *hypertonic*. In such a situation, whenever possible, water leaves the cell by osmosis in an attempt to equalize the two concentrations. *Osmosis* is defined as the movement of a solvent (*e.g.,* water) through a permeable membrane, from a solution having a lower concentration of solute to a solution having a higher concentration of solute. If the cell is a human cell, such as a red blood cell (erythrocyte), the loss of water causes the cell to shrink; this shrinkage is called *crenation* and the cell is said to be *crenated*. If the cell is a bacterial cell, having a rigid cell wall, the

Isotonic solution	Hypotonic solution	Hypertonic solution

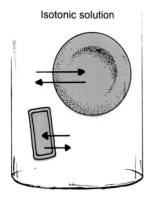 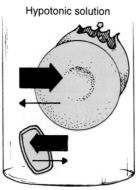 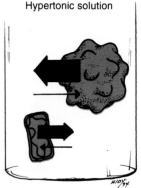

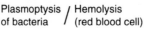

Plasmoptysis / Hemolysis
of bacteria / (red blood cell)   Plasmolysis / Crenation
of bacteria / of red blood cell

FIGURE 7-1. Changes in osmotic pressure. No change in pressure occurs inside the cell in an isotonic solution; pressure is increased in a hypotonic solution; and pressure is decreased in a hypertonic solution. Arrows indicate direction of water flow. The larger the arrow, the greater the amount of water flowing in that direction.

cell does not shrink. Instead, the cell membrane and cytoplasm shrink away from the cell wall; this condition, known as *plasmolysis,* inhibits bacterial cell growth and multiplication. Salts and sugars are added to certain foods as a way of preserving them. Bacteria entering such hypertonic environments will die.

When the concentration of solutes outside a cell is less than the concentration of solutes inside the cell, the solution in which the cell is suspended is said to *hypotonic.* In such a situation, whenever possible, water enters the cell in an attempt to equalize the two concentrations. If the cell is a human cell, such as an erythrocyte, the increased water within the cell causes the cell to swell. If sufficient water enters, the cell will burst (or lyse). In the case of erthyrocytes, this bursting is called *hemolysis.* If a bacterial cell is placed in a hypotonic solution (such as distilled water), the cell may not burst (due to the rigid cell wall), but the fluid pressure within the cell increases greatly. This increased pressure occurs in cells having rigid cell walls such as plant cells and bacteria. If the pressure becomes so great that the cell ruptures, the escape of cytoplasm from the cell is referred to as *plasmoptysis.*

When the concentration of solutes outside a cell equals the concentration of solutes inside the cell, the solution is said to be *isotonic.* In an isotonic environment, excess water neither leaves nor enters the cell and, thus, no plasmolysis or plasmoptysis occurs; the cell has normal turgor (fullness). Refer to Figure 7-1 for a comparison of the effects of various solution concentrations on bacteria and red blood cells.

Sugar solutions for jellies and pickling brines (salt solutions) for meats preserve these foods by inhibiting the growth of microorganisms. However, many types of molds and some types of bacteria can survive and even grow in a salty environment. Organisms capable of surviving salty environments—such as *Staphylococcus aureus*—are referred to as being *haloduric.* Those microbes that actually prefer salty environments (such as the concentrated salt water found in the Great Salt Lake and solar salt evaporation ponds) are called *halophilic;* "halo" for "salt" and "philic" meaning "to love." Microbes that live in the ocean, such as *Vibrio cholerae* and other *Vibrio* species, are halophilic.

pH

The term pH refers to the acidity or alkalinity of a solution (see Chapter 4). Most microorganisms prefer a neutral growth medium (about pH 7), but *acidophilic* microbes (*acidophiles*), such as those that can live in the stomach and in pickled foods, prefer a pH of 2 to 5. Acidophiles thrive in highly acidic environments such as those produced by the production of sulfurous gases in hydrothermal vents and hot springs, and in the debris produced from coal mining. *Alkaliphiles* prefer an alkaline environment (above pH 8.5), such as is found inside the intestine (about pH 9), in soils laden with carbonate, and in so-called soda lakes.

Barometric Pressure

Most bacteria are not affected by minor changes in barometric pressure. Some thrive at normal atmospheric pressure, and some, known as *barophiles* ("baro" refers to "pressure"), thrive deep in the ocean and in oil wells, where the atmospheric pressure is very high.

Autoclaves and home pressure cookers kill microbes by a combination of high pressure and high temperature (steam under pressure). The increase in pressure raises the temperature above the temperature of boiling water (100°C). At a pressure of 15 pounds per square inch (p.s.i.), the temperature of boiling water is 121°C. However, home canning done without the use of a pressure cooker does not destroy the endospores of bacteria, notably the anaerobe, *Clostridium botulinum*. Occasionally, local newspapers report cases of food poisoning resulting from the ingestion of *C. botulinum* toxins (poisons) in improperly canned vegetables and meats.

STUDY AID The suffix "phile" means to love something. Thus, acidophiles are organisms that love acidic conditions; they live in acidic environments. Alkaliphiles live in alkaline environments. Halophiles live in salty environments. Barophiles live in environments where there is high barometric pressure, such as the bottom of the ocean. Thermophiles prefer hot temperatures. Mesophiles prefer moderate temperatures. Psychrophiles prefer cold temperatures.

Gases

The types of gases present in a particular environment and their concentrations determine which species of microbes are able to live there. Most microbes grow best in an atmosphere containing oxygen, but obligate anaerobes die in its presence. For this reason, oxygen is sometimes forced into wound infections caused by anaerobes. For instance, wounds that may contain tetanus bacteria (*Clostridium tetani*) are lanced (opened) to expose them to the air. Another example is gas gangrene; this is a deep wound infection that is often treated by placing the patient in a hyperbaric (meaning increased pressure) oxygen chamber or in a room with high oxygen pressure, because the causative bacteria, most often *Clostridium perfringens,* are anaerobic and cannot live in the presence of oxygen.

ANTIMICROBIAL METHODS

The methods used to destroy or inhibit microbial life are either physical or chemical, and sometimes both types are used. The effectiveness of any antimicrobial procedure depends on (1) the length of time it is applied, (2) the temperature, (3) its concentration, (4) the nature and number of microbes and spores present (bioburden), and

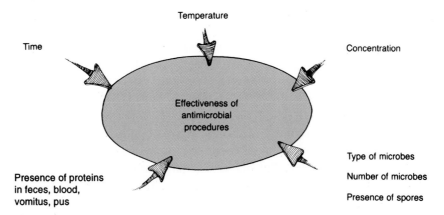

FIGURE 7-2. Factors that determine the effectiveness of any antimicrobial procedure: time, temperature, concentration, the type and number of microbes present, the presence of spores, and the presence of proteinaceous materials.

(5) the presence of organic matter, such as proteins in feces, blood, vomitus, and pus, on the materials being treated (Fig. 7-2)

Physical Antimicrobial Methods

The physical methods commonly used in hospitals, clinics, and laboratories to destroy or control pathogens are heat, pressure, desiccation, radiation, sonic disruption, and filtration.

HEAT

Heat is the most practical, efficient, and inexpensive method of disinfection and sterilization of those inanimate objects and materials that can withstand high temperatures. Because of these advantages, it is the means most frequently employed.

Two factors, *temperature* and *time,* determine the effectiveness of heat for sterilization. There is considerable variation from organism to organism in susceptibility to heat; pathogens usually are more susceptible than nonpathogens. Also, the higher the temperature, the shorter the time required to kill the organisms. The *thermal death point* (TDP) of any particular species of microorganism is the lowest temperature that will kill all the organisms in a standardized pure culture within a specified period of time. The *thermal death time* (TDT) is the length of time necessary to sterilize a pure culture at a specified temperature.

In practical applications of heat for sterilization, one must consider the material in which a mixture of organisms and their spores may be found. Pus, feces, vomitus, mucus, and blood contain proteins that serve as a protective coating to insulate the pathogens; when these substances are present on bedding, bandages, surgical instruments, and syringes, very high heat is required to destroy vegetative (growing) microorganisms and spores. In practice, the most effective procedure is to wash away the protein debris with strong soap, hot water, and a disinfectant, and then sterilize the equipment with heat.

Heat applied in the presence of moisture, as boiling or steaming, is more effective than dry heat because moist heat causes proteins to coagulate. Because cellular enzymes are proteins, they are also inactivated. This is exactly what happens when an egg is hard-boiled: the combination of heat and moisture causes the proteins to coagulate. Moist heat sterilization is faster than dry heat sterilization and can be done at a lower temperature; thus, it is less destructive to many materials that otherwise would be damaged by higher temperatures.

The vegetative forms of most pathogens are quite easily destroyed by boiling; however, bacterial endospores are particularly resistant to heat and drying. The autoclave, which combines heat and pressure, offers the most effective yet inexpensive means of destroying spores. Two examples of sporeformers are *Clostridium tetani,* the causative agent of tetanus, and *C. botulinum,* which causes a severe form of food poisoning; their spores are usually found in contaminated dirt and dust. Botulism food poisoning is preventable by properly washing and pressure cooking (autoclaving) the food.

Certain viruses are remarkably resistant to heat. A case in point is the hepatitis virus, which is frequently transferred from one person to another by the reuse of contaminated syringes and needles that have not been adequately sterilized. It is recommended that all equipment used in the transfer of blood be sterilized in an autoclave at 121°C for 20 minutes or boiled for 30 minutes or baked in an oven at 180°C for 1 hour.

Dry Heat.   Dry heat baking in a thermostatically controlled oven provides effective sterilization of metals, glassware, some powders, oils, and waxes. These items must be baked at 160° to 165°C for 2 hours or at 170° to 180°C for 1 hour. An ordinary oven of the type found in most homes may be used if the temperature remains constant. The effectiveness of dry heat sterilization depends on how deeply the heat penetrates throughout the material, and the items to be baked must be placed so that the hot air circulates freely among them.

Incineration, or burning, is an effective means of destroying contaminated disposable materials. An incinerator must never be overloaded with moist or protein-laden materials, such as feces, vomitus, or pus, because the contaminating microorganisms within these moist substances may not be destroyed if the heat does not readily penetrate and burn them. Flaming the surface of heat-resistant material is an effective way to kill microorganisms on forceps and bacteriological loops and, for many years, was a common laboratory procedure. Flaming is accomplished by holding the end of the loop or forceps in the yellow portion of a gas flame (Fig. 7-3). Open flames are dangerous and, therefore, rarely used in modern laboratories, where either sterile, disposable, plastic inoculating loops are used or heat sterilization of wire inoculating loops is accomplished using electrical heating devices (Fig 7-3).

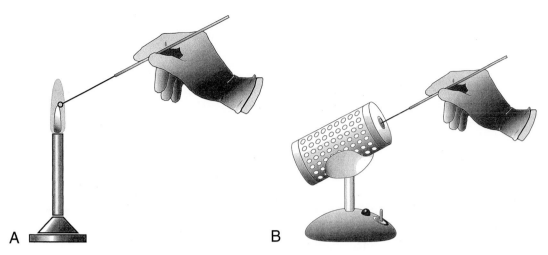

FIGURE 7-3. (*A*) The technique for flaming a wire inoculating loop. (*B*) Heat-sterilization of a wire inoculating loop using an electrical heating device.

Moist Heat. As previously stated, moist heat causes cellular proteins (including enzymes) in the microorganisms to become inactivated, and the cells die. Boiling water and steam are favored in disinfection, because no expensive equipment is necessary and the required time is short. Most pathogens die after 10 minutes of steaming at 70°C; also, boiling for 10 to 30 minutes at 90° to 100°C, depending on the altitude, destroys most viable bacteria, fungi, and viruses. Clean articles made of metal and glass, such as syringes, needles, and simple instruments, may be disinfected by boiling for 30 minutes. However, this technique is not always effective, because heat-resistant bacterial endospores, mycobacteria, and viruses may be present. As mentioned in Chapter 2, the endospores of the bacteria that cause anthrax, tetanus, gas gangrene, and botulism, as well as hepatitis viruses, are notably heat resistant and often survive normal disinfection procedures. Because the temperature at which water boils is lower at higher altitudes, water should always be boiled for longer times at high altitudes.

An effective way to disinfect clothing, bedding, and dishes is to use hot water, above 60°C, with detergent or soap and to agitate the solution around the items. This combination of heat, mechanical action, and chemical inhibition is deadly to most pathogens.

Pressurized Steam. An *autoclave* is a large metal pressure cooker that uses steam under pressure to completely destroy all microbial life. Pressure raises the temperature of the steam and shortens the time necessary to sterilize materials that can tolerate the high temperature and moisture. Autoclaving at a pressure of 15 p.s.i. at a temperature of 121.5°C for 20 minutes kills viable microorganisms, viruses, and exposed bacterial endospores, if they are not protected by pus, feces, vomitus, blood, or other proteinaceous substances. Some types of equipment and certain materials, such as rubber, which may be damaged by high temperatures, can be autoclaved at lower temperatures for longer periods. The timing must be carefully determined based on the contents and compactness of the load. All articles must be properly packaged and arranged within the autoclave to allow steam to penetrate each package. Cans should be open, bottles covered loosely with foil or cotton, and instruments wrapped in cloth. Sealed containers should not be autoclaved. Pressure-sensitive autoclave tape (Fig. 7-4) and commercially available solutions containing bacterial spores can be used as quality control measures to ensure that autoclaves are functioning properly.

COLD

Most microorganisms are not killed by cold temperatures and freezing, but their metabolic activities are slowed, greatly inhibiting their growth. Thus, freezing is a microbistatic method of preservation in which the microorganisms are in a state similar to suspended animation. When the temperature is raised above the freezing point, the metabolic reactions speed up and the organisms slowly begin to reproduce again. Refrigeration merely slows the growth of microorganisms; it does not altogether inhibit them. Many foods, biologic specimens, and bacterial cultures are preserved by rapid freezing to very low temperatures, using liquid nitrogen. It should be noted that slow freezing causes ice crystals to form within cells and may rupture the cell membranes and cell walls of some bacteria; hence, if it is important to preserve a pure culture of bacteria, such slow freezing should be avoided.

Persons who are involved in the preparation and preservation of foods must be aware that thawing to room temperature allows bacterial spores to germinate and microorganisms to resume growth. Consequently, refreezing of thawed foods is an unsafe practice, because it preserves the millions of microbes that might be present and the food deteriorates quickly when it is rethawed. Also, if bacterial endospores

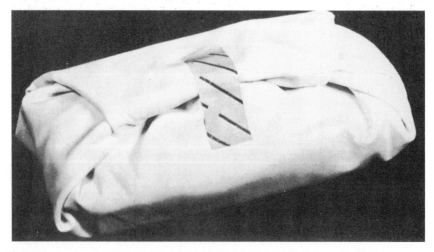

FIGURE 7-4. Pressure-sensitive autoclave tape shows dark stripes after sterilization.

of *C. botulinum* or *C. perfringens* were present, the viable bacteria would begin to produce toxins that would cause food poisoning.

DRYING

For many centuries, foods have been preserved by drying. When moisture and nutrients are lacking, many dried microorganisms remain viable, although they cannot reproduce. Foods, antisera, toxins, antitoxins, antibiotics, and pure cultures of microorganisms are often preserved by lyophilization (discussed previously).

In the hospital or clinical environment, healthcare professionals should keep in mind that dried viable pathogens may be lurking in dried matter, including blood, pus, fecal material, and dust that are found on floors, in bedding, on clothing, and in wound dressings. Should these dried materials be disturbed, such as by dry dusting, the microbes would be easily transmitted through the air or by contact. They would then grow rapidly if they settled in a suitably moist, warm, nutrient environment such as a wound or a burn. Therefore, important precautions must be observed, including the following: wet mop and damp dust floors and furniture, roll the bed linens and towels carefully, and properly dispose of wound dressings.

RADIATION

The sun is not a particularly reliable disinfecting agent because it kills only those microorganisms that are exposed to direct sunlight. The rays of the sun include the long infrared (heat) rays, the visible light rays, and the shorter ultraviolet (UV) rays. The UV rays, which do not penetrate glass and building materials, are effective only in the air and on the surface of equipment. They do, however, penetrate cells and, thus, can cause damage to DNA. When this occurs, genes may be so severely damaged that the cell dies (especially unicellular microorganisms) or is drastically changed.

In practice, a UV lamp is useful for reducing the number of microorganisms in the air. A UV lamp is often called a germicidal lamp. Its main component is a low-pressure mercury vapor tube. Such lamps are found in newborn nurseries, operating rooms, elevators, entry ways, cafeterias, and classrooms, where they are incorporated into louvered ceiling fixtures designed to radiate across the top of the room without striking persons in the room. The sterility of an area may also be maintained by having a UV lamp placed in a hood or cabinet containing instruments, paper and cloth equipment, liquid, and other inanimate articles. Many biologic materials, such as sera, antisera, toxins, and vaccines, are sterilized with UV rays.

Those whose work involves the use of UV lamps must be particularly careful not to expose their eyes or skin to the rays, because they can cause serious burns and cellular damage. Because UV rays do not penetrate cloth, metals, and glass, these materials may be used to protect persons working in a UV environment. It has been shown that skin cancer can be caused by excessive exposure to the UV rays of the sun; thus, extensive suntanning is harmful.

X-rays and gamma and beta rays of certain wavelengths from radioactive materials may be lethal or cause mutations in microorganisms and tissue cells, because they damage DNA and proteins within those cells. Studies performed in radiation research laboratories have demonstrated that these radiations can be used for the prevention of food spoilage, sterilization of heat-sensitive surgical equipment, preparation of vaccines, and treatment of some chronic diseases such as cancer, all of which are very practical applications for laboratory research. The Food and Drug Administration approved the use of gamma rays (from cobalt-60) to process chickens and red meat in 1992 and 1997, respectively. Since then, gamma rays have been used by some food processing plants to kill pathogens (like *Salmonella* and *Campylobacter* spp.) in chickens, which are labeled "irradiated" and bear the green international symbol for radiation. When radiations are used in the treatment of disease, care must be taken to focus the rays precisely on the specific area being treated to minimize damage to surrounding normal cells.

ULTRASONIC WAVES

In hospitals and clinics, ultrasonic waves are a frequently used means of cleaning and sterilizing delicate equipment. Ultrasonic cleaners consist of tanks filled with liquid solvent (usually water); the short sound waves are then passed through the liquid. The sound waves mechanically dislodge organic debris on instruments and glassware.

Glassware and other articles that have been cleansed in ultrasonic equipment must be then washed to remove the dislodged particles and solvent, and then sterilized by another method before they are used.

FILTRATION

Filters of various pore sizes are used to filter or separate cells, larger viruses, bacteria, and certain other microorganisms from the liquids or gases in which they are suspended. The filtered solution (filtrate) is not necessarily sterile, because small viruses may not be filtered out. The variety of filters is large and includes sintered glass (in which uniform particles of glass are fused), plastic films, unglazed porcelain, asbestos, diatomaceous earth, and cellulose membrane filters. Small quantities of liquid can be filtered through a syringe; large quantities require larger apparatuses.

A cotton plug in a test tube, flask, or pipette is a good filter for preventing the entry of microorganisms. Dry gauze and paper masks prevent the outward passage of microbes from the mouth and nose, at the same time protecting the wearer from inhaling airborne pathogens and foreign particles that could damage the lungs. Biological safety cabinets and laminar flow hoods contain high efficiency particulate air (HEPA) filters to protect workers from contamination.

Chemical Antimicrobial Methods

Chemical disinfection means the use of chemical agents to inhibit the growth of microorganisms, either temporarily or permanently. The effectiveness of a chemical disinfectant depends on many factors: the concentration of the chemical; the time allowed for the chemical to work (called contact time); the pH of the solution; the temperature; and the presence of proteins, blood, pus, feces, mucous secretions, and vomitus. Directions for the preparation and dilution of the disinfectant must be carefully followed, and the proper concentration, pH, and temperature must be maintained for the specified time period to ensure the best results. The items to be disinfected must first be washed to remove any proteinaceous material in which pathogens may be hidden. Although the washed article may then be clean, it is not safe to use until it has been properly disinfected. Healthcare personnel need to understand an important limitation of chemical disinfection: many disinfectants that are effective against pathogens in the controlled conditions of the laboratory become ineffective in the actual hospital or clinical environment. Furthermore, the stronger and more effective antimicrobial chemical agents are of limited usefulness because of their destructiveness to human tissues and certain other substances.

Almost all bacteria in the vegetative state as well as fungi, protozoa, and most viruses are susceptible to many disinfectants, although the mycobacteria that cause tuberculosis and leprosy, bacterial endospores, pseudomonads, fungal spores, and hepatitis viruses are notably resistant. Therefore, chemical disinfection should never be attempted when it is possible to use proper physical sterilization techniques.

The disinfectant most effective for each situation must be carefully chosen. Chemical agents used to disinfect respiratory therapy equipment and thermometers must destroy all pathogenic bacteria, fungi, and viruses that may be found in sputum and saliva. One must be particularly aware of the oral and respiratory pathogens, including *Mycobacterium tuberculosis* species of *Pseudomonas, Staphylococcus,* and *Streptococcus,* the various fungi that cause candidiasis, blastomycosis, coccidioidomycosis, and histoplasmosis, and all of the respiratory viruses.

Because most disinfection methods do not destroy all bacterial endospores that are present, any instrument or dressing used in the treatment of an infected wound or a disease caused by spore-formers must be autoclaved or incinerated. Gas gangrene, tetanus, and anthrax are examples of diseases caused by spore-formers that require the health worker to take such precautions. Formaldehyde and ethylene oxide, when properly used, are highly destructive to spores, mycobacteria, and viruses. Certain articles are heat sensitive and cannot be autoclaved or safely washed before disinfection; such articles are soaked for 24 hours in a strong detergent and disinfectant solution, washed, and then sterilized in an ethylene oxide autoclave. The use of disposable equipment whenever possible in these situations helps to protect patients and healthcare team members.

The effectiveness of a chemical agent depends to some extent on the physical characteristics of the article on which it is used. A smooth, hard surface is readily disinfected, whereas a rough, porous, or grooved surface is not. Thought must be given to selection of the most suitable germicide for cleaning patient rooms and all other areas where patients are treated.

The most effective antiseptic or disinfectant should be chosen for the specific purpose, environment, and pathogen or pathogens likely to be present. The characteristics of a good chemical antimicrobial agent are as follows:

- It must kill pathogens within a reasonable period and in specified concentrations.
- It must be nontoxic to human tissues and noncorrosive and nondestructive to materials on which it is used.
- It must be soluble in water and easy to apply. If a tincture (*e.g.*, alcohol-water solution) is used, the proper concentration must be used. Evaporation of the alcohol solvent can cause a 1% solution to increase to a 10% solution, and at this concentration, it may cause tissue damage.
- It should be inexpensive and easy to prepare for use with simple, specific directions.
- It must be stable in the dissolved or solid form so that it can be shipped and stored for a reasonable period.
- It should be stable to pH and temperature changes within reasonable limits.

ANTISEPSIS

Most antimicrobial chemical agents are too irritating and destructive to be applied to mucous membranes and skin. Those that may be safely used on human tissues are called *antiseptics*. An antiseptic merely reduces the number of organisms on a surface but does not penetrate the pores and hair follicles to destroy microorganisms residing there. To remove organisms lodged in pores and folds of the skin, health personnel use an antiseptic soap and scrub with a brush. To prevent resident indigenous microflora from contaminating the surgical field, surgeons wear sterile gloves on freshly scrubbed hands and masks and hoods to cover face and hair. Also, an antiseptic is applied at the site of the surgical incision to destroy local microorganisms.

HOW ANTIMICROBIAL CHEMICALS WORK

Injury of Cell Membranes.   Soap and detergents are referred to as surfactants; this means that they are surface-active agents that help to disperse bacteria, allowing them to more readily be rinsed away. These agents concentrate on the surface and, thus, reduce the surface tension; this characteristic makes them good wetting and dispersing agents. Some agents, such as Dial® and Safeguard® soaps, contain disinfectants, which also aid in killing bacteria. Certain concentrations of weak acids such as acetic and benzoic acids may also be used in disinfectant soaps.

Inactivation of Enzymes.   Alcohols, such as ethyl and isopropyl, are good skin antiseptics at 70% solution. Ethyl alcohol has a low toxicity for humans; hence, it is frequently used to disinfect clinical thermometers and other instruments. However, when taken internally, isopropyl alcohol causes severe gastrointestinal upset and methanol causes brain damage. Alcohols are tuberculocidal (destructive to tuberculosis-causing organisms), but not sporicidal (destructive to spores).

The phenolics including phenol, carbolic acid, xylenols, orthophenylphenol, and cresol are used as disinfectants in hospitals and laboratories. However, they are too irritating and toxic to be used on skin. The commercial mixture of phenolics, Lysol®, is an effective germicide because it works in the presence of organic material and remains active on hard surfaces for extended periods. These chemicals are tuberculocidal, but not sporicidal.

The effectiveness of phenol was demonstrated by Joseph Lister in 1867, when it was used to reduce the incidence of infections following surgical procedures. The effectiveness of other disinfectants is compared with that of phenol using the *phenol coefficient test*. To perform this test, a series of dilutions of phenol and the experimental disinfectant are inoculated with the test bacteria, *Salmonella typhi* and *Staphylococcus aureus,* at 37° C. The highest dilutions (lowest concentrations) that kill the bacteria after 10 minutes are used to calculate the phenol coefficient.

Salts of heavy metals such as mercury chloride (Merthiolate®, Mercurochrome®, Metaphen®—generic names: thimerosal, merbromin, nitromersol, respectively) and silver nitrate (Argyrol®, Protargol®) are bacteriostatic antiseptics, but they are not sporicidal and are ineffective against many pathogens. In the past, silver nitrate in low concentrations was used in the eyes of newborns to kill *Neisseria gonorrhoeae,* thus preventing gonococcal eye infections which could cause blindness.

Chemical oxidizing agents are useful disinfectants. Two of these are hydrogen peroxide and sodium perborate, which destroy bacteria and tissue debris and prevent growth of anaerobes in damaged tissues. A third, potassium permanganate, is used in weak solutions to treat urethral infections and fungal infections of the skin. Another agent in this group is ethylene oxide (Carboxide®, Cryoxide®, Oxygume®), which is used as a sterilant in gas autoclaves to sterilize heat-sensitive materials. Although it is a good microbicide and sporicide, this gas must be used with great care because it is flammable and toxic to humans.

The many compounds of chlorine, iodine, bromine, and fluorine are also useful disinfectants. For example, chlorine compounds (Clorox®, Halozone®, hypochlorites, Warexin®) are used to disinfect water and sewage and for sanitization of dishes, floors, and plumbing fixtures. It has been found that human immunodeficiency virus (HIV) can be destroyed on syringes and needles by soaking them in a 1:10 Clorox® (chlorine bleach) solution for 10 minutes. Iodine compounds, such as Wescodyne®, Betadine®, Isodine®, and tincture of iodine, are effective skin antiseptics and disinfectants. However, these compounds can be dangerous. If an alcohol solution of iodine is left open to the air, allowing the alcohol to evaporate, the solution may become too concentrated, and an iodine burn may result if it is used on skin. Most compounds of bromine and fluorine are too toxic at the effective concentrations to be used as antiseptics. All of these compounds are viricidal, bactericidal, and tuberculocidal; however, none is sporicidal.

Damage to Genetic Material.   The DNA of cells is inactivated by caustic compounds such as formalin. Formalin is a 37% aqueous solution of gaseous formaldehyde that inactivates proteins and nucleic acids. It is one of the few antimicrobial agents that are also sporicidal; however, it is so irritating to skin and mucous membranes that it cannot be used on living tissues. Frequently, it is used in the laboratory to preserve tissue specimens.

Basic aniline dyes also inactivate nucleic acids. This group includes gentian violet and crystal violet, which are useful in the treatment of fungal skin infections (ringworm) and vaginal infections caused by yeasts (*Candida*) and Gram-positive bacteria, as well as intestinal roundworm infections. Pyridium is a dye in the same group. It is sometimes prescribed for urinary tract infections (UTIs) caused by Gram-negative enteric organisms.

CHEMOTHERAPY

Chemotherapeutic agents are chemical substances (drugs) used to treat diseases, including infectious diseases. For thousands of years, people have been finding and using herbs and chemicals to cure diseases. Native witch doctors in Central and South America long ago discovered that the herb, ipecac, aided in the treatment of dysentery and that a quinine extract of cinchona bark was effective in treating malaria. During the 16th and 17th centuries, the alchemists of Europe searched for ways to cure smallpox, syphilis, and many other diseases that were rampant during that period of history. Many of the mercury and arsenic chemicals that were used frequently caused more damage to the patient than to the pathogen. Chemotherapeutic agents used to treat infectious diseases are called *antimicrobial agents.*

Major Discoveries

The true beginning of modern chemotherapy was in the late 1800s when Paul Ehrlich, a German chemist, began his search for chemicals that would destroy bacteria, yet would not damage normal body cells. By 1909, he had tested and discarded more than 600 chemicals. Finally, in that year, he discovered an arsenic compound that proved effective against syphilis. Because this was the 606th compound Ehrlich had tried, he called it "compound 606." The technical name for it is arsphenamine and the trade name became Salvarsan®. Until the purification of penicillin in 1938, arsphenamine was used to treat syphilis.

In 1928, Alexander Fleming, a Scottish bacteriologist, noted that a substance produced by a mold, *Penicillium notatum*, inhibited the growth of staphylococci on an agar plate (Fig. 7-5). He also found that broth cultures of the mold were nontoxic to his laboratory animals and destroyed staphylococci and other bacteria. During World War II, two biochemists, Sir Howard Walter Florey and Ernst Boris Chain, purified penicillin and demonstrated its effectiveness in the treatment of various bacterial infections. By 1942, the U.S. drug industry was able to produce sufficient penicillin for human use, and the search for other antibiotics began. In 1935, a chemist named Gerhard Domagk discovered that the red dye, Prontosil®, was effective against streptococcal infections in mice. Further research demonstrated that

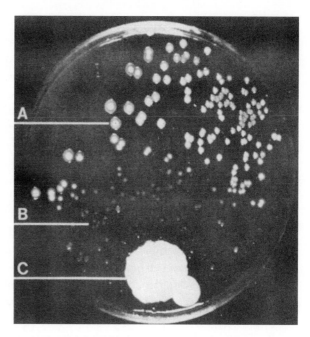

FIGURE 7-5. The discovery of penicillin by Fleming. (A) Colonies of *Staphylococcus aureus* are growing well in this area of the plate. (B) Colonies are poorly developed in this area of the plate due to an antibiotic being produced by the colony of *Penicillium notatum* shown at (C). (This photograph originally appeared in the *British Journal of Experimental Pathology* in 1929.)

Prontosil® was degraded or broken down in the body into sulfanilamide, and that sulfanilamide was the effective agent. For their outstanding contributions to medicine, these investigators—Ehrlich, Fleming, Florey, Chain, and Domagk—were all Nobel Prize recipients at various times.

Characteristics of Antimicrobial Agents

An *antimicrobial agent* is any chemical (drug) used to treat an infectious disease by inhibiting or killing pathogens *in vivo* (in the living animal). Some antimicrobial agents are antibiotics. An *antibiotic* is a substance produced by a microorganism that is effective in killing other microorganisms. Antibiotics are produced by certain molds and bacteria, usually those that live in soil. Penicillin is an example of an antibiotic produced by a mold and bacitracin is an example of an antibiotic produced by a bacterium. Although originally produced by microorganisms, many antibiotics are now synthesized or manufactured in pharmaceutical laboratories. Also, many antibiotics have been chemically modified to kill a wider variety of pathogens; these modified antibiotics are called semisynthetic antibiotics.

The ideal antimicrobial agent should (1) kill or inhibit the growth of pathogens, (2) cause no damage to the host, (3) cause no allergic reaction in the host, (4) be stable when stored in solid or liquid form, (5) remain in specific tissues in the body long enough to be effective, and (6) kill the pathogens before they mutate and become resistant to it. However, most antimicrobial agents have some side effects, produce allergic reactions, or permit development of resistant mutant pathogens.

How Antimicrobial Agents Work

To be acceptable, an antimicrobial agent must inhibit or destroy the pathogen without damaging the host. To accomplish this, the agent must target a metabolic process or structure possessed by the pathogen, but not possessed by the host. The following examples illustrate this principle.

Sulfonamide drugs inhibit production of folic acid in those bacteria that require para-aminobenzoic acid (PABA) to synthesize folic acid. Folic acid (a vitamin) is essential to these bacteria. Because the sulfonamide molecule is similar in shape to the PABA molecule, bacteria attempt to metabolize sulfonamide to produce folic acid (Fig. 7-6). However, the enzymes that convert PABA to folic acid cannot produce folic acid from the sulfonamide molecule. Without folic acid, bacteria cannot produce certain essential proteins and finally die. Sulfa drugs, therefore, are called competitive inhibitors; that is, by competing with an enzyme that metabolizes an essential nutrient, they inhibit growth of microorganisms. They are, therefore, bacteriostatic. Cells of humans and animals do not synthesize folic acid from PABA; they get folic acid from the food they eat. Consequently, they are unaffected by sulfa drugs.

In most Gram-positive bacteria, including streptococci and staphylococci, penicillin interferes with the synthesis of the peptidoglycan that is required in bacterial cell walls. Thus, by inhibiting cell wall synthesis, penicillin destroys the bacteria. Why doesn't penicillin also destroy human cells? Because human cells do not have a cell wall.

There are other antimicrobial agents that have similar action; they inhibit a specific step that is essential to the microorganism's metabolism and thereby cause its destruction. Antibiotics are used in this way against bacteria and they are highly effective. Some specifically destroy Gram-positive bacteria; others specifically destroy

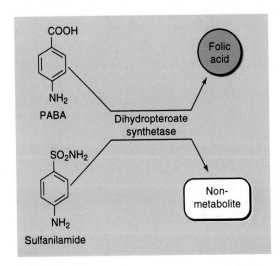

FIGURE 7-6. The effect of sulfonamide drugs.

Gram-negative bacteria. Those that are destructive to both Gram-positive and Gram-negative bacteria are called broad-spectrum antibiotics. Examples of broad-spectrum antibiotics are ampicillin, chloramphenicol, and tetracycline. Table 7-2 lists some of the antimicrobial drugs most frequently used against many common pathogens.

Antimicrobial agents work well against bacterial pathogens because the bacteria (being procaryotic) have different cellular structures and metabolic pathways that can be disrupted or destroyed by drugs that do not damage the host's (eucaryotic) cells. Bactericidal agents kill bacteria, whereas bacteriostatic agents stop them from growing and dividing. Bacteriostatic agents should only be used in patients whose host defense mechanisms (Chapter 11) are functioning properly. Some of the mechanisms by which antibacterial agents kill or inhibit bacteria are shown in Table 7-3.

Frequently, a single antimicrobial agent is not sufficient to destroy all the pathogens that develop during the course of a disease; thus, two or more drugs may be used simultaneously to kill all of the pathogens and to prevent resistant mutant pathogens from emerging. In tuberculosis, for example, four drugs (isoniazid, rifampin, pyrazinamide, and either ethambutol or streptomycin) are routinely prescribed, and as many as 12 drugs may be required for especially resistant strains. Many urinary, respiratory, and gastrointestinal infections respond particularly well to a combination of trimethoprim and sulfamethoxazole, a combination referred to as co-trimoxazole.

It is much more difficult to use antimicrobial drugs against fungal and protozoal pathogens because they are eucaryotic cells; thus, the drugs are much more toxic to the host. Most antifungal agents work in one of three ways: (1) by binding with cell membrane sterols; *e.g.*, nystatin and amphotericin B; (2) by interfering with sterol synthesis; *e.g.*, clotrimazole and miconazole; or (3) by blocking mitosis or nucleic acid synthesis; *e.g.*, griseofulvin and 5-flucytosine. Antiprotozoal drugs are usually quite toxic to the host and work by (1) interfering with DNA and RNA synthesis; *e.g.*, chloroquine, pentamidine, and quinacrine; or (2) interfering with protozoal metabolism; *e.g.*, metronidazole (Flagyl®).

Antiviral chemotherapeutic agents are particularly difficult to develop and use because viruses are produced within host cells. A few drugs have been found to be effective in certain viral infections; these work by inhibiting viral replication within cells. Some antiviral chemotherapeutic agents are listed in Table 7-4.

In cancer, in which malignant cells are dividing more rapidly than normal cells, chemical agents that inhibit DNA and RNA synthesis can be used, provided the dosage and total period of administration are carefully controlled. Cancer drugs interfere with normal DNA function in rapidly dividing cells, regardless of whether the cells are normal or malignant. Thus, normal cells that are rapidly dividing, including skin cells, erythroblasts that later become red blood cells, and sperm cells, are damaged along with the malignant cells. This is why blood cell counts are performed frequently in cancer patients; the physician must be able to determine at

(text continues on page 199)

TABLE 7-2. Examples of Antimicrobial Agents Used to Treat Important Infectious Diseases

Pathogen	Disease(s)	Antimicrobial Agent(s)
Bacteria		
Bacillus anthracis	Anthrax	Penicillin, tetracyclines, erythromycin
Bordetella pertussis	Whooping cough	Erythromycin, trimethoprim-sulfamethoxazole
Brucella abortus and *B. melitensis*	Brucellosis, undulant fever	Tetracyclines, streptomycin, gentamicin
Chlamydia trachomatis	Trachoma, urethritis, lymphogranuloma venereum	Sulfonamides, tetracyclines, erythromycin, azithromycin
Clostridium perfringens	Gas gangrene, wound infections	Penicillin, clindamycin, metronidazole, imipenem
Clostridium tetani	Tetanus (lockjaw)	Penicillin, tetracyclines, tetanus immune globulin
Corynebacterium diphtheriae	Diphtheria	Antitoxin, penicillin, erythromycin
Escherichia coli	Urinary tract infections	Cefotaxime, ceftizoxime, ceftriaxone, etc.
Francisella tularensis	Tularemia	Streptomycin, tetracyclines, gentamicin, chloramphenicol
Haemophilus ducreyi	Chancroid	Azithromycin, ceftriaxone, ciprofloxacin
Haemophilus influenzae	Meningitis, pneumonia, ear infections, epiglottitis	Cefotaxime, ceftriaxone, trimethoprim-sulfamethoxazole
Klebsiella pneumoniae	Pneumonia, urinary tract infections	Cefotaxime, ceftizoxime, ceftriaxone, etc.
Legionella pneumophilia	Legionellosis	Erythromycin, clarithromycin, azithromycin, doxycycline
Mycobacterium leprae	Leprosy	Dapsone, rifampin, clofazimine, minocycline
Mycobacterium tuberculosis	Tuberculosis	Isoniazid, streptomycin, pyrazinamide, rifampin, ethambutol
Mycoplasma pneumoniae	Atypical pneumonia	Tetracyclines, erythromycin, clarithromycin, azithromycin
Neisseria gonorrhoeae	Gonorrhea	Ceftriaxone, cefixime, ciprofloxacin, ofloxacin
Neisseria meningitidis	Nasopharyngitis, meningitis	Penicillin, cefotaxime, ceftizoxime, ceftriaxone
Proteus vulgaris and *Morganella morganii*	Urinary tract infections	Cefotaxime, ceftizoxime, ceftriaxone, etc.
Pseudomonas aeruginosa	Urinary tract infections	Ciprofloxacin, carbenicillin, ticarcillin, etc.

Pathogen	Disease(s)	Antimicrobial Agent(s)
Rickettsia rickettsii	Rocky Mountain spotted fever	Tetracyclines, chloramphenicol, a fluoroquinolone
Salmonella typhi	Typhoid fever	A fluoroquinolone, ceftriaxone, chloramphenicol, etc.
Salmonella spp.	Gastroenteritis (salmonellosis)	Cefotaxime, ceftriaxone, a fluoroquinolone
Shigella spp.	Shigellosis (bacillary dysentery)	A fluoroquinolone, azithromycin, etc.
Staphylococcus aureus	Pneumonia, septicemia, etc.	Penicillin G or V (if not producing penicillinase), cloxacillin, dicloxacillin, nafcillin, oxacillin, etc.
Streptococcus pyogenes	Strep throat, scarlet fever, rheumatic fever, septicemia	Penicillin, a cephalosporin, erythromycin, clindamycin, etc.
Streptococcus pneumoniae	Pneumonia, meningitis, ear infections	Penicillin, a cephalosporin, erythromycin, azithromycin, etc.
Treponemia pallidum	Syphilis	Penicillin, tetracyclines, ceftriaxone
Vibrio cholerae	Cholera	Trimethoprim-sulfamethoxazole, tetracyclines, a fluoroquinolone
Yersinia pestis	Plague	Streptomycin, tetracyclines, chloramphenicol, gentamicin
Fungi		
Dermatophytes	Tinea ("ringworm") infections	Miconazole, clotrimazole, econazole, ketoconazole, griseofulvin
Candida	Mucosal candidiasis	Fluconazole, ketoconazole, itraconazole
Blastomyces *Histoplasma* *Cryptococcus* *Coccidioides* *Candida*	Systemic mycosis	Amphotericin B, fluconazole, itraconazole, flucytosine
Pneumocystis carinii	Pneumonia	Pentamidine, trimethoprim-sulfamethoxazole, etc.
Protozoa		
Trichomonas	Trichomoniasis	Metronidazole, tinidazole
Giardia lamblia	Giardiasis	Metronidazole, tinidazole, furazolidone, paromomycin
Entamoeba histolytica	Amebiasis	Metronidazole, tinidazole
Toxoplasma	Toxoplasmosis	Pyrimethamine + sulfadiazine, spiramycin
Plasmodium spp.	Malaria	Chloroquine, mefloquine, doxycycline, primaquine

(continued)

Pathogen	Disease(s)	Antimicrobial Agent(s)
Viruses		
Herpes spp.	Orolabial herpes	Penciclovir
	Genital herpes	Acyclovir, famciclovir, valacyclovir
	Encephalitis	Acyclovir
Influenza A	Influenza	Rimantadine, amantadine
Human immunodeficiency virus (HIV)	AIDS	Reverse transcriptase inhibitors, protease inhibitors

Note: The above listed antimicrobial agents were recommended for use at the time this table was prepared. Recent issues of *The Medical Letter* should be consulted for current treatment recommendations. Whenever possible, antimicrobial susceptibility testing should be performed on isolates.

TABLE 7-3. Antibacterial Agents

Mode of Action	Agent	Source	Bactericidal or Bacteriostatic
Inhibition of cell wall synthesis	Bacitracin	*Bacillus subtilis* (a bacterium)	Bactericidal
	Cephalosporins	*Cephalosporium* spp. (molds)	Bactericidal
	Penicillins	*Penicillium* spp. (molds)	Bactericidal
	Beta-lactamase-resistant penicillins: methicillin, nafcillin, oxacillin, cloxacillin	These are modified penicillins	Bactericidal
	Semisynthetic penicillins: ampicillin, carbenicillin, amoxicillin	These are modified penicillins	Bactericidal
	Vancomycin	*Streptomyces* sp. (fungus-like bacteria)	Bactericidal
Inhibition of protein synthesis	Chloramphenicol	*Streptomyces venezuelae* (fungus-like bacteria)	Bacteriostatic
	Erythromycin	*Streptomyces erythreus* (fungus-like bacteria)	Bacteriostatic (usually)
	Streptomycin and other aminoglycosides	*Streptomyces* spp. (fungus-like bacteria)	Bactericidal
	Tetracycline	*Streptomyces rimosus* (fungus-like bacteria)	Bacteriostatic
Disruption of cell membranes	Polymyxin B and polymyxin C (colistin)	*Bacillus* spp. (bacteria)	Bactericidal
Inhibition of enzyme activity	Sulfonamides	Synthetic	Bacteriostatic
	Trimethoprim	Synthetic	Bacteriostatic

TABLE 7-4. Examples of Antiviral Agents		
Antiviral Agent	Viral Infection(s) the Drug is Used to Treat	Mechanism of Action
Acyclovir and virarabine	Herpes simplex infections, shingles, chickenpox	Interferes with viral DNA synthesis
Amantadine	Influenza A (prevention)	Inhibits penetration, uncoating and assembly of viruses
Antiviral creams (e.g., idoxuridine, trifluridine, acyclovir)	Herpes blisters	Interferes with viral DNA synthesis
Reverse transcriptase inhibitors (e.g., zidovudine [AZT], didanosine, zalcitabine, stavudine, lamivudine, nevirapine, delavirdine)	HIV	Inhibits HIV reverse transcriptase
Ribavirin	RSV infection in infants	Inhibits viral mRNA synthesis
Rimantadine	Influenza A (prevention)	Inhibits penetration, uncoating and assembly of viruses
Viral protease inhibitors (e.g., saquinavir, ritonavir, indinavir)	HIV	Inhibits HIV protease (thus inhibiting viral protein synthesis)

Note: Of the eleven antiretroviral drugs licensed for use in the U.S. at the time this table was prepared, only two viral gene products (reverse transcriptase and protease) were targeted by these drugs.

what point the chemotherapy must be discontinued to avoid critical damage to the patient's normal cells.

How Bacteria Become Resistant to Antimicrobial Agents

It is common to hear about drug-resistant bacteria or "superbugs" these days (see Insight Boxes), but how do bacteria become resistant to drugs? Some bacteria are naturally resistant to a particular antimicrobial agent because they lack the specific target site for that drug (*e.g.*, mycoplasmas have no cell walls and are, therefore, resistant to any drugs that interfere with cell wall synthesis) or the drug is unable to cross the organism's cell wall or cell membrane and, thus, cannot reach its site of action (*e.g.*, ribosomes); such resistance is known as *intrinsic resistance*.

It is also possible for bacteria that were once susceptible to a particular drug to become resistant to it; this is called *acquired resistance*. Bacteria usually become resistant to antibiotics and other antimicrobial agents by one of four mechanisms: (1) by alteration of drug binding sites, (2) by alteration of membrane permeability, (3) by

INSIGHT
"Superbugs," Part I—What Are They?

Undoubtedly, you've heard much about drug-resistant bacteria, or "superbugs," as they've been labeled by the press. "Superbugs" are microorganisms, mainly bacteria, that have become resistant to one or more antimicrobial agents. The worst of them are multi-resistant; that is, resistant to a variety of antimicrobial agents. Microorganisms become resistant by chromosomal mutations or by inheriting genes, most frequently via conjugation and the transfer of plasmids. Common ways by which bacteria become resistant include (1) a surface alteration that prevents the drug from binding to the cell, (2) a membrane alteration that prevents the drug from entering the cell, (3) by developing the ability to produce an enzyme that destroys the drug, and (4) by developing mechanisms by which drugs are pumped out of the cell before they can damage the cell. Let's examine some examples of "superbugs."

Mycobacterium tuberculosis—the etiologic agent of tuberculosis (TB)—is the organism that's received the most publicity in the past few years. Tuberculosis remains one of the biggest killers worldwide, with about 8 million new cases of active TB every year and approximately 3 million deaths annually. In the United States, there are about 27,000 new cases and approximately 2000 deaths each year. Outbreaks of multi-drug-resistant TB have been reported in most states. One particular strain, designated "strain W," is resistant to most of the antitubercular drugs such as isoniazid, rifampin, streptomycin, and ethambutol; some strains cannot be treated with any drug or combination of drugs. Patients infected with these strains may have to have a lung or section of lung removed, as in pre-antibiotic days, and many will die.

MRSA and MRSE: Another ever-growing problem concerns the multi-drug-resistant *Staphylococcus aureus* (MRSA) and *Staphylococcus epidermidis* (MRSE). Today, over 90% of *S. aureus* strains are penicillin-resistant, and many strains (the MRSA) have developed resistance to methicillin, nafcillin, oxacillin, and cloxacillin, as well as other antimicrobial agents. *Staphylococcus aureus* is the second most common cause of nosocomial infections, including skin and wound infections, bac-teremia, and lower respiratory infections. Today, 40% of nosocomial *S. aureus* infections are due to MRSA. Coagulase-negative staphylococci, like *S. epidermidis,* are the most frequent cause of infections related to intravenous catheters and prosthetic devices, and also cause urinary tract infections and endocarditis. Depending upon the particular hospital, anywhere from 60% to 90% of coagulase-negative staphylococci are methicillin resistant. Vancomycin, a very expensive and potentially toxic agent, must be used to treat infections with MRSA and MRSE. Some strains of *Staphylococcus* have developed resistance to vancomycin, and scientists worry that the frequency with which vancomycin is used will ultimately lead to the emergence of vancomycin-resistant MRSA and MRSE. Should that occur, there will be no drugs left to treat the numerous types of infections caused by these staphylococci.

Other "superbugs" include **vancomycin-resistant *Enterococcus* spp., penicillin-resistant strains of *Neisseria gonorrhoeae, Haemophilus influenzae,* and *Streptococcus pneumoniae.*** *Enterococcus* is the third most common cause of nosocomial infections, including wound infections, urinary tract infections, septicemia, and endocarditis. *Haemophilus influenzae* is a common cause of bacterial meningitis and pneumonia and causes about one-third of the cases of ear infections. *Streptococcus pneumoniae* is the most common cause of bacterial pneumonia (about 500,000 cases in the U.S. per year), is a major cause of bacterial meningitis (about 6000 cases in the U.S. per year), causes about one-third of the cases of ear infection (about 6 million cases in the U.S. per year), and causes about 55,000 cases of bacteremia in the U.S. per year. *Streptococcus pneumoniae* causes about 40,000 deaths in the U.S. every year. [Other multi-drug-resistant organisms that cause especially severe and sometimes untreatable infections are *Pseudomonas aeruginosa* and *Pseudomonas cepacia. Pseudomonas aeruginosa* is a common cause of nosocomial infections, and both of these organisms are frequently isolated from the lungs of cystic fibrosis patients. Additional "superbugs" include certain strains of *Escherichia coli, Salmonella,* and *Shigella.*]

developing the ability to produce an enzyme that destroys or inactivates the antimicrobial agent, or (4) by developing the ability to produce multidrug resistance pumps which pump drugs out of the cells before the drugs can damage or kill the cells.

Before a drug can enter a bacterial cell, molecules of the drug must first bind to proteins on the surface of the cell; these protein molecules are called binding sites. A chromosomal mutation can result in an alteration in the structure of the binding sites, so that the drug is no longer able to bind to the cell. If the drug cannot bind to the cell, it cannot enter the cell, and the organism is, therefore, resistant to the drug.

To enter a bacterial cell, a drug must be able to pass through the cell wall and cell membrane. A chromosomal mutation can result in an alteration in the structure of the cell membrane, which, in turn can change the permeability of the membrane. If the drug is no longer able to pass through the cell membrane, it cannot reach its target (*e.g.,* a ribosome or the DNA of the cell) and, if the drug cannot reach its target, the organism is now resistant to the drug.

Another way in which bacteria become resistant to a certain drug is by developing the ability to produce an enzyme that destroys or inactivates the drug. Because enzymes are coded for by genes, a bacterial cell would have to acquire a new gene to produce an enzyme that it never before produced. The primary way in which bacteria acquire new genes is by conjugation. Often, a plasmid containing such a gene, is transferred from one bacterial cell (the donor cell) to another bacterial cell (the recipient cell) during conjugation (see Chapter 6). Many bacteria have become resistant to penicillin because they have acquired the gene for penicillinase production during conjugation. Penicillinase is an enzyme that destroys part of the structure of the penicillin molecule, rendering the penicillin molecule ineffective (see following section on beta-lactamases). A bacterium that produces penicillinase is, thus, resistant to penicillin. A plasmid containing multiple genes for drug resistance is called a *resistance factor* or *R-factor*. Bacteria can also acquire new genes by transduction and transformation (refer back to Chapter 6, if necessary).

A fourth way in which bacteria become resistant to drugs is by developing the ability to produce multi-drug resistance (MDR) pumps. An MDR pump enables the cell to pump drugs out of the cell before the drugs can damage or kill the cell. The genes encoding these pumps are often located on plasmids that bacteria receive during conjugation. Bacteria receiving such plasmids become multi-drug resistant; *i.e.,* they become resistant to several drugs.

Thus, bacteria can acquire resistance to antimicrobial agents as a result of chromosomal mutation or the acquisition of new genes by transduction, transformation, and, most commonly, by conjugation.

Beta-Lactamases.   At the heart of every penicillin and cephalosporin molecule is a double-ringed structure, which in penicillins resembles a "house and garage" (Fig. 7-7).

The "garage" is called the beta-lactam ring. Some bacteria produce enzymes that destroy the beta-lactam ring; these enzymes are known as *beta-lactamases*. When the

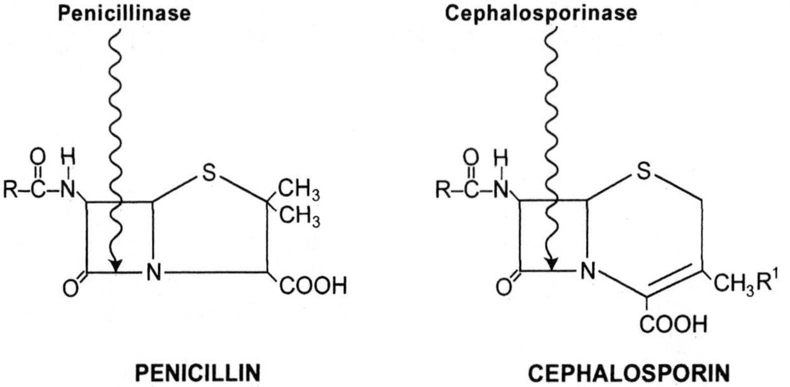

FIGURE 7-7. Sites of beta-lactamase attack on penicillin and cephalosporin molecules. (See text for details.)

beta-lactam ring is destroyed, the antibiotic no longer works. Thus, an organism that produces a beta-lactamase is resistant to antibiotics containing the beta-lactam ring (collectively referred to as beta-lactam antibiotics or beta-lactams).

There are two types of beta-lactamases: penicillinases and cephalosporinases. Penicillinases destroy the beta-lactam ring in penicillins; thus, an organism that produces penicillinase is resistant to penicillins. Cephalosporinases destroy the beta-lactam ring in cephalosporins; thus, an organism that produces cephalosporinase is resistant to cephalosporins. Some bacteria produce both types of beta-lactamases.

Side Effects of Chemotherapeutic Agents

There are many reasons why chemotherapeutic drugs should not be used indiscriminately:

- Whenever antimicrobial agents are administered to a patient, susceptible organisms die, but resistant ones survive. The resistant organisms then multiply, become dominant, and can be transferred to other people. As stated earlier, microorganisms can become resistant by chromosomal mutation or by acquiring new genes via transduction, transformation, and conjugation. To prevent the overgrowth of resistant organisms, sometimes several drugs, each with a different mode of action, are administered simultaneously.
- The patient may become allergic to the agent. Penicillin G in low doses, for example, often sensitizes those who are prone to allergies; when these persons receive a second dose of penicillin at some later date, they may have a severe reaction known as anaphylactic shock, or they may break out in hives. Allergic reactions are described in more detail in Chapter 11.
- Many antimicrobial agents are toxic to humans, and some are so toxic that they are administered only for serious diseases for which no other agents are

INSIGHT
"Superbugs," Part 2—What Can Be Done About Them?

"Superbugs" or multi-drug-resistant microorganisms are an enormous public health problem. What can be done about them?

In the past, we merely relied on drug companies to come up with "bigger and better" drugs. However, due to a variety of factors, there are relatively few drugs ready for introduction today. If we can no longer rely on drug companies to solve the problems, what *can* be done?

Education is crucial—education of health care professionals and, in turn, education of patients. The following are some of the ways in which "superbugs" can be controlled:

- Patients must stop demanding antibiotics every time they are sick or have a sick child. Most sore throats and many respiratory infections are caused by viruses, and viruses are unaffected by antibiotics. Consequently, antibiotics should not be prescribed for viral infections. According to the *Journal of the American Medical Association* (*JAMA*), instead of demanding antibiotics from the physician, consumers/patients should be asking why one *is* being prescribed.

- Physicians must not let themselves be pressured by patients. They should prescribe antibiotics only when they are warranted—only when there is a demonstrated need for them. Whenever possible, physicians should collect a specimen for culture and have the clinical microbiology laboratory perform antimicrobial susceptibility testing to determine *which* antimicrobial agents are apt to be effective. Physicians should prescribe an inexpensive, narrow spectrum drug whenever the laboratory results demonstrate that such a drug effectively kills the pathogen. According to *JAMA*, by some estimates, at least half of current antibiotic use in the United States is inappropriate—antibiotics are either not indicated at all, or they are incorrectly prescribed as the wrong drug, the wrong dosage, or the wrong duration. Another study demonstrated that colds, upper respiratory infections, and bronchitis accounted for 20% of all antibiotic prescriptions despite the fact that these conditions typically do not benefit from antibiotics.

- Patients must take their antibiotics in the exact manner in which they are prescribed. Health professionals should emphasize this to patients and do a better job explaining exactly how medications are to be taken.

- Patients must take *all* their pills—even after they are feeling better. Again, this must be explained and *emphasized*. As stated in *JAMA,* if treatment is cut short, there is selective killing of only the most susceptible members of a bacterial population. The more resistant variants are left behind to regrow into a new infection having a greater number of more resistant members. Shorter treatment schedules might lead to better patient compliance.

- Never keep antibiotics in your medicine cabinet and never give them to anyone else. Unless prescribed by a physician, never use antibiotics in a prophylactic manner—such as to avoid "traveler's diarrhea" when traveling to a foreign country. Taking antibiotics in that manner will actually increase your chances of developing traveler's diarrhea. The antibiotics kill off some of your indigenous intestinal flora, making it easier for pathogens to gain a foothold.

- As always, practice good infection prevention and control procedures. Frequent and proper hand-washing is essential to prevent the transmission of pathogens from one patient to another. Monitor for important pathogens, such as methicillin-resistant *Staphylococcus aureus* (MRSA). Hospital personnel with skin infections are potential sources of MRSA. Isolate patients that are infected with MRSA.

- Vaccines also help. They cut down on the number of infections with drug-resistant organisms such as ampicillin-resistant strains of *Haemophilus influenzae* and penicillin-resistant strains of *Streptococcus pneumoniae.*

available. One such drug is chloramphenicol (Chloromycetin®), which, if given in high doses for a long period, may cause a very severe type of anemia called aplastic anemia. Another is streptomycin, which can damage the auditory nerve and cause deafness. Other drugs are hepatotoxic or nephrotoxic, causing liver or kidney damage, respectively.

- With prolonged use, broad-spectrum antibiotics may destroy the normal flora of the mouth, intestine, or vagina. The person no longer has the protection of the indigenous microflora and, thus, becomes much more susceptible to infections caused by opportunists or secondary invaders. The resultant overgrowth by such organisms is referred to as a *superinfection*. An example of such an infection is diarrhea, which can result from prolonged antibiotic therapy owing to the loss of the normal protective microbes. *Clostridium difficile*-associated diseases such as antibiotic-associated diarrhea and pseudomembranous colitis are also the result of superinfections. A vaginal yeast infection often follows antibacterial therapy because many bacteria of the vaginal flora were destroyed, allowing the indigenous yeast (*Candida albicans*) to overgrow. Superinfections are discussed more fully in Chapter 8.

Therefore, antimicrobial agents, including antibiotics, should be taken only when prescribed and only under a physician's supervision. Also, the proper dosage must be administered for the recommended period to prevent resistant organisms from gaining a foothold.

In recent years, microorganisms have developed resistance at such a rapid pace that many people, including some scientists, are beginning to fear that pathogens are becoming so resistant to antibiotics that the human race may be destroyed (see Insight Box). Some problems already have arisen. A case in point is pneumonia. In the more-developed countries of the world, the types of pneumonia currently occurring are different from those that were seen 30 years ago and sometimes they are highly resistant to treatment. It is hoped that antibiotics will be used with greater care and that effective vaccines may become available. *Haemophilus influenzae* and *Streptococcus pneumoniae* vaccines are available to protect against some types of pneumonia, but not all. Also, resistant strains of gonococci (*e.g.,* penicillinase-producing *Neisseria gonorrhoeae* or PPNG) have developed. Scientists hope to someday see vaccines against gonorrhea, syphilis, and AIDS. The difficulty then will be to convince the American public to receive the vaccines to protect themselves against these pathogens.

■ REVIEW OF KEY POINTS

- Controlling growth of certain microbes is important to prevent and control infectious diseases in humans, animals, and plants; preserve food; prevent contaminating microbes from interfering with certain industrial processes; and prevent contamination of pure culture research.

- Avoidance, inhibition of, and destruction of potentially pathogenic microorganisms is crucial to control diseases.
- Some disinfectants destroy cell walls, others attack cell membranes, and still others destroy enzymes, structural proteins or nucleic acids.
- The effectiveness of a chemical disinfectant depends on its concentration, time of contact, pH, temperature, and the presence of organic matter (*e.g.*, proteins).
- Growth of microorganisms is affected by temperature, moisture, osmotic pressure, pH, barometric pressure, atmosphere, radiation, chemicals, and other microbes.
- Lyophilization (freeze-drying) will not sterilize or kill microbes, but it will prevent microbial growth. It is a method of preserving microbes for future use.
- The types and concentrations of gases like oxygen and carbon dioxide in an environment determine the types of microbes that will grow there.
- Physical methods used to destroy or control microorganisms include heat, cold, barometric pressure, desiccation, radiation, sonic disruption, and filtration.
- Bacterial endospores and certain viruses are very resistant to heat and desiccation.
- Washing with hot water, detergent, and agitation is very effective against most pathogens.
- Freezing is a microbistatic method of preservation in which the microorganisms are in a state similar to suspended animation. Thawing at room temperature allows bacterial spores to germinate and vegetative microbes to resume growth.
- Antiseptics reduce the number of microbes on the skin by injuring microbial cell membranes and inactivating essential enzymes.
- The types of chemotherapeutic agents used to treat infectious diseases are called antimicrobial agents (some of which are antibiotics), which work by destroying the pathogen with minimal damage to the host.
- Side effects of antimicrobial agents include microbial mutation to become resistant to the drug, patients becoming allergic to the agent, toxicity and damage to humans, destruction of human indigenous microflora of the mouth, vagina, and intestine, and superinfections.
- Ways in which bacteria become resistant to antimicrobial agents include alteration of drug binding sites, alteration of membrane permeability, developing the ability to produce an enzyme that destroys or inactivates the drug, and developing multi-drug resistance pumps.
- Beta-lactamases (penicillinases and cephalosporinases) are enzymes that destroy the beta-lactam ring in antibiotics containing a beta-lactam ring (*e.g.*, penicillins and cephalosporins). When the beta-lactam ring is destroyed, the drug no longer works. Bacteria that produce penicillinases are resistant to penicillins and those that produce cephalosporinases are resistant to cephalosporins.
- Healthcare professionals must take special care not to transfer potentially pathogenic microbes from patient to patient, from themselves to patients, or from patients to themselves by using physical and chemical methods to control pathogens.

Problems and Questions

1. Discuss why it is necessary to control microbial growth.
2. Define sterilization, disinfection, pasteurization, and lyophilization.
3. What are the differences between sterile, aseptic, and antiseptic techniques? In what circumstances might each be used?
4. What are the characteristics of bacteria indicated by the following terms? Psychrophilic, mesophilic, thermophilic, halophilic, haloduric, alkaliphilic, acidophilic.
5. List six characteristics of a good antimicrobial agent.
6. List some effective physical and chemical means of controlling microbial growth.
7. How do chemotherapeutic agents destroy microbes without also harming the patient?
8. Discuss why antibiotics should be used discriminantly.

Self Test

After you have read Chapter 7, reviewed the chapter outline, examined the objectives, studied the new terms, and answered the problems and questions above, complete the following self test.

MATCHING EXERCISES

Complete each statement from the list of words provided with each section.

Terms

disinfection	fungistatic agent	sterile technique
sterilization	sepsis	aseptic technique
pasteurization	asepsis	
fungicidal agent	antiseptic technique	

1. A chemical that kills fungi is a _____.
2. A chemical that inhibits growth of the fungus that causes athlete's foot is a _____.
3. The growth of pathogens in living tissues is _____.
4. The surgical technique of using disinfectants and antiseptics to cleanse skin, instruments, and so on, is _____.
5. The lack of pathogens on living tissues is _____.
6. The surgical technique that eliminates and avoids pathogens is _____.
7. The process of opening a sterile packet without exposing the sterile equipment to any microorganisms is the _____.
8. A process during which all microorganisms are killed is called _____.
9. The process of destroying the pathogens that are on or in nonliving objects is called _____.
10. Heating milk to destroy pathogens is called _____.
11. To spray the base of the shower stall to kill the fungi growing there, one could use Lysol®, a _____.

12. When you apply iodine or Merthiolate® to a cut or abrasion, you are using _____ technique.
13. The microbes that spoil beer and wine can be destroyed by _____.
14. Washing table tops with an antimicrobial agent is an example of _____.

Microbial Types

thermophiles	aerobes	obligate anaerobes
mesophiles	acidophiles	facultative
psychrophiles	alkaliphiles	anaerobes
halophiles	barophiles	

1. Organisms that can thrive deep within the ocean, in a high barometric pressure, are _____.
2. Organisms that can live in the colon (where it is anaerobic) and in the air are _____.
3. Microbes that die in the presence of oxygen are _____.
4. Microbes that thrive best in air are considered _____.
5. Organisms that are not inhibited by chlorine or iodine disinfectants and tolerate a high salt concentration are called _____.
6. Microbes that can live in the acid environment of the stomach are _____.
7. Microbes that prefer the alkaline environment of the intestine are _____.
8. Microbes found living in an iceberg are _____.
9. Organisms that can live in hot springs are _____.
10. Most pathogens are _____ because they grow best at body temperature.
11. When you gargle with salt water, most microbes are inhibited; exceptions are the staphylococci, which are _____.
12. *Escherichia coli* and other enteric bacteria would be considered _____.
13. Pathogens of the genus *Clostridium* are all sporeformers, and they grow best in a closed wound or jar where there is no oxygen; they are therefore called _____.
14. The pathogen that causes botulism is one of the _____.

Physical Antimicrobial Methods

ultraviolet rays	desiccation	osmotic pressure
x-rays	autoclaving	
filtration	sonic waves	

1. Microorganisms of various sizes may be removed from a solution by a process called _____, so that the remaining solution may be sterile.
2. The use of concentrated salt and sugar solutions inhibits the growth of bacteria by changing the _____.
3. An excellent method of cleaning and sterilizing delicate instruments is the use of _____.

4. The type of radiation used to keep certain areas sterile, such as cabinets containing sterile instruments and equipment, is _____.
5. When pressurized steam is used to kill microorganisms as well as spores, the process is called _____.
6. When all moisture is removed, microbes are inhibited by _____.
7. The rays of the sun that cause suntans and may cause skin cancer are called _____.

Chemical Antimicrobial Methods

ethyl alcohol	phenolics	formalin
isopropyl alcohol	hydrogen	ethylene oxide
detergent or soap	peroxide	mercury salts

1. A type of alcohol that is toxic when taken internally, but is good for rubbing on the skin is _____.
2. The type of alcohol that is the best antiseptic is _____.
3. An oxidizing agent that is frequently used to cleanse wounds and remove pus is _____.
4. A solution that is sporicidal but is too caustic to use on living tissues is _____.
5. Lysol® and carbolic acid are _____.
6. Merthiolate® is one of the _____.
7. Before surgery, the surgeon scrubs well with a _____ to destroy the surface bacteria on the skin.
8. A toxic, flammable gas that is sporicidal and is used in gas autoclaves is _____.

Chemotherapy

antibiotic	Salvarsan®	amphotericin B
penicillin	broad-spectrum	antibiotics
sulfanilamide	Prontosil®	

1. The antibiotic that was discovered by Fleming, isolated from a mold, and purified by Florey and Chain is _____.
2. Antibiotics that are effective against many gram-positive and gram-negative pathogens are known as _____.
3. The drug that worked against *in vivo* streptococcal infections but was not effective *in vitro* was _____.
4. A drug that is effective against fungal infections is _____.
5. When Prontosil® was broken apart, one half of the molecule was effective against gram-positive infections *in vitro*. This drug is _____.
6. A chemotherapeutic drug that is produced by a living organism (a bacterium or fungus) is called an _____.
7. The antimicrobial agent that Ehrlich discovered after 605 unsuccessful attempts was _____.

8. The first antimicrobial agent drug found to be effective against syphilis was _____.

9. The antimicrobial agent used to treat syphilis today is _____.

TRUE OR FALSE (T OR F)

___ 1. Sulfonamide drugs inhibit production of the essential vitamin, folic acid, in all fungi.

___ 2. Penicillin interferes with the synthesis of nuclei in bacteria.

___ 3. Antimicrobial agents work by inhibiting a specific essential step in the metabolism of a pathogen, which is different from the metabolism of a human cell.

___ 4. Drugs that destroy viral infections generally also destroy host cells.

___ 5. Drugs that destroy cancer cells affect all rapidly dividing cells.

___ 6. Many pathogens can mutate to become resistant to chemotherapeutic drugs by changing their metabolic pathways.

___ 7. The concentration of a disinfectant is not important; it will be effective at any concentration.

___ 8. The mycobacteria that cause leprosy and tuberculosis are among the most resistant pathogens.

___ 9. A rough, porous surface is more easily disinfected than a smooth, hard surface.

___ 10. Formaldehyde and ethylene oxide are not effective against spores.

___ 11. Pasteurization kills all the bacteria present in milk.

___ 12. All bacteria must be destroyed because they all cause disease.

___ 13. An infectious disease is any disease caused by the growth of microorganisms.

___ 14. An antiseptic is a mild disinfectant used on the skin.

___ 15. Antiseptic technique was developed by Lister in the late 1800s.

MULTIPLE CHOICE

1. It would be necessary to use a tuberculocidal agent to kill a particular species of
   a. *Pseudomonas*
   b. *Clostridium*
   c. *Staphylococcus*
   d. *Mycobacterium*
   e. *Streptococcus*

2. Pasteurization is a type of
   a. surgical asepsis
   b. disinfection
   c. sterilization
   d. antiseptic technique

3. The combination of freezing and drying is known as
   a. desiccation
   b. tyndallization
   c. pasteurization
   d. sterilization
   e. lyophilization

4. Organisms that live around thermal vents at the bottom of the ocean are
   a. halophilic, alkaliphilic, and psychrophilic
   b. halophilic, thermophilic, and barophilic
   c. acidophilic, psychrophilic, and halophilic
   d. halophilic, psychrophilic, and barophilic

5. When placed into a hypertonic solution, a bacterial cell will
   a. swell
   b. shrink
   c. lyse
   d. hemolyze

6. To prevent *Clostridium* infections in a hospital setting, the disinfectant used should be
   a. virucidal
   b. fungicidal
   c. sporicidal
   d. tuberculocidal
   e. pseudomonicidal

7. Sterilization can be accomplished by use of
   a. antiseptics
   b. disinfectants
   c. an autoclave
   d. medical asepsis

8. The goal of medical asepsis is to kill _____, whereas the goal of surgical asepsis is to kill _____.
   a. all microorganisms, pathogens
   b. nonpathogens, pathogens
   c. bacteria, bacteria and viruses
   d. pathogens, all microorganisms
   e. pathogens, nonpathogens

9. Autoclaves are usually set at
   a. 160°C, 20 p.s.i., 20 minutes
   b. 100°C, 15 p.s.i., 15 minutes
   c. 121.5°C, 15 p.s.i., 20 minutes
   d. 100°C, 20 p.s.i., 20 minutes
   e. 121.5°C, 20 p.s.i., 15 minutes

10. Antiseptics are used to kill bacteria on
    a. surgical instruments
    b. floors
    c. skin
    d. thermometers
    e. door handles

CHAPTER 8

Microbial Ecology

INTERACTIONS BETWEEN HUMANS AND MICROBES
Indigenous Microflora
  Microflora of the Skin
  Microflora of the Mouth
  Microflora of the Ear and Eye
  Microflora of the Respiratory Tract
  Microflora of the Urogenital Area

Microflora of the Gastrointestinal Tract
Beneficial Roles of Indigenous Microflora
Symbiotic Relationships
  Mutualism
  Commensalism
  Neutralism and Antagonism
  Parasitism

Pathogenic Relationships
Nonpathogenic Microbes
MICROBES IN AGRICULTURE
The Role of Microbes in Elemental Cycles
BIOTECHNOLOGY
BIOREMEDIATION
BIOLOGICAL WARFARE AGENTS

OBJECTIVES

After studying this chapter, you should be able to:

- *Define ecology, human ecology, and microbial ecology*
- *Discuss the importance of indigenous microflora and where it is found*
- *List three types of symbiotic relationships*
- *Differentiate between mutualism and commensalism and give an example of each*
- *Cite an example of a parasitic relationship*
- *Discuss factors related to the pathogenicity of microbes*

- *Describe ecological interrelationships of plants, animals, and microorganisms*
- *Define biotechnology and cite four examples of how microbes are used in industry*
- *Name 10 foods that require microbial activity for their production*
- *Define bioremediation*
- *Describe the role of microbes in the nitrogen cycle*

NEW TERMS

Ammonification
Bioremediation
Biotechnology
Biotherapeutic agents
Candidiasis
Carrier

Commensalism
Denitrifying bacteria
Ecology
Ectoparasite
Endoparasite
Endosymbiont

Host
Infestation
Lysozyme
Microbial antagonism
Microbial ecology
Mutualism

Neutralism	Parasitism	Symbiosis
Nitrifying bacteria	Superinfection	Synergism
Nitrogen-fixing bacteria	Symbiont	

The science of *ecology* is the systematic study of the interrelationships that exist between organisms and their environment. Application of the ecological approach to the study of humans and human societies is referred to as human ecology. *Microbial ecology* is the study of the numerous interrelationships between microorganisms and the world around them; how microbes interact with other microbes, organisms other than microbes, and the nonliving world around them. Interactions between microorganisms and animals, plants, other microbes, soil, and our atmosphere have far-reaching effects on our lives. We are all aware of the diseases caused by pathogens (Chapter 12), but this is just one example of the many ways that microbes interact with humans. Most relationships between humans and microbes are beneficial rather than harmful.

INTERACTIONS BETWEEN HUMANS AND MICROBES

Microorganisms interact with humans in many ways and at many levels. The most intimate association that we have with microorganisms is their presence both on and within our bodies. Additionally, microbes play important roles in agriculture, various industries, sewage treatment, and water purification. Microbes are essential in the fields of bioremediation, genetic engineering, and gene therapy.

Indigenous Microflora

The *indigenous microflora* or *indigenous microbiota* (referred to in the past as "normal flora") of a person includes all the microbes that reside on or within that person. These microorganisms include bacteria, fungi, protozoa, and viruses. A fetus has no indigenous microflora. During and after delivery, a newborn is exposed to many microorganisms from its mother, food, air, and everything that touches the infant. Both harmless and helpful microbes take up residence on the skin, at all body openings, and in mucous membranes that line the digestive tract (mouth to anus) and the urogenital tract. These moist, warm environments provide excellent conditions for growth. Conditions for proper growth (moisture, pH, temperature, oxygen supply, nutrients) vary throughout the body; thus, the types of resident flora differ from one anatomical site to another. Blood, lymph, spinal fluid, and most internal tissues and organs should be free of microorganisms; *i.e.*, they should be sterile. See Table 8-1 for a list of the microorganisms frequently found on and within the human body.

Relatively few types of microbes establish themselves as indigenous microflora because most organisms in our external environment do not find the body to be a suitable host. A *host* is defined as a living organism that harbors another living organism. In addition to the resident microflora, transient microflora take up temporary residence on and within humans. The body is constantly exposed to the flow

TABLE 8-1. Locations of Microorganisms Usually Found on and in Humans

Organism	Skin	Eye	Ear	Mouth	Nose	Respiratory Tract	Intestine	Urogenital Tract
Bacteria								
Bacillus spp.	+	−	−	+	−	−	+	−
Bacteroides	+	−	−	+	+	+	+	+
Borrelia spp.	−	−	−	+	−	−	−	+
Clostridium spp.	+	−	−	−	−	−	+	+
Coliforms	+	−	−	−	−	+	+	+
Escherichia coli	−	−	−	−	−	−	+	+
Corynebacterium spp.	+	+	+	+	+	+	+	+
Fusobacterium spp.	−	−	−	+	−	−	+	−
Haemophilus influenzae	−	+	+	−	+	+	−	−
Klebsiella pneumoniae	−	−	−	−	+	−	+	−
Lactobacillus spp.	+	−	+	+	−	−	+	+
Leptotrichia	−	−	−	+	−	−	−	+
Micrococcus spp.	+	−	−	+	−	−	+	−
Mycobacterium spp.	−	−	+	−	−	−	−	+
Mycoplasmas	−	−	−	+	+	+	+	+
Neisseria spp.	−	+	−	+	+	+	−	−
Proteus spp.	−	−	−	−	−	−	+	+
Pseudomonas aeruginosa	−	−	+	−	−	−	+	−
Staphylococci	+	+	+	+	+	+	+	+
S. aureus	+	−	+	+	+	−	−	−
S. epidermidis	+	+	+	+	+	−	−	−
Streptococci	+	+	+	+	+	+	+	+
S. mitis	+	+	−	+	+	−	−	−
S. pneumoniae	−	+	+	+	+	−	−	−
S. pyogenes	+	+	+	−	−	−	−	−
Veillonella spp.	−	−	−	+	−	−	+	−
Fungi								
Actinomyces spp.	−	−	−	+	−	−	−	−
Candida albicans	+	+	−	+	−	−	+	+
Cryptococcus spp.	+	−	−	−	−	−	−	−
Protozoa	−	−	−	+	−	−	+	+
Viruses	+	−	−	+	+	+	+	−

+, present; −, absent

of microorganisms from the external environment, and these transient microbes frequently are attracted to the moist body areas. These microbes are only temporary for many reasons: they may be washed from external areas by bathing; they may not be able to compete with the resident microflora; they may fail to survive in the acid or alkaline environment of the site; or they may be flushed out in excretions or secretions, such as urine, feces, tears, and perspiration.

Destruction of the resident microflora disturbs the delicate balance established between the host and its microorganisms. For example, prolonged therapy with certain antibiotics often destroys many of the intestinal microflora. Diarrhea is usually the result of such an imbalance, which, in turn, leaves the body more susceptible to secondary invaders. When the number of usual resident microbes is greatly reduced, opportunistic invaders can more easily establish themselves within those areas. One important opportunist usually found in small numbers near body openings is the yeast, *Candida albicans,* which, in the absence of sufficient numbers of other resident microflora, may grow unchecked in the mouth, vagina, or lower intestine, causing the disease *candidiasis* (also known as moniliasis). Such an overgrowth of an organism usually present in low numbers is referred to as a *superinfection.*

MICROFLORA OF THE SKIN

The resident microflora of the skin consists primarily of bacteria and fungi. The most common bacteria on skin are species of *Staphylococcus* (especially *S. epidermidis* and other coagulase-negative staphylococci), *Micrococcus, Corynebacterium, Propionibacterium, Brevibacterium,* and *Acinetobacter.* Yeasts in the genus *Pityrosporum* are frequently present, and *Candida albicans* is present on the skin of some people. The number and species of microorganisms present on the skin depends on many factors. Moist, warm conditions in hairy areas where there are many sweat and oil glands, such as under the arms and in the groin area, stimulate the growth of many different microorganisms. Dry, calloused areas of skin have few bacteria, whereas moist folds between the toes and fingers support many bacteria and fungi. The surface of the skin near mucosal openings of the body (the mouth, eyes, nose, anus, and genitalia) is inhabited by bacteria present in various excretions and secretions.

Frequent washing with soap and water removes most of the potentially harmful transient microorganisms harbored in sweat, oil, and other secretions from moist body parts. All persons involved in patient care must be particularly careful to keep their skin and clothing as free of transient microbes as possible, to help prevent personal infections, and to avoid transferring pathogens to patients. Such persons should always remember that most infections following burns, wounds, and surgery result from the growth of resident or transient skin microflora in these susceptible areas. A cross-section of skin is shown in Figure 12-1 in Chapter 12.

MICROFLORA OF THE MOUTH

The mouth and throat have an abundant and varied population of microorganisms. These areas provide moist, warm mucous membranes that furnish excellent conditions for microbial growth. Bacteria thrive especially well in particles of food and in the debris of dead epithelial cells around the teeth. The peculiar anatomy of the oral cavity and the throat affords shelter for numerous anaerobic and aerobic bacteria. Anaerobic microorganisms flourish in gum margins, cervices between the teeth, and deep folds (crypts) on the surface of the tonsils.

The list of microbes that have been isolated from healthy human mouths reads like a manual of the main groups of microorganisms. It includes Gram-positive and Gram-negative cocci, Gram-positive and Gram-negative bacilli, spirochetes, and sometimes yeasts, mold-like organisms, protozoa, and viruses. The first such list was made by Leeuwenhoek in 1690 (see Chapter 1).

Most microorganisms found in the healthy mouth and throat are beneficial (or at least harmless); these include diphtheroids, lactobacilli, and micrococci. Others, such as certain streptococci and staphylococci, are potentially pathogenic opportunists and are frequently associated with disease. Some people carry virulent pathogens in their nasal passages or throats, but do not have the diseases associated with them, such as diphtheria, meningitis, pneumonia, and tuberculosis. These people are healthy *carriers* who are resistant to these pathogens, but who can transmit them to susceptible persons.

Food remaining on and between teeth provides a rich nutrient medium for growth of the many oral bacteria. Carelessness in dental hygiene allows growth of these bacteria, with development of dental caries (tooth decay), gingivitis (gum disease), and more severe periodontal diseases. These bacteria include species of *Actinomyces, Bacteroides, Fusobacterium, Lactobacillus, Streptococcus, Neisseria*, and *Veillonella*. Many areas within the mouth are anaerobic, thus supporting growth of a variety of obligate, aerotolerant, and facultative anaerobes.

The most common organisms in the indigenous microflora of the mouth and throat are various species of alpha-hemolytic (α-hemolytic) streptococci. When Group A, beta-hemolytic (β-hemolytic) streptococci (*Streptococcus pyogenes*) are present, however, the person should be treated with an antibiotic to destroy these pathogens which may cause strep throat and its serious complications (*e.g.,* scarlet fever, rheumatic fever, and glomerulonephritis). The anatomy of the mouth is shown in Figure 12-7 in Chapter 12.

MICROFLORA OF THE EAR AND EYE

The middle ear and inner ear are usually sterile, whereas the outer ear and the auditory canal contain the same types of microorganisms as are found on moist areas, such as the mouth and the nose. When a person coughs, sneezes, or blows his or her nose, these microbes are carried along the eustachian tube and into the middle ear where they can cause infection. Infection can also develop in the middle ear when the eustachian tube does not open and close properly to maintain correct air pressure within the ear. The anatomy of the ear is shown in Figure 12-9 in Chapter 12.

Many microorganisms are found in the external opening of the eye, the conjunctiva that lines the eyelid, and tears. But these microbes are not a frequent cause of disease because the intact membranes serve as a barrier. These mucous membranes are constantly flushed by tears, which contain an enzyme called *lysozyme* that destroys bacteria. The indigenous microflora of the eye area includes species of *Staphylococcus, Streptococcus,* and *Corynebacterium,* as well as *Moraxella catarrhalis.* The anatomy of the eye is shown in Figure 12-5 in Chapter 12.

MICROFLORA OF THE RESPIRATORY TRACT

The respiratory tract consists of the nose, pharynx (throat), larynx (voice box), trachea, bronchi, bronchioles, and alveoli. The lower respiratory tract, below the larynx, is usually free of microbes because the mucous membranes and lungs have defense mechanisms that efficiently remove invaders (described in Chapter 11). Thus, staphylococci, streptococci, *Pseudomonas* species, or yeasts found in sputum specimens indicate either an infectious disease of the lungs or specimen contamination by indigenous microflora of the upper respiratory tract.

The membranes of the upper part of the tract, including the nasopharynx and the oropharynx, provide a suitable environment for the growth of many species of *Streptococcus, Staphylococcus, Neisseria, Corynebacterium,* yeasts, and other microorganisms. In susceptible persons, many of these opportunists can cause disease. The respiratory system is shown in Figure 12-10 in Chapter 12.

MICROFLORA OF THE UROGENITAL AREA

The healthy kidney, ureters, and urinary bladder are sterile. However, the distal urethra (that part of the urethra furthest from the urinary bladder) and the external opening of the urethra house many microflora, such as nonpathogenic *Neisseria* species, staphylococci, streptococci, enterococci, diphtheroids, mycobacteria, mycoplasmas, enteric (intestinal) Gram-negative rods, some anaerobic bacteria, yeasts, and viruses. As a rule, these organisms do not invade the bladder, as the urethra is periodically flushed by acidic urine; however, persistent, recurring urinary infections can develop following obstruction or narrowing of the urethra and with infrequent urination, which allows the invasive organisms to multiply and cause urinary tract infections. Chlamydias and mycoplasmas are frequent causes of nongonococcal urethritis (NGU; urethritis not caused by *Neisseria gonorrhoeae*). All of these organisms are easily introduced into the urethra by sexual intercourse.

The reproductive systems of both men and women are usually sterile, with the exception of the vagina; here the microflora varies with the stage of sexual development. During puberty and following menopause, vaginal secretions are alkaline, supporting the growth of various diphtheroids, streptococci, staphylococci, and coliforms (*E. coli* and closely related enteric Gram-negative rods). Through the childbearing years, vaginal secretions are acidic, encouraging the growth mainly of lactobacilli, along with a few α-hemolytic streptococci, staphylococci, diphtheroids, and yeasts.

Neisseria species, particularly *Neisseria gonorrhoeae* (also called gonococci), can survive the acidic environment of the vagina and the penis; hence, they may be harbored by infected persons who show no symptoms of gonorrhea. This disease, which is readily transmitted through sexual contact, is asymptomatic (causing no symptoms) in 80% of infected women and 20% of infected men. The urinary tract and male and female reproductive systems are shown in Figures 12-15 and 12-16 in Chapter 12.

MICROFLORA OF THE GASTROINTESTINAL TRACT

The acidic environment of the stomach prevents growth of indigenous microflora. However, a few microbes, protected by foods, manage to pass through the stomach during periods of low acid concentration. Also, when the amount of acid is reduced in the course of diseases such as stomach cancer, certain bacteria may be found in this site.

Usually, in the upper portion of the small intestine (the duodenum) few microflora exist because bile inhibits their growth, but many are found in the lower part of the small intestine. The most abundant organisms include many species of *Staphylococcus, Lactobacillus, Streptobacillus, Veillonella,* and *Clostridium perfringens.*

The colon, or large intestine, contains the largest number of microorganisms of any colonized area of the body. It has been estimated that more than 400 species—primarily bacteria—live there. Because the colon is anaerobic, the bacteria living there are obligate, aerotolerant, and facultative anaerobes. The obligate anaerobes include Gram-positive *Clostridium* and *Peptostreptococcus* species and Gram-negative *Bacteroides* and *Fusobacterium* species. The facultative anaerobes include members of the family *Enterobacteriaceae* (*e.g., Escherichia coli* and species of *Klebsiella, Enterobacter,* and *Proteus*), and species of *Enterococcus, Pseudomonas, Streptococcus, Lactobacillus,* and *Mycoplasma.* Also, many fungi, protozoa, and viruses can live in the colon. Many of the microflora of the colon are opportunists, causing disease only when they lodge in the other areas of the body or when the balance among the microorganisms is upset. The anatomy of the digestive system is shown in Figure 12-14 in Chapter 12.

Beneficial Roles of Indigenous Microflora

Many benefits are derived by humans from the symbiotic relationship established with their indigenous microflora. Some nutrients, particularly vitamins K, B_{12}, pantothenic acid, pyridoxine, and biotin are obtained from secretions of the coliform bacteria.

Evidence also indicates that these indigenous microbes provide a constant source of irritants and antigens to stimulate the immune system. This activity causes the immune system to respond more readily by producing antibodies to foreign invaders and substances, which, in turn, enhances the body's protection against pathogens. It appears that by merely occupying a place and using the nutrients present, these resident microflora prevent other microorganisms that may be pathogenic from "gaining a foothold" and establishing a site of infection. This protection is maintained by competition for food, controlled pH and oxygen levels, and antibiotic production by certain resident microbes.

When the delicate balance of the various species in the population of indigenous microflora is upset by antibiotic or other chemotherapy, many complications may result. Certain microorganisms may flourish out of control, such as *Candida albicans,* the cause of yeast vaginitis. Also, diarrhea and pseudomembranous colitis may occur. Cultures of *Lactobacillus* in yogurt or present in medications may be prescribed to reestablish and stabilize the microbial balance. Bacteria and yeasts used in this manner are called *biotherapeutic agents*. Other microorganisms that have been used as biotherapeutic agents include *Bifidobacterium,* nonpathogenic *Enterococcus* and *Saccharomyces* (a yeast).

Symbiotic Relationships

The relationships between the indigenous microflora and the human host are excellent examples of *symbiosis,* a term meaning the living together or close association of two dissimilar organisms (usually two different species). The organisms that live together in such a relationship are called *symbionts*. The relationship may be beneficial, harmless, or harmful to one or both of the symbionts. The types of relationships outlined in the following sections demonstrate that the pathogenicity of a microbe can be represented as a balance between the virulence (disease-causing ability) of the pathogen and the resistance of the host. The dynamic symbiotic balance may shift toward the parasitic/pathogenic disease state if host defenses are reduced with an accompanying rise in host susceptibility. Recovery from disease occurs with a shift toward mutualism and commensalism (described below). These factors are further discussed in later chapters. Various symbiotic relationships are illustrated in Figure 8-1.

MUTUALISM
In the symbiotic relationship called *mutualism,* both symbionts benefit; the relationship is mutually beneficial. An example of this is the intestinal bacterium *Escherichia coli,* which obtains nutrients from food materials ingested by the host and produces vitamin K which is used by the host. Vitamin K is a blood-clotting factor that is essential to humans. Also, some members of our indigenous microflora maintain conditions that prevent colonization by pathogens and overgrowth by opportunistic pathogens.

As another example of a mutualistic relationship, consider the protozoa that live in the intestine of termites. These protozoa enable the termites to digest the wood they eat by breaking cellulose down into nutrients to be absorbed and used by the termites. In turn, the termite provides food and a warm, moist place for the protozoa to live. Without these protozoa, the termites would die of starvation. Lichens that you see as colored patches on rocks and tree trunks are further examples of mutualism. A lichen is composed of an alga (or a cyanobacterium) and a fungus, living so closely together that they appear to be one organism. The fungus uses some of the energy that the alga produces by photosynthesis, and the chitin in the fungal cell walls protects the alga from desiccation. Thus, both symbionts benefit from the relationship.

In some mutualistic relationships, two organisms work together to produce a result that neither could accomplish alone. This is referred to as *synergism,* or a synergistic relationship. Oral fusobacteria and spirochetes, which together cause the disease "trench mouth," represent such a relationship. Also, nitrogen-fixing bacte-

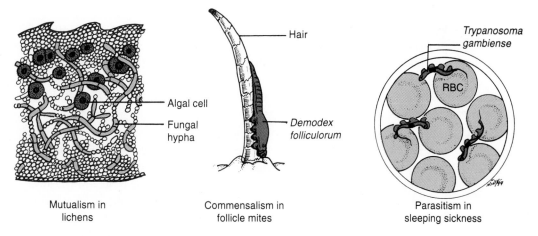

Mutualism in
lichens

Commensalism in
follicle mites

Parasitism in
sleeping sickness

FIGURE 8-1. Various symbiotic relationships.

ria and the roots of legumes where they exist have a true synergistic relationship, because each depends on the other for nutrients.

COMMENSALISM

A symbiotic relationship that is beneficial to one symbiont and neither beneficial nor harmful to the other is called *commensalism.* Most of the indigenous microflora of humans are considered to be commensals, in that the microbes are provided nutrients and "housing" with no effect on the host. However, as mentioned above, we do have a mutualistic relationship with certain members of our indigenous flora, and keep in mind that some species are opportunistic pathogens.

NEUTRALISM AND MICROBIAL ANTAGONISM

Indifference, or *neutralism,* exists when organisms occupy the same niche but do not affect each other, as with many bacteria that live in the human mouth and intestines. However, sometimes the waste products of one microorganism can destroy certain neighboring microbes. This situation is called *microbial antagonism.* For example, *Penicillium* mold growing on a culture plate of certain strains of staphylococci can inhibit growth of the staphylococci by producing the antibiotic, penicillin (see Fig. 7-5).

PARASITISM

The relationship in which an organism benefits at the expense of the host organism is called *parasitism.* Depending on the parasite and the circumstances, the damage may range from none to fatal. The "wise" parasite does not kill its host, but rather takes only the nutrients it needs to exist. Intestinal worms (such as pinworms and tapeworms) and external parasites (such as mites, lice, and ticks) usually cause only minor damage in humans. The presence of *ectoparasites* (parasites that live on the outside of the body) is referred to as an *infestation,* whereas the presence of *endoparasites* (parasites that live inside the body) constitutes an actual infection.

PATHOGENIC RELATIONSHIPS

When microorganisms cause damage (pathology) to the host during the infection process, a pathogenic relationship exists. The pathogen may be only a displaced commensal; for example, the staphylococci that usually inhabit the skin can cause an infection when the skin is wounded or burned. It may be a highly virulent airborne pathogen, such as the common cold virus, or it may be carried in food and water such as the dysentery pathogens. Many opportunistic pathogens (opportunists) cause disease only in hosts who are physically impaired or debilitated, because the hosts' usual defenses against disease are weakened. Pneumonia developing in a bedridden patient is another example of such a pathogenic relationship. In the usual course of events, the opportunist is harmless; it moves in to cause damage when an abnormal situation develops, such as a wound, a burn, or the destruction of the indigenous microflora by antibiotic therapy. Opportunists can also cause disease in otherwise healthy persons, if they gain access to the blood, urinary bladder, lungs, or other organs and tissues. Pathogenic relationships are discussed more fully in Chapter 9.

NONPATHOGENIC MICROBES

Microbes that never cause disease are referred to as nonpathogens. We know now that many microorganisms, thought at one time to be nonpathogens, can cause serious illness in immunosuppressed patients, such as AIDS patients. Thus, many microorganisms originally classified as nonpathogens have been reclassified as opportunistic pathogens.

MICROBES IN AGRICULTURE

There are many uses for microorganisms in agriculture. They are used extensively in the field of genetic engineering, which can be used to create new or genetically altered plants. Such genetically engineered plants might grow larger, be better tasting, or be more resistant to insects, plant diseases, or extremes in temperature. Some microorganisms are used as pesticides. Many microorganisms are decomposers, which return minerals and other nutrients to soil. In addition, microorganisms play major roles in elemental cycles, such as the carbon, oxygen, nitrogen, phosphorus, and sulfur cycles.

The Role of Microbes in Elemental Cycles

Bacteria are exceptionally adaptable and versatile. They are found on the land, in all waters, in every animal and plant, and even inside of other microorganisms (in which case they are referred to as *endosymbionts*). Some bacteria and fungi serve a valuable function by recycling back into the soil the nutrients from dead, decaying animals and plants, as discussed in Chapter 1 (see Figures 1-3 and 1-4). Free-living fungi and bacteria that decompose dead organic matter into inorganic materials are called saprophytes. The inorganic nutrients that are returned to the soil are used by chemotrophic bacteria and plants for synthesis of biological molecules necessary for their growth. The plants are eaten by animals, which eventually die and are recycled again with the aid of saprophytes. The cycling of elements by microorganisms is sometimes referred to as biogeochemical cycling.

Good examples of the cycling of nutrients in nature are the nitrogen, carbon, oxygen, sulfur, and phosphorus cycles, in which microorganisms play very important roles. In the nitrogen cycle (Fig. 8-2), free atmospheric nitrogen gas (N_2) is converted by *nitrogen-fixing bacteria* and cyanobacteria into ammonia (NH_3) and the ammonium ion (NH_4^+). Then chemolithotrophic soil bacteria, called *nitrifying bacteria,* convert ammonium ions into nitrite ions (NO_2^-) and nitrate ions (NO_3^-). Plants then use the nitrates to build plant proteins; these proteins are eaten by animals, which then use them to build animal proteins. Excreted nitrogen-containing animal waste products (such as urea in urine) are converted by certain bacteria to ammonia by a process known as *ammonification.* Also, dead plant and animal nitrogen-containing debris and fecal material are transformed by saprophytic fungi and bacteria into ammonia, which, in turn, is converted into nitrites and nitrates for recycling by plants. To replenish the free nitrogen in the air, a group of bacteria called *denitrifying bacteria* convert nitrates to atmospheric nitrogen gas (N_2). Thus, the cycle continues.

Nitrogen-fixing bacteria are of two types: free-living and symbiotic. Symbiotic bacteria in the genera *Rhizobium* and *Bradyrhizobium* live in and near the root nodules of plants called legumes, such as alfalfa, clover, peas, soy beans, and peanuts (Fig.

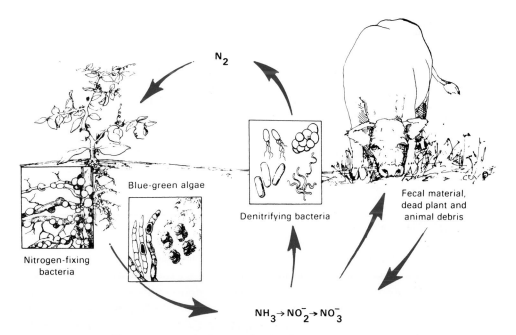

FIGURE 8-2. The nitrogen cycle. See text for details.

FIGURE 8-3. Nodules on the roots of a legume. These root nodules contain nitrogen-fixing bacteria, such as *Rhizobium* species.

8-3). These plants are often used in crop-rotation techniques by farmers to return nitrogen compounds to the soil and thus avoid the loss of nutrients. Nitrifying soil bacteria include *Nitrosomonas, Nitrosospira, Nitrosococcus, Nitrosolobus* and *Nitrobacter* species. Denitrifying bacteria include certain species of *Pseudomonas* and *Bacillus*.

The types and amounts of microorganisms living in soil depend on many factors: amount of decaying organic material, available nutrients, moisture content, amount of oxygen available, acidity, temperature, and the presence of waste products of other microbes. Likewise, the types and number of harmless microbes that live on and within the human body depend on pH, moisture, nutrients, antibacterial factors, and the presence of other microorganisms at the site they are colonizing.

BIOTECHNOLOGY

The United States Congress defines *biotechnology* as "any technique that uses living organisms or substances from those organisms, to make or modify a product, to improve plants or animals, or to develop microorganisms for specific uses." Microbes are used in a variety of industries, including the production of certain foods and beverages, food additives, amino acids, enzymes, chemicals, vitamins (such as vitamins B_{12} and C), vaccines, and antibiotics, and the mining of ores such as copper and uranium.

Microorganisms are used in the production of foods such as acidophilus milk, bread, butter, buttermilk, cocoa, coffee, cottage cheese, cream cheese, fish sauces, green olives, kimchi (from cabbage), pickles, poi (fermented taro root), sauerkraut, sour cream, soy sauce, various cheeses (*e.g.,* cheddar, Swiss, Limburger, Camembert, Roquefort and other blue cheeses), vinegar, and yogurt. Microbes are also used in the production of beer, ale, wine, sake (rice wine), brandy, rum, vodka, and whiskey. Two amino acids produced by microbes are used in the artificial sweetener called aspartame (NutraSweet®). Microbes are also used in the commercial production of amino acids (*e.g.,* alanine, aspartate, cysteine, glutamate, glycine, histidine, lysine, methionine, tryptophan) that are used in the food industry.

Microbial enzymes used in industry include cellulase, collagenase, lipase, pectinase, and protease. Microbes are used in the large-scale production of chemicals such as acetic acid, acetone, butanol, citric acid, ethanol, formic acid, glycerol, isopropanol, lactic acid, and methane. They are also used in the mining of arsenic, cadmium, cobalt, copper, nickel, uranium, and other metals in a process known as leaching or bioleaching.

Many antibiotics are produced in pharmaceutical company laboratories by fungi and bacteria. Examples include amphotericin B, bacitracin, cephalosporins, chloramphenicol, cycloheximide, cycloserine, erythromycin, griseofulvin, kanamycin, lincomycin, neomycin, novobiocin, nystatin, penicillin, polymyxin B, streptomycin, and tetracycline. Genetically engineered bacteria and yeasts are used in the production of human insulin, human growth hormone, interferon, hepatitis B vaccine, and other important substances (see Chapter 6).

BIOREMEDIATION

The term *bioremediation* refers to the use of microorganisms to clean up industrial wastes (including toxic wastes) and other pollutants, such as herbicides and pesticides. Some of the microbes used in this manner have been genetically engineered to digest specific wastes. For example, genetically engineered, petroleum-digesting bacteria were used to clean up the 11-million gallon oil spill in Prince William Sound, Alaska, in 1989. In addition, microbes are used extensively in sewage treatment and water purification (see Chapter 10).

BIOLOGICAL WARFARE AGENTS

Sad to say, pathogenic microorganisms sometimes find themselves in the hands of mentally deranged people who want to use them to cause harm to others. In times of war, the use of microorganisms in this manner is called biological warfare, and the microbes are referred to as biological warfare agents or bw agents. But the danger doesn't just exist during times of war. There is always a possibility that members of terrorist or radical hate groups might use pathogens to create fear, chaos, illness, and death.

Four of the pathogens most often discussed as potential biological weapons are *Bacillus anthracis, Clostridium botulinum,* smallpox virus, and *Yersinia pestis,* the etiologic agents of anthrax, botulism, smallpox, and plague, respectively. If disseminated in some type of aerosol, either *B. anthracis* spores or *Y. pestis* bacilli could result in numerous, severe, and potentially fatal pulmonary infections. In addition, entry of *B. anthracis* into wounds could cause cutaneous anthrax, and ingestion of the organisms could result in intestinal anthrax. Anthrax infections involve significant hemorrhage and serous effusions in various organs and body cavities and are frequently fatal.

Clostridium botulinum spores could be added to water supplies or food. Botulinal toxin is odorless and tasteless, and only a tiny quantity of the toxin need be ingested to cause potentially fatal cases of botulism. Since 1980, when the World Health Organization (WHO) announced that smallpox had been eradicated, civilians no longer receive smallpox vaccinations. Thus, throughout the world, huge numbers of people are highly susceptible to the virus. Although there are no reservoirs for smallpox virus in nature, preserved samples of the virus exist in a few medical research laboratories throughout the world. There is always the danger that smallpox virus, or any of the other pathogens mentioned here, could fall into the wrong hands. Other pathogens viewed as potential bw agents are the etiologic agents of brucellosis, Q fever, tularemia, viral encephalitis, and viral hemorrhagic fevers.

An instance of biological terrorism occurred in a small Oregon town in 1984. Members of a religious cult purposely contaminated salad bars in 10 area restaurants with *Salmonella typhimurium* in an attempt to influence the outcome of a local election. They also contaminated the drinking water of two county commissioners. Over 750 people became ill, including the two commissioners, but no deaths occurred.

To minimize the danger of potentially deadly microorganisms falling into the wrong hands, the U.S. Antiterrorism and Effective Death Penalty Act of 1996 makes the Centers for Disease Control and Prevention (CDC) responsible for controlling shipment of those pathogens and toxins deemed most likely to be used as bw agents. Authorities must constantly be on the alert for possible theft of these pathogens from biological supply houses and legitimate laboratories. In addition, vaccines, antitoxins, and other antidotes must be available wherever the threat of the use of these biological agents is high (*e.g.,* in various potential war zones).

■ REVIEW OF KEY POINTS

- Microbial ecology is the study of the interrelationships among microorganisms and the living and nonliving world around them.
- Most relationships between humans and microbes are beneficial rather than harmful.

- Microbes play important roles in agriculture, industrial processes, sewage treatment and water purification, as well as the fields of genetic engineering and bioremediation.

- Relatively few types of microbes become human indigenous microflora because the human body is not a suitable host for most environmental microorganisms.

- Destruction of the resident microflora disturbs the delicate balance established between the host and its microorganisms.

- Frequent washing with soap and water removes most of the potentially harmful transient microbes found in sweat, oil, and other human body secretions.

- Many benefits are derived by humans from the symbiotic relationships established with their indigenous microflora.

- A mutualistic relationship is of benefit to both parties (symbionts), whereas a commensalistic relationship is of benefit to one symbiont but of no consequence to the other (*i.e.,* neither beneficial nor harmful). A parasitic relationship is of benefit to the parasite and detrimental to the host. Although many parasites cause disease, others do not.

- Synergism is a mutualistic relationship in which two organisms work together to produce a result that neither could accomplish alone.

- A pathogenic relationship exists when microorganisms cause damage to the host during the infection process.

- A usually harmless opportunist may cause complications when an abnormal situation occurs, such as entry of the organism into a wound, the bloodstream, or an organ (*e.g.,* the urinary bladder), or following destruction of much of the indigenous microflora by antibiotic therapy.

- Inorganic nutrients, returned to the soil by saprophytes, are used by chemotrophic bacteria and plants for synthesis of biological molecules necessary for growth. The plants are eaten by animals, which eventually die and are recycled again with the aid of saprophytes.

- Biotechnology includes the industrial use of microbes in the production of certain foods and beverages, food additives, chemicals, amino acids, enzymes, vitamins B_{12} and C, and antibiotics, as well as in the refining of ores to obtain copper, uranium and gold.

- Bioremediation includes the use of microbes to destroy industrial and toxic wastes and other environmental pollutants, such as pesticides, herbicides, and petroleum spills. Many of the microbes used in bioremediation are found in nature, but others are genetically engineered to digest specific wastes.

Problems and Questions

1. What types of relationships exist between humans and their indigenous microflora?
2. Where would you find symbiotic relationships in your environment?
3. What are the differences between mutualism, commensalism, neutralism, antagonism, and parasitism?
4. Why are the microbial decomposers so necessary for life on earth?
5. What factors control the number of microorganisms in the soil and on the human body?
6. In what industries do microbes play a role?

Self Test

After you have read Chapter 8, reviewed the chapter outline, examined the objectives, studied the new terms, and answered the problems and questions above, complete the following self test.

MATCHING EXERCISES

Complete each statement from the list of words provided with each section.

Symbiotic Relationships

symbiotic	pathogen	neutralism
mutualism	opportunist	parasite
synergism	symbionts	infestation
antagonism	commensalism	infection

1. A parasitic microorganism that causes damage to its host is called a _____.
2. An organism that lives on or within a host organism is called a _____.
3. If a parasite is an endoparasite, like pinworms, the host has an _____.
4. If the parasite is an ectoparasite, the host has an _____.
5. When two dissimilar organisms live together they live in a _____ relationship.
6. The two organisms that live together are called _____.
7. If both organisms benefit from the relationship, they live in a state of _____.
8. If the two organisms work together to produce an effect, they live in a state of _____.
9. If one organism secretes a material that damages or repels another organism, the two organisms live in a state of _____.
10. The secretion of penicillin by a *Penicillium* mold in an area where bacteria are established would be an example of _____.
11. When two organisms live together without harming or benefiting each other, the relationship is termed _____.
12. If the relationship is beneficial to one party, but of no consequence to the other, the relationship is one of _____.

TRUE OR FALSE (T OR F)

___ 1. All indigenous microflora are nonpathogens.

___ 2. Newborn infants acquire their first resident microflora organisms as they pass through the birth canal.

___ 3. Saprophytes aid in the cycling of nutrients, which provides plants with proper carbon, oxygen, sulfur, phosphorus, and nitrogen sources.

___ 4. Nitrogen-fixing protozoa change free nitrogen from the air into ammonia.

___ 5. Legumes, like alfalfa and clover, help fertilize the soil because of the bacteria that live on and in their roots.

___ 6. The yeast *Candida albicans* is an indigenous microflora organism that is an opportunist.

___ 7. The cool, dry areas of the skin support the growth of most indigenous microflora of the skin.

___ 8. Wound and burn infections are frequently caused by resident microflora.

___ 9. Careless dental hygiene encourages dental caries and gingivitis.

___ 10. Beta-hemolytic streptococci are responsible for strep throat, scarlet fever, and rheumatic fever.

___ 11. The lysozyme found in saliva and tears helps to destroy bacteria.

MULTIPLE CHOICE

1. Symbionts might also be
   a. commensals
   b. opportunists
   c. indigenous microflora
   d. endosymbionts
   e. all of the above

2. Indigenous microflora are found in or on the
   a. skin
   b. mouth
   c. gastrointestinal tract
   d. nasal passages
   e. all of the above

3. *Escherichia coli* living in the colon would be considered a/an
   a. opportunist
   b. endosymbiont
   c. member of the indigenous microflora
   d. symbiont in a mutualistic relationship
   e. all of the above

4. Which of the following sites does not have indigenous microflora?
   a. the distal urethra
   b. the vagina
   c. the colon
   d. the bloodstream
   e. the skin

5. Harmless members of the indigenous microflora are called
   a. opportunists
   b. parasites
   c. commensals
   d. pathogens
   e. opportunistic pathogens

6. Which of the following would be present in high numbers in the indigenous microflora of the mouth?
   a. beta-hemolytic streptococci
   b. alpha-hemolytic streptococci
   c. *Staphylococcus aureus*
   d. *Staphylococcus epidermidis*
   e. *Candida albicans*

7. Which of the following would be present in high numbers in the indigenous microflora of the skin?
   a. *Pseudomonas aeruginosa*
   b. *Escherichia coli*
   c. *Enterococcus* spp.
   d. coagulase-negative staphylococci
   e. *Candida albicans*

8. A symbiotic relationship that is of benefit to both parties is
   a. commensalism
   b. neutralism
   c. mutualism
   d. parasitism

9. The relationship that humans have with the vitamin-producing bacteria in their intestinal tract is an example of
   a. commensalism
   b. neutralism
   c. mutualism
   d. parasitism

10. The microorganisms most likely to be causing tooth decay are
    a. beta-hemolytic streptococci
    b. alpha-hemolytic streptococci
    c. *Staphylococcus aureus*
    d. *Staphylococcus epidermidis*
    e. *Candida albicans*

CHAPTER 9

Microbial Pathogenicity and Epidemiology

MICROBIAL PATHOGENICITY
Why Infection Does Not Always
    Occur
The Development of an
    Infectious Disease
The Disease Process
Virulence and Virulence
    Factors
Virulence

*Bacterial Structures Associated
    with Virulence*
*Enzymes Associated with
    Virulence*
Toxins Associated with Virulence
Pathogenicity and Virulence
**EPIDEMIOLOGY AND DISEASE
TRANSMISSION**
Endemic Diseases

Epidemic Diseases
Pandemic Diseases
Sporadic and Nonendemic
    Diseases
RESERVOIRS OF INFECTION
**MODES OF DISEASE
TRANSMISSION**
CONTROL OF EPIDEMIC DISEASES

OBJECTIVES

After studying this chapter, you should be able to:

- *Differentiate between infectious, communicable, and contagious diseases*
- *List the six components of the chain of infection*
- *List six reasons why an infection may not occur even though a pathogen is present*
- *Discuss the disease process*
- *Define acute and chronic diseases*
- *State the difference between primary and secondary diseases*
- *State the difference between local and generalized infections*

- *List three factors associated with the virulence of a pathogen*
- *List and discuss eight factors that affect the pathogenicity of bacteria*
- *Define epidemiology and the following types of diseases: epidemic, endemic, pandemic, sporadic, and nonendemic diseases*
- *List three factors that contribute to an epidemic*
- *List six reservoirs of infection*
- *List five modes of disease transmission*
- *Discuss the procedure for stopping an epidemic*

NEW TERMS

Active carrier	Bacteriocins	Collagen
Acute disease	Carrier	Collagenase
Asymptomatic disease	Chronic disease	Communicable disease
Avirulent	Coagulase	Contagious disease

Convalescent carrier
Edema
Endemic disease
Endotoxin
Enterotoxin
Epidemic disease
Epidemiology
Erythrocytes
Erythrogenic toxin
Exfoliative toxin
Exotoxin
Exudate
Fomites
Generalized infection
Hemolysin
Hyaluronic acid
Hyaluronidase
Incubatory carrier
Kinase

Latent infection
Lecithin
Lecithinase
Leukocidin
Leukocytes
Local infection
Microbial antagonism
Neurotoxin
Nonendemic disease
Pandemic disease
Passive carrier
Parenteral injection
Pathogenicity
Plasma
Primary infection
Pyrogens
Reservoirs of infection
Secondary infection
Septicemia

Septic shock
Serum (pl. *sera*)
Shock
Signs of a disease
Sporadic disease
Staphylokinase
STD
Streptokinase
Symptomatic disease
Symptoms of a disease
Toxigenicity
Toxin
Vasodilation
Vectors
Virulence
Virulence factors
Virulent
Zoonosis (pl. *zoonoses*)

MICROBIAL PATHOGENICITY

There are many diseases that are not caused by pathogens. Included among them are those caused by malfunction of an organ, such as diabetes and hyperthyroidism; those caused by a vitamin deficiency, such as scurvy and rickets; those caused by an allergic response, such as asthma and hay fever; and those caused by uncontrolled cell growth, such as tumors and cancer. However, this chapter focuses on the many infectious diseases caused by the growth of pathogens on and in living tissues.

Virulence is a measure or degree of *pathogenicity* (the ability to cause disease). Microbes that cause infectious diseases—pathogens—vary in their ability to cause disease. They are said to vary in virulence, with some pathogens being more virulent than others. For example, it may take as few as 10 *Shigella* cells to cause shigellosis (a diarrheal disease), but it takes 100 to 1000 *Salmonella* cells to cause salmonellosis (also a diarrheal disease). *Shigella* is said to be more virulent than *Salmonella* because it takes fewer *Shigella* cells to cause disease. Within a particular species, some strains may be *virulent* (capable of causing disease) and other strains may be *avirulent* (incapable of causing disease), as determined by the genetic characteristics of the organisms. For example, encapsulated strains of a particular species may be virulent, but nonencapsulated strains of that species are avirulent. Likewise, piliated strains of some species are virulent, but nonpiliated strains of the same species are avirulent. As you can see, the terms virulent and avirulent are frequently used as synonyms for pathogenic and nonpathogenic, respectively.

Whether a disease results when a person is exposed to a pathogen depends on many factors, including the degree of susceptibility or resistance of the host and the degree of virulence of the pathogen. The 6 components in the infectious disease process (also known as the chain of infection) are shown in Figure 9-1; they are discussed in detail later in this chapter.

The potential for infections always exists when organisms live in a parasitic relationship. When a pathogen finds the appropriate places and conditions in which to grow and cause damage, it produces an infectious disease.

An infection occurs when a pathogen multiplies in the organ or tissue where it lodges. An infectious disease starts with the growth of a pathogen. As stated earlier, all infectious diseases are caused by pathogens. A *communicable disease* is an infectious disease that can be transmitted from one person to another, as with measles, gonorrhea, and diphtheria. A *contagious disease* is a communicable disease that is

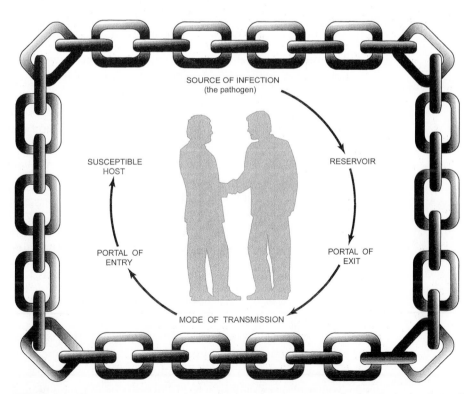

FIGURE 9-1. The six components of the infectious disease process; also known as the chain of infection.

STUDY AID   The term "infection" can be especially confusing because it is used in different ways by different people. Physicians and many other health care personnel use "infection" as a synonym for infectious disease. Thus, when a physician says "My patient has an ear infection," it is the same as saying "My patient has an infectious disease of the ear." Microbiologists, on the other hand, tend to use "infection" to mean colonization by a pathogen. In other words, when a pathogen enters a person's body and remains there and multiplies, the person is "infected" with that pathogen. The pathogen may or may not be causing illness. If it is, then the person has an infectious disease. As the word "infection" is used by microbiologists, it is possible to have an infection without having an infectious disease.

easily transmitted from person to person, for example via droplets in the air, such as occurs with the common cold and influenza.

The severity of an infectious disease and the amount of damage it causes are determined by the host's ability to resist invasion by the pathogen and to neutralize the damaging enzymes and *toxins* (poisons) produced by the pathogen.

Why Infection Does Not Always Occur

Many people who are exposed to pathogens do not get sick for a variety of reasons:

- The microbe may land in the wrong place and may, thus, be unable to multiply. For example, when a respiratory pathogen falls on the skin, it may be unable to grow there because the skin lacks the necessary warmth, moisture, and nutrients required for growth of that particular microorganism.
- Many pathogens must be able to attach to specific receptor sites (described later) before they can multiply and cause damage.
- Antibacterial factors that destroy or inhibit the growth of microbes (*e.g.,* the lysozyme that is present in tears, saliva, and perspiration) may be present at the site where a pathogen lands.
- The indigenous microflora of that site (*e.g.,* the mouth, vagina, or intestine) may inhibit growth of the foreign microorganism by occupying space and using up the available nutrients. This is a type of *microbial antagonism,* where one microbe or group of microbes wards off another.
- The microbes already present at the site may produce antibacterial factors (called *bacteriocins*) that destroy the newly arrived pathogen. This is also a type of microbial antagonism.
- The person may be immune to that particular pathogen, perhaps as a result of prior infection with that pathogen or having been vaccinated against that pathogen. Immunity and vaccination are discussed in Chapter 11.
- Phagocytes present in the blood and other tissues may engulf and destroy the invader.

The Development of an Infectious Disease

A typical sequence in the development of an infectious disease is as follows: the organism enters the host, attaches to a tissue, multiplies, invades deeper into the tissue, and causes damage to the tissue. Not all pathogens follow this sequence, however. Some pathogens are able to cause disease even though they are unable to attach. Others cause disease even though they are not invasive.

The body often responds to an infectious process via the inflammatory response (Fig. 9-2). The symptoms of inflammation are redness, heat, swelling (*edema*), and pain. One of the first events in inflammation is *vasodilation*—an increase in the diameter of small blood vessels (capillaries) at the site. Vasodilation causes additional blood to flow to the site, resulting in redness and heat. Additional heat is due to increased metabolic activity of cells in the area. When capillaries are dilated, the endothelial cells that line the capillaries are stretched apart, allowing some of the liquid portion of the blood (plasma) to escape from the capillaries. The accumulation of plasma at the site results in edema; the area becomes edematous (swollen). Edema exerts pressure on nerve endings, resulting in pain. Additional pain may be due to substances being produced by the invading pathogens. Sometimes the area becomes so edematous that there is a loss of function; for example, a patient may be unable to bend an inflamed, swollen finger.

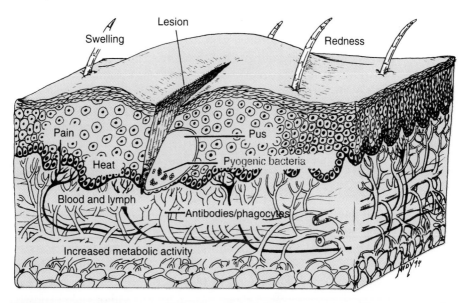

FIGURE 9-2. The development of infection and the inflammatory process. The symptoms of inflammation include redness, heat, swelling, and pain. Pus may also be present whether or not pyogenic bacteria are present.

The increased blood flow also brings phagocytic white blood cells (phagocytes) to the inflamed area to phagocytize (ingest) the invaders. Pus may be formed at the site of inflammation. Pus consists primarily of white blood cells, but also contains tissue fluids, dead tissue cells, and sometimes pathogens. Pus that is oozing from the site is called an *exudate*. Infections due to certain pathogens are always associated with pus production; such pathogens (like staphylococci and streptococci) are referred to as pyogenic (pus-producing) pathogens.

Various other activities occur as the body attempts to fight off the infection. A clot or connective tissue wall may be formed around the pathogens in an attempt to keep them contained and prevent them from invading deeper into the tissues. The body will produce special proteins called antibodies to attack the pathogens, but it takes about 10–14 days for these antibodies to be produced. Inflammation, antibody production, and other infection-fighting activities are described in Chapter 11.

The Disease Process

If the body wins the battle against the invading pathogens at the site of inflammation, the local infection is stopped. As a result of the infection, the person will usually have some antibodies to protect the body against a later similar infection. If all the pathogens have not been destroyed, the person may become a carrier. A *carrier* is a person who is colonized with a particular pathogen, but the pathogen is not causing disease in that person. However, the pathogen can be transmitted from the carrier to others, who may become sick. Some people become carriers after having strep throat, for example. These people continue to harbor the pathogen—*Streptococcus pyogenes*—and can transmit it to others, who then develop strep throat.

Clinical disease occurs when the body's primary defenses lose the battle with the pathogen. The disease may remain localized to one site, in which case it is referred to as either a *local infection* or *localized infection*. Pimples, boils, and abscesses are examples of localized infections. But if the pathogens are not stopped at the local level, they may be carried to other parts of the body by way of lymph, blood, or, in some cases, by phagocytes. When the infection has spread throughout the body, it is referred to as either a *generalized infection* or a *systemic infection*. For example, the bacterium that causes tuberculosis—*Mycobacterium tuberculosis*—may spread to many internal organs, a condition known as miliary tuberculosis.

A disease may be described as being acute, subacute, or chronic. An *acute disease* has a rapid onset, followed by a relatively rapid recovery; measles, mumps, influenza are examples. A *chronic disease* has an insidious (slow) onset, and lasts a long time, such as tuberculosis, leprosy (Hansen's disease), and syphilis. Sometimes a disease having a sudden onset can develop into a long-lasting disease. Some diseases, such as bacterial endocarditis, come on more suddenly than a chronic disease, but less suddenly than an acute disease; they are referred to as subacute diseases.

A *symptom of a disease* is defined as some evidence of a disease that is experienced or perceived by the patient; something that is subjective. Examples of symptoms include any type of ache or pain, a ringing in the ears, blurred vision, nausea, dizziness, and chills. Diseases, including infectious diseases, may either be symptomatic or asymptomatic. A *symptomatic disease* (or clinical disease) is a disease where the patient is experiencing symptoms. An *asymptomatic disease* (or subclinical disease) is a disease that the patient is unaware of because he or she is not experiencing any symptoms.

A *sign of a disease* is defined as some type of objective evidence of a disease. For example, while palpating a patient, a physician might discover a lump or an enlarged liver or spleen. Other signs of disease would be abnormal heart or breath sounds, abnormal laboratory results, or abnormalities that show up on x-rays, ultrasound studies, or CAT-scans.

In its early stages, especially in female patients, gonorrhea may be asymptomatic. Only after several months, during which the organism has caused damage, scarring, and destruction of the female reproductive organs, is pain experienced by the infected person. Gonorrhea is especially difficult to control because people are often unaware that they are infected, and unknowingly transmit the pathogen—*Neisseria gonorrhoeae*—to others during sexual activities.

An infectious disease may go from being apparent (symptomatic) to inapparent (asymptomatic) and, then some time later, back to being apparent. Such diseases are referred to as *latent infections.* Herpes virus infections, such as cold sores (fever blisters), genital herpes infections, and chickenpox are examples of latent infections. Cold sores occur intermittently, but the patient continues to harbor the herpes virus between cold sore episodes. The virus remains dormant within cells of the nervous system until some type of stress acts as a trigger. The stressful trigger may be a fever, sunburn, extreme cold, or emotional stress. A person who had chickenpox as a child may harbor the virus throughout his or her lifetime and then, later in life, as the immune system weakens, that person may develop shingles. Shingles, a painful infection of the nerves, is considered a latent manifestation of chickenpox.

If not successfully treated, syphilis progresses through primary, secondary, latent, and tertiary stages (Fig. 9-3). During the primary stage, the patient has an open lesion called a chancre, which contains the spirochete *Treponema pallidum.* A few weeks after the spirochete enters the bloodstream, the chancre disappears and the symptoms of the secondary stage arise, including rash, fever, and mucous membrane lesions. These symptoms also disappear after a few weeks and the disease enters a latent stage, which may last from 1 to 50 years, during which the patient has few or no symptoms. In tertiary syphilis, the spirochetes cause destruction of the organs in which they have been "hiding"—the brain, heart, and bone tissue.

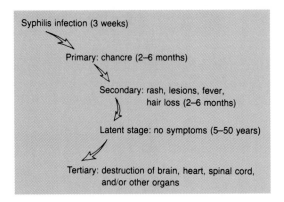

FIGURE 9-3. Stages of syphilis.

One infectious disease may commonly follow another, in which case the first disease is referred to as a *primary infection* and the second is referred to as a *secondary infection*. For example, serious cases of bacterial pneumonia frequently follow relatively mild viral respiratory infections. During the primary infection, the virus causes damage to the ciliated epithelial cells that line the respiratory tract. The function of these cells is to move foreign materials up out of the respiratory tract into the throat where they can be swallowed. While coughing, the patient may inhale some saliva, containing an opportunistic pathogen, such as *Streptococcus pneumoniae* or *Haemophilus influenzae*. Because the ciliated epithelial cells were damaged by the virus, they are unable to clear the bacteria from the lungs. The bacteria multiply and cause pneumonia. In this example, the viral infection is the primary infection and the bacterial pneumonia is the secondary infection.

Often, the primary infection leaves the patient in a debilitated (weakened) condition, and more susceptible to infection. Complications frequently follow measles, for example, because the patient's body defenses are lowered as a result of the viral infection. Thus, the measles virus (Rubeola virus) and possibly opportunistic bacteria can invade deeper to cause pneumonia, ear and sinus infections, or encephalitis (infection of the brain).

Once exposure to a pathogen has occurred, the course of an infectious disease has four periods or phases (Fig. 9-4): (1) the incubation period, during which the pathogen multiplies, but the patient experiences no symptoms; (2) the prodromal period, during which the patient feels "out of sorts," but is not yet experiencing actual symptoms of the disease; (3) the period of illness, during which the patient experiences the typical symptoms associated with that particular disease; and (4) the convalescent period, disability, or death. Fortunately, this final stage is usually the convalescent period, during which the patient recovers. For certain infectious diseases, especially respiratory diseases, the convalescent period can be quite long. Communicable diseases are most easily transmitted during the third period.

Although the patient may recover from the illness itself, permanent damage

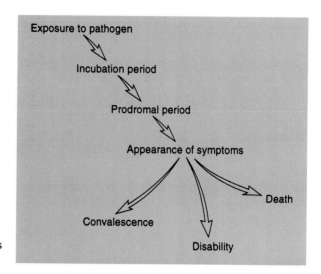

FIGURE 9-4. The course of an infectious disease.

may be caused by destruction of tissues in the affected area. For example, brain damage may follow encephalitis or meningitis; paralysis may follow poliomyelitis; deafness may follow ear infections.

Virulence and Virulence Factors

VIRULENCE

The capability of a pathogen to cause disease (its *pathogenicity*) is related to its abilities to (1) infect the host, (2) protect itself from the body's defenses, (3) to invade and multiply in tissues, and (4) to cause damage to or destruction of tissue. As defined earlier, virulence is a measure or degree of pathogenicity; some pathogens are more virulent than others.

Each species of pathogenic microbe has specific characteristics, including its unique metabolism, that determine its pathogenicity and virulence. Properties or characteristics that contribute to the virulence of a pathogen are called *virulence factors* (Fig. 9-5). It would be impossible to list every individual pathogen and its virulence factors in this book; in fact, the various mechanisms are frequently not completely understood. Keep these limitations in mind as we discuss some of the factors associated with virulence and pathogenicity.

The ability of a pathogen to settle onto a susceptible tissue and to survive the shock of landing is largely a matter of chance. However, once the pathogen finds itself in a moist, warm environment, it must be able to attach to the site (so that it is not flushed away) and it must be able to resist the body's bacteriolytic enzymes, antibodies, and phagocytes. It should also be noted that certain pathogens are able to cause disease without the need to attach and/or invade body tissues.

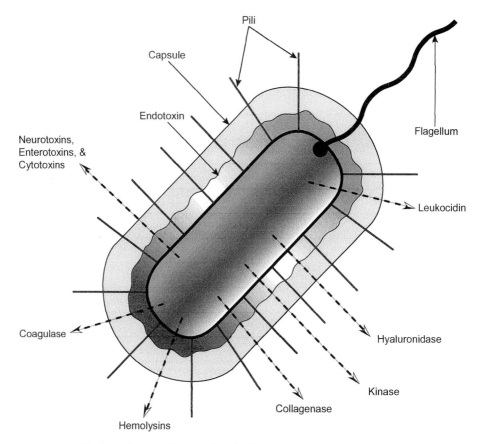

FIGURE 9-5. Virulence factors. (See text for details.)

BACTERIAL STRUCTURES ASSOCIATED WITH VIRULENCE

Some structural features of pathogens enable them to reach and attach to tissues in certain areas of the host, where they multiply and cause an infection. Such structures include flagella, capsules, and pili.

Flagella. Possession of flagella is considered a virulence factor because flagella enable bacteria to invade aqueous areas of the body. Motile bacteria are able to reach areas that nonmotile organisms are unable to reach.

Capsules. Capsules are a type of glycocalyx. They are considered a portion of the cell envelope, which includes the cell membrane, cell wall, and outer glycocalyx. Capsular constituents vary among the different species of bacteria. Some capsules consist only of polysaccharides, whereas others consist of polysaccharides and proteins. Although many of the bacteria that infect the body have some type of capsule, the capsules are frequently not observable when the organisms are grown on artificial media in the laboratory. They seem to lose the ability to produce capsules when grown in vitro (in the laboratory), perhaps because the organisms no longer need capsules to protect themselves.

Capsules serve as virulence factors in two ways: (1) they enable bacteria to attach to tissues, and (2) they serve an antiphagocytic function; *i.e.,* they protect encapsulated bacteria from being phagocytized by certain white blood cells (called phagocytes). Some bacteria secrete polysaccharide fibers (glycocalyx) to increase their adherence to teeth and mucous membranes. Virulent strains of *Streptococcus pneumoniae* are encapsulated (Fig. 9-6), whereas avirulent strains lack capsules. Because the capsule protects the encapsulated organisms from being phagocytized, they are able to survive, multiply, invade, and cause disease.

Pili (Fimbriae). Because pili (fimbriae) enable bacteria to adhere to tissue, piliated pathogens are able to cause infections in areas where nonpiliated pathogens cannot. For example, piliated *Neisseria gonorrhoeae* cells are able to attach to the inner walls of the urethra, where they multiply and cause urethritis. Because they are unable to attach, nonpiliated strains of *N. gonorrhoeae* are flushed away by urination and unable to cause infection. Likewise, piliated strains of *Escherichia coli* are able to anchor themselves to the inner walls of the urethra and urinary bladder. Thus, piliated strains of *E. coli* are able to cause cystitis, whereas nonpiliated strains are flushed away by urination. Enterotoxigenic strains of *E. coli,* with their many pili, are able to attach to cells in the intestine where they multiply and secrete the exotoxin that causes gastroenteritis. The pili of group A, β-hemolytic streptococci (*Streptococcus pyogenes*) contains a certain antigenic protein (M-protein) that enables these bacteria to adhere to pharyngeal cells. In addition, M-protein serves an antiphagocytic function. (See Insight Box in Chapter 12 for more about *S. pyogenes.*) Other bacterial pathogens possessing pili are *Vibrio cholerae, Salmonella* spp., *Shigella* spp., *Pseudomonas aeruginosa,* and *Neisseria meningitidis.*

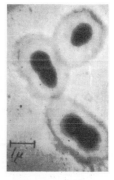

FIGURE 9-6. Electron micrograph of *Streptococcus pneumoniae,* type 1. The capsule has been treated with a specific antibody to enhance its visibility; this is known as a Quellung reaction.

ENZYMES ASSOCIATED WITH VIRULENCE

Some pathogens excrete enzymes (called exoenzymes) that enable them to evade host defense mechanisms, invade, or cause damage to body tissues. These exoenzymes include coagulase, kinases, hyaluronidase, collagenase, hemolysins, and lecithinase. Note that many enzymes end in "-ase." You will recall that enzymes are proteins that catalyze (speed up) particular chemical reactions.

Coagulase. *Staphylococcus aureus* is noted for its production of the exoenzyme, *coagulase.* In the body, coagulase enables these organisms to clot plasma and thereby to form a sticky coat of fibrin around themselves for protection from phagocytes and other body defense mechanisms. In the laboratory, the coagulase test is used to differentiate *S. aureus* (which is coagulase-positive) from all other species of staphylococci (which are coagulase-negative).

Kinases. *Kinases* (also known as *fibrinolysins*) have the opposite effect of coagulase. Sometimes the host will cause a fibrin clot to form around pathogens in an attempt to wall them off and prevent them from invading deeper into body tissues. Kinases are enzymes that lyse (dissolve) clots; therefore, pathogens that produce kinases are able to escape from clots. *Streptokinase* is a kinase produced by streptococci and *staphylokinase* is a kinase produced by staphylococci. *Staphylococcus aureus* produces both coagulase and staphylokinase; thus, it not only produces clots, but it also can dissolve them.

Hyaluronidase. The spreading factor, as *hyaluronidase* is sometimes called, enables pathogens to spread through connective tissue by breaking down *hyaluronic acid,* the polysaccharide "cement" that holds tissue cells together. Hyaluronidase is secreted by several pathogenic species of *Staphylococcus, Streptococcus,* and *Clostridium.*

Collagenase. The enzyme *collagenase* breaks down *collagen,* the supportive protein found in tendons, cartilage, and bones. *Clostridium perfringens,* a major cause of gas gangrene, spreads deeply within the body by secreting both collagenase and hyaluronidase.

Hemolysins. *Hemolysins* (from "hemo," meaning "blood," and "lysis," meaning "breakdown or dissolution") are enzymes that cause damage to the host's red blood cells (*erythrocytes*). Not only does the lysis of red blood cells harm the host, but it also provides the pathogens with a source of iron. In the laboratory, the effect an organism has on the red blood cells in blood agar enables differentiation between alpha- (α-) hemolytic and beta- (β-) hemolytic bacteria. The hemolysins produced by α-hemolytic bacteria convert hemoglobin (which is red) to methemoglobin (which is green), so that colonies of α-hemolytic organisms are surrounded by a green zone. The hemolysins produced by β-hemolytic bacteria cause complete lysis of the red blood cells, so that colonies of β-hemolytic organisms are surrounded by clear zones. (Refer to Color Figure 22.) Hemolysins are produced by many pathogenic bacteria, but the type of hemolysins produced by an organism is of most importance when attempting to identify (speciate) a *Streptococcus* species.

Lecithinase. *Clostridium perfringens,* the major cause of gas gangrene (myonecrosis), is able to rapidly destroy large areas of tissue, especially muscle tissue. One of the enzymes produced by *C. perfringens* is called *lecithinase,* which breaks down phospholipids collectively referred to as *lecithin.* This enzyme is destructive to cell membranes of red blood cells and other tissues.

TOXINS ASSOCIATED WITH VIRULENCE

The ability of pathogens to damage host tissues and cause disease may depend on the production and release of various types of poisonous substances, referred to as toxins. *Endotoxin* ("endo," meaning "within"), which is an integral part of the cell wall of Gram-negative bacteria, can cause a number of adverse physiologic effects. *Exotoxins* ("exo," meaning "outside"), on the other hand, are toxins that are produced within cells, and then released from the cells.

Endotoxin. *Septicemia* (also called sepsis) is a very serious disease. It consists of chills, fever, prostration (extreme exhaustion), and the presence of bacteria and/or their toxins in the blood stream. A blood stream infection with Gram-negative bacteria, sometimes referred to as Gram-negative sepsis, is an especially serious type of septicemia. The cell walls of Gram-negative bacteria contain lipopolysaccharide (LPS), the lipid portion of which is called Lipid-A or endotoxin. Endotoxin can cause serious, adverse, physiological effects, such as fever and shock. Substances that cause fever are known as *pyrogens.*

    Shock is a life-threatening condition, resulting from very low blood pressure and an inadequate blood supply to body tissues and organs, especially the kidneys and brain. The type of shock that results from Gram-negative sepsis is known as *septic shock.* Symptoms include reduced mental alertness, confusion, rapid breathing, chills, fever, and warm, flushed skin. As shock worsens, several organs begin to fail, including the kidneys, lungs, and heart. Blood clots may form within blood vessels. More than 500,000 cases of sepsis occur annually in the United States; approximately half of these are caused by Gram-negative bacteria. There is a 30%–35% mortality rate associated with Gram-negative sepsis.

Exotoxins. Bacterial exotoxins are proteins secreted by living (vegetative) pathogens. Exotoxins are often named for the target organs they affect. The most toxic exotoxins are *neurotoxins;* they affect the central nervous system. The neurotoxins produced by *Clostridium tetani* and *Clostridium botulinum*—tetanospasmin and botulinal toxin—cause tetanus and botulism, respectively. Tetanospasmin affects control of nerve transmission, leading to a spastic, rigid type of paralysis, where the patient's muscles are contracted. Botulinal toxin also blocks nerve impulses, but by a different mechanism, leading to a generalized flaccid type of paralysis, where the patient's muscles are relaxed. Both diseases are often fatal.

Enterotoxins are toxins that affect the gastrointestinal tract, often causing diarrhea, and sometimes vomiting. Examples of pathogens that secrete enterotoxins are *Vibrio cholerae, Salmonella* spp., *Shigella* spp., *Clostridium difficile, Clostridium perfringens, Bacillus cereus*, certain serotypes of *E. coli*, and some strains of *Staphylococcus aureus*. In addition to secreting an enterotoxin, *C. difficile* also produces a cytotoxin that damages the lining of the colon, leading to a condition known as pseudomembranous colitis.

Symptoms of toxic shock syndrome are caused by exotoxins secreted by certain strains of *Staphylococcus aureus* and, less commonly, *Streptococcus pyogenes*. *Exfoliative toxins* of *S. aureus* cause the epidermal layers of skin to slough away, and *erythrogenic toxin*, produced by some strains of *S. pyogenes*, causes scarlet fever. The exotoxin produced by *Corynebacterium diphtheriae* inhibits protein synthesis in many cell types and causes diphtheria. As you learned in Chapter 6, it is actually a bacteriophage gene that codes for diphtheria toxin. *Leukocidins* are toxins that destroy white blood cells (*leukocytes*). Thus, leukocidins (which are produced by some staphylococci and streptococci) cause destruction of the very cells that the body sends to the infection site to ingest and destroy pathogens.

Pathogenicity and Virulence

All of the terms presented thus far in this chapter concerning the pathogenicity and virulence of pathogens are useful in discussing general concepts of disease causation. Although the terms pathogenicity and virulence are sometimes used synonymously, virulence actually refers to the degree of pathogenicity.

A pathogen is a species of microorganism capable of causing disease. However, some avirulent strains may exist within a pathogenic species. For instance, some avirulent strains of *Escherichia coli, Neisseria gonorrhoeae, Corynebacterium diphtheriae*, and *Streptococcus pneumoniae* have been identified. Nonpiliated strains of *E. coli* and *N. gonorrhoeae*, nontoxigenic strains of *C. diphtheriae*, and nonencapsulated strains of *S. pneumoniae* are avirulent.

Virulence depends on infectivity, invasiveness, and *toxigenicity*. Pathogens that are highly infective but low in invasiveness might produce a large number of carriers. This large group of pathogens would include the opportunists found among the indigenous microflora. If the infectivity of a pathogen was low but its invasiveness and toxigenicity were high, sporadic outbreaks of the disease would occur.

The production of toxic products by pathogens is an important factor in their pathogenicity. Toxin-producing bacteria differ greatly in the amount of exotoxin they produce, depending on growth conditions. For example, virulent strains of some pathogens become avirulent when grown in vitro (in the laboratory) under artificial conditions. Some virulent encapsulated pathogens lose their ability to produce capsules when grown on artificial media in the laboratory, and become avirulent. Also, some toxin producers cease producing exotoxins when grown on artificial media.

These changes can be regarded as the pathogen's response to different environments. On a completely defined, artificial medium in the incubator where a constant temperature is maintained, optimal conditions exist for the microorganisms. An ample supply of nutrients is present; temperature and humidity are ideal; no overcrowding occurs from other microbes; and no phagocytes or antibodies exist. Thus, pathogens have no need to arm themselves with capsules and toxins, as they do when attempting to grow in human tissues.

It should also be noted that virulence usually increases in pathogens that are transmitted from animal to animal and human to human, which explains why virulence may increase during an epidemic. This factor may also account for the relatively high virulence of pathogens in a hospital environment, where there is a continuously changing population of susceptible persons. Rampaging hospital staphylococcal infections are caused by virulent staphylococci that continue to mutate and to become resistant to the antibiotics used within hospitals in ever greater varieties and frequency to fight infections.

EPIDEMIOLOGY AND DISEASE TRANSMISSION

Epidemiology is the study of the factors that determine the occurrence of diseases in human populations. These factors include the virulence of various pathogens; susceptibility of various populations because of overcrowding, lack of immunization, nutritional status, inadequate sanitation procedures, etc.; and the various modes of transmission of infectious diseases.

Terms like endemic, epidemic, pandemic, sporadic, and nonendemic are used to describe the prevalence of a particular disease in a given population over a specific period of time or at a particular point in time. Each of these terms is discussed in this section.

Endemic Diseases

Endemic diseases are diseases that are constantly present in a population, community, or country. The number of cases increases and decreases at various times, but the disease never dies out completely. Endemic infectious diseases that occur regularly in the United States include tuberculosis, staphylococcal and streptococcal infections, sexually transmitted diseases like gonorrhea and syphilis, and viral diseases, such as the common cold, influenza, chickenpox, and mumps. Tuberculosis remains endemic throughout the world, particularly among the poor, who live in crowded conditions. Although the number of tuberculosis deaths has been greatly reduced over the years by drug therapy, strains of multi-drug resistant *Mycobacterium*

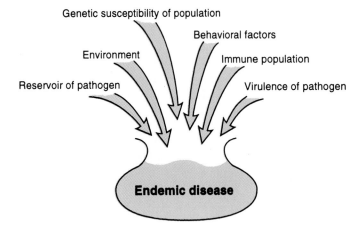

FIGURE 9-7. Factors influencing an endemic disease.

tuberculosis are now common in the United States and other countries. In some parts of the United States, plague (caused by a bacterium called *Yersinia pestis*) is endemic among rats, prairie dogs, and other rodents, but is not endemic among humans. Plague in humans is only occasionally observed in the United States. The actual incidence of an endemic disease at any particular time depends on a balance among several factors: environment, the genetic susceptibility of the population, behavioral factors, the number of people who are immune, the virulence of the pathogen, and the reservoir or source of infection (Fig. 9-7).

Epidemic Diseases

Infections that are usually endemic may on occasion become epidemic. An *epidemic* is defined as a greater than normal number of cases of a disease in a particular region within a short period of time. Two epidemics are represented graphically in Figure 9-8.

Epidemics in the United States include the 1976 epidemic of Legionnaires' disease or legionellosis (Fig. 9-9) during an American Legion convention in Philadelphia, Pennsylvania; the 1993 epidemic of hantavirus pulmonary syndrome in the four-corner area (where Colorado, New Mexico, Arizona, and Utah all meet); and the 1993 epidemic of cryptosporidiosis (described below) in Milwaukee, Wisconsin. Other such epidemics have been identified through constant surveillance and accumulation of data by the U.S. Centers for Disease Control and Prevention (CDC). Epidemics usually follow a specific pattern in which the number of cases of a disease increases to a maximum and then decreases rapidly, because the number of susceptible and exposed individuals is limited.

Epidemics may occur in communities that have not been previously exposed to a particular pathogen. People from populated areas who travel into isolated communities frequently introduce a new virulent pathogen to susceptible natives of that

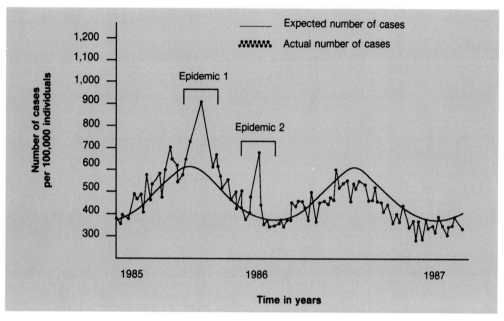

FIGURE 9-8. Graph illustrating two epidemics.

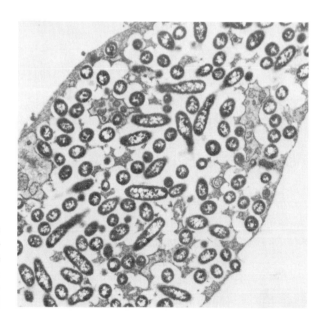

FIGURE 9-9. *Legionella pneumophila* cultured in human embryonic lung fibroblasts. This gram-negative bacillus causes a respiratory disease called Legionnaires' disease or legionellosis. (Courtesy of Mae C. Wong, Centers for Disease Control and Prevention, Atlanta, Georgia)

community; then the disease spreads like wildfire. There have been many such examples described in history. The syphilis epidemic in Europe in the early 1500s might have been caused by a highly virulent spirochete carried back from the West Indies by Columbus' men in 1492. Also, measles and tuberculosis introduced to Native Americans by early explorers and settlers almost destroyed many tribes. Recently, we have observed devastating outbreaks of measles and other contagious diseases in Australia, Africa, Greenland, and other relatively isolated areas.

In communities where normal sanitation practices are relaxed, allowing fecal contamination of water supplies and food, epidemics of typhoid, cholera, giardiasis, and dysentery frequently occur. Visitors to these communities should be aware that they are especially susceptible to these diseases, because they never developed a natural immunity by being exposed to them during childhood.

In the 1970s, gonorrhea reached epidemic proportions in the United States and most of the world, in part because no immunity remains after the disease is cured (Fig. 9-10). This disease runs rampant in a promiscuous, mobile society. The

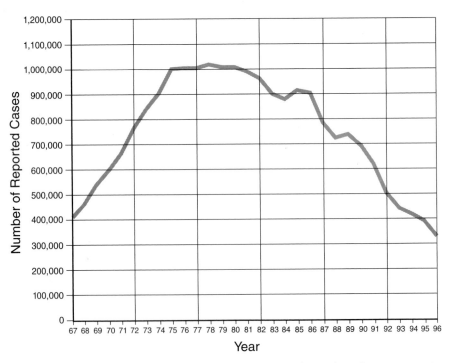

FIGURE 9-10. Gonorrhea in the United States. The graph shows the number of cases reported annually to the Centers for Disease Control and Prevention between 1967 and 1996. (Based on data from Morbidity and Mortality Weekly Report, Vol. 45, No. 53, 1996. U.S. Department of Health and Human Services, Public Health Service, Centers for Disease Control and Prevention, Atlanta, GA)

use of birth control pills makes women more susceptible by changing the pH and microbial flora of the vagina. Also, the careless use of penicillin allowed the gonococcus to mutate into penicillin-resistant strains (penicillinase-producing-*Neisseria gonorrhoeae* or PPNG), making control by penicillin more difficult. An additional complicating factor is that this disease is *asymptomatic* (without symptoms) in 80% of infected women and 20% of infected men; that is, the gonococcus lives and multiplies in the urogenital areas, yet produces no symptoms. An effective vaccine might stimulate the body to produce antibodies against pili to prevent the pathogen from adhering to mucous membranes.

Diseases such as the various types of influenza occur in many areas during certain times of the year and involve most of the population because the immunity developed is usually temporary. Thus, the disease recurs each year among those who are not revaccinated or naturally resistant to the infection.

In the spring of 1993, a water-borne epidemic of cryptosporidiosis (a diarrheal disease) affected over 400,000 people in Milwaukee, Wisconsin. The oocysts of *Cryptosporidium parvum* (a protozoan) were present in the city's drinking water. Although the water had been treated, the tiny oocysts passed through the filters that were being used at that time. The disease caused the death of some immunosuppressed individuals. Ebola virus has caused several epidemics in Africa. In the summer of 1995, an Ebola virus epidemic occurred in Kikwit, Zaire, resulting in over 300 cases within a few weeks and over 200 deaths. The source of the virus is not yet known.

In a hospital setting, a relatively small number of infected patients can constitute an epidemic. If a larger than usual number of patients on a particular ward should suddenly become infected by a particular pathogen, this would constitute an epidemic, and the situation must be brought to the attention of the Hospital Infection Control Committee (discussed in Chapter 10).

Pandemic Diseases

A *pandemic* is a worldwide epidemic of a specific disease. Pre-1900 pandemics of influenza ("flu") occurred in 1729, 1732, 1781, 1830, 1833, and 1889, but it was the influenza pandemic of 1918—1919 that was the most devastating pandemic of the twentieth century. That pandemic killed 21 million people worldwide, including 500,000 in the United States. Almost every nation on Earth was affected. Recent influenza pandemics are often named for the point of origin or first recognition, such as the Taiwan flu, Hong Kong flu, London flu, Port Chalmers flu, and the Russian flu.

A current example of an important pandemic is the expanding HIV infection/AIDS pandemic (Fig. 9-11). Various theories exist regarding the origin of the etiologic agent. Because HIV is similar to simian immunodeficiency virus (SIV), some scientists believe that HIV came from primates, perhaps African monkeys. Although

New U.S. AIDS Cases: 1984-1996

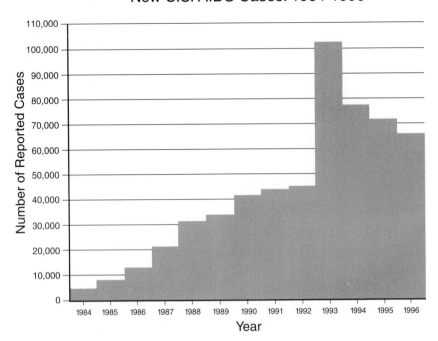

FIGURE 9-11. Acquired immunodeficiency syndrome (AIDS) in the United States. The graph shows the number of cases reported annually to the Centers for Disease Control and Prevention between 1984 and 1996. The sharp rise in 1993 was the result of an expanded AIDS surveillance case definition in that year. (Based on data from Morbidity and Mortality Weekly Report, Vol. 45, No. 53, 1996. U.S. Department of Health and Human Services, Public Health Service, Centers for Disease Control and Prevention, Atlanta, GA)

the first documented evidence of HIV infection in humans can be traced to an African serum sample collected in 1959, it is possible that humans were infected with HIV prior to that date. The AIDS epidemic began in the United States around 1979, but the epidemic was not detected until 1981, and it wasn't until 1983 that the virus that causes AIDS was discovered. The virus was originally called human T-cell lymphotrophic virus, type 3 (HTLV—III) by one group of investigators and lymphadenopathy-associated virus (LAV) by another; it was later named human immunodeficiency virus (HIV) by international agreement.

The World Health Organization (WHO) estimated that, as of the end of 1997, about 30 million people worldwide were infected with HIV, and that between 2 and 2.5 million people had died of AIDS. Of the 30 million infected people, approximately half were living in sub-Saharan Africa, compared to about 750,000 living in North America. Approximately 68,500 new United States AIDS cases were reported to the CDC during 1996, bringing the total number of U.S. cases to 573,800 since it was first reported in 1981. It has been estimated that as many as

16,000 people per day were becoming infected worldwide during 1997. The WHO estimates that as many as 40 million people worldwide will be infected with HIV by the year 2000. Most people who develop AIDS will probably die, unless new treatments are developed. This sexually transmitted or blood-borne virus has a long incubation period during which it destroys the CD4$^+$ T-helper lymphocytes, thus, crippling the immune system so that most victims succumb to secondary infections (see Chapter 12). One significant outcome of the AIDS pandemic is the heightened awareness of other sexually transmitted diseases (STDs) like gonorrhea, syphilis, and genital herpes infections.

Sporadic and Nonendemic Diseases

Certain diseases follow neither the endemic nor the epidemic pattern but occur only occasionally. In the United States, these *sporadic diseases* include botulism, gas gangrene, plague, and tetanus. Diseases that are controlled as a result of immunization and sanitation procedures are termed *nonendemic diseases*. These include smallpox, poliomyelitis, and diphtheria (in most parts of the world). Occasionally, outbreaks of these controlled diseases occur where vaccination programs and other public health programs have been neglected.

The communicability of pathogens relies entirely on the survival of infectious agents during their transfer from one host to another. Thus, health professionals should be thoroughly familiar with sources of potential pathogens and pathways of transfer. For instance, a hospital staphylococcal epidemic may begin when aseptic conditions are relaxed and a *Staphylococcus aureus* carrier introduces the organism to the many susceptible patients (babies, surgical patients, and debilitated persons). Such an epidemic may quickly spread from one person throughout the entire hospital population.

RESERVOIRS OF INFECTION

The sources of microorganisms that cause infectious diseases are many and varied. They are known as *reservoirs of infection* or simply *reservoirs*. A reservoir of infection is any site where the pathogen can multiply or merely survive until it is transferred to a host. Reservoirs of infection may be living hosts or inanimate objects or materials (Fig. 9-12).

Living animal reservoirs include humans, cats, dogs, farm animals, wild animals, insects, arachnids like ticks and mites, and many others. Humans may acquire pathogens from other humans and animals, which may or may not be diseased; through direct contact and bites; by eating meat or other products from diseased animals; or through the bites of mosquitoes, flies, fleas, mites, ticks, and other arthropod vectors.

FIGURE 9-12. Reservoirs of infection include soil, dust, contaminated water, contaminated foods, insects, and infected humans, domestic animals, and wild animals. (Illustration from *Principles and Practice of Clinical Anaerobic Bacteriology* by P.G. Engelkirk et al., 1992; reproduced courtesy of Star Publishing Company, Belmont, CA)

Infectious diseases that humans acquire from animal sources are called *zoonotic diseases* or *zoonoses*. Some of the more than 200 known zoonoses are listed in Table 9-1. Zoonoses are acquired by direct contact, inhalation, ingestion, or injection of the pathogen. The most prevalent zoonotic infection in the U.S. is Lyme disease, one of many arthropod-borne zoonoses. Other zoonoses that are endemic in the U.S. include anthrax, brucellosis, campylobacteriosis, cryptosporidiosis, echinococcosis, ehrlichiosis, hantavirus pulmonary syndrome (HPS), leptospirosis, pasteurellosis, plague, psittacosis, Q fever, rabies, ringworm, Rocky Mountain spotted fever, salmonellosis, toxoplasmosis, tularemia, and various viral encephalitides (*e.g.,* Western equine encephalitis, Eastern equine encephalitis, St. Louis encephalitis, California encephalitis). A variant form of Creutzfeldt-Jakob (CJ) disease in humans may be transmitted by ingestion of prion-infected beef from cows with bovine spongiform encephalopathy ("mad cow disease"). Measures for the control of zoonotic diseases include the use of personal protective equipment, proper use of pesticides, isolation or destruction of infected animals, and proper disposal of animal carcasses and waste products. For a discussion of nosocomial zoonoses, see the Insight Box in Chapter 10.

The most important reservoirs of human infections, however, are other humans. Many human pathogens can only cause disease in humans because they are

(text continues on page 253)

TABLE 9-1. Examples of Zoonotic Diseases

Category	Disease	Pathogen	Animal Reservoir(s)	Mode of Transmission
Viral diseases	Ebola disease	Ebola virus	Unknown (possibly monkeys or rodents)	Unknown
	Equine encephalitis	Various arboviruses	Birds, small mammals	Mosquito bite
	Hantavirus pulmonary syndrome	Hantaviruses	Rodents	Inhalation of contaminated dust or aerosols
	Lassa fever	Lassa virus	Wild rodents	Inhalation of contaminated dust or aerosols
	Marburg disease	Marburg virus	Monkeys	Contact with blood or tissues from infected monkeys
	Rabies	Rabies virus	Rabid dogs, cats, skunks, foxes, wolves, raccoons, coyotes, bats	Animal bite or inhalation
	Yellow fever	Yellow fever virus	Monkeys	*Aedes aegypti* mosquito bite
Bacterial diseases	Anthrax	*Bacillus anthracis*	Cattle, sheep, goats	Inhalation, ingestion, entry through cuts, contact with mucous membranes
	Bovine tuberculosis	*Mycobacterium bovis*	Cattle	Ingestion
	Brucellosis	*Brucella* spp.	Cattle, swine, goats	Inhalation, ingestion of contaminated milk, entry through cuts, contact with mucous membranes
	Campylobacter infection	*Campylobacter* spp.	Wild mammals, cattle, sheep, pets	Ingestion of contaminated food and water
	Cat-scratch disease	*Bartonella henselae*	Domestic cats	Cat scratch, bite, or lick
	Ehrlichiosis	*Ehrlichia* spp.	Deer, mice	Tick bite
	Endemic typhus	*Rickettsia typhi*	Rodents	Flea bite
	Leptospirosis	*Leptospira* spp.	Cattle, rodents, dogs	Contact with contaminated animal urine

(continued)

Category	Disease	Pathogen	Animal Reservoir(s)	Mode of Transmission
	Lyme disease	*Borrelia burgdorferi*	Deer, rodents	Tick bite
	Pasteurellosis	*Pasteurella multocida*	Oral cavities of animals	Bites, scratches
	Plague	*Yersinia pestis*	Rodents	Flea bite
	Psittacosis (ornithosis, parrot fever)	*Chlamydia psittaci*	Parrots, parakeets, other pet birds, pigeons, poultry	Inhalation of contaminated dust and aerosols
	Relapsing fever	*Borrelia* spp.	Rodents	Tick bite
	Rickettsial pox	*Rickettsia akari*	Rodents	Mite bite
	Rocky Mountain spotted fever	*Rickettsia rickettsii*	Rodents, dogs	Tick bite
	Salmonellosis	*Salmonella* spp.	Poultry, livestock, reptiles	Ingestion of contaminated food, handling reptiles
	Scrub typhus	*Rickettsia tsutsugamushi*	Rodents	Mite bite
	Tularemia	*Francisella tularensis*	Wild mammals	Entry through cuts, inhalation, tick or deer fly bite
	Q fever	*Coxiella burnetii*	Cattle, sheep, goats	Tick bite, air, milk, contact with infected animals
Fungal diseases	Tinea (ringworm) infections	Various dermatophytes	Various animals including dogs	Contact with infected animals
Protozoal diseases	African trypanosomiasis	Subspecies of *Trypanosoma brucei*	Cattle, wild game animals	Tsetse fly bite
	American trypanosomiasis (Chagas' disease)	*Trypanosoma cruzi*	Numerous wild and domestic animals, including dogs, cats, wild rodents	Trypomastigotes in the feces of reduviid bug are rubbed into bite wound
	Babesiosis	*Babesia microti*	Deer, mice, voles	Tick bite
	Leishmaniasis	*Leishmania* spp.	Rodents, dogs	Sandfly bite
	Toxoplasmosis	*Toxoplasma gondii*	Cats, pigs, sheep, rarely cattle	Ingestion of oocysts in cat feces or cysts in raw or undercooked meat
Helminth diseases	Echinococcosis (hydatid disease)	*Echinococcus granulosis*	Dogs	Ingestion of eggs

Category	Disease	Pathogen	Animal Reservoir(s)	Mode of Transmission
	Dog tapeworm infection	*Dipylidium caninum*	Dogs, cats	Ingestion of flea containing the larval stage
	Rat tapeworm infection	*Hymenolepis diminuta*	Rodents	Ingestion of beetle containing the larval stage

species-specific. A species-specific pathogen can cause disease in only one species of animal. Some people may harbor a pathogen and transmit it to susceptible people, even though they themselves show no symptoms of the infectious disease caused by that pathogen. These people "carry" the pathogen and are called *carriers*. There are several types of carriers. An *incubatory carrier* is a person who is capable of transmitting a pathogen during the incubation period of a particular infectious disease. *Convalescent carriers* harbor and can transmit a particular pathogen while recovering from an infectious disease; *i.e.*, during the convalescence period. *Active carriers* have completely recovered from the disease, but continue to harbor the pathogen indefinitely. *Passive carriers* carry the pathogen without ever having had the disease. Usually, respiratory secretions and intestinal or urinary excretions are the vehicles by which the pathogen is transferred, either directly to a susceptible individual or via food or water. Human carriers are very important in the spread of staphylococcal and streptococcal infections as well as in the spread of hepatitis, diphtheria, dysentery, meningitis, and STDs.

Inanimate reservoirs of infection include air, soil, food, milk, water, as well as fomites (see Insight Box entitled "Pathogens in Our Kitchens, Part 1"). *Fomites* include articles of clothing, bedding, eating and drinking utensils, and hospital equipment, such as bedpans, and urinals, that are easily contaminated by pathogens from the respiratory tract, intestinal tract, and the skin of patients. Air is contaminated by dust, smoke, and respiratory secretions of humans expelled into the air by breathing, blowing, sneezing, and coughing. The most highly contagious diseases include colds and influenza, in which the respiratory viruses can be transmitted through the air on droplets of respiratory tract secretions or by the hands to another person. Dust particles can carry spores of certain bacteria and dried bits of human and animal excretions that contain pathogens. Bacteria cannot multiply in the air but can be easily transported via airborne particles to a warm, nutrient site for growth. All personnel in hospitals and other facilities where housekeeping is conducted for large numbers of people must be especially aware of air currents that carry dust and pathogens throughout the facility. Great care must be taken by the hospital staff to prevent transmission of pathogens to patients.

INSIGHT
Pathogens in Our Kitchens, Part I

Are there bacteria in the milk you drink? Pasteurization is designed to kill *pathogens*. The process does not kill *all* bacteria. According to accepted standards, raw milk may not have more than 75,000 bacteria per mL *before* pasteurization and must have less than 15,000 per mL *after* pasteurization. Let's say that the milk you are drinking contains 10,000 bacteria per mL. One fluid ounce equals approximately 29.6 mL. Therefore, an 8-ounce glass of that milk contains 2,368,000 bacteria. You will be chug-a-lugging over two million bacteria, assumed to be nonpathogens.

Are there bacteria in the food you eat? Bacteria and fungi occur in most foods, but vary in quantity from one type of food to another. Assuming the food has been stored correctly (refrigeration, for example), there are usually fewer than 100,000 per gram or mL, depending upon the type of food. The number may be *much* higher if the food has not been stored properly. Also, the way the food is prepared will influence the number of live organisms that are present. For example, a thoroughly cooked ("well done") hamburger may not contain any live bacteria. A rare or medium-rare hamburger, on the other hand, will contain *many* live bacteria. If some of those bacteria are pathogens (such as *E. coli* 0157:H7), you could develop severe gastrointestinal disease.

Are there bacteria in your drinking water? There are many types of bacteria in the water you drink, but hopefully not too many pathogens. Ideally, drinking water should not contain any *coliforms* (gram-negative bacilli, like *E. coli*, that live in the gastrointestinal tract and are present in feces). The presence of coliforms in drinking water represents contamination of the water with human or animal feces. The extent of fecal contamination of water can be determined by performing a coliform count. A "satisfactory" coliform count is one colony (a colony is derived from one organism) or less per 100 mL of water. If the coliform count is satisfactory, the water is considered potable (drinkable). It usually takes many coliforms per mL to cause disease in humans.

MODES OF DISEASE TRANSMISSION

There are four principal modes by which transmission of pathogens occurs: airborne, vehicular, contact (either direct or indirect), and vectors (Fig. 9-13 and Table 9-2). Each of these modes will be discussed in detail.

Communicable diseases, those that are transmitted from person to person, are usually transmitted in the following ways:

- Direct skin-to-skin contact. For example, the common cold virus is frequently transmitted from the hand of someone who just blew his/her nose to another person by hand shaking.
- Direct mucous membrane-to-mucous membrane contact by kissing or sexual intercourse. Many STDs are transmitted in this manner.
- Indirectly via airborne droplets of respiratory secretions, as a result of sneezing or coughing.
- Indirectly via contamination of food and water by fecal material. Many infectious diseases are transmitted by restaurant food handlers who fail to wash their hands after using the rest room.
- Indirectly via fomites that become contaminated by respiratory secretions, blood, urine, feces, vomitus, or exudates from an ill person.

- Indirectly via transfusion of contaminated blood or blood products from an ill person or by *parenteral injection* (injection directly into the bloodstream) using nonsterile syringes and needles.

Most diseases transmitted by direct contact are those in which the causative organism can be carried on the skin; usually the pathogen is transferred by the hands and face. Many viruses and opportunistic bacteria are thus transferred, causing colds, influenza (flu), staphylococcal and streptococcal infections, pneumonia, polio, and diphtheria. In hospitals, this mode of transfer is particularly prevalent. Even the dysentery organisms, *Salmonella* and *Shigella,* can be transferred by fecal material on the hands of one person to the hands, mouth, or food of another.

A wide variety of diseases is transmitted via the mucous membrane-to-mucous membrane mode, by kissing or sexual contact. STDs include syphilis, gonorrhea, and chlamydia and herpes infections, and HIV infections. Chlamydial genital infections are the most common nationally notifiable infectious diseases in the United States. Nationally notifiable diseases are those diseases that must be reported to the CDC.

Most contagious airborne diseases are due to respiratory pathogens carried in droplets of respiratory secretions to susceptible people. Some respiratory pathogens may dry as they settle on dust particles and be carried long distances through the air

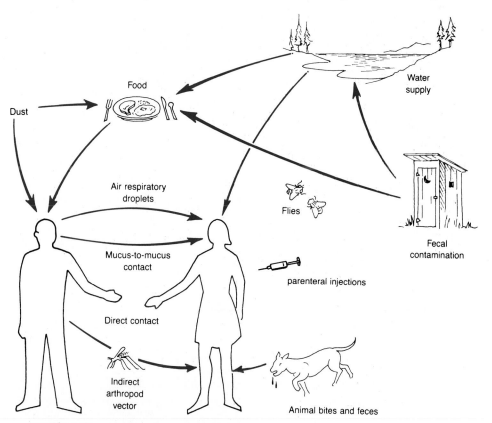

FIGURE 9-13. Modes of disease transmission.

and into a building's ventilation or air conditioning system (as observed in the Legionnaires' disease epidemic). Improperly cleaned inhalation therapy equipment can easily transfer these pathogens from one patient to another. Diseases that may be transmitted in this manner include colds, influenza, measles, mumps, chickenpox, smallpox, and pneumonia. Psittacosis or parrot fever is a respiratory infection that may be acquired from infected birds, usually parakeets and parrots. Also, some fungal respiratory diseases (*e.g.,* histoplasmosis) are frequently transferred via dried bird feces.

Indirect transfer of organisms to a susceptible person frequently occurs through contamination of foods and water. Human and animal fecal matter from outhouses, cesspools, and feed lots often is carried into water supplies. Improper disposal of sewage and inadequate treatment of drinking water contribute to the spread of fecal and soil pathogens. Food and milk may be contaminated by careless handling, which allows pathogens to enter from dust particles, dirty hands, hair, and respiratory secretions. If these pathogens and bacterial spores are not destroyed by proper processing and cooking, food poisoning can develop. Diseases frequently transmitted via foods and water are botulism, staphylococcal food poisoning, diarrhea caused by *Salmonella* and *Shigella* species, typhoid fever (caused by *Salmonella typhi*), infectious hepatitis (caused by Hepatitis A virus), amebiasis (caused by the ameba, *Entamoeba*

TABLE 9-2. Common Routes of Transmission

Route of Exit	Route of Transmission or Entry	Disease
Skin	Skin discharge → air → respiratory tract	Chickenpox, colds, influenza, measles, staph and strep infections
	Skin to skin	Impetigo, eczema, boils, warts, syphilis
Respiratory	Aerosol droplet inhalation Nose or mouth → hand or object → nose	Colds, influenza, pneumonia, mumps, measles, chickenpox, tuberculosis
Gastrointestinal	Feces → hand → mouth Stool → soil → food or water → mouth	Gastroenteritis, hepatitis, salmonellosis, shigellosis, typhoid fever, cholera, giardiasis, amebiasis
Salivary	Direct salivary transfer	Herpes cold sore, infectious mononucleosis, strep throat
Genital secretions	Urethral or cervical secretions Semen	Gonorrhea, herpes, *Chlamydia* infection Cytomegalovirus infection, AIDS, syphilis, warts
Blood	Tranfusion or needle stick injury	Hepatitis B; cytomegalovirus infection; malaria, AIDS
	Insect bite	Malaria, relapsing fever
Zoonotic	Animal bite	Rabies
	Contact with animal carcasses	Tularemia, anthrax
	Arthropod	Rocky Mountain spotted fever; Lyme disease, typhus, viral encephalitis, yellow fever, malaria, plague

INSIGHT
Pathogens in Our Kitchens, Part 2

Many of the foods that we bring into and work with in our kitchens are contaminated with pathogens. For example, gastrointestinal pathogens such as *E. coli* 0157:H7, *Salmonella*, and *Campylobacter* are often present on poultry, ground beef, and other meat products. *Salmonella* and *Campylobacter* may also be present within and on the surface of eggs. Protozoan parasites also gain access to our kitchens via contaminated foods. *Toxoplasma gondii* cysts may be present in meat, especially pork or mutton, and *Cyclospora* outbreaks in 1996 and 1997 were associated with imported raspberries.

Assuming that the meat and poultry that you serve to your family are properly and thoroughly cooked, the pathogens that are present on the surface of, or within, these foods are usually killed. They are not the problem. The real problem concerns the handling of these foods *prior* to cooking them. As we handle and prepare foods in the kitchen, pathogens from the foods get on our hands, counter tops, plates and cutting boards, knives, and almost anything else that we touch in the kitchen.

Here's a typical scenario: You place a package of chicken breasts on a plate and then unwrap it. You then place the chicken breasts, one at a time, on a cutting board and trim away the excess fat with a knife. By the time the chicken breasts are placed into the oven, the plate, the cutting board, the knife, your hands, and anything that you have touched with your hands have become contaminated. Now you take a head of lettuce from the refrigerator, place it on the cutting board, and proceed to chop it up for use in a salad. It is quite possible that any pathogens that

were present on your hands, the knife, or the cutting board are now in the salad. It's not likely that you'll be cooking the salad, so later when you eat it, you and your family will be ingesting live pathogens.

It is very important that, as you prepare foods in the kitchen, you remain aware of the presence of pathogens, and take steps to eliminate contamination of yourself, your kitchen, and other foods with those pathogens. Wash your hands often, using hot water and antibacterial soap. Don't merely rinse them. In the kitchen, just as in the hospital setting, frequent and thorough handwashing is the most important way to prevent the transmission of pathogens.

Using hot water, thoroughly rinse poultry and meat blood from plates, and then place them in the dishwasher. Don't use them for anything else until after they've been washed. Always wash knives before reusing them. After working with poultry and meat, be sure to thoroughly wash counter tops and cutting boards with hot, soapy water. Because bacteria can get between the slats in wooden cutting boards, consider replacing them with smooth plastic cutting boards. Use an antibacterial kitchen spray to clean counter tops, refrigerator and oven handles, and anything else that you touched as you prepared the food. Remember to follow the manufacturer's directions regarding the length of time to leave the disinfectant in place before wiping it off. Since wet sponges and dishcloths are havens for pathogens, be sure to wash or replace them often. People should place their sponges or dishcloths in the dishwasher each time they wash dishes.

histolytica), giardiasis (diarrhea caused by the protozoan, *Giardia lamblia*), and trichinosis (caused by *Trichinella spiralis* worms in pork). The Insight Boxes contain additional information about pathogens in our kitchens.

Blood is normally sterile; thus, the discovery of organisms in the blood may indicate the presence of an infection. A *vector* is often necessary to carry pathogens from the blood of one person to another. Common vectors are arthropods which bite an infected person or animal and then transfer the pathogens to a healthy individual. Included among the arthropod vectors are insects such as mosquitoes, flies, fleas, and lice and arachnids such as mites and ticks.

To better understand this mode of transmission, consider the tick. (If you own a dog, you probably already have an appropriate concern about ticks.) *Rickettsia rickettsii,* which causes Rocky Mountain spotted fever (see Chapter 12), is widespread throughout the animal population and their ticks. If an infected tick bites a person, the pathogen may be injected, causing an infection. Ticks may also carry the pathogens of Q fever, typhus, Lyme disease, babesiosis, tularemia, and the more recently described erhlichial diseases, human granulocytic ehrlichiosis (HGE) and human monocytic ehrlichiosis (HME). In similar fashion, body lice and head lice are the vectors of epidemic typhus, trench fever, and relapsing fever. Blood-sucking fleas carry the pathogens of plague and endemic typhus. Deer flies transmit tularemia, tsetse flies carry African sleeping sickness, and certain species of mosquitoes transmit malaria, viral encephalitis, and yellow fever. Appendix C contains additional information about arthropods and arthropod-borne diseases.

Many pets and other animals are important reservoirs of zoonoses. Dogs, cats, bats, skunks, and other animals are known vectors of rabies, transmitting the rabies

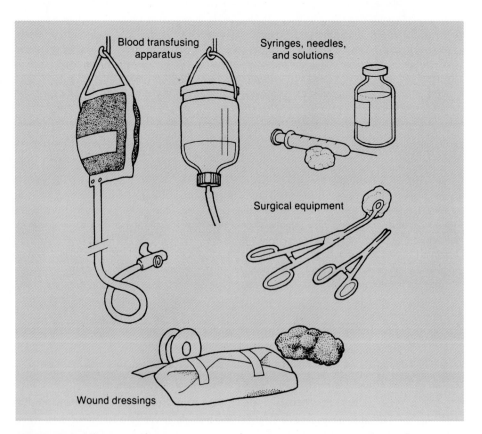

FIGURE 9-14. Various medical instruments and pieces of apparatus that may serve as inanimate vectors of infection (fomites).

virus to a human via the saliva that is injected when they bite. Salmonellosis is frequently acquired from the feces of turtles, other reptiles, and poultry. Cat and dog bites easily transfer *Pasteurella, Staphylococcus,* and *Streptococcus* species into tissues where severe infections may result. Toxoplasmosis, a protozoan disease, may cause severe brain damage to the fetus when contracted by a woman during the first 3 months of pregnancy. This pathogen can be contracted from cat feces in cat litter boxes as well as from infected raw or undercooked meats.

The instruments and devices handled by medical personnel are common inanimate vectors of blood and other internal infections. These most often include nonsterile syringes, needles, and solutions, as well as blood-processing equipment such as kidney dialyzers and blood-transfusing apparatus (Fig. 9-14). One reason why disposable sterile tubes, syringes, and various other types of single-use hospital equipment have become very popular is that they are effective in preventing blood infections that result from re-use of equipment. Any blood-borne disease can be transferred by improperly sterilized instruments and equipment. Hepatitis, syphilis, malaria, AIDS, and systemic staphylococcal infections are the diseases most often transmitted in this manner. Individuals using illegal intravenous drugs commonly transmit these diseases to each other by sharing needles and syringes, which easily become contaminated with the blood of an infected person.

CONTROL OF EPIDEMIC DISEASES

The WHO and the U.S. Public Health Service constantly strive to prevent epidemics and to identify and eliminate any that do occur. One way in which health personnel and community workers participate in this massive program is by reporting cases of communicable diseases to the proper agencies. They also can help by educating the public by describing how diseases are transmitted and explaining proper sanitation procedures, by identifying and attempting to eliminate reservoirs of infection, by carrying out measures to isolate diseased persons, by participating in immunization programs, and by helping to treat sick persons. Through measures like these, diphtheria, poliomyelitis, and smallpox have been totally or nearly eliminated in most parts of the world. Everyone in our society should contribute in whatever way possible to eliminate infectious diseases from the human environment. (Additional information about the WHO and the CDC is contained in Chapter 10.)

■ REVIEW OF KEY POINTS

- When individuals are exposed to pathogens, these microorganisms may or may not cause disease, depending on a number of factors, including the natural resistance of the individual and the virulence, infectivity, invasiveness, and toxicity of the pathogen.
- An infectious disease starts with the growth of a pathogen after it lands on and attaches to a suitable organ or tissue. The severity of the ensuing infectious disease

and the amount of damage it causes depends on the host's ability to resist invasion by the pathogen and to neutralize the damaging enzymes and toxins produced by the pathogen.

■ The inflammatory response (inflammation) is the body's normal response to the infectious process, with symptoms of redness, heat, swelling and pain, and sometimes accompanied by loss of movement at the site. Pyogenic pathogens are those that produce pus at the infected site.

■ When the body loses the battle with the pathogen, clinical disease results, accompanied by characteristic signs and symptoms. An infection may be acute, subacute, or chronic; localized or systemic; and symptomatic or asymptomatic. As the disease progresses it may change from one stage to another.

■ A primary infection may set the stage for a secondary infection caused by another pathogen. The four phases of an infectious disease are the incubation period, prodromal period, period of illness, and convalescent period. Sometimes the period of illness is followed by disability or death.

■ Pathogenicity is the ability of a microbe to cause disease, usually by infecting the host, protecting itself from the body's defenses, and by invading, multiplying and damaging host tissues. Virulence is a measure or degree of pathogenicity. Different species or even different strains of the same species vary in their ability to cause disease; thus, some are more virulent than others. Virulence depends on infectivity, invasiveness, and toxigenicity of a particular strain of pathogen.

■ Virulence factors are the phenotypic characteristics of a microorganism that enable it to be pathogenic (*i.e.,* that explain how it causes disease). Some virulence factors are structural features (*e.g.,* capsules, flagella, pili) that enable pathogens to avoid phagocytosis and reach and attach to various tissues within the host. Some virulence factors are exoenzymes (*e.g.,* coagulase, kinases, hyaluronidase, collagenase, hemolysins, and lecithinase), which enable pathogens to evade host defenses and invade and cause damage to body tissues. Endotoxins and exotoxins are virulence factors associated with such disease manifestations as tissue destruction, paralysis, diarrhea, fever, and septic shock.

■ Epidemiology is the study of the frequency and distribution of diseases and contributing factors (*e.g.,* virulence of pathogens; susceptibility of a population because of overcrowding, lack of immunization, or inadequate sanitation; reservoirs of infection; and various modes of transmission). Epidemic, endemic, pandemic, sporadic, and nonendemic diseases are epidemiological terms used to describe the prevalence of a disease in an area at a particular time.

■ The sources of pathogens are known as reservoirs of infection; they may be living hosts (*e.g.,* humans, arthropods, or other animals) or nonliving (*e.g.,* air, soil, dust, food, water, or inanimate objects found in the home, office, or hospital). The principal modes of transmission of pathogens are airborne, vehicular, contact, and vector transfer.

■ To eradicate certain diseases and prevent epidemics, epidemiologists must consider the virulence of the pathogens, susceptibility of the population, sanitation practices, reservoirs of infection, and ways in which pathogens are transmitted.

Problems and Questions

1. Define infectious, communicable, and contagious disease. Why are some diseases more contagious than others?
2. List several ways that the human body resists infection by foreign invaders.
3. Define and illustrate the following terms used to describe infectious diseases: acute, chronic, latent, primary, secondary, local, systemic.
4. Name four phases in the course of an infectious disease.
5. On what general properties do virulence and pathogenicity depend?
6. Compare exotoxins with endotoxins. List some diseases produced by each.
7. Name some general types of exotoxins.
8. Define and give examples of epidemic, endemic, pandemic, sporadic, and nonendemic diseases.
9. In what way do carriers influence epidemics?
10. Discuss how reservoirs of infection could be reduced or destroyed.
11. How could each of the four major modes of transmission be interrupted to prevent diseases transmitted in that manner?
12. If you were called to a distant island to stop an epidemic, what measures would you take?

Self Test

After you have read Chapter 9, reviewed the chapter outline, examined the objectives, studied the new terms, and answered the problems and questions above, complete the following self test.

MATCHING EXERCISES

Complete each statement from the list of words provided with each section.

communicable	infectious	virulence
contagious	inflammation	

1. The ability of a pathogen to produce disease may depend on the pathogen's _____.
2. The body tissues respond to damage or infections by a process called _____.
3. A disease that results from dietary deficiencies, such as rickets, is not a/an _____ disease.
4. A disease resulting from the presence, growth, and destructive effects of microorganisms is a/an _____ disease.
5. Diseases that are transmissible from person to person are called _____ diseases.
6. Diseases that are very easily transmitted from person to person, such as colds, are called _____ diseases.
7. Localized symptoms including heat, redness, swelling, and pain are indications of a/an _____.

Types of Disease States

local	chronic	carrier
generalized	primary	systemic
latent	secondary	
acute	asymptomatic	

1. When an infection is confined to a single area it is called a/an _____ disease.
2. During the course of a disease, the pathogen may be walled off or hidden in certain organs; then the infection is referred to as a/an _____ disease.
3. If the pathogen is present and detectable but there are no symptoms of the disease, it is called a/an _____ disease.
4. A disease that has a rapid onset and a rapid recovery period is said to be a/an _____ disease.
5. If the disease is characterized by slow onset and long duration, it is said to be a/an _____ disease.
6. A person who has recovered from a disease but who still harbors the pathogens and may transmit them to others is known as a _____.
7. When a pathogen has invaded the bloodstream and caused an infection throughout the body, the disease is referred to as a _____ or _____ disease.
8. An initial or _____ infection, such as influenza, may leave the individual in a weakened condition so that a _____ infection may then occur because the body defenses are at a low level.

Disease Causation

capsule	hyaluronidase	endotoxin
coagulase	collagenase	hemolysins
streptokinase	exotoxins	
fibrinolysin	leukocidin	

1. Very toxic proteins that are secreted by living pathogens and cause diseases such as botulism, tetanus, and diphtheria are called _____.
2. The protective coating outside the cell wall of an organism, which may protect it from phagocytosis, is a _____.
3. An enzyme secreted by a pathogen that enables it to spread through connective tissue by breaking down hyaluronic acid is called _____.
4. Some gram-negative bacteria cause disease because their cell walls are very toxic to humans. This toxic material, which is part of the bacterial cell, is referred to as _____.
5. An enzyme secreted by some bacteria that clots plasma around the bacterial cell is _____.
6. Streptococci can break down a fibrin blood clot by secreting _____, which could also be called a _____.
7. Red blood cells (erythrocytes) are lysed by _____.

8. Some streptococci and staphylococci are able to destroy white blood cells (leukocytes) by secreting _____.
9. An enzyme that destroys collagen in tendons and cartilage is called _____.

Epidemiology

epidemic	pandemic	nonendemic
endemic	sporadic	

1. An epidemic that spreads throughout the world is then called a/an _____.
2. When a greater than usual number of cases of a disease occur in a specific area during a certain time frame, the disease has reached _____ proportions.
3. Those diseases, such as smallpox, that do not occur in most countries are called _____ diseases.
4. Certain diseases, such as botulism, that occur only occasionally are referred to as _____ diseases.
5. Some diseases are usually present in an area, but the numbers of cases vary with the season and month of the year. These diseases are said to be _____.
6. In most areas, colds and influenza are considered _____ diseases.
7. In the United States, diphtheria has been controlled as a result of immunization programs; thus, it is said to be a/an _____ disease.
8. In recent years, gonorrhea has become _____ in the United States and _____ throughout the world.

Reservoirs of Infection

reservoirs	vectors	respiratory
carriers	arthropod	secretions
STDs	fomites	fecal material

1. Arthropods such as mosquitoes, ticks, fleas, mites, and lice that transfer pathogens by their bites are considered to be _____.
2. Pathogens may be transferred to a susceptible patient via bed clothes, urinals, towels, or sheets of another sick patient. These items are referred to as _____.
3. Food, milk, and water supplies are easily contaminated with human _____ containing enteric pathogens.
4. Contaminated syringes, needles, and other hospital equipment frequently serve as inanimate _____ of disease agents.
5. The most contagious diseases are usually transferred by droplets of _____ in the air and by dust.
6. Persons who transmit pathogens to other susceptible people but who have no symptoms of the disease themselves are called _____.
7. Examples of pathogens that must be transmitted by mucous membrane-to-mucous membrane contact are those that cause _____, such as syphilis and gonorrhea.

8. Rocky Mountain spotted fever and tularemia are transmitted from wildlife to humans by the bite of infected ticks, which are a type of _____.

9. Contaminated food and water are considered to be _____ of infection.

TRUE OR FALSE (T OR F)

___ 1. Infectious diseases are those caused by the growth of pathogenic microorganisms.

___ 2. The virulence of a pathogen depends on its ability to infect, invade, and damage the host.

___ 3. A communicable disease is one that can be transmitted from one person to another.

___ 4. When a pathogen lands on the skin, it immediately starts to grow and cause illness.

___ 5. A carrier is a person who harbors a pathogen and may transmit it to others but does not have symptoms of the disease caused by that pathogen.

___ 6. An acute disease, such as gonorrhea, may become a chronic one.

___ 7. In the latent stage, herpes cold sores are not apparent.

___ 8. Gonorrhea is never asymptomatic.

___ 9. Pneumonia and ear infections are usually primary diseases.

___ 10. Streptokinase helps staphylococci break down blood clots.

___ 11. The actual number of cases of an endemic disease depends on many factors, including the environment.

___ 12. Hospital staphylococcal epidemics are often caused by the hospital personnel.

___ 13. Gonorrhea has never been epidemic in the United States.

___ 14. Living hosts or inanimate objects may be reservoirs of infection.

___ 15. Humans are the most important reservoir of infection for human diseases.

___ 16. All arthropod vectors are insects.

___ 17. Body lice may be vectors of epidemic typhus, trachoma, and impetigo.

___ 18. Improperly sterilized syringes, needles, and blood transfusion and dialysis equipment can be the sources of blood infections such as systemic staphylococcus infections and hepatitis.

___ 19. An endemic disease can become an epidemic.

___ 20. *Legionella pneumophila* has been and continues to be the cause of epidemics.

___ 21. Toxoplasmosis is an example of a zoonosis.

MULTIPLE CHOICE

1. Which of the following best describes gonorrhea and syphilis?
   a. sporadic diseases in the U.S.A.
   b. nonendemic diseases in the U.S.A.
   c. communicable diseases
   d. contagious diseases

2. Which of the following virulence factors enables bacteria to attach to tissues?
   a. neurotoxins
   b. endotoxin
   c. capsules
   d. flagella
   e. pili

3. Which of the following pathogens produce neurotoxins?
   a. *Staphylococcus aureus* and *Streptococcus pyogenes*
   b. *Clostridium difficile* and *Clostridium perfringens*
   c. *Clostridium botulinum* and *Clostridium tetani*
   d. *Pseudomonas aeruginosa* and *Mycobacterium tuberculosis*

4. Which of the following pathogens produce enterotoxins?
   a. *Staphylococcus aureus* and *Streptococcus pyogenes*
   b. *Clostridium difficile* and *Clostridium perfringens*
   c. *Clostridium botulinum* and *Clostridium tetani*
   d. *Pseudomonas aeruginosa* and *Mycobacterium tuberculosis*

5. Which of the following pathogens could release endotoxin?
   a. *Staphylococcus aureus* and *Streptococcus pyogenes*
   b. *Clostridium difficile* and *Clostridium perfringens*
   c. *Neisseria gonorrhoeae* and *Escherichia coli*
   d. *Staphylococcus aureus* and *Mycobacterium tuberculosis*

6. Which of the following are considered reservoirs of infection?
   a. contaminated food
   b. contaminated drinking water
   c. carriers
   d. rabid animals
   e. all of the above

7. Infectious diseases are usually most contagious during the
   a. incubation period
   b. prodromal period
   c. period of illness
   d. period of convalescence

8. The usual mode of transmission of Lyme disease, Rocky Mountain spotted fever, and ehrlichiosis is
   a. contaminated drinking water
   b. aerosols produced by coughing patients
   c. mosquito bites
   d. flea bites
   e. tick bites

9. An example of a sporadic disease in the U.S. is
   a. AIDS
   b. tuberculosis
   c. gonorrhea
   d. plague
   e. all of the above

10. Enterotoxins affect cells in the
    a. respiratory tract
    b. genitourinary tract
    c. gastrointestinal tract
    d. central nervous system
    e. cardiovascular system

CHAPTER 10

Preventing the Spread of Communicable Diseases

PREVENTION OF HOSPITAL-ACQUIRED INFECTIONS
How Hospital-Acquired Infections Develop
General Control Measures
Prevention of Airborne Contamination
Handling Food and Eating Utensils
Handling of Fomites
Handwashing
Infection Control Procedures
Medical and Surgical Asepsis
Standard Precautions
Transmission-Based Precautions
Reverse Isolation
Hospital Infection Control
Medical Waste Disposal

SPECIMEN COLLECTING, PROCESSING, AND TESTING
Role of Healthcare Professionals
Proper Collection of Specimens
Types of Specimens Usually Required
Blood
Urine
Cerebrospinal Fluid
Sputum
Mucous Membrane Swabs
Feces
Shipping Specimens
The Pathology Department ("the Lab")

Identification and Antimicrobial Susceptibility Testing of Pathogens
Types of Tests Used for Identification
Molecular Diagnostic Procedures
Antimicrobial Susceptibility Testing
Quality Control in the Laboratory
ENVIRONMENTAL DISEASE CONTROL MEASURES
Public Health
Water Supplies and Sewage Disposal
Sources of Water Contamination
Water and Sewage Treatment

OBJECTIVES

After studying this chapter, you should be able to:

- List six factors that have contributed to an increase in hospital-acquired (nosocomial) infections
- List areas in the hospital where nosocomial infections are most probable
- List several types of patients who are extremely vulnerable to infectious diseases
- Write a brief description of reverse isolation and source isolation
- Briefly describe important procedures to follow in standard precautions

- Discuss the role of healthcare professionals in the collection of clinical specimens
- List types of specimens that usually must be collected from patients
- Discuss general precautions that must be observed during collection and handling of specimens
- Describe proper procedures for obtaining specimens
- Describe the organization of the Pathology Department and the Clinical Microbiology Laboratory

- *List ten phenotypic characteristics of value when attempting to identify bacterial pathogens*
- *Discuss the importance of quality control in a microbiology laboratory*
- *List sources of water contamination*

- *Describe how water and sewage are usually treated*
- *Discuss how epidemics are controlled and prevented*

NEW TERMS

Airborne precautions
Bacteremia
Bacteriuria
Community-acquired
  infection
Contact precautions
Droplet precautions

Hospital-acquired infection
Iatrogenic infection
Medical asepsis
Medical aseptic techniques
Nosocomial infection
Pathologist
Pathology

Reverse isolation
Septicemia
Standard precautions
Surgical asepsis
Surgical aseptic techniques
Transmission-based
  precautions

PREVENTION OF HOSPITAL-ACQUIRED INFECTIONS

How Hospital-Acquired Infections Develop

The importance of microbiology to those who work in health-related occupations can never be overemphasized. Whether working in a hospital, nursing home, medical or dental clinic, or caring for sick persons in their homes, all healthcare professionals must follow the same procedures to prevent the spread of communicable diseases.

Thoughtless or careless actions when giving patient care can cause preventable infections. Infections associated with hospitalization are of two types: *community-acquired infections* and *hospital-acquired infections;* the latter are also called *nosocomial infections.* According to the Centers for Disease Control and Prevention (CDC), community-acquired infections are those present or incubating at the time of hospital admission. All other hospital infections are considered nosocomial, including those that erupt within 14 days of hospital discharge. *Iatrogenic* (literally meaning "physician-induced") *infections* or diseases are the result of medical or surgical treatment and are, thus, caused by surgeons, other physicians, or other healthcare personnel.

A nosocomial infection often adds several weeks to the patient's hospital stay and may cause serious complications and even death. Cross-infections transmitted by hospital personnel, including physicians, are all too common; this is particularly true when hospitals and clinics are overcrowded and the staff is overworked. However, these infections *can* be avoided through proper care and the disciplined use of aseptic techniques and precautions.

Nurses and physicians play major roles in preventing nosocomial infections; nevertheless, respiratory therapists, physical therapists, laboratory personnel, phlebotomists, occupational therapists, radiologic technologists, dental hygienists, dentists, and all others who deal with patients must be equally knowledgeable about methods of preventing the spread of pathogens. Members of the hospital housekeeping staff and central supply department; those who prepare, dispense, and dispose of food; administrative personnel; and those who dispose of medical wastes can help to prevent cross-contamination among patients or, through their carelessness, may contribute to the spread of diseases. Of course, sick and debilitated (weakened) hospitalized patients are much more susceptible to even minor opportunistic pathogens than are the healthy people who are caring for them.

In the United States, the number of nosocomial infections has risen sharply in the past 20 years to approximately 2 million cases per year despite the availability of new disinfectants and antibiotics. There are many reasons for this situation:

- Indiscriminate use of broad-spectrum antibiotics, which may allow opportunistic pathogens to mutate and become resistant to antibiotics.
- A false sense of security about the effectiveness of antibiotics, with a corresponding neglect of aseptic techniques and precautions.
- Lengthy, more complicated types of surgery.
- Overcrowding of hospitals and shortages of staff.
- Increased numbers and types of hospital workers who are often not aware of the importance of routine infection control and aseptic and sterile techniques.
- Increased use of anti-inflammatory and immunosuppressant agents, such as radiation, steroids, anticancer chemotherapy, and antilymphocytic serum.
- Use of indwelling medical devices.

The presence of even one of these factors is sufficient to enable opportunistic pathogens to rush in and create problems; taken together, they are the cause of approximately 60% of hospital-acquired infections. The seven most common pathogens involved in nosocomial infections (in no particular order) are the Gram-positive bacteria, *Staphylococcus aureus,* coagulase-negative staphylococci, and *Enterococcus* spp.; and the Gram-negative bacteria, *Escherichia coli, Pseudomonas aeruginosa, Enterobacter* spp., and *Klebsiella* spp. In the United States, during the period 1990 to 1996, the three Gram-positive pathogens caused 34% of nosocomial infections, and the four Gram-negative pathogens caused 32%. Collectively, these seven pathogens caused 66% of all nosocomial infections (see Insight Box).

Medical devices that support or monitor basic body functions contribute greatly to the success of modern medical treatment. However, by bypassing normal defensive barriers, these devices provide microorganisms access to normally sterile body fluids and tissues. The risk of bacterial or fungal infection is related to the degree of debilitation of the patient and the design and management of the device.

INSIGHT
Nosocomial Infections

A nosocomial infection is an infection that a person acquires while they are hospitalized. The person was not infected at the time they entered the hospital, but became infected during his or her hospital stay. Of the approximately 36 to 38 million patients who are hospitalized in the United States per year, an estimated 2 million (about 5%) acquire nosocomial infections. In 1995, approximately 88,000 deaths were related to nosocomial infections, and nosocomial infections added an estimated $4.5 billion to the cost of health care in the United States that year.

The hospital setting harbors many pathogens and potential pathogens. They live on and in healthcare professionals, other hospital employees, and patients themselves. Some live in dust, whereas others live in wet or moist areas like sink drains, shower heads, whirlpool baths, mop buckets, flower pots, and even food from the kitchen.

The seven most common causes of nosocomial infections are *Staphylococcus aureus,* coagulase-negative staphylococci, *Enterococcus* spp., *Escherichia coli, Pseudomonas aeruginosa, Enterobacter* spp., and *Klebsiella* spp., which collectively cause approximately two-thirds of all nosocomial infections in the United States. Although some of these pathogens come from the external environment, most come from the patients themselves—their own indigenous microflora that enter a surgical incision or otherwise gain entrance to the body. Urinary catheters, for example, provide a "superhighway" for organisms to gain access to the urinary bladder. In fact, urinary tract infections (UTIs) are the most common type of nosocomial infections, followed by surgical wound infections, lower respiratory tract infections, and bacteremia.

About 70% of nosocomial infections involve drug-resistant bacteria, which are common in the hospital environment as a result of all the antimicrobial agents that are used there. The drugs have put selective pressure on the microbes, ensuring that only those that are resistant to the drugs will survive. *Pseudomonas* infections are especially hard to treat, as are infections caused by vancomycin-resistant *Enterococcus* species (VRE), and methicillin-resistant strains of *Staphylococcus aureus* (MRSA) and *Staphylococcus epidermidis* (MRSE).

Healthcare professionals and other hospital employees must be aware of the problem of nosocomial infections and must take appropriate measures to minimize the number of such infections that occur within the hospital. Hand-washing remains the most effective means of control; hands must be washed before and after working with every patient. Other means of reducing the incidence of nosocomial infections include sterilization techniques, air filtration, use of ultraviolet lights, isolating especially infectious patients, and wearing gloves, masks, and gowns whenever appropriate.

The most common nosocomial infections are urinary tract infections associated with the use of urinary catheters. Thus, it is advisable to discontinue the use of urinary catheters, vascular catheters, respirators, and hemodialysis on individual patients as soon as medically feasible.

The emergency room, operating room, delivery room, nursery, and the central supply area are the most critical areas for disease transmission. In the emergency room, many patients who are carrying unknown pathogens are rushed in and treated quickly to halt a life-threatening situation; in haste, the emergency room attendants may neglect to protect themselves and others by failing to use recommended precautions. In operating and delivery rooms, particular care must be taken because portals of entry into the body are readily accessible to pathogens. The nursery is critical because newborns have very little resistance to disease. The central supply department must take great care not to distribute contaminated materials or supplies, as these might expose patients to a myriad of pathogens.

The most vulnerable patients in a hospital are:

- premature infants and newborns
- women in labor and delivery
- surgical and burn patients
- diabetic and cancer patients
- patients receiving treatment with steroids, anticancer drugs, antilymphocyte serum, and radiation
- patients with a deficient immune response; *e.g.,* patients with AIDS
- patients who are paralyzed or are undergoing renal dialysis or catheterization; these patients' normal defense mechanisms are not working properly.

The greatest risk to healthcare professionals who handle blood and body fluids is the transmission of hepatitis B virus (HBV) and/or the human immunodeficiency virus (HIV). Control of these viruses requires meticulous attention to procedures, such as the use of gloves and gowns, that prevent direct contact with blood and other body substances.

Nosocomial viral infections are frequently ignored because most hospitals lack adequate viral diagnostic laboratories. Influenza virus, respiratory syncytial virus (RSV), and other respiratory viruses have been shown to spread in hospitals via direct contact or droplet inhalation. Other highly infectious viruses that cause measles and chickenpox may cause outbreaks among susceptible patients and hospital staff. Nosocomial zoonoses are a recently recognized problem in hospitals (see Insight Box).

General Control Measures

A general state of cleanliness must be maintained throughout the hospital or health care institution using general principles of sanitation, disinfection, and sterilization. Each area, from the trash disposal area to the operating rooms, must be thoroughly cleaned and disinfected to prevent the growth and spread of microorganisms.

Hospitals vary greatly in their specific requirements for the maintenance of medical asepsis (the exclusion of pathogens; to be discussed later), but there is general agreement about sanitary methods for handling food and eating utensils, proper cooking and storage of food, proper disposal of waste products and contaminated materials, proper handwashing and personal hygiene of hospital personnel, proper use of gowns and masks in isolation rooms, proper washing and sterilizing of hospital equipment, proper use of disposable equipment, and proper disinfection of a room after a patient has been discharged.

INSIGHT
Nosocomial Zoonoses

Literally hundreds of diseases are transmissible from animals to humans; these are collectively known as zoonoses or zoonotic diseases (see Table 9–1). Transmission from animals to humans occurs by many routes, including direct contact, scratch, bite, inhalation, contact with urine or feces, and ingestion. Fortunately, most zoonoses have no association with hospitals or hospitalized patients. There have been reports, however, of nosocomial zoonoses—nosocomial infections transmitted directly or indirectly from live animals. How does this happen?

Wild rodents such as rats and mice might enter hospitals in which they can transmit diseases such as leptospirosis, rat-bite fever, and rickettsial pox. Droppings from wild birds can enter air vents and air-conditioning systems, causing diseases such as histoplasmosis and cryptococcosis. Pathogens from laboratory animals could enter air ventilating systems and be conveyed to patient care facilities. Pet therapy is becoming increasingly popular in nursing homes, in which pets provide companionship for the residents. Such pets can be the source of pathogens. If animal tissues and organs are used for transplantation, there is always the danger of undetected microorganisms and other infectious agents (*e.g.,* viruses and prions) being present in the transplanted material.

Healthcare professionals can also be the source of zoonotic pathogens, especially those who have domestic pets or farm animals where they live. Should these workers wear their uni-

forms while playing with or caring for their animals, the uniforms could be contaminated with zoonotic pathogens. Or, if they should fail to wash their hands after touching their animals, pathogens could be carried from home to hospital on their hands.

Pathogens that could potentially cause nosocomial zoonoses include cutaneous fungi (such as *Microsporum canis* and others that cause ringworm infections); bacteria such as staphylococci, streptococci, *Pseudomonas*, *Salmonella*, and *Campylobacter*; yeasts; and parasites such as *Cryptosporidium*. Several outbreaks in intensive care nurseries have resulted from *Malassezia furfur* and *M. pachydermatis,* fungi that were transmitted from pets to low-birth-weight neonates via the hands and clothing of healthcare professionals.

Prevention of nosocomial zoonoses involves keeping medical facilities free of rodents, preventing birds from nesting near air-conditioners and air vents, and ensuring that laboratory animal ventilation systems are not linked to ventilation systems in patient care areas. In addition, laboratory coats worn in animal facilities should never be worn in patient care areas, and hospital uniforms should not be worn to and from work. Animals used in pet therapy should be vaccinated and in good health. And the most important way to prevent nosocomial infections of any kind is frequent and proper handwashing. Healthcare professionals must always wash their hands after handling animals and prior to providing patient care.

PREVENTION OF AIRBORNE CONTAMINATION

Respiratory infections are most often transmitted through the air. The following measures should be taken to decrease the number of pathogens transmitted by this means:

- Cover the mouth and nose when coughing or sneezing.
- Limit the number of persons in a room.
- Remove the dirt and dust from the floor and furniture by dusting with a damp cloth.
- Open the room to fresh air and sunlight whenever possible.

- Roll linens together carefully to prevent dispersal of microbes in the air.
- Remove bacteria from the air with a filtered air-conditioning system.

HANDLING FOOD AND EATING UTENSILS

Contaminated food provides an excellent environment for the growth of pathogens. Most often, human carelessness—such as failing to take precautions when handling fecal material, flies, insects, dust, dirt, and domestic animals and pets, as well as neglecting the practice of handwashing—is responsible for this contamination. The pathogens most frequently carried in food are *Staphylococcus* species (from skin and dust); *Clostridium botulinum* (from dust and dirt); *Clostridium perfringens* (from dust, dirt, and hands); *Salmonella, Shigella, Campylobacter* and *Proteus* species (from feces, hands, flies, and/or pets); and *Pseudomonas* species (from dirt, hands, and contaminated equipment).

Regulations for safe handling of food and eating utensils are not difficult to follow. They include:

- using high quality fresh food;
- properly refrigerating and storing food;
- properly washing, preparing, and cooking food;
- properly disposing of uneaten food;
- thoroughly washing hands and fingernails before handling food and after visiting a rest room;
- properly disposing of nasal and oral secretions in tissues and then thoroughly washing hands and fingernails;
- covering hair and wearing clean clothes and aprons;
- providing periodic health examinations for kitchen workers;
- prohibiting anyone with a respiratory or gastrointestinal disease from handling food or eating utensils;
- keeping all cutting boards and other surfaces scrupulously clean; and
- rinsing and then washing cooking and eating utensils in a dishwasher in which the water temperature is above 80°C

HANDLING OF FOMITES

As previously described, fomites are any articles or substances other than food that may harbor and transmit microbes. Examples of fomites include eating utensils, bedpans, urinals, thermometers, washbasins, bed linen, and clothing and other personal patient items. Transmission of pathogens by these items may be prevented by observing the following rules:

- Use disposable equipment and supplies wherever possible.
- Disinfect or sterilize equipment as soon as possible after use.
- Use individual equipment for each patient.
- Use electronic or glass thermometers fitted with one-time use, disposable

covers or disposable, single-use thermometers; electronic and glass thermometers must be cleaned or sterilized on a regular basis, following manufacturer's instructions.

- Empty bedpans and urinals, wash them in hot water, and store them in a clean cabinet between uses.
- Place bed linen and soiled clothing in bags to be sent to the laundry.

HANDWASHING

Handwashing is the single most important measure to reduce the risks of transmitting microorganisms from one person to another or from one site to another on the same patient. Washing hands as promptly and as thoroughly as possible between patient contacts and after contact with blood, body fluids, secretions, excretions, and equipment or articles contaminated by these body substances is an important component of infection control and isolation precautions. Hands should be washed with an antimicrobial or antiseptic agent during hospital epidemics with *Staphylococcus aureus* or other pathogens. Everyday, common-sense, handwashing guidelines that one should always follow include:

- Washing your hands before you . . .
  - Prepare or eat food
  - Treat a cut or wound or tend to someone who is sick
  - Insert or remove contact lenses
- Washing your hands after you . . .
  - Use the rest room
  - Handle uncooked foods, particularly raw meat, poultry or fish
  - Change a diaper
  - Cough, sneeze, or blow your nose
  - Touch a pet, particularly reptiles and exotic animals
  - Handle garbage
  - Tend to someone sick or injured
- Washing your hands in the following manner:
  - Use warm or hot, running water
  - Use soap (preferably an antibacterial soap)
  - Wash all surfaces thoroughly, including wrists, palms, back of hands, fingers and under fingernails (preferably with a nail brush)
  - Rub hands together for at least 10 to 15 seconds
  - When drying, begin with your forearms and work toward your hands and finger tips, and pat your skin rather than rubbing to avoid chapping and cracking

The above guidelines were originally published by the Bayer Corporation and the American Society for Microbiology. Additional guidelines regarding handwashing are presented in a following section entitled "Standard Precautions."

Infection Control Procedures

From the discoveries and observations of Semmelweis and Lister in the 19th century, we know that wound contamination is not inevitable and that we must prevent microorganisms from reaching susceptible areas, a concept referred to as *asepsis*. Asepsis, which literally means "without infection," includes any activities that prevent infection or break the chain of infection. There are two types of asepsis: medical asepsis and surgical asepsis. The techniques used to achieve asepsis depend on the site, circumstances, and environment.

MEDICAL AND SURGICAL ASEPSIS

Once basic cleanliness is achieved, it is not difficult to maintain asepsis. *Medical asepsis,* or clean technique, involves procedures and practices that reduce the number and transfer of pathogens. Medical asepsis includes all the precautionary measures necessary to prevent direct transfer of pathogens from person to person and indirect transfer of pathogens through the air or onto instruments, bedding, equipment, and other inanimate objects (fomites). *Medical aseptic techniques* include frequent and thorough handwashing; personal grooming; proper cleaning of supplies and equipment; disinfection; proper disposal of needles, contaminated materials and infectious waste; and sterilization.

Surgical asepsis, or sterile technique, includes practices used to render and keep objects and areas sterile (*i.e.,* free of microorganisms). Note the differences between medical and surgical asepsis: (1) medical asepsis is a clean technique, whereas surgical asepsis is a sterile technique; (2) the goal of medical asepsis is to exclude pathogens, whereas the goal of surgical asepsis is to exclude all microorganisms.

Surgical aseptic techniques are practiced in operating rooms, labor and delivery areas, certain areas of the hospital laboratory, and at the patient's bedside. Invasive procedures, such as drawing blood, injecting medications, urinary catheter insertion, cardiac catheterization, and lumbar punctures, for example, must be performed using strict surgical aseptic precautions. Other surgical aseptic techniques include scrubbing hands and fingernails before entering the operating room; using sterile gloves, masks, gowns, and shoe covers; using sterile solutions and dressings; using sterile drapes and creating a sterile field; and using heat-sterilized surgical instruments.

The surgical site of the patient's skin must be shaved and thoroughly cleansed and scrubbed with soap and antiseptic. If the surgery is to be extensive, the surrounding area is covered with a sterile plastic film or sterile cloth drapes so that a sterile surgical field is established. The surgeon and all surgical assistants must scrub for 10 minutes with a disinfectant soap and cover their clothes, mouth, and hair, as these areas might shed microorganisms onto the operative site. These coverings include sterile gloves, gowns, caps, masks, and shoe covers (Figs. 10-1, 10-2). All instruments, sutures, and dressings must be sterilized; as soon as they become contaminated, they

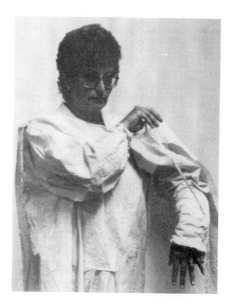

FIGURE 10-1. Nurse donning a sterile gown.

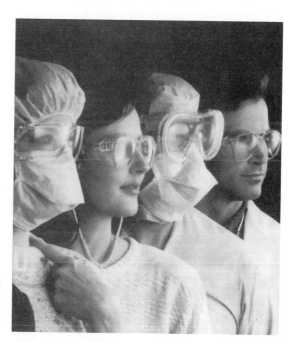

FIGURE 10-2. Various pieces of personal protective equipment, including masks, goggles, hair protection, and disposable gowns. (Photo courtesy of 3M Health Care.)

must be discarded and replaced with sterile ones. Any instruments or equipment that cannot be autoclaved must be properly disposed of. All needles, syringes, and other "sharps" must be placed in appropriate puncture-proof containers.

Floors, walls, and all equipment in the operating room must be thoroughly cleaned and disinfected before and after each use. Proper ventilation must be maintained to ensure that fresh, filtered air is circulated throughout the room at all times.

STANDARD PRECAUTIONS

In a hospital setting, one is not always aware of which patients are infected with HIV or hepatitis B virus (HBV) or other transmissible pathogens. Thus, *Standard Precautions* (as defined by the Centers for Disease Control and Prevention in 1996[1]) are used for the care of *all* hospitalized patients, regardless of their diagnosis or presumed infection status. Standard Precautions incorporate the major features of Universal Precautions (which were instituted in 1985 to reduce the risk of transmission of blood-borne pathogens) and Body Substance Isolation (instituted in 1987 to reduce the risk of transmission of pathogens from moist body substances). They are designed to reduce the risk of transmission of blood-borne and other pathogens in hospitals, and apply to blood; all body secretions and excretions except sweat, regardless of whether they contain visible blood; non-intact skin; and mucous membranes. Standard Precautions provide guidelines regarding handwashing; wearing of gloves, masks, eye protection, and gowns; cleaning of patient-care equipment, environmental control (including cleaning and disinfection); handling of soiled linens; handling and disposal of used needles and other "sharps"; resuscitation devices; and patient placement. Standard Precautions will protect healthcare professionals from becoming infected with HIV, HBV, and many other pathogens.

Handwashing. Hands must be washed after touching blood, body fluids, secretions, excretions, and contaminated items, regardless of whether gloves are worn. Hands must be washed immediately after gloves are removed, between patient contacts, and whenever necessary to avoid transfer of microorganisms to other patients or environments. In addition, it may be necessary to wash hands between tasks and procedures on the same patient to prevent cross contamination of different body sites. A plain (nonantimicrobial) soap may be used for routine handwashing, but an antimicrobial or antiseptic agent should be used in certain circumstances (*e.g.,* to control outbreaks within the hospital).

Gloves. Gloves must be worn when touching blood, body fluids, secretions, excretions, and contaminated items, and just before touching mucous membranes or non-intact skin. Gloves must be changed between tasks and procedures on the same patient whenever there is risk of transferring microorganisms from one body site to

---

[1]Information in this chapter on Standard and Transmission-Based Precautions is from *Guideline for Isolation Precautions in Hospitals.* Centers for Disease Control and Prevention, Atlanta, GA, 1996.

another. Always remove gloves promptly after use and before going to another patient. Thoroughly wash your hands immediately after removing gloves.

Masks, Eye Protection, Face Shields, Gowns. Always wear a mask and eye protection or a face shield during procedures and patient-care activities that are likely to generate splashes or sprays of blood, body fluids, secretions, or excretions. Always wear a gown during procedures and patient-care activities that are likely to generate splashes or sprays of blood, body fluids, secretions, or excretions, or cause soiling of clothing. Remove a soiled gown as quickly as possible and thoroughly wash your hands immediately after removing the gown.

Patient-Care Equipment. Handle soiled patient-care equipment in a manner that prevents contaminating yourself or your clothing and prevents transfer of microorganisms to other patients and areas. Ensure that reusable equipment is not used for the care of another patient until it has been appropriately cleaned, disinfected, or sterilized. Properly dispose of single-use items.

Environmental Control. The hospital must have and follow adequate procedures for the routine care, cleaning, and disinfection of environmental surfaces, beds, bedrails, bedside equipment, and other frequently touched surfaces.

Linens. Soiled linens must be handled, transported, and processed in a manner that prevents contaminating yourself or your clothing and prevents transfer of microorganisms to other patients and areas.

Occupational Health and Blood-Borne Pathogens. Needle-stick injuries and injuries resulting from broken glass and other "sharps" are the primary manner in which healthcare professionals become infected with such pathogens as HIV and hepatitis B virus. Thus, Standard Precautions includes guidelines regarding the safe handling of such items. Never recap used needles, unless a device is used that eliminates the danger of sticking yourself with the needle. Do not remove used needles from disposable syringes by hand and do not attempt to bend or break used needles. Place used disposable syringes, needles, scalpel blades, broken glass, and other "sharps" in appropriate puncture-resistant containers. Such containers should be located in areas where such items are likely to be used.

Patient Placement. Whenever possible, use private rooms for patients who contaminate the hospital environment or cannot assist in maintaining appropriate hygiene or environmental control.

TRANSMISSION-BASED PRECAUTIONS

The five main routes of transmission are airborne, droplet, contact, common vehicle, and vector-borne. Within a hospital, microorganisms are transmitted by three major routes: airborne, droplet, and contact. *Transmission-Based Precautions* are designed for patients known or suspected to be infected with highly transmissible or epidemiologically important pathogens for which additional precautions beyond Standard Precautions are required to interrupt transmission within hospitals. There are three types of Transmission-Based Precautions, which may be used either singularly or in combination: Airborne Precautions, Droplet Precautions, and Contact Precautions. Please note that these Transmission-Based Precautions are to be used in addition to the Standard Precautions already being employed.

Airborne Precautions. Airborne transmission involves either airborne droplet nuclei or dust particles containing an infectious agent. Airborne droplet nuclei are small-particle residue (5 μm or smaller) of evaporated droplets containing microorganisms; they remain suspended in air for long periods of time. *Airborne Precautions* apply to patients known or suspected to be infected with epidemiologically important pathogens that can be transmitted by the airborne route (*e.g., Mycobacterium tuberculosis,* rubella virus, varicella virus). In addition to Standard Precautions, the patient is placed in a private room, having negative pressure and from which air is either discharged outdoors or passed through high efficiency particulate air (HEPA) filters. If a private room is not available, the patient may be placed in a room with a patient having active infection with the same pathogen, but with no other infection. Persons entering the patient's room must wear respiratory protection, unless they are known to be immune to the pathogen. A surgical mask is placed on the patient whenever it is necessary to transport the patient from the room. Special air handling and ventilation are required to prevent airborne transmission. Pathogens transmitted via airborne transmission are listed in Table 10-1.

Droplet Precautions. Technically, droplet transmission is a form of contact transmission. However, in droplet transmission, the mechanism of transfer is quite different than either direct- or indirect-contact transmission. Droplets are produced primarily as a result of coughing, sneezing, and talking, and during such procedures as suctioning and bronchoscopy. Transmission occurs when droplets (larger than 5 μm in size) containing microorganisms are propelled a short distance through the air and become deposited on another person's conjunctivae, nasal mucosa, or mouth. *Droplet Precautions* must be used for patients known or suspected to be infected with microorganisms transmitted by droplets that can be generated in the ways previously mentioned. In addition to Standard Precautions, the patient is placed in a private room. Special air handling and ventilation are not required to prevent droplet transmission. If a private room is not available, the patient may be placed in a room with a patient having active infection with the same pathogen, but with no other infection. Persons working within three feet of the patient must wear a mask. A surgical mask is placed on the patient whenever it is necessary to transport the patient from the room. Pathogens transmitted via droplet transmission are listed in Table 10-1.

TABLE 10-1. Infectious Diseases Requiring Transmission-Based Precautions

Type of Transmission-Based Precautions	Infectious Diseases
Airborne Precautions	Chickenpox, disseminated shingles or shingles in immunocompromised patients; measles (rubeola); pulmonary or laryngeal tuberculosis.
Droplet Precautions	Adenovirus infection in infants and young children; adenovirus pneumonia; epiglottitis caused by *Haemophilus influenzae;* German measles; Group A streptococcal infections in infants and children; *H. influenzae* or *Neisseria meningitidis* meningitis or pneumonia; influenza; meningococcemia; mumps; *Mycoplasma* pneumonia; parvovirus B19 infections; pertussis (whooping cough); pharyngeal diphtheria; pharyngitis, pneumonia, or scarlet fever in infants and young children; pneumonic plague.
Contact Precautions	Acute viral (hemorrhagic) conjunctivitis, adenovirus infection in infants and young children; adenovirus pneumonia; cellulitis with uncontrolled drainage, chickenpox, *Clostridium difficile* infections, congenital rubella, cutaneous diphtheria; disseminated shingles or shingles in immunocompromised patients; enterohemorrhagic O157:H7 *E. coli,* hepatitis A, *Shigella,* or rotavirus infections in diapered or incontinent patients; enteroviral infections in infants and young children; gastrointestinal, respiratory, skin, wound, or burn infections or colonization with multidrug-resistant organisms; hemorrhagic fevers (*e.g.,* Lassa and Ebola viruses); impetigo; lice (pediculosis); major draining abscesses, major infected decubitus ulcer; Marburg virus disease; major staphylococcal or Group A streptococcal skin, wound, or burn infections; neonatal or mucocutaneous herpes simplex infections; parainfluenza virus respiratory infection in infants and young children; RSV infection in infants, young children, and immunocompromised adults; scabies; staphylococcal furunculosis in infants and young children.

Contact Precautions. Contact transmission is the most important and frequent mode of transmission of nosocomial infections. Contact transmission is divided into two subgroups: direct-contact transmission (transfer of microorganisms by body surface-to-body surface contact) and indirect-contact transmission (transfer of microorganisms via a contaminated intermediate object, such as instruments, needles, and dressings). *Contact Precautions* are used for patients known or suspected to be infected or colonized with epidemiologically important microorganisms that can be transmitted by direct or indirect contact. In addition to Standard Precautions, the patient is placed in a private room. If a private room is not available, the patient may be placed in a room with a patient having active infection with the same pathogen, but with no other infection. In addition to wearing gloves as outlined under Standard Precautions, wear gloves when entering the patient's room. Change gloves after having contact with infective material that may contain high concentration of pathogens (*e.g.,* fecal material and wound drainage). Remove gloves before leaving the room and wash hands immediately with an antimicrobial or antiseptic agent. In addition to wearing a gown as outlined in Standard Precautions, wear a gown when entering the patient's room if you anticipate that your clothing will have substantial contact with the patient, environmental surfaces, or items in the patient's room, or if the patient is incontinent, or has diarrhea, an ileostomy, a colostomy, or wound drainage not contained by a dressing. Remove the gown before leaving the room. If the patient is transported out of the room, ensure that precautions are maintained to minimize the risk of transmission of pathogens to other patients and contamination of environmental surfaces or equipment. When possible, dedicate the use of noncritical patient-care equipment to a single patient to avoid sharing between patients. If this is not possible, then such equipment must be adequately cleaned and disinfected before use for another patient. Pathogens transmitted via contact transmission are listed in Table 10-1.

REVERSE ISOLATION

Certain patients are especially vulnerable to infection; among them are patients with severe burns, those who have leukemia, patients who have received a transplant, immunodeficient persons, those receiving radiation treatments, and leukopenic patients (those having abnormally low white blood cell counts). Premature babies are also highly susceptible. All such patients are protected through an isolation procedure known as *reverse isolation* (also referred to as protective or neutropenic isolation). Vented air entering the room is passed through HEPA filters and the room is under positive pressure to prevent hallway air from entering when the door is opened. The room must be thoroughly cleaned and disinfected before the patient is admitted. Those entering the room must wear sterile gowns and masks to prevent introducing microorganisms into the room from their clothes or respiratory tracts. Proper handwashing procedures must be followed before entering the room.

HOSPITAL INFECTION CONTROL

All hospitals are required by the regulatory agencies to have a formal infection control program. Although its functions vary slightly from one type of hospital to another, it is usually under the jurisdiction of the hospital's Infection Control Committee (ICC) or Epidemiology Service. The ICC is composed of representatives from most of the hospital's departments, including medical and surgical services, pathology, nursing, hospital administration, housekeeping, food services, and central supply. The chairperson is usually an Infection Control Professional (see Insight Box), such as a physician (*e.g.,* an epidemiologist or infectious disease specialist), an infection control nurse, a microbiologist, or some other person knowledgeable

INSIGHT
Infection Control Professionals

Individuals wishing to combine their interest in detective work with a career in medicine might consider a career as an infection control professional (ICP). ICPs include physicians (infectious disease specialists or epidemiologists), nurses, clinical laboratory scientists (medical technologists), and microbiologists. Most ICPs are nurses, many having baccalaureate degrees and some with master's degrees. In addition to having strong clinical skills, ICPs require knowledge and expertise in such areas as epidemiology, microbiology, infectious disease processes, statistics, and computers. To be effective, they must be part detective, part diplomat, part administrator, and part educator. In addition, ICPs function as role models, patient advocates, and consultants.

Within the hospital, ICPs provide valuable services that minimize the risks of infection and spread of disease, thereby aiding patients, healthcare professionals, and visitors. The ICP is the key person in implementing and facilitating the institution's infection control program. The ICP is often the head of the hospital's Infection Control Committee (ICC) and, as such, is responsible for scheduling, organizing and conducting ICC meetings. At these meetings, medical records are reviewed of all patients suspected of having incurred a nosocomial infection since the previous meeting. The committee discusses possible or known causes of such infections and ways to prevent them from occurring in the future. The ICP receives timely information from the clinical microbiology laboratory concerning possible outbreaks of infection within the hospital, and is responsible for rapidly organizing a team to investigate these outbreaks. ICPs are also responsible for educating healthcare personnel about infection risk, prevention, and control.

(Additional information about ICPs can be obtained from the Association for Professionals in Infection Control and Epidemiology, Inc. 1016 Sixteenth Street NW, Sixth Floor, Washington, DC 20036-5703.)

about infection control. The ICC periodically reviews the hospital's infection control program and the incidence of nosocomial infections. It is a policy-making and review body that may take drastic action (*e.g.*, instituting quarantine measures) when epidemiologic circumstances warrant. Other ICC responsibilities include patient surveillance, environmental surveillance, investigation of outbreaks/epidemics, and education of the hospital staff regarding infection control.

Although every department of the hospital endeavors to maintain aseptic conditions, the total environment is constantly bombarded with microbes from outside the hospital. These must be controlled for the protection of the patients.

Hospital personnel (usually ICPs) entrusted with this aspect of health care diligently and constantly work to maintain the proper environment. Clinical microbiology laboratory (CML) personnel cooperate by monitoring the types and numbers of pathogens isolated from hospitalized patients. Should an unusual pathogen or an unusually high number of isolates of a common pathogen be detected, CML personnel notify the ICP, who, in turn, initiates an investigation of the outbreak. Environmental samples, including samples from hospital employees, will be collected from within the affected ward(s), and processed in the CML, in an attempt to pinpoint the exact source of the pathogen. In the event of an epidemic, the ICP notifies city, county, and state health authorities so that they can assist in ending the epidemic.

MEDICAL WASTE DISPOSAL

General Regulations. According to Occupational Safety and Health Administration (OSHA) Standards, medical wastes must be disposed of properly. These standards include the following:

- Any receptacle used for decomposable solid or liquid waste or refuse must be constructed so that it does not leak and must be maintained in a sanitary condition. This receptacle must be equipped with a solid, tight-fitting cover, unless it can be maintained in a sanitary condition without a cover.
- All sweepings, solid or liquid wastes, refuse, and garbage shall be removed to avoid creating a menace to health and shall be removed as often as necessary to maintain the place of employment in a sanitary condition.
- The infection control program must address the handling and disposal of potentially contaminated items.

Sharp Instruments and Disposables. These items must be disposed of in the following manner:

- Needles shall not be recapped, purposely bent or broken by hand, removed from disposable syringes, or otherwise manipulated by hand.
- After use, disposable syringes, scalpel blades, and other sharp items must be placed in puncture-resistant containers for disposal of "sharps."
- These containers must be easily accessible to all personnel needing them and must be located in all areas where needles are commonly used, as in areas where blood is drawn, including patient rooms, emergency rooms, intensive care units, and surgical suites.
- The containers must be constructed so that the contents will not spill if knocked over and will not cause injuries.

Laboratory Specimens. Healthcare professionals who collect and transport clinical specimens should exercise extreme caution during collection and transport not to stick themselves with needles, cut themselves with other types of "sharps," or come in contact with any type of specimen. According to the National Committee for Clinical Laboratory Standards, after collection, "all specimens should be placed into a leakproof primary container having a secure closure. Care should be taken by the person collecting the specimen not to contaminate the outside of the primary container. Before being transported to the laboratory, the primary container should be placed into a second container, which will contain the specimen if the primary container breaks or leaks in transit to the laboratory." Within the laboratory, all specimens are handled carefully, following Standard Precautions, and disposed of as infectious waste.

SPECIMEN COLLECTING, PROCESSING, AND TESTING

It would not be feasible in a book of this size to provide a complete discussion of collecting, handling, processing, and testing of clinical specimens. Only a few important concepts are discussed here.[2]

Role of Healthcare Professionals

Extreme care must be taken by those involved in collecting, handling, and processing specimens that are to be examined for the presence of microorganisms. High quality specimens are required to achieve accurate, clinically relevant results. The three components of specimen quality are (1) proper specimen selection (*i.e.,* the correct type of specimen must be submitted), (2) proper specimen collection, and (3) proper transport of the specimen to the laboratory. Whenever possible, specimens must be collected in a manner that will eliminate or minimize contamination of the specimen with indigenous microflora. Certain types of specimens must be rushed to the laboratory. Some require transit on ice, whereas others must never be placed on ice. The laboratory must provide written guidelines regarding specimen selection, collection, and transport. Copies of this "Floor Manual" must be available on every ward and in every clinic. Furthermore, the laboratory is responsible for ensuring that proper specimen collection and transport devices are available.

The Three Components of Specimen Quality
Proper selection of the specimen (*i.e.,* the proper specimen)
Proper collection of the specimen
Proper transport of the specimen to the laboratory

When specimens are improperly collected and handled, (1) the etiologic (causative) agent may not be found or may be destroyed, (2) overgrowth by indigenous microflora may mask the pathogen, and (3) contaminants may interfere with the identification of pathogens and the diagnosis of the patient's infectious disease.

A close working relationship among the members of the healthcare team is essential for the proper identification of pathogens. When the attending physician recognizes the clinical symptoms of a possible infectious disease, certain specimens and

---

[2]For more detailed information about specimen collection, refer to the *Manual of Clinical Microbiology,* 6th Ed, 1995, and *Cumitech 9: Collection and Processing of Bacteriological Specimens,* 1979, both of which were published by the American Society for Microbiology, Washington, D.C.

clinical tests may be requested. The clinical microbiologist who performs the laboratory microbial analysis must provide adequate collection materials and instructions for their proper use. The doctor, nurse, medical technologist, or other qualified healthcare professional must perform the collection procedure properly, and then the specimen must be transmitted properly to the laboratory where it is cultured, stained, and analyzed. Laboratory findings must then be conveyed to the attending physician as quickly as possible to facilitate the prompt diagnosis and treatment of the infectious disease.

Proper Collection of Specimens

When collecting specimens, these general precautions should be taken:

- All specimens should be placed or collected into a sterile container to prevent contamination of the specimen by indigenous microflora and airborne microbes. Appropriate types of collection devices and specimen containers should be specified in the "Floor Manual."
- The material should be collected from a site where the suspected pathogen is most likely to be found and where the least contamination is likely to occur.
- Whenever possible, specimens should be obtained before antimicrobial therapy has begun. If this is not possible, the laboratory should be informed as to which antimicrobial agent(s) the patient is receiving.
- The acute stage of the disease (when the patient is experiencing the signs and symptoms of the disease) is the appropriate time to collect most specimens. Some viruses, however, are more easily isolated during the onset stage of disease.
- Specimen collection should be performed with care and tact to avoid harming the patient, causing discomfort, or causing undue embarrassment. If the patient is to collect the specimen, such as sputum or urine, the patient must be given clear and detailed collection instructions.
- A sufficient quantity of the specimen must be obtained to provide enough material for all required diagnostic tests. The amount of specimen to collect should be specified in the "Floor Manual."
- Specimens should be protected from heat and cold and promptly delivered to the laboratory so that the results of the analyses will validly represent the number and types of organisms present at the time of collection. If delivery to the laboratory is delayed, some delicate pathogens might die, *e.g.*, obligate anaerobes die when exposed to air. Any indigenous microflora in the specimen may overgrow, inhibit, or kill pathogens. Delay of delivery considerably decreases the chances of isolating pathogens. Certain types of specimens must be placed on ice during delivery to the laboratory, whereas other specimens should never be refrigerated or placed on ice due to the fragile nature of the pathogens. Specimen transport instructions should be contained in the "Floor Manual."

- Hazardous specimens must be handled with even greater care to avoid contamination of the courier, patients, and healthcare professionals. Such specimens must be placed in a sealed plastic bag for immediate and careful transport to the laboratory.
- Whenever possible, sterile, disposable specimen containers should be used. If reusable containers are used, they should be cleaned, sterilized, and properly stored to avoid contamination of the specimen by microbes and potentially harmful chemicals.
- The specimen container must be properly labeled and accompanied by an appropriate request slip containing adequate instructions. At minimum, labels must contain the patient's name and identification number, specific source of specimen, the date and time of collection, and the collector's initials. The laboratory should always be given sufficient clinical information to aid in performing appropriate analyses. The request slip that accompanies a wound specimen, for example, should state the specific type of wound (*e.g.,* burn wound, dog bite wound, postsurgical wound infection, etc.).
- Specimens should be collected and delivered to the laboratory as early in the day as possible to give the technologists sufficient time to process the material, especially when the hospital or clinic does not have 24-hour laboratory service.

Types of Specimens Usually Required

Special techniques in collection and handling are required to obtain specific types of specimens.

BLOOD

Blood is usually sterile. Bacteria in the bloodstream (*bacteremia*) may indicate a disease, although temporary or transient bacteremias may occur following oral surgery, tooth extraction, etc. To prevent contamination of the blood specimen with indigenous skin flora, extreme care must be taken to use sterile technique when collecting blood for culture. After locating a suitable vein, disinfect the skin with 70% isopropyl alcohol and then with an iodophor. When disinfecting the site, use a concentric swabbing motion, starting at the point you intend to insert the needle, and working outward from that point. Allow the iodophor to dry. Apply a tourniquet and withdraw 10 ml to 20 ml of blood with a 21-gauge needle into a sterile blood culture bottle, containing an anticoagulant. After venipuncture, remove the iodophor from the skin with alcohol. The blood culture bottle(s) should be transported promptly to the laboratory for incubation at 37°C.

Bacteremia may occur during certain stages of many infectious diseases. These diseases include bacterial meningitis, typhoid fever and other salmonella infections, pneumococcal pneumonia, urinary infections, endocarditis, brucellosis, tularemia,

The suffix "-emia" refers to the bloodstream; often, the presence of something in the bloodstream. Toxemia refers to the presence of toxins in the bloodstream; bacteremia, the presence of bacteria; fungemia, the presence of fungi; viremia, the presence of viruses; parasitemia, the presence of parasites. Septicemia, however, is an actual disease; quite often, a serious, life-threatening disease. Septicemia is defined as chills, fever, prostration (extreme fatigue), and the presence of bacteria and/or their toxins in the bloodstream.

plague, anthrax, syphilis, and wound infections caused by β-hemolytic streptococci, staphylococci, and other invasive bacteria. *Septicemia* is a serious disease characterized by chills, fever, prostration, and the presence of bacteria and/or their toxins in the bloodstream. The most severe types of septicemia are those caused by Gramnegative bacilli, due to the endotoxin that is released from their cell walls. Endotoxin can induce fever and septic shock, which can be fatal. To diagnose either bacteremia or septicemia, it is recommended that at least three blood cultures be collected over a 24-hour period.

URINE

Urine is ordinarily sterile while it is in the urinary bladder. However, during urination, it becomes contaminated by indigenous microflora of the distal urethra (the section of the urethra furthest from the bladder). Contamination can be reduced by collecting a "clean-catch, midstream urine" (CCMS urine). "Clean-catch" refers to the fact that the area around the external opening of the urethra is cleansed by washing with soap and rinsing with water before urination. This cleansing removes the indigenous microflora that live in the area. "Midstream" refers to the fact that the initial portion of the urine stream is directed into a toilet or bedpan, and then the urine stream is directed into in a sterile container. Thus, the microorganisms that live in the distal urethra are flushed out of the urethra by the initial portion of the urine stream, into the toilet or bedpan rather than into the specimen container. In some circumstances, the physician may prefer to collect a catheterized specimen or to use the suprapubic needle aspiration technique to obtain a sterile sample of urine. In the latter technique, a needle is inserted through the abdominal wall into the urinary bladder and a syringe is used to withdraw urine from the bladder. To prevent continued bacterial growth, all urine specimens must be processed within an hour or refrigerated at 4°C until they can be analyzed (within 5 hours).

A urinary tract infection (UTI) is indicated if the number of bacteria in a CCMS urine equals or exceeds 100,000 (1×10^5) organisms per milliliter (expressed as colony-forming units per mL [CFU/mL]). A CCMS urine collected from someone who does not have a UTI usually (but not always) contains fewer than 10,000 CFU/mL. The presence of two or more bacteria per ×1000 microscopic field of a Gram-stained urine smear indicates *bacteriuria* (bacteria in the urine) with 100,000 or more CFU per mL.

CEREBROSPINAL FLUID

Meningitis, encephalitis, and meningoencephalitis are rapidly fatal diseases that can be caused by a variety of microbes, including bacteria, fungi, protozoa, and viruses. To diagnose these diseases, spinal fluid specimens must be collected into a sterile tube by a lumbar puncture ("spinal tap") under surgically aseptic conditions (Fig.10-3). This difficult procedure is performed by a physician. Because *Neisseria meningitidis* (meningococci) are susceptible to cold temperatures, the specimen must be cultured immediately and not refrigerated. Specimens to be further examined for viruses may be kept frozen at −20°C.

SPUTUM

Sputum (pus that accumulates in the lungs) may be collected by allowing the patient to spit the coughed-up specimen into a sterile wide-mouthed bottle with a lid, after warning the patient not to contaminate the sputum with saliva. If proper mouth hygiene is maintained, the sputum will not be severely contaminated with oral flora. If tuberculosis is suspected, extreme care in collecting and handling the specimen should be exercised because one could easily be infected with the pathogens. Usually, sputum specimens may be refrigerated for several hours without loss of the pathogens.

The physician may wish to obtain a better quality specimen by bronchial aspiration through a bronchoscope or by a process known as transtracheal aspiration. Needle biopsy of the lungs may be necessary for diagnosis of *Pneumocystis carinii*

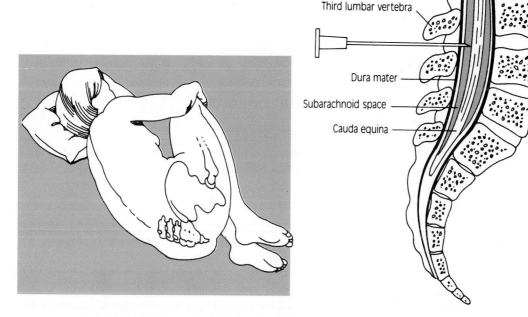

Third lumbar vertebra

Dura mater

Subarachnoid space

Cauda equina

FIGURE 10-3. Technique of lumbar puncture.

pneumonia (as in AIDS patients) and for certain other pathogens. Although once classified as a protozoan, *P. carinii* is currently considered to be a fungus.

MUCOUS MEMBRANE SWABS

Sterile polyester swabs are used to collect specimens of exudates and secretions of the throat, nose, ear, eye, urethra, rectum, wounds, operative sites, and ulcerations. Cotton swabs are no longer used because fatty acids in the cotton inhibit the growth of some microorganisms. Handy, sterile, disposable collection units can be obtained from many medical supply companies. Each unit contains a sterile polyester swab and transport medium in a sterile tube. By using this set-up, pathogens are kept alive and protected during transportation to the laboratory.

When attempting to diagnose gonorrhea, vaginal, cervical, and urethral swabs should be inoculated immediately onto Thayer-Martin chocolate agar plates and incubated in a CO_2 (carbon dioxide) environment. Alternatively, they should be inoculated into a tube or bottle (*e.g.*, Transgrow®) that contains an appropriate culture medium and CO_2, while the bottle is held in an upright position to prevent loss of the CO_2. These cultures should be incubated at 37°C overnight, and then shipped to a public health diagnostic facility for positive identification of gonococci.

FECES

Ideally, fecal specimens should be collected at the laboratory and processed immediately to prevent a decrease in temperature, which allows the pH to drop, causing the death of many *Shigella* and *Salmonella* species; or the specimen may be placed in a container with a preservative that maintains a pH of 7.

Because the colon is anaerobic, fecal bacteria are obligate, aerotolerant, and facultative anaerobes. However, fecal specimens are cultured anaerobically only when *Clostridium difficile*-associated disease is suspected or for diagnosing clostridial food poisoning. In intestinal infections, the pathogens frequently overwhelm the normal microflora so that they are the predominant organisms seen in smears and cultures. A combination of culture, direct microscopic examination, and immunological tests may be performed to identify Gram-negative and Gram-positive bacteria (*e.g.*, enteropathogenic *E. coli*, *Salmonella* spp., *Shigella* spp., *Clostridium perfringens*, *Clostridium difficile*, *Vibrio cholerae*, *Campylobacter* spp., and *Staphylococcus* spp.), fungi (*Candida*), intestinal protozoa (*Giardia*, *Entamoeba*), and intestinal helminths.

Shipping Specimens

Occasionally, an etiologic agent or specimen must be sent to a laboratory in another city for identification or processing. Cultures and specimens must be shipped in accordance with regulations established by the U.S. Public Health Service. The primary container (usually a test tube or vial) must be fitted with a watertight cap and surrounded by sufficient packing material to absorb any fluid material should a leak occur. The primary

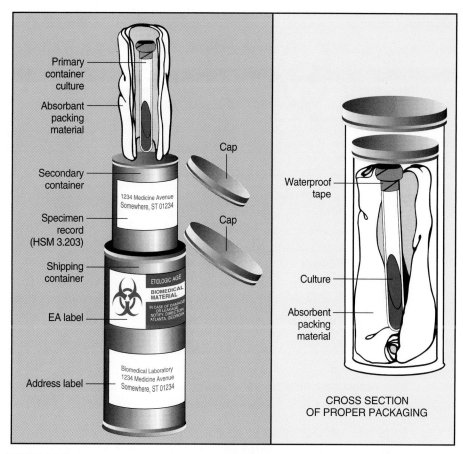

FIGURE 10-4. Proper method of packaging cultures, specimens, and other biological hazardous materials. (CDC Lab. Manual, DHEW publication No. [CDC] 74–8272, Atlanta, GA, Centers for Diesease Control, 1974.)

container is placed into a secondary container, preferably metal, fitted with a screw-cap lid. The secondary container is then packed in an outer shipping carton, constructed of corrugated fiberboard, cardboard, or Styrofoam. (See Fig. 10-4). The carton must bear a red and white etiologic agents/biomedical label picturing the biohazard symbol (Fig. 10-5). If dry ice is used as a refrigerant, the package must be appropriately marked that it contains dry ice and it must allow the escape of carbon dioxide gas.

The Pathology Department ("The Lab")

Virtually all healthcare personnel will interact in some way(s) with the hospital laboratory, which is technically called the Pathology Department, but more commonly referred to as "the Lab." As shown in Figure 10-6, the Pathology Department is divided into two major areas: anatomical pathology and clinical pathology.

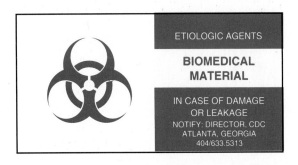

FIGURE 10-5. Etiologic agent/bio-medical material shipping label.

Most *pathologists* (physicians with specialized training in *pathology,* the study of disease) work in anatomical pathology, where they perform autopsies (in the morgue) and examine stained tissue sections and cytology specimens. Healthcare professionals employed in anatomical pathology include pathology assistants and cytopathology, cytogenetics, and histopathology technologists.

The Chemistry Laboratory is where various chemistry procedures are performed on blood and urine specimens, such as determining cholesterol, glucose, and protein levels. The Hematology Laboratory is where blood specimens are examined to determine if the patient has too many or too few red blood cells or white blood cells. Differential blood cell counts ("diffs") are also performed in the Hematology Laboratory to determine if the patient's blood contains the proper types of white blood cells in the appropriate ratios. The Immunology Laboratory is where various immunological procedures are performed to diagnose

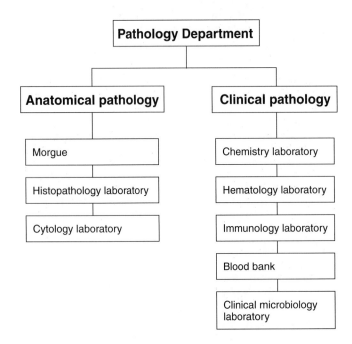

FIGURE 10-6. Organization of the Pathology Department.

infectious diseases (immunodiagnostic procedures) and to determine tissue compatibility for organ and tissue transplants. The Blood Bank is where individuals go to donate blood and where that blood is then processed, tested, and stored for patient use.

The primary function of the Clinical Microbiology Laboratory (CML) is to assist physicians in the diagnosis of infectious diseases. Healthcare professionals employed in the CML include microbiologists (persons having masters or doctoral degrees in microbiology), clinical laboratory scientists/medical technologists (having bachelors degrees), and clinical laboratory technicians/medical laboratory technicians (having associate degrees).

The four, major, day-to-day responsibilities of the CML are (1) to process the various clinical specimens that are submitted to the CML, (2) to isolate pathogens from those specimens, (3) to identify pathogens, and (4) to perform antimicrobial susceptibility testing when appropriate to do so. The exact steps in the processing of clinical specimens vary from one specimen type to another, but usually include (1) a macroscopic examination of the specimen, (2) a microscopic examination of the specimen, and (3) inoculation of appropriate culture media. The CML is sometimes called upon to assume an additional responsibility—the processing of environmental samples. Such samples are collected from appropriate hospital sites (*e.g.,* floors, sink drains, shower heads, whirlpool baths, respiratory

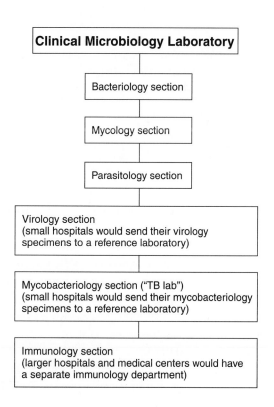

FIGURE 10-7. Organization of the Clinical Microbiology Laboratory.

INSIGHT
Specimen Quality and Clinical Relevance

Microbiology laboratory results are clinically relevant if they reveal information about the patient's infectious disease—if they provide the physician with useful information that can be used to diagnose infectious diseases, monitor their progress, and guide therapy. To provide clinically relevant information, the microbiology laboratory must receive high quality clinical specimens. The quality of the results can be no better than the quality of the specimen. If a poor quality specimen is submitted to the laboratory, in all likelihood, the results obtained using that specimen will not be clinically relevant. In fact, results obtained from poor quality specimens might very well be harmful to the patient.

What constitutes a high quality clinical specimen? The best quality specimen is one that has been selected, collected, and transported properly. First, it must be an appropriate specimen—the correct type of specimen required to diagnose the suspected disease. Next, the specimen must be collected in a manner that will minimize its contamination with indigenous microflora. And finally, it must be transported to the laboratory in the proper manner—rapidly, if necessary; on ice, if necessary; anaerobically, if necessary; with the proper preservative, if necessary; etc. A specimen labeled "sputum," for example, must contain sputum—not merely saliva. A urine specimen

submitted for culture must be a clean catch, midstream specimen. Adequate care must be taken to adequately disinfect the phlebotomy site when blood is drawn for culture, to minimize the chance of contamination of the specimen with indigenous skin flora.

Those people who are responsible for submitting specimens to the laboratory are responsible for the quality of the specimens they submit. But, how do these people know which specimen to submit, or how to collect it, or the proper way to transport it to the laboratory? If they have not been taught such procedures in their course of study, they should consult the "Floor Manual," which contains such information. A copy of the "Floor Manual," which may be called the "Laboratory Procedures Manual" or some other name, should be present on each ward or clinic, and readily available for reference.

It is the laboratory's responsibility to publish and distribute such a manual. If the laboratory demands high quality specimens, as it should, then it must take the time to educate healthcare professionals as to what constitutes an appropriate specimen for the diagnosis of each infectious disease. Only in this way will the highest quality of service be assured and only then will the microbiology laboratory's results be clinically relevant.

therapy equipment) and employees whenever there is an outbreak or epidemic within the hospital, in an attempt to locate the source of the pathogen involved.

As shown in Figure 10-7, the CML is divided into various sections which generally correspond to the various categories of microorganisms. The role of the Bacteriology Section is to assist in the diagnosis of bacterial diseases. It is here that bacterial pathogens are isolated from clinical specimens, identified, and tested for susceptibility/resistance to various antibacterial drugs. The role of the Mycology Section is to assist in the diagnosis of fungal diseases (mycoses). Here, fungal pathogens (yeasts and molds) are isolated from clinical specimens and then identified. In the Parasitology Section, specimens are examined to diagnose diseases caused by parasitic protozoa and helminths (parasites are described in Appendix C).

If the CML is part of a large hospital or medical center, it will also have a Virology Section and a Mycobacteriology Section (or "TB Lab"). If, however, the hospital is small, it is far too expensive to maintain these sections and virology and my-cobacteriology specimens will be forwarded to a reference laboratory. In smaller hospitals, immunological or serological procedures will be performed in the Im-munology Section of the CML. Larger hospitals will have an Immunology De-partment which is separate from the CML.

It is extremely important that the results sent out by the CML be clinically rel-evant; *i.e.,* they must provide information about the patient's infectious disease. Nothing contributes more significantly to clinical relevancy than high quality clin-ical specimens. Various aspects of specimen quality are described in the Insight Box. Those hospital personnel (primarily physicians and nurses) who collect clinical spec-imens are responsible for the quality of the specimens being sent to the laboratory. It is the Pathology Department's responsibility to provide written guidelines for the proper selection, collection, and transport of clinical specimens. A copy of these guidelines, often referred to as the "Floor Manual", must be present on every ward and clinic in the hospital.

In emergency situations when laboratory personnel are not available (*e.g.,* at night or on weekends in small hospitals), nurses and other healthcare professionals are sometimes required to perform certain relatively simple microbiological proce-dures, such as inoculation of culture media, Gram staining of clinical specimens, wet mounts, potassium hydroxide preparations (KOH preps), and India ink preps. These procedures are described in Appendix D.

Identification and Antimicrobial Susceptibility Testing of Pathogens

After a pathogen has been isolated from a specimen, the main remaining tasks are to identify the organism (*i.e.,* determine what organism it is) and determine its sus-ceptibility (sensitivity) to antimicrobial agents.

TYPES OF TESTS USED FOR IDENTIFICATION

In some ways, microbiology laboratory personnel are like detectives, gathering "clues" (phenotypic characteristics) about a pathogen until they have enough clues to identify it. Some pathogens can be identified to the species level by using relatively few clues, but others require a large number of clues. Important phenotypic charac-teristics can be determined by examining a Gram-stained smear of the organism, in-cluding its Gram reaction, basic shape, and morphological arrangement. Additional clues can be learned by examining colonies of the organism on solid medium, such as the overall shape, height, consistency, size, color, and odor of the colonies. The organism's atmospheric requirements (*e.g.,* its relationships with oxygen and carbon dioxide) are also important clues. Additional phenotypic characteristics can be determined using certain selective and differential media, such as MacConkey agar, and a few relatively simple biochemical tests, such as those for catalase or oxidase, or the ability of the organism to catabolize certain substrates. Once the organism has been identified to species, it is sometimes necessary to determine its serotype (as in the case of *E. coli* and *Salmonella* species). Different serotypes of a particular species possess different surface molecules on their capsules (called K antigens), cell walls (called O antigens), and/or flagella (called H antigens). For example, *E. coli* O157:H7 has an antigen designated "O157" on its cell walls and an antigen designated "H7" on its flagella.

Phenotypic Characteristics of Value in the Identification of Bacterial Pathogens

Gram reaction (positive or negative)

Basic cell shape (cocci, bacilli, curved, or spiral-shaped)

Morphological arrangement of cells (*e.g.,* pairs, tetrads, chains, clusters)

Colony morphology

Atmospheric requirements

Presence or absence of capsules

Presence or absence of flagella and, if present, the number and location of the flagella

Presence or absence of spores

Type of hemolysis the organism produces (alpha, beta, or gamma)

Presence or absence of certain enzymes, such as catalase, coagulase, oxidase

Ability to catabolize certain carbohydrates and amino acids

Many laboratories make use of miniaturized, multiple-biochemical test systems available from Analytab Products, Inc.; Roche Diagnostics; BBL-Bioquest; and many other companies (see Color Figures 17, 18). Such "mini-systems" are very popular because they provide accurate results, are relatively inexpensive, are easy to inoculate, and take up little incubator space. In some cases, mini-system test results can be determined by electrical instruments, within which internal computers then analyze the pattern of test results and print out the identity of the organism.

MOLECULAR DIAGNOSTIC PROCEDURES

Traditionally, the major criticism of the CML is speed—that it takes too long to get results from the lab. This is primarily due to the fact that it takes a while, days or even weeks in some cases, to isolate pathogens in the laboratory; *i.e.,* to get them growing in pure culture, in large numbers, so that there is sufficient inoculum for the biochemical-based identification procedures. Two approaches to the diagnosis of infectious diseases have greatly reduced the time it takes to diagnose infectious diseases. One approach is the use of immunodiagnostic procedures, which are described in Chapter 11. The other approach is the use of molecular diagnostic procedures.

Molecular diagnostic procedures (MDPs), such as DNA probes, detect pathogen-specific DNA or RNA in clinical specimens. Because it is not necessary for the pathogens to be alive, there is less emphasis on the need for fresh specimens. Also, because of what are known as amplification procedures, only small amounts of pathogen DNA or RNA need be present in the specimen. One example of an amplification procedure is the polymerase chain reaction (PCR), where DNA polymerase is used to make many additional copies of existing segments of DNA. MDPs typically provide results in as short as 4 to 6 hours after receipt of the specimen and they tend to be more sensitive and specific than traditional methods of culture and identification. A variety of DNA probes are currently commercially available for the diagnosis of many different infectious diseases, including bacterial, viral, fungal, and protozoal diseases.

ANTIMICROBIAL SUSCEPTIBILITY TESTING

Not only does the CML isolate and identify the pathogens that are causing patients' infections, but it also tests the pathogens to determine what drugs will kill them and what drugs seem to have no effect. This is called antimicrobial susceptibility testing (AST) and, in the CML, AST is performed only on bacterial pathogens. Techniques are available to test fungi and protozoa, but these tests are performed only in research and reference laboratories.

Four different methods are available for performing AST in the CML: (1) the agar dilution method (which uses many agar plates), (2) the macro broth dilution method (which uses many test tubes), (3) the micro broth dilution method (which uses small plastic, microtiter trays), and (4) the disk-diffusion method. Although agar dilution is the most accurate method (referred to as the "gold standard"), it is not practical for use in the CML. Likewise, the macro broth dilution method is also impractical for use in the CML. The most practical and most popular methods for use in the CML are the micro broth dilution and disk-diffusion methods.

Disk-diffusion antimicrobial susceptibility testing is also referred to as the "Kirby-Bauer method," named for two microbiologists—Drs. Kirby and Bauer—who, in 1966, described a standardized technique for performing disk-diffusion AST. In this method, a pure culture of the organism in Mueller-Hinton broth is inoculated onto the entire surface of a Mueller-Hinton agar plate. Small, filter-paper disks containing various antimicrobial agents are then placed on the agar surface and the plate is incubated for 18 to 24 hours at 35°C in a non-CO_2 incubator (see Figs. 10-8 and 10-9).

For Gram-positive bacteria, disks of amoxicillin/clavulanic acid, ampicillin, ampicillin/sulbactam, cephalosporins, chloramphenicol, ciprofloxacin, clindamycin, erythromycin, gentamicin, imipenem, nitrofurantoin, oxacillin, penicillin,

FIGURE 10-8. An antimicrobial disk dispenser. (Photograph courtesy Difco Laboratories, Detroit, Michigan.)

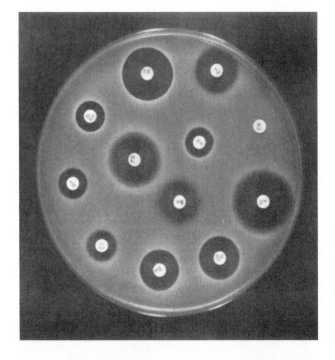

FIGURE 10-9. Paper disks impregnated with various antimicrobial agents are placed on the surface of an inoculated plate of Mueller-Hinton agar. The agents diffuse into the medium, inhibiting growth if the organism is sensitive to them. The sizes of the zones of inhibition of growth determine if the organism is susceptible or resistant to the various agents.

tetracycline, trimethoprim/sulfamethoxazole, and vancomycin might be used. For Gram-negative bacteria, disks of amikacin, amoxicillin/clavulanic acid, ampicillin, ampicillin/sulbactam, azlocillin, aztreonam, carbenicillin, cephalosporins, chloramphenicol, ciprofloxacin, gentamicin, imipenem, mezlocillin, nitrofurantoin, piperacillin, tetracycline, ticarcillin, ticarcillin/clavulanic acid, trimethoprim/sulfamethoxazole, and tobramycin, as well as other newer antibiotics, might be used. After 18 to 24 hours of incubation, the diameter of the zone of inhibition around each disk is measured in millimeters. These measurements must then be compared with zone sizes listed on published charts, to determine whether the organism is susceptible or resistant to the various drugs tested. The test procedure (which must be strictly followed) and charts for interpretation of zone sizes are published by the National Committee for Clinical Laboratory Standards (NCCLS).

QUALITY CONTROL IN THE LABORATORY

A quality control (QC) program is necessary in any laboratory to monitor the reliability and the quality of the work performed there. In the microbiology laboratory, all test procedures, media, reagents, and staining solutions should be evaluated frequently for effectiveness. Quality control organisms (organisms known to give positive and negative test results) are maintained in the laboratory and used for this purpose. Also, all equipment should be properly maintained and monitored for performance. Refrigerators, freezers, incubators, and water baths should be checked periodically for accuracy of temperature control. By constant surveillance and frequent checking, the efficiency and reliability of the laboratory work can be maintained so that other members of the hospital team have confidence in the information supplied to them by the laboratory.

ENVIRONMENTAL DISEASE CONTROL MEASURES

Public Health

Massive networks involving the World Health Organization (WHO), public health agencies at all levels, and community groups work together to coordinate preventive health programs and to maintain constant surveillance of sources and causes of epidemics. The WHO, a specialized agency of the United Nations, was founded in 1948. Its missions are to promote technical cooperation for health among nations, carry out programs to control and eradicate diseases, and improve the quality of human life. When an epidemic strikes, such as the 1995 Ebola outbreak in Kikwit, Zaire, teams of epidemiologists are sent to the site to investigate the situation and assist in bringing the outbreak under control. Because of this assistance, many countries have been successful in their fight to control smallpox, diphtheria, malaria, trachoma, and numerous other diseases. At one time, smallpox killed about 40% of those infected, and caused scarring and blindness in many others. In 1980, the WHO announced that smallpox had been completely eradicated from the face of the earth; hence, smallpox vaccination is no longer required. More recently, the WHO has been attempting to eradicate polio and dracunculiasis (Guinea worm infection); to eliminate leprosy, neonatal tetanus, and Chagas' disease; and to control onchocerciasis ("river blindness"). WHO definitions of eradication, elimination, and control of disease are presented in Table 10-2.

In the United States, a Federal Agency called the U.S. Department of Health and Human Services administers the Public Health Service and the Centers for Disease Control and Prevention (CDC), which assist state and local health departments in the application of all aspects of epidemiology. When the CDC was first established as the Communicable Disease Center in Atlanta, GA, in 1946, its focus was communicable diseases (*i.e.,* infectious diseases transmissible from person to person). The two most important infectious diseases in the United States at that time were malaria and typhus. Since then, the CDC's scope has been expanded greatly, and it now consists of 11 Centers, Institutes and Offices, one of which is the National Center for Infectious Diseases (NCID). The CDC's overall mission is "to promote health and quality of life by preventing and controlling diseases, injury, and disability." Highlights in the history of the CDC are illustrated in Table 10-3. The NCID's mission is "to prevent illness, disability, and death caused by infectious diseases in the United States and around the world." Certain infectious diseases, referred to as nationally notifiable diseases, must be

TABLE 10-2. Definitions of Epidemiologic Terms Relating to Infectious Diseases	
Term	**Definition**
Eradication	The WHO defines eradication of an infectious disease as achieving a status where no further cases of that disease occur anywhere, and where continued control measures are unnecessary
Elimination	The WHO defines elimination of an infectious disease as the reduction of case transmission to a predetermined very low level (e.g., to a level below one case per million population)
Control	The WHO defines control of an infectious disease as ongoing operations or programs aimed at reducing the incidence and/or prevalence of that disease
Incidence	The number of new cases of a disease in a defined population over a specific period of time (e.g., the number of new cases of hantavirus pulmonary syndrome in the United States during 1997)
Prevalence	The number of cases of a disease existing in a given population at a specific period of time (period prevalence; e.g., the total number of cases of gonorrhea that existed in the U.S. population during 1997) or at a particular moment in time (point prevalence; e.g., the number of cases of malaria in the U.S. population right now)

TABLE 10-3. Highlights in the History of the CDC

Year	Event
1946	The Communicable Disease Center (CDC) opened in downtown Atlanta, GA
1951	The Epidemic Intelligence Service (EIS) was established; it becomes the nation's (and world's) response team for a wide range of health emergencies; national disease surveillance systems also began
1960	The CDC moved to its present location at 1600 Clifton Road in Atlanta
1961	The CDC took over publication of the *Morbidity and Mortality Weekly Report*
1970	The National Nosocomial Infections Surveillance System was established to monitor trends in hospital-acquired infections; the CDC was renamed the Center for Disease Control
1978	The CDC opened an expanded, maximum-containment laboratory enabling the agency to work with viruses and other pathogens too dangerous to work with in ordinary laboratories
1980	The CDC was renamed the Centers for Disease Control
1985	The CDC co-sponsored the first International Conference on AIDS
1992	The CDC was renamed the Centers for Disease Control and Prevention; the initials "CDC" were still used

reported to the CDC by all 50 states. The 10 most common nationally notifiable infectious diseases in the United States are listed in Table 10-4.

Through the efforts of these agencies, working with local physicians, nurses, other healthcare professionals, educators, and community leaders, many diseases are no longer endemic in the United States. Some of the diseases that no longer pose a serious threat to U.S. communities include cholera, diphtheria, malaria, polio, smallpox, and typhoid fever.

TABLE 10-4. Top Ten Notifiable Infectious Diseases in the U. S. (1997)

Ranking	Disease	No. of U.S. Cases Reported
1	Genital chlamydial infections	526,671
2	Gonorrhea	324,907
3	Chickenpox (varicella)	98,727
4	AIDS	58,492
5	Syphilis	46,540
6	Salmonellosis	41,901
7	Hepatitis A	30,021
8	Shigellosis	23,117
9	Tuberculosis	19,851
10	Lyme disease	12,801

The prevention and control of epidemics is a never-ending community goal. To be effective, it must include measures to:

- increase host resistance through development and administration of vaccines that induce active immunity and maintain it in susceptible persons;
- ensure that persons who have been exposed to a pathogen are protected against the disease; *e.g.,* injections of gamma globulin or antisera are effective against outbreaks of diphtheria;
- segregate, isolate, and treat those who have contracted a contagious infection to prevent the spread of pathogens to others; and
- identify and control potential reservoirs and vectors of infectious diseases; this control may be accomplished by prohibiting healthy carriers from working in restaurants, hospitals, nursing homes, and other institutions where they may transfer pathogens to susceptible people, and by instituting effective sanitation measures to control diseases transmitted through water supplies, sewage, and food (including milk).

Water Supplies and Sewage Disposal

Water is the most essential resource necessary for the survival of humanity. The main sources of community water supplies are surface water from rivers, natural lakes, and reservoirs, as well as groundwater from wells. However, two types of water pollution are present in our society that are making it increasingly difficult to provide safe water supplies; these come from chemical and biological sources.

Chemical pollution of water occurs when industrial installations dump waste products into local waters without proper pretreatment, when pesticides are used indiscriminately, and when chemicals are expelled in the air and carried to earth by rain ("acid rain"). The main source of biological pollution is waste products of humans—fecal material and garbage—that swarm with pathogens. The etiologic agents of cholera, typhoid fever, bacterial and amebic dysentery, giardiasis, cryptosporidiosis, infectious hepatitis, and poliomyelitis can all be spread through contaminated water.

Waterborne epidemics today are the result of failure to make use of the available existing knowledge and technology. In those countries that have established safe sanitary procedures for water purification and sewage disposal, outbreaks of typhoid fever, cholera, and dysentery only rarely occur.

SOURCES OF WATER CONTAMINATION

Rainwater falling over a large area collects in lakes and rivers and, thus, is subject to contamination by soil microbes and raw fecal material. For example, an animal feed lot located near a community water supply source harbors innumerable

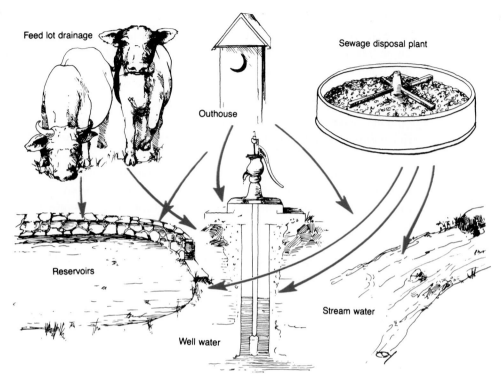

FIGURE 10-10. Sources of water contamination.

pathogens, which are washed into lakes and rivers. A city that draws its water from a local river, processes it, and uses it, but then dumps inadequately treated sewage into the river at the other side of town may be responsible for a serious health problem in another city downstream on the same river. The city downstream must then find some way to rid its water supply of the pathogens. In many communities, untreated raw sewage and industrial wastes are dumped directly into local waters; also, a storm or a flood may result in contamination of the local drinking water with sewage (Fig. 10-10).

Groundwater from wells also can become contaminated. To prevent such contamination, the well must be dug deep enough to ensure that the surface water is filtered before it reaches the level of the well. Outhouses, septic tanks, and cesspools must be situated in such a way that surface water passing through these areas does not carry fecal microbes directly into the well water. With the growing popularity of trailer homes, a new problem has arisen because of trailer sewage disposal tanks that are located too near the water supply. In some very old cities in which water pipes are cracked and sewage pipes leak, sewage has easy access to water pipes wherever there is a break in the pipe at a point just before it enters a dwelling.

WATER AND SEWAGE TREATMENT

Water must be properly treated to make it safe for human consumption. It is interesting to trace the many steps involved in such treatment (Fig. 10-11). The water first is filtered to remove large pieces of debris such as twigs and leaves. Next, it is held in sedimentation ponds where the addition of alum coagulates bacteria and organic materials, which then settle more rapidly. The water is then filtered through sand filters to remove the remaining bacteria and other small particles. Finally, chlorine gas or sodium hypochlorite is added to a final concentration of 0.2 to 1.0 part per million (ppm); this kills most remaining bacteria.

Small communities in rural areas may be financially unable to construct water treatment plants that incorporate all of the above-mentioned steps. Some may rely on chlorination alone. Unfortunately, the levels of chlorine routinely used for water treatment do not kill some pathogens, such as *Giardia lamblia* cysts and *Cryptosporidium parvum* oocysts. Other communities use all of the water treatment steps but fail to use filters having a small enough pore size to trap tiny pathogens such as *Cryptosporidium* oocysts (about 4 to 5 μm in diameter).

In the laboratory, water can be tested for fecal contamination by checking for the presence of coliform bacteria (*E. coli* and closely-related bacteria). These bacteria normally live in the human intestine; thus, their presence in drinking water is a sure sign that the water was fecally contaminated.

If one is unsure about the purity of drinking water, boiling it for 20 minutes destroys most pathogens that are present. It can then be cooled and used. Boiling

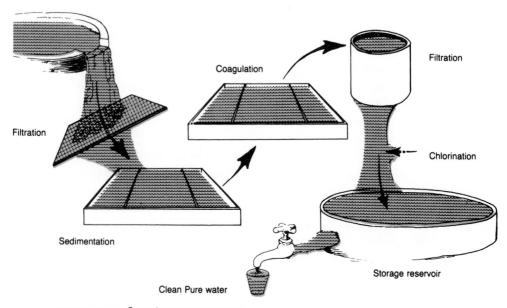

FIGURE 10-11. Steps in water treatment.

will kill *Giardia* cysts and *Cryptosporidium* oocysts, but there are some bacterial spores and viruses that can withstand long periods of boiling.

When sewage is adequately treated in a disposal plant, the water it contains can be returned to lakes and rivers to be recycled. Raw sewage consists mainly of water, fecal material including intestinal pathogens, and garbage and bacteria from the drains of human habitations. In the sewage disposal plant, large debris is first filtered out, and then the bacteria break down some of the organic material. Next, the activated sludge, which includes solid matter and bacteria, is settled out in a settling tank. In some communities, the sludge is heated to kill bacteria, then dried and used as fertilizer. The remaining liquid is filtered and chlorinated so that the effluent water can be returned to rivers or oceans. In some desert cities where water is in short supply, the effluent water from the sewage disposal plant is distilled so that it can be returned directly to the drinking water system. In some other cities, effluent water is used to irrigate lawns; however, it is expensive to install a separate water system for this purpose.

■ REVIEW OF KEY POINTS

- Infections that are acquired in the hospital are called nosocomial infections; those that are acquired elsewhere are called community-acquired infections. Iatrogenic infections or diseases are the result of medical or surgical treatment by surgeons, other physicians, and other healthcare personnel.
- Nosocomial infections occur all too frequently because of widespread indiscriminate use of broad-spectrum antibiotics, neglect of aseptic techniques and safety precautions, lengthy complicated surgeries, overcrowding of hospitals, shortage of hospital staff, lack of awareness of routine infection control measures, the use of anti-inflammatory and immunosuppressant agents, and the use of indwelling medical devices.
- The four most common causes of nosocomial infections are *Staphylococcus aureus*, *Escherichia coli*, *Enterococcus* species, and *Pseudomonas* species.
- The patients most susceptible to nosocomial infections are women in delivery, newborns, and surgical, cancer, diabetic and paralyzed patients. Reverse isolation techniques are designed to protect the most vulnerable patients, such as those with severe burns, leukemia, transplants, those who are undergoing radiation treatments, and other immunodeficient persons.
- All healthcare personnel must follow the same procedures to prevent the spread of communicable diseases. They must prevent cross-infections from themselves to susceptible patients, from hospitalized, contagious patients to susceptible patients, and from hospitalized, contagious patients to themselves. Healthcare personnel must use precautions that will protect them from bloodborne pathogens such as hepatitis B virus and HIV.
- Standard Precautions must be used for the care of *all* patients. Such precautions

are designed to reduce the risk of transmission of blood-borne and other pathogens; and apply to mucous membranes, non-intact skin, blood and all body secretions and excretions except sweat, regardless of whether they contain visible blood.

■ The five major routes of pathogen transmission are contact, droplet, airborne, common vehicle, and vector-borne. Within the hospital setting, Transmission-based Precautions (Airborne Precautions, Droplet Precautions, and Contact Precautions) are used in addition to Standard Precautions to protect healthcare personnel and hospital patients from airborne, droplet, and contact modes of pathogen transmission.

■ Medical aseptic techniques are designed to eliminate pathogens, whereas surgical aseptic techniques are designed to eliminate all microorganisms.

■ Medical aseptic techniques include sanitary methods for handling food and eating utensils; proper cooking and storage of food; proper disposal of waste products and contaminated materials; proper handwashing and personal hygiene of hospital personnel; proper use of gowns and masks in isolation rooms; proper washing and sterilizing of hospital equipment; proper use of disposable equipment, and proper disinfection of a room after a patient has been discharged.

■ To avoid becoming infected, extreme care must be taken by those involved in collecting, handling, and processing clinical specimens, particularly, blood, urine, cerebrospinal fluid, sputum, mucous membranes, and feces.

■ The primary function of the Clinical Microbiology Laboratory (CML) is to assist physicians in the diagnosis of infectious diseases. The major responsibilities of those employed in the CML are (1) processing clinical specimens, (2) isolating pathogens from specimens, (3) identifying pathogens, and (4) performing antimicrobial susceptibility testing.

■ A quality control program is necessary in the laboratory to monitor the reliability and quality of the work performed there; all equipment, test procedures, media, reagents, and staining solutions must be evaluated frequently to ensure that accurate results are being obtained.

■ The World Health Organization and public health and community groups at all levels must work together to coordinate preventive health programs and maintain constant surveillance of sources and causes of epidemics.

■ Prevention and control of epidemics includes measures to increase host resistance by immunizations; protect people from exposure to pathogens; segregate, isolate, and treat those with contagious infections to prevent the spread of pathogens to others; identify and control potential reservoirs and vectors of infectious diseases; and institute effective sanitation measures to control diseases transmitted through water supplies, sewage, and food.

Problems and Questions

1. How do nosocomial infections differ from community-acquired infections?
2. How can hospital-acquired infections be eliminated or reduced?
3. Which patients are most susceptible to nosocomial infections?
4. Which pathogens are most frequently associated with nosocomial infections?
5. Why are Standard Precautions necessary and how do they differ from Transmission-Based Precautions?
6. What are the differences between surgical and medical asepsis?
7. Who must be notified to help control an epidemic?
8. How can you be hurt by collecting and handling laboratory specimens?
9. In what ways can a specimen become contaminated?
10. What precautions must be followed when collecting a blood specimen for culture?
11. Which pathogens are most apt to be found in a blood sample?
12. How would you prepare a pathogenic specimen for shipment to a distant laboratory?
13. How would you determine which antimicrobial agent would be most effective against a certain pathogen?
14. How could you determine which pathogen you had isolated from a throat culture?
15. How is water treated to make it safe for human consumption?
16. If a laboratory test showed that *E. coli* was present in a sample of drinking water, what would this indicate?

Self Test

After you have read Chapter 10, reviewed the chapter outline, examined the objectives, studied the new terms, and answered the problems and questions above, complete the following self test.

MATCHING EXERCISES

Complete each statement from the list of words provided

aseptic   sterile   medical
surgical   sanitation   disinfection
sterilization   reverse isolation

1. The technique used to avoid *all* microorganisms is the _____ technique and is accomplished by _____.
2. The _____ technique is used to exclude *all* microorganisms from surgical areas to maintain _____ asepsis.
3. To maintain _____ asepsis, _____ technique is employed to exclude *all* pathogens from the area.
4. _____ of dressings and _____ of the skin is used to maintain _____ asepsis while dressing a wound.

5. The practical application of sanitary measures and cleanliness is termed _____.

6. Patients with severe burns and organ transplants are protected from hospital infections by _____.

TRUE OR FALSE (T OR F)

___ 1. Hospital infections can be avoided by proper awareness and the use of aseptic techniques.

___ 2. Sick and debilitated patients are much more susceptible to opportunistic pathogens than are healthy individuals.

___ 3. Because of the new disinfectants and antibiotics, the incidence of hospital-acquired infections has decreased.

___ 4. Many health-care workers are not adequately aware of the importance of aseptic and sterile techniques and universal precautions.

___ 5. The operating and delivery rooms are always clean, so no special precautions are necessary to protect the patient.

___ 6. Patients receiving steroids, anticancer drugs, antilymphocyte sera, and radiation treatments are usually resistant to hospital-acquired infections.

___ 7. Medical asepsis includes all precautionary measures necessary to prevent transfer of pathogens from person to person, including indirect transfer of pathogens through the air or on inanimate objects.

___ 8. Contaminated foods provide an excellent growth medium for pathogens.

___ 9. *Staphylococcus aureus* is one of the main pathogens spread by the hands and nasal secretions of hospital workers.

___ 10. Isolation techniques are used to protect everyone in the hospital from contagious diseases.

___ 11. Damp or wet masks are just as effective as dry ones.

___ 12. The microbiology laboratory should constantly monitor aseptic conditions in the hospital.

___ 13. Smallpox is thought to be totally eradicated due to the efforts of the public health authorities.

___ 14. All pathogens can be identified in the clinical laboratory, even in the presence of many contaminants.

___ 15. If the specimen is improperly collected, unnecessary contamination may mask the etiologic agent.

___ 16. Laboratory findings must be conveyed to the attending physician as soon as possible to aid in the diagnosis and treatment of infectious diseases.

___ 17. The specimens most likely to yield pathogens are collected before antimicrobial therapy begins.

___ 18. Aerobic microbes die when exposed to the air.

___ 19. Dangerous specimens should be placed in a sealed container for immediate transport to the laboratory.

___ 20. The label on the specimen need only identify the patient.

___ 21. Staphylococci are a part of the indigenous microflora of the blood and spinal fluid.

___ 22. Bacteremia and septicemia are different names for the same condition.

___ 23. Sputum specimens may be refrigerated for several hours without loss of pathogens.

___ 24. Genital swabs for gonorrhea must be inoculated immediately onto Thayer-Martin media and incubated in the presence of carbon dioxide.

___ 25. Only people who work in hospitals are responsible for the prevention and control of epidemics.

MULTIPLE CHOICE QUESTIONS

1. A nosocomial infection is one that
   a. the patient has at the time of hospital admission
   b. is acquired in the community
   c. a patient develops during hospitalization
   d. affects only the nose
   e. none of the above

2. An example of a "fomite" would be
   a. a drinking glass used by a patient
   b. bandages from an infected surgical site
   c. a contaminated bedpan
   d. soiled bed linens
   e. all of the above

3. Reverse isolation would be appropriate for
   a. a patient with tuberculosis
   b. a patient with a diarrheal disease
   c. a leukopenic patient
   d. a patient with pneumonic plague

4. The microorganism most apt to be the cause of a nosocomial infection is
   a. *Yersinia pestis*
   b. *Staphylococcus aureus*
   c. *Streptococcus pyogenes*
   d. *Clostridium perfringens*

5. Housekeeping and central supply personnel contribute to hospital asepsis by
   a. restricting their contact with patients
   b. protecting themselves from infection
   c. using techniques to prevent cross-contamination
   d. all of the above

6. Which of the following is not part of Standard Precautions?
   a. handwashing between patient contacts
   b. wearing gloves, masks, eye protection, and gowns when appropriate to do so
   c. cleaning and reprocessing reusable equipment before it is used for the care of another patient
   d. placing a patient in a private room having negative air pressure
   e. proper disposal of needles, scalpels, and other "sharps"

7. Assuming that a clean-catch midstream urine was processed, which of the following is indicative of a urinary tract infection?
   a. <100 CFU/mL
   b. <1000 CFU/mL
   c. >1000 CFU/mL
   d. >10,000 CFU/mL
   e. >100,000 CFU/mL

8. Which of the following statements is not true about the disk-diffusion method of antimicrobial susceptibility testing?
   a. It is also known as the "Kirby-Bauer test."
   b. A pure culture of the organism is required.
   c. The plate should be incubated in a CO_2 incubator for 12 hours.
   d. The test should be performed in the manner described by the NCCLS.

9. Spinal fluid specimens
   a. are easy to obtain
   b. are used to diagnose serious conditions such as meningitis and encephalitis
   c. should always be refrigerated
   d. are usually collected by nurses

10. All specimens must be
   a. properly selected
   b. properly collected
   c. properly transported to the laboratory
   d. all of the above

CHAPTER 11

Human Defenses Against Infectious Diseases

NONSPECIFIC MECHANISMS OF DEFENSE
First Line of Defense
Second Line of Defense
Fever Production
Iron Balance
Cellular Secretions
Blood Proteins
Phagocytosis
Inflammation
IMMUNE RESPONSE TO DISEASE: THIRD LINE OF DEFENSE

Immunity
Acquired Immunity
Immunology
Antigens
Antibodies
The Immune System
T Cells
B Cells
Immune Responses
Humoral Immunity
Cell-Mediated Immunity
NK and K Cells

Antibody Structure and Function
Monoclonal Antibodies
Hypersensitivity
The Allergic Response
Localized Anaphylaxis
Systemic Anaphylaxis
Latex Allergy
Skin Testing and Allergy Shots
Autoimmune Diseases
Immune Deficiency
Immunodiagnostic Procedures

OBJECTIVES

After studying this chapter, you should be able to:

- List and describe the nonspecific defenses of the human body
- Define the first and second lines of defense
- Define phagocytosis
- List the various types of phagocytic cells
- Describe the process of inflammation
- Describe the immune response, or third line of defense
- Define antigen, antibody, and immunoglobulin
- Differentiate between active and passive immunity

- Compare natural and artificial immunity
- List three ways in which vaccines are prepared
- Draw a graph representing the primary and secondary antibody responses to antigens
- Differentiate between immediate and delayed hypersensitivity
- Define autoimmunity and give examples
- List five serological tests used to determine the presence of a specific antibody in human serum

NEW TERMS

Acquired immunity
Active acquired immunity
Agammaglobulinemia
Agglutination
Allergen
Anamnestic response

Anaphylactic shock
Anaphylaxis
Antibody
Antigen
Antigenic determinant
Antigenic variation
Antiserum
Antitoxins
Atopic person
Attenuated vaccines
Autogenous vaccine
Autoimmune disease
B cell
Bacteriocins
Basophil
Beta-lysin (β-lysin)
Blocking antibodies
Cell-mediated immunity
Chemotactic agents
Chemotaxis
Colicin
Complement
Cytokines
Cytotoxic
Edema/edematous
Eosinophil

Eosinophilia
Epitope
Erythema/erythematous
Exudate
Fibronectin
Granulocyte
Hapten
Histamine
Histiocyte/ histocyte
Humoral immunity
Hybridoma
Hypersensitivity
Hypogammaglobulinemia
Immunity
Immunodiagnostic
   procedures
Immunogen
Immunoglobulin
Immunology
Inflammation
Interferon
Interleukins
Lymphokines
Macrophage
Mast cell
Microbial antagonism

Monoclonal antibodies
Monocyte
Natural killer cell
Neutrophil
Opsonins
Opsonization
Passive acquired
   immunity
Phagolysosome
Phagosome
Plasma cell
Precipitate
Precipitin tests
Prostaglandins
Pus
Pyogenic
Pyrogen/pyrogenic
Reticuloendothelial system
   (RES)
Serology
Serum (pl. *sera*)
T cell
Toxoid
Vasodilation
Wandering
   macrophages

NONSPECIFIC MECHANISMS OF DEFENSE

Humans and lower animals have survived on earth for millions of years because they have many built-in mechanisms of defense against harmful microorganisms and the infectious diseases they cause. The ability of any animal to resist these invaders and recover from disease is due to many complex interacting functions within the body.

Humans have three lines of defense against bacteria, viruses, fungi, and other pathogens. The first two lines of defense are nonspecific; these are ways in which the body attempts to destroy *all* types of substances that are foreign to it. The third line of defense, the immune response, is very specific. In the third line of defense, special proteins called *antibodies* are formed in response to the presence of particular foreign substances. These foreign substances are called *antigens* because they cause the production of specific antibodies; they are *"antibody generating"* substances. The immune response is discussed in more detail later in this chapter.

Nonspecific defense mechanisms are general in nature and serve to protect the body against many harmful substances. One of the nonspecific defenses is the innate, or inborn, resistance observed among some species of animals, some races of humans, and some persons who have a natural resistance to certain diseases. Innate or inherited characteristics make these people and animals more resistant to some diseases than to others. Resistance to cholera is an example of species resistance. Human beings can contract human cholera but not chicken cholera. The exact factors that produce this innate resistance are not well understood but are probably due to chemical, physiological, and temperature differences between the species, as well as the general state of physical and emotional health of the person and environmental factors that affect certain races and not others.

Although we are usually unaware of it, our bodies are, more or less, constantly in the process of defending against microbial invaders. Nonspecific defense mechanisms include such things as mechanical and physical barriers to invasion, chemical factors, microbial antagonism by our indigenous microflora, phagocytic host cells, fever, and the inflammatory response (inflammation) (Fig. 11-1).

First Line of Defense

One of the ways in which the body is protected from invasion by foreign microorganisms and other substances is the presence of mucous membranes at the openings to the respiratory, digestive, urinary, and reproductive systems. Mucus produced at mucous membranes entraps invaders.

Another way in which the body is protected is by intact skin providing a complete external covering for all other parts of the body. The unbroken skin acts as a

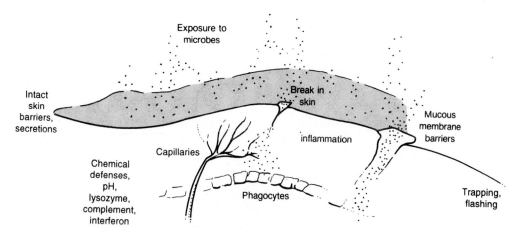

FIGURE 11-1. Nonspecific defenses against microbial invasion.

physical or mechanical barrier to pathogens; only when it is cut, abraded (scratched), or burned can they gain entrance. There are several factors that account for the skin's ability to resist pathogens. One is the skin's normal secretions, which destroy bacteria or inhibit their growth on its surface. These bactericidal secretions include acidic perspiration from sweat glands and fatty acids from oil glands. In addition, pathogens may not be able to establish themselves and multiply because of the many microorganisms already living in the pores and in moist parts of the head, underarms, hands, feet, and perianal and urethral regions.

The respiratory system would be particularly accessible to invaders that could ride on dust or other particles inhaled with each breath, were it not for the hair, mucous membranes, and irregular chambers of the nose that serve to trap much of the inhaled debris. The cilia (mucociliary covering) of the epithelial (surface) cells of the posterior nasal membranes, nasal sinuses, bronchi, and trachea sweep the trapped dust and microbes toward the throat, where they are swallowed or expelled by sneezing and coughing. Phagocytic white blood cells (phagocytes) in the mucous membranes may also be involved in this mucociliary clearance mechanism. Lysozyme and other enzymes that lyse or destroy bacteria are present in nasal secretions, saliva, and tears. The microflora usually residing in the nose and mouth may also serve a protective function.

To a large degree, the digestive system is protected by the process of digestion, digestive enzymes, acidity of the stomach, and alkalinity of the intestines. Bile, which is secreted from the liver into the intestine, lowers the surface tension and causes chemical changes in bacterial cell walls and membranes that make bacteria more digestible. Many invading microorganisms are trapped in the mucous lining of the digestive tract where they may be destroyed by bactericidal enzymes and phagocytes. The indigenous microflora that use available nutrients and occupy space in the intestines include *Escherichia coli, Enterobacter aerogenes, Enterococcus faecalis,* and many other anaerobic or facultatively anaerobic enteric (intestinal) bacteria. Peristalsis and the expulsion of feces serve to remove bacteria from the intestine, as well.

The urinary tract is usually sterile in healthy persons; therefore, there should be no microorganisms present in urine drawn from the urinary bladder in an aseptic manner (*i.e.,* via needle and syringe). Also, the reproductive system of the male and most of the reproductive organs in the female lack an indigenous microflora. However, indigenous microflora (including opportunistic pathogens) from the anus and perianal skin may enter the vagina and invade the urethra. Microorganisms are continually flushed from these areas by frequent urination and expulsion of mucous secretions. Many bladder infections occur simply as a result of infrequent urination, especially failure to urinate after intercourse. Usually, the acidic urine and vaginal secretions also inhibit microbial growth. Many women who are taking certain oral contraceptives are particularly susceptible to some infections because those chemicals increase the pH of the vagina.

The prevention of colonization of potential microbial pathogens by the indigenous microflora of a given anatomical site is called *microbial antagonism;* this is another example of a nonspecific defense mechanism. The inhibitory capability of these microflora has been attributed to a competition for nutrients and the production of certain inhibitory substances. Examples of such inhibitory substances (proteins) include the *colicins* produced by certain strains of *E. coli*. Similar substances are produced by some strains of *Pseudomonas* and *Bacillus* species, as well as by other bacteria. Collectively these antibacterial substances, produced by bacteria, are known as *bacteriocins*. The effectiveness of microbial antagonism is frequently decreased following prolonged administration of broad-spectrum antibiotics. The antibiotics reduce or eliminate certain members of the microflora (*e.g.*, the vaginal and gastrointestinal flora) and permit overgrowth by bacteria and/or fungi that are resistant to the antibiotic being administered. This overgrowth of organisms, by *Candida albicans* in the vagina or *Clostridium difficile* in the colon, is called a *superinfection*.

Second Line of Defense

The nonspecific cellular and chemical responses to microbial invasion are considered the second line of defense. Virulent pathogens that penetrate the first line of defense usually are destroyed by a series of defense mechanisms, including the inflammatory response. A complex sequence of events develops involving fever production, iron balance, cellular secretions (interferon, fibronectin, β-lysin, interleukins, prostaglandins, histamine), activation of blood proteins (complement, properdin), chemotaxis, phagocytosis, neutralization of toxins, and clean-up and repair of damaged areas. Each of these responses is discussed in this section.

FEVER PRODUCTION
Normal body temperature fluctuates between 36.2°C and 37.5°C, with an average of about 37°C (see Appendix E for information about Fahrenheit temperatures). A body temperature over 38°C is generally considered to be a fever. Substances that stimulate the production of fever are called *pyrogens* or *pyrogenic substances*. Pyrogens may come either from outside or inside the body. Those from outside the body include pathogens and various pyrogenic substances that they produce and/or release. Interleukin-1 (IL-1; discussed in following sections) is an example of a pyrogen that is produced within the body. The resulting increased body temperature (fever) is considered to be a nonspecific host defense mechanism. It augments the host's defenses by (1) stimulating white blood cells (leukocytes) to deploy and destroy invaders; (2) reducing available free plasma iron, which limits the growth of pathogens that require iron for replication and synthesis of toxins; and (3) inducing the production of IL-1, which causes the proliferation, maturation, and activation of lymphocytes in the immunological response. Elevated body temperatures also slow down the rate of growth of certain pathogens, and can even kill some especially fastidious pathogens.

IRON BALANCE

The virulence of many bacteria is enhanced in the presence of free iron, which is used for the synthesis of exotoxins. In response to pathogenic invasion, some of the host's leukocytes produce IL-1. Among other functions, this substance induces the release of lactoferrin, which stimulates iron storage in the liver and, thus, reduces the amount of free iron available for the pathogen. Interleukin-1 is one of many different proteins that are produced by cells to control various aspects of cell growth and functioning; collectively, these proteins are referred to as *cytokines*.

CELLULAR SECRETIONS

Interferons. *Interferons* are small, anti-viral proteins produced by virus-infected cells. The three known types of interferon, referred to as α-, β-, and γ-(gamma) interferons, are induced by different stimuli, including viruses, tumors, bacteria, and other foreign cells. The interferons produced by a virus-infected cell are unable to save that cell from destruction, but, when released from that cell, they attach to the membranes of surrounding cells and prevent viral replication from occurring in those cells. Thus, the spread of the infection is inhibited, allowing other body defenses to fight the disease more effectively. In this way, many viral diseases (*e.g.,* colds, influenza, and measles) are self-limiting in duration. Similarly, the acute phase of herpes simplex cold sores is of limited duration. The virus then enters a latent phase and hides in nerve ganglion cells where it is protected until the person's defenses are down; thus, the cycle of disease is repeated.

Interferons are not virus-specific, meaning that they are effective against a variety of viruses, not just the particular type of virus that stimulated their production. Interferons are species-specific, however, meaning that they are effective only in the species of animal that produced them. Thus, rabbit interferons are only effective in rabbits, and could not be used to treat viral infections in humans. Human interferons are industrially produced by genetically engineered bacteria (with interferon genes inserted) and are used experimentally to treat certain viral infections (*e.g.,,* warts, herpes simplex, hepatitis B and C) and cancers (*e.g.,* leukemias, lymphomas, Kaposi's sarcoma in AIDS patients). In addition to interfering with viral multiplication, interferons also activate certain lymphocytes (natural killer cells) to kill virus-infected cells. Natural killer (NK) cells are discussed in a later section.

Fibronectin. *Fibronectin* is an epithelial tissue secretion (glycoprotein) that enables cells to bind with collagen and other components of the extracellular matrix. It can also interact with certain bacteria (*e.g.,* staphylococci, streptococci) to block their attachment to epithelial cells; thus, it aids in clearing and flushing these pathogens from the body.

ß-lysin. *ß-lysin* is a polypeptide that is released from blood platelets during an infection. It destroys Gram-positive bacteria by disrupting their plasma membranes, causing lysis. It is also found inside phagocytes where it aids in the digestion of microbes.

Interleukins. *Interleukins* (IL-1, IL-2, IL-3) are polypeptides that are secreted by antigen-stimulated macrophages (large, phagocytic white blood cells) and lymphocytes. They enhance T lymphocyte activation, proliferation, and activity during the immune response.

BLOOD PROTEINS

Complement. *Complement* is actually a group of approximately 30 different proteins (including proteins designated as C1 through C9) found in normal blood plasma. They constitute what is called "the complement system"—so named because it is "complementary" to the action of the immune system. The proteins of the complement system interact with each other in a step-wise manner, known as the complement cascade. Activation of the complement system is considered a nonspecific defense mechanism because it assists in the destruction of many different pathogens. The major consequences of complement activation are the (1) initiation and amplification of inflammation, (2) attraction of phagocytes to the sites to which they are needed (chemotaxis; discussed later), (3) activation of leukocytes, (4) lysis of bacteria and other foreign cells, and (5) increased phagocytosis by phagocytic cells (opsonization). *Opsonization* is a process by which phagocytosis is facilitated by the deposition of *opsonins* (*e.g.,* antibodies or complement components) onto the surface of particles or cells. In some cases, phagocytes are unable to ingest certain particles or cells until opsonization occurs.

Activation of the complement system by antigen-antibody complexes (immune complexes) is known as the classical pathway. Certain microbial surface molecules, microbial secretions (*e.g.,* endotoxin and proteases), and aggregated immunoglobulins can also activate the complement system; this is known as the alternative pathway. Complement components C1, C2, and C4 do not participate in the alternative pathway. Instead, plasma protein factors B, D, and P (known as properdin) work in tandem with complement components C3 and C5 through C9 to attract phagocytes and enhance phagocytosis, inflammation, and the destruction of bacteria and certain viruses.

Prostaglandins. *Prostaglandins* are membrane-associated lipids that act much like local hormones. They are biologically reactive in controlling platelet aggregation, immune response, inflammation, increased capillary permeability, pain production, diarrhea, autoimmune responses, and many other conditions in health and disease. Some anti-inflammatory drugs, such as aspirin, function by inhibiting the production of prostaglandins.

Prostaglandins also play a role in fever, as illustrated by the following scenario for fever production:

1. A patient has septicemia due to Gram-negative bacteria ("Gram-negative sepsis").

2. The bacteria release endotoxin into the patient's bloodstream. (Recall that endotoxin is part of the cell wall structure of Gram-negative bacteria.)
3. Phagocytic white blood cells (phagocytes) ingest (phagocytize) the endotoxin.
4. The ingested endotoxin stimulates the phagocytes to produce IL-1 (also known as endogenous pyrogen).
5. The IL-1 stimulates the hypothalamus (a part of the brain, referred to as the body's "thermostat") to produce prostaglandins.
6. The prostaglandins cause the hypothalamus to raise the body temperature, resulting in a fever.

PHAGOCYTOSIS

Phagocytic white blood cells (leukocytes) are called phagocytes and the process by which phagocytes surround and engulf (ingest) foreign material is called phagocytosis. The two most important phagocytes in the human body are macrophages and neutrophils (Fig. 11-2); they are sometimes called "professional phagocytes" because phagocytosis is their major function. Phagocytes serve as a "clean-up crew" to rid the body of unwanted and often harmful substances, such as dead cells, unused cellular secretions, debris, and microorganisms.

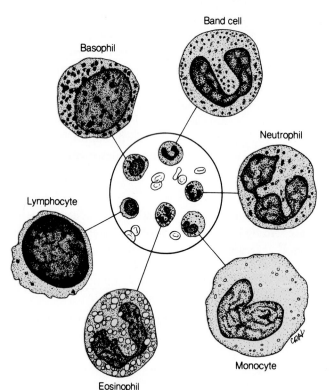

FIGURE 11-2. Types of blood cells. When blood smears are stained with Wright's stain, eosinophils have reddish-orange granules, basophils have purple granules, and neutrophil granules are colorless. In this way, the different granulocytes can be distinguished from one another.

Cellular Elements of the Blood

Erythrocytes (red blood cells)

Thrombocytes (platelets)

Leukocytes (white blood cells)

    Granulocytes

        Basophils

        Eosinophils

        Neutrophils

    Monocytes/Macrophages

    Lymphocytes

        B cells

        Helper T cells (T_H cells)

        Suppressor T cells (T_S cells)

        Cytotoxic T cells (T_C cells)

        Delayed hypersensitivity T cells (T_H cells)

        Natural killer (NK) cells

Granulocytes are named for the prominent cytoplasmic granules that they possess. Phagocytic granulocytes include *neutrophils* and *eosinophils*. Neutrophils (also known as polymorphonuclear cells, polys, and PMNs) are much more efficient at phagocytosis than are eosinophils, and represent the body's most efficient and abundant phagocytes. An abnormally high number of eosinophils in the peripheral bloodstream is known as *eosinophilia;* examples of conditions that cause eosinophilia are allergies and helminth infections. A third type of granulocyte, *basophils,* (discussed later in the chapter) are also involved in allergic and inflammatory reactions, although they are not phagocytes.

Macrophages develop from a type of leukocyte called *monocytes* during the inflammatory response to infections. Those that leave the bloodstream and migrate to infected areas are called *wandering macrophages. Fixed macrophages,* or *histiocytes,* remain in tissues and organs and serve to trap foreign debris. Macrophages are extremely efficient phagocytes. They are found in tissues of the *reticuloendothelial system* (RES); this nonspecific defensive system includes cells in the liver (Kupffer cells), spleen, lymph nodes, and bone marrow, as well as the lungs (dust cells), blood vessels, intestines, and brain (microglia). The principal function of the entire RES is the engulfment and removal of foreign and useless particles, living or dead, such as excess cellular secretions, dead and dying leukocytes, erythrocytes, and tissue cells, as well as foreign debris and microorganisms that gain entrance to the body.

Phagocytosis begins when phagocytes move to the site where they are needed. This directed migration is called *chemotaxis,* and is the result of chemical attractants

 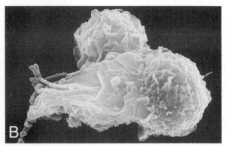

FIGURE 11-3. Scanning electron micrographs showing phagocytosis of *Giardia* trophozoites by rat leukocytes. (SEMs by S. Erlandsen and P. Engelkirk)

called *chemotactic agents*. The phagocytes move along a concentration gradient, meaning that they move from areas of low concentrations of chemotactic agents to the area of highest concentration. The area of highest concentration is the site where the chemotactic agents are being produced or released—often the site of inflammation. Different types of chemotactic agents attract different types of leukocytes; some attract monocytes, others attract neutrophils, and still others attract eosinophils.

The next step is attachment of the phagocyte to the object (*e.g.*, a bacterium) to be ingested. Phagocytes can only ingest objects that they can attach to. The phagocyte then surrounds the object with pseudopodia, which fuse together, and the object is ingested (also called endocytosis) (see Figures 11-3 and 11-4). Within the cytoplasm of the phagocyte, the object is contained within a membrane-bound vesicle called a *phagosome*. The phagosome next fuses with nearby lysosomes to form a digestive vacuole (*phagolysosome*), within which killing and digestion occur. Lysosomal enzymes, including lysozyme and β-lysin, digest and degrade carbohydrates, lipids, proteins, and nucleic acids (see Figure 11-5).

Other mechanisms also participate in the destruction of phagocytized microorganisms. In neutrophils, for example, a membrane-bound enzyme called NADPH oxidase reduces oxygen to very destructive products such as superoxide anions, hydroxyl radicals, hydrogen peroxide and singlet oxygen. These highly reactive reduction products assist in the destruction of the ingested microbes.

Four Steps in Phagocytosis
1. Chemotaxis
2. Attachment
3. Ingestion
4. Digestion

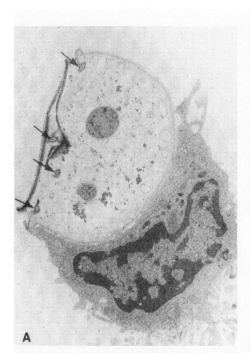

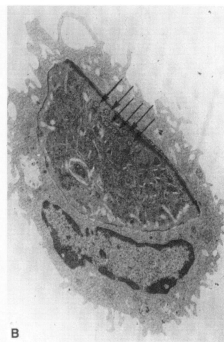

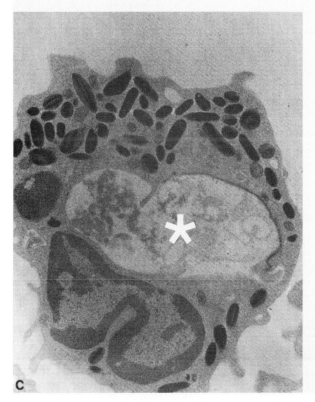

FIGURE 11-4. Transmission electron micrographs showing phagocytosis of *Giardia* trophozoites by rat leukocytes. (*A*) Attachment. (*B*) Ingestion. (*C*) Digestion. Note the cross sections of flagella (arrows) in *A* and *B*, and the phagolysosome (*) and darkly-stained granules in the eosinophil at *C*. (TEMs by S. Koester and P. Engelkirk)

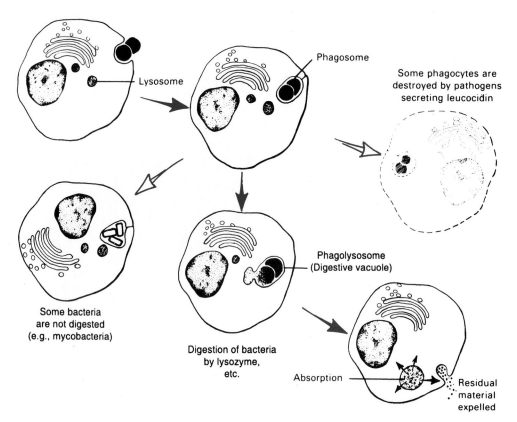

FIGURE 11-5. Phagocytosis. Not all pathogens are destroyed by phagocytes. Encapsulated bacteria can escape phagocytosis and bacteria that secrete leukocidin can destroy phagocytes. Some bacteria (e.g., *Mycobacterium tuberculosis*) can survive within the phagolysosome.

During the initial phases of infection, capsules serve an antiphagocytic function, protecting encapsulated bacteria from being phagocytized. Other bacteria produce an exoenzyme called leukocidin, which kills phagocytes (discussed in Chapter 9). It should also be noted that not all bacteria engulfed by phagocytes are destroyed within phagolysosomes. Waxes in the cell wall of *Mycobacterium tuberculosis,* for example, protect the organism from digestion. The bacteria are even able to multiply within the phagocytes and be transported by them to other parts of the body. Other pathogens that are able to survive within phagocytes include bacteria such as *Rickettsia rickettsii, Legionella pneumophila, Brucella abortus, Coxiella burnetii, Listeria monocytogenes,* and *Salmonella,* and protozoan parasites such as *Toxoplasma gondii, Trypanosoma cruzi,* and *Leishmania* spp. The mechanism by which each pathogen evades digestion by lysosomal enzymes differs from one pathogen to another and, in some cases, the mechanism is not understood. These pathogens may remain dormant within phagocytes for months or years before they escape to cause disease. Thus, these types of virulent pathogens usually win the battle with phagocytes.

Unless antibodies and complement components are present to aid in the destruction of these pathogens, the infection may progress unchecked.

Ehrlichia spp., closely related to rickettsias, are obligate, intracellular, Gram-negative bacteria that live within leukocytes; *i.e.,* they are intraleukocytic parasites. *Ehrlichia* spp. cause two endemic, tick-borne diseases in the United States. In human monocytic ehrlichiosis (HME), the bacteria infect monocytic phagocytes, whereas in human granulocytic ehrlichiosis (HGE), they infect granulocytes. The bacteria are somehow able to prevent the fusion of lysosomes with phagosomes.

INFLAMMATION

The body normally responds to any local injury, irritation, microbial invasion, or bacterial toxins by a complex series of events collectively referred to as *inflammation*. The purposes of the inflammatory response are to (1) localize an infection, (2) prevent the spread of microbial invaders, (3) neutralize toxins, and (4) aid in the repair of damaged tissue (Fig. 11-6). In this process, many nonspecific defense mechanisms come into play. These interrelated physiological reactions result in the characteristic signs and symptoms of inflammation: redness, heat, swelling (edema), pain, often pus formation, and occasionally a loss of function of the damaged area.

A complex series of physiological events occurs immediately after the initial damage to the tissue. One of the initial events is *vasodilation* (an increase in the diameter of blood vessels) in the area of the injury. This allows more blood to flow to the site, bringing redness and heat. Additional heat results from increased metabolic activities in the tissue cells at the site. Vasodilation causes the endothelial cells that line the capillaries to stretch and separate, causing increased permeability; plasma escapes from the vessels into the surrounding area, causing the area to become edematous (swollen). Sometimes the swelling is severe enough to interfere with the bending of a particular joint (*e.g.,* knuckle, elbow, knee, ankle), leading to a loss of function.

A variety of chemotactic agents are produced at the site of inflammation, leading to an influx of phagocytes. The pain or tenderness that accompanies inflammation

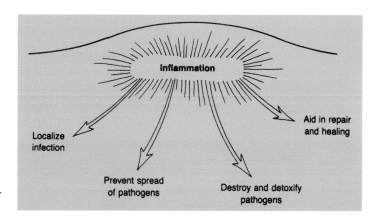

FIGURE 11-6. The purposes of inflammation.

may result from actual damage of the nerve fibers because of the injury, irritation by microbial toxins or other cellular secretions (such as prostaglandins), or increased pressure on nerve endings due to the edema (Fig. 11-7).

The accumulation of fluid, cells, and cellular debris at the inflammation site is referred to as the inflammatory exudate. If the exudate is thick and greenish-yellow, containing many live and dead leukocytes, it is known as a *purulent exudate* or *pus*. However, it should be noted that in many inflammatory responses, such as arthritis or pancreatitis, there is no exudate and no invading microorganisms. When *pyogenic* (pus-producing) microorganisms, such as some staphylococci and streptococci, are present, even more pus is produced as a result of the killing effect of the bacterial toxins on phagocytes and tissue cells. Although most pus is greenish-yellow, the exudate is often bluish-green in infections caused by *Pseudomonas aeruginosa*. This is due to the bluish-green pigment (pyocyanin) produced by this organism.

When the inflammatory response is over and the body has won the battle, the phagocytes clean up the area and help to restore order. The cells and tissues can then repair the damage and begin to function normally again in a homeostatic (equilibrated) state, although some permanent damage and scarring may have occurred.

The lymphatic system—including the lymph, lymphatic vessels, lymph nodes, and lymphatic organs (tonsils, spleen, and thymus gland)—also plays an important role in defending the body against invaders. The primary functions of this system include draining and circulating intercellular fluids from the tissues and transporting digested fats from the digestive system to the blood. Also, macrophages, B lymphocytes (B cells), and T lymphocytes (T cells) in the lymph nodes serve to filter the lymph by removing foreign matter and microbes and by producing antibodies and other factors to aid in the destruction and detoxification of any invading microorganisms.

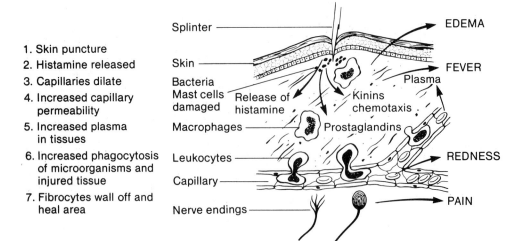

1. Skin puncture
2. Histamine released
3. Capillaries dilate
4. Increased capillary permeability
5. Increased plasma in tissues
6. Increased phagocytosis of microorganisms and injured tissue
7. Fibrocytes wall off and heal area

FIGURE 11-7. The inflammatory response.

The body continually wages war against damage, injury, malfunction, and microbial invasion. The outcome of each battle depends on the person's age, hormonal balance, genetic resistance, and overall state of physical and mental health, as well as the virulence of the pathogens involved.

IMMUNE RESPONSE TO DISEASE: THIRD LINE OF DEFENSE

The immune response is the third line of defense against pathogens. Usually, in this protective type of defense mechanism, special glycoproteins called *antibodies* are produced by lymphocytes to recognize, bind with, inactivate, and destroy specific microorganisms. These humoral (circulating) antibodies are normally found in blood plasma, lymph, and other body secretions where they readily protect against the specific pathogens that stimulated their production. Thus, a person is immune to a particular infectious disease because of the presence of specific protective antibodies that are effective against the causative (etiologic) agent of that disease. In addition, there are protective cell-mediated immune responses where antibodies play only a minor role.

Although the immune system protects against disease and aids in fighting cancer, it may also cause damage to its host, as you will learn from the upcoming discussions on cell-mediated hypersensitivity and autoimmunity.

Immunity

Immunity is the condition of being immune or resistant to a particular infectious disease, usually as a result of the presence of protective antibodies that are directed against the etiologic agent of that disease. The innate, or native, resistance to disease found in certain individuals, races, and species of animals is not a type of immunity conferred by antibodies but, rather, is a resistance resulting from natural nonspecific factors. A person who is susceptible to a disease usually has inadequate levels of protective antibodies and/or insufficient nonspecific defenses, which, in some cases, may simply reflect a very poor state of health or the presence of an immunodeficiency disease.

ACQUIRED IMMUNITY

Immunity that results from the active production or receipt of antibodies is called *acquired immunity*. If the antibodies are actually produced within the person's body, the immunity is called *active acquired immunity;* such protection is usually long-lasting. In *passive acquired immunity,* the person receives antibodies that were produced by another person or, in some cases, by an animal; such protection is usually only temporary. The various types of acquired immunity are summarized in Table 11-1.

TABLE 11-1. Types of Acquired Immunity

ACTIVE		PASSIVE	
Natural	**Artificial**	**Natural**	**Artificial**
Clinical or subclinical disease	Vaccines: Inactivated (killed) pathogens Attenuated (weakened) pathogens Extracts (parts of pathogens) Toxoids	Congenital (across placenta) Colostrum	Antiserum Antitoxin Gamma globulin

Active Acquired Immunity. People who have had a specific infection usually have some resistance to reinfection by the causative pathogen because of the presence of antibodies and stimulated lymphocytes. This is called *natural active acquired immunity.* Symptoms of the disease may or may not be present when these antibodies are formed. Such resistance to reinfection may be permanent (as with mumps, measles, smallpox, diphtheria, whooping cough, poliomyelitis, plague, and typhoid fever), or it may be only temporary (as with pneumonia, influenza, gonorrhea, and streptococcal and staphylococcal infections). There is no immunity to reinfection following recovery from gonorrhea, syphilis and tuberculosis; although antibodies are produced against the etiologic agents of these diseases, they are not protective antibodies (*i.e.,* the antibodies that are produced do not protect the person from being reinfected).

Artificial active acquired immunity is the second type of actively acquired immunity. This type of immunity results when a person receives a vaccination—the administration of a vaccine that stimulates the production of specific protective antibodies. The vaccine contains sufficient antigens of a pathogen to enable the person to form antibodies against that pathogen. An ideal vaccine is one that (1) contains enough antigens to protect against infection by the pathogen; (2) contains antigens from all of the strains of the pathogen that cause that disease (*e.g.,* three strains of virus cause polio); (3) is not too toxic; and (4) does not cause disease in the vaccinated person. A timeline for vaccine development is shown in Table 11-2.

Because so many different types of viruses cause colds, a successful vaccine for colds has not been developed. Maintaining a successful vaccine for influenza is also difficult because the viruses continually change by mutation—a phenomenon known as *antigenic variation.*

Vaccines are made from living or dead (inactivated) pathogens or from certain toxins they excrete (Fig. 11-8). In general, vaccines made from living organisms are most effective, but they must be prepared from harmless organisms that are antigenically closely related to the pathogens or from weakened pathogens that have been genetically changed so they are no longer pathogenic. The process of weakening pathogens is called attenuation, and the vaccines are referred to as *attenuated vaccines*. The smallpox (variola) vaccine is derived from the cowpox virus (vaccinia), which causes a mild pox infection in cattle and humans. Most other live vaccines are avirulent (nonpathogenic) mutant strains of pathogens that have been derived from the virulent (pathogenic) organisms; this is accomplished by growing them for many generations under various conditions or by exposing them to mutagenic chemicals or radiation. Some attenuated live vaccines are shown in Table 11-3.

Vaccines made from dead pathogens, which have been killed by heat or chemicals, can be produced faster and more easily, but they are less effective than live vaccines. This is because the antigens on the dead cells are usually less effective and produce a shorter period of immunity. Inactivated vaccines are safer to use during the experimental phase of vaccine production because of the remote possibility that an avirulent strain of a living organism may revert to the virulent strain before the strain's stability has been established. For example, the first poliomyelitis virus vaccine, developed by Jonas Salk, contained the inactivated virions of three strains of

TABLE 11-2. Timeline for Vaccine Development

Time Period	Development
1500s	Asian physicians used dried smallpox crusts to vaccinate people by scratching the material into their skin or by inhalation.
1796	Edward Jenner, a British physician, used cowpox virus to vaccinate against smallpox.
1885	Louis Pasteur developed a rabies vaccine.
1890–1904	Vaccines for diphtheria and tetanus were developed.
1904–1914	Typhoid fever and cholera vaccines were produced.
1914	Tetanus vaccine became available.
1920s	Tuberculosis vaccine (BCG) was made. BCG stands for Bacillus of Calmette and Guérin. It is a live (attenuated) strain of *Mycobacterium bovis*.
1930s	Diphtheria and yellow fever vaccines were produced.
1940s	Influenza and pertussis (whooping cough) vaccines were created.
1955–1960	Jonas Salk and Alfred Sabin developed different forms of poliomyelitis vaccine. Salk's dead (inactivated) injectable vaccine was first introduced in 1955. Several years later, Sabin developed a live (attenuated) oral vaccine.
1960s	Measles and rubella vaccines were made.
1968	Mumps vaccine was created.
1970s	Meningococcal and chickenpox vaccines were produced.
1978	Pneumococcal pneumonia vaccine was developed.
1980s	Hepatitis B and M-M-R combination vaccines became available.
1990	Vaccine for *Haemophilus influenzae* meningitis was introduced.

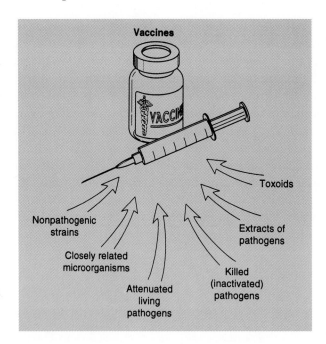

FIGURE 11-8. Sources of vaccines.

polioviruses, which had to be injected to produce adequate immunity against the disease. A few years later, it was found that the live (attenuated) virus vaccine, developed by Sabin and others, was safe when taken orally. Some killed (inactivated) vaccines are shown in Table 11-3. The use of such vaccines illustrates a very important and practical application of the principles of microbiology and immunology.

A subunit vaccine (or acellular vaccine) is one that uses antigenic (antibody-stimulating) portions of pathogens, rather than using the whole pathogen. For example, a vaccine containing pili of *Neisseria gonorrhoeae* could theoretically stimulate the body to produce antibodies that would attach to *N. gonorrhoeae* pili, thus, preventing the bacteria from adhering to cells. If *N. gonorrhoeae* cannot adhere to cells that line the urethra, then they cannot cause urethritis.

Another type of vaccine used to prevent tetanus and diphtheria, called a *toxoid* vaccine, is prepared from exotoxins that have been inactivated or made nontoxic by heat or chemicals. Toxoids can be injected safely to stimulate the production of antibodies that are capable of neutralizing the exotoxins of pathogens, such as those that cause tetanus and diphtheria. Antibodies that neutralize toxins are called *antitoxins* and a serum containing such antitoxins is referred to as an *antiserum*. Serum is the liquid portion of blood that has been allowed to clot (*i.e.*, the liquid remaining after the clotting factors have been removed from plasma).

As microbiologists made further studies of the characteristics of vaccines, they found that it was practical to vaccinate against several diseases by combining specific vaccines in a single injection. Thus, the diphtheria-tetanus-pertussis (DTP) vaccine contains toxoids to prevent diphtheria and tetanus and portions of killed bacteria (*Bordetella pertussis*) to prevent whooping cough (pertussis). Another example is the measles-mumps-rubella (MMR) vaccine. According to the Centers for Diseases Control and Prevention (CDC), U.S. children should receive the following vaccines between birth and entry into school:

- Diphtheria-tetanus-pertussis (DTP) vaccine
- Poliovirus vaccine
- Measles-mumps-rubella (MMR) vaccine
- *Haemophilus influenzae* type b (Hib) vaccine
- Hepatitis B (HepB) vaccine
- Varicella (chickenpox) vaccine
- Rotavirus vaccine

(From: *Morbidity and Mortality Weekly Report,* Vol. 48, No. 1, January 15, 1999)

An *autogenous vaccine* is prepared from bacteria isolated from a localized infection, such as a staphylococcal boil. The pathogens are killed, then injected into the same person to induce production of more antibodies.

Passive Acquired Immunity. Passive immunity differs from active immunity in that antibodies formed in one person are transferred to another to protect the latter from infection. Because the person receiving the antibodies did not actively produce them, the immunity is temporary, lasting only about 3 to 6 weeks. The antibodies of passive immunity may be transferred naturally or artificially.

In *natural passive acquired immunity,* small antibodies (like IgG, which is described later in this chapter) present in the mother's blood cross the placenta to reach the fetus while it is in the uterus (*in utero*). Also, colostrum, the thin, milky fluid secreted by mammary glands a few days before and after delivery, contains maternal antibodies to protect the infant during the first months of life.

TABLE 11-3. Types of Vaccines

Type of Vaccine	Examples
Live (attenuated) viruses	Adenovirus, measles (rubeola), mumps, German measles (rubella), polio (oral), smallpox (varicella), yellow fever
Dead (inactivated) viruses or viral antigens	Hepatitis B, influenza, Japanese encephalitis, polio (subcutaneous), rabies
Live (attenuated) bacteria	BCG (for protection against tuberculosis), typhoid fever (oral)
Dead (inactivated) bacteria	Anthrax, cholera, pertussis, plague, typhoid fever (subcutaneous)
Bacterial capsular antigens	Hib (for protection against *Haemophilus influenzae* type b), meningococcal, pneumococcal
Bacterial toxoids	Diphtheria, tetanus

Artificial passive acquired immunity is accomplished by transferring antibodies from an immune person to a susceptible person. After a patient has been exposed to a disease, the length of the incubation period usually does not allow sufficient time for vaccination to be an effective preventive measure. This is because a span of about 2 weeks is needed before sufficient antibodies are formed to protect the exposed person. To provide temporary protection in these situations, the patient is given human gamma globulin or "pooled" immune serum globulin (ISG); that is, antibodies taken from the blood of many immune people. In this manner, the patient receives some antibodies to all of the diseases to which the donors are immune. The ISG may be given to provide temporary protection against measles, mumps, polio, diphtheria, and hepatitis in people, especially infants, who are not immune and have been exposed to these diseases.

Hyperimmune serum globulin (or specific immune globulin) has been prepared from the serum of persons with high antibody levels (titer) against certain diseases. For example, hepatitis B immune globulin (HBIG) is given to protect those who have been, or are apt to be, exposed to hepatitis B virus; tetanus immune globulin (TIG) is used for nonimmunized patients with deep, dirty wounds; and rabies immune globulin (RIG) may be given following a bite by a rabid animal. Other examples include chickenpox immune globulin, measles immune globulin, pertussis immune globulin, poliomyelitis immune globulin, and zoster immune globulin. In potentially lethal cases of botulism, antitoxin antibodies are used to neutralize the toxic effects of the botulinal toxin. Remember that passive acquired immunity is always temporary because the antibodies are not actively produced by the lymphocytes of the protected person.

Immunology

Immunology is the scientific study of immune responses. It is a huge and complex field of study, and only the basic fundamentals can be presented in this book. The topics briefly discussed in this chapter include active and passive immunity to infectious agents, processes involved in antibody production, cell-mediated immune responses, allergies and other types of hypersensitivity reactions, autoimmunity, and immunodiagnostic procedures for detecting antibodies and antigens in clinical specimens.

ANTIGENS

An *antigen* (or *immunogen*) can be any foreign organic substance that is large enough to stimulate the production of antibodies; in other words, it is an *anti*body-*gen*erating substance. Such a substance is said to be *antigenic* or *immunogenic*. Antigens may be proteins of more than 10,000 daltons molecular weight, polysaccharides larger than 60,000 daltons, large molecules of DNA or RNA, or any combination of biochemical molecules (*e.g.,* glycoproteins, lipoproteins, and nucleoproteins) that are cellular components of either microorganisms or macroorganisms. Foreign proteins are the best antigens. On the surface of a bacterial cell are many molecules capable of stimulating the production of antibodies; these individual molecules or antigenic sites are known as *antigenic determinants* (or *epitopes*). The important point is that antigens must be *foreign* materials that the human body does not recognize as *self* antigens. Certainly, all microorganisms fall into this category. Some small molecules called *haptens* may act as antigens if they are coupled with a large carrier molecule such as a protein. Then the antibodies formed against the antigenic determinant(s) of the hapten may combine with the hapten molecules when they are not coupled with the carrier protein. As an example, penicillin and other low-molecular-weight chemical molecules may act as haptens, causing some people to become allergic (or hypersensitive) to them.

ANTIBODIES

Antibodies are glycoproteins produced by lymphocytes in response to the presence of an antigen. (As will be described later, the antibody-producing cells are a specific type of lymphocyte called B lymphocytes [or B cells], which often work in coordination with T lymphocytes [T cells] and macrophages.) A bacterial cell has numerous antigenic determinants on its cell membrane, cell wall, capsule, and flagella that stimulate the production of many different antibodies. Usually, an antibody is considered to be "specific" in that it will recognize and bind to only the antigenic determinant that stimulated its production. Occasionally, an antibody will bind to an antigenic determinant that is similar, but not identical, in structure to the antigenic determinant that stimulated its production; in this case, it is referred to as a cross-reacting antibody.

STUDY AID Note that the definitions of antigens and antibodies are cyclical; *i.e.*, one is defined in terms of the other. Antigens are substances that stimulate the body to produce antibodies. Antibodies are glycoproteins produced by our bodies in response to antigens.

All antibodies are in a class of proteins called *immunoglobulins* (Ig); they are globular glycoproteins in the blood that participate in the immune reactions. We usually use the term antibodies to refer to immunoglobulins with particular specificity for an antigen. In addition to being found in blood, immunoglobulins are found in lymph, tears, saliva, and colostrum (Fig. 11-9). Colostrum contains a large number of antibodies and some lymphocytes from the mother that serve to protect the newborn during the first few months of life (natural passive acquired immunity).

The amount and type of antibodies produced by a given antigenic stimulation depend on the nature of the antigen, the site of antigenic stimulus, the amount of antigen, and the number of times the person is exposed to the antigen. Figure 11-10 shows that after the first exposure to an antigen (such as a vaccine), there is a

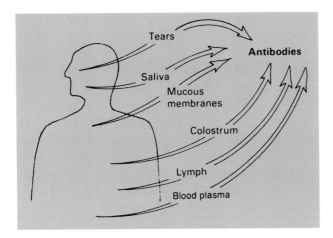

FIGURE 11-9. Body fluids and sites where antibodies can be found.

delayed primary response in the production of antibodies. During this lag phase, the antigen is processed by macrophages, T cells and B cells, or by B cells only. The majority of antigens are *T-dependent antigens,* the processing of which requires the involvement of all three cell types. Other antigens are *T-independent antigens,* the processing of which requires only B cells. Ultimately, small B cells develop into large B cells (or *plasma cells*) that are capable of producing antibodies by protein synthesis. This initial immune response to a particular antigen is called the *primary response.* When the antigen is used up, the number of antibodies in the blood declines as the plasma cells die off. Other antigen-stimulated B cells become "memory cells," which are small lymphocytes that can be stimulated to rapidly produce large quantities of antibodies when later exposed to the same antigens. This increased production of antibodies following the second exposure to the antigen (*e.g.,* a booster shot) is called the *secondary response, anamnestic response,* or *memory response.* A sec-

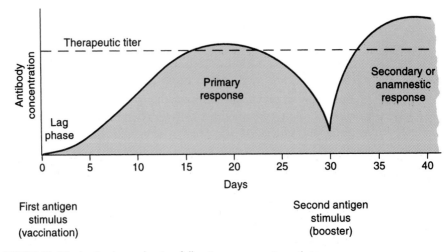

FIGURE 11-10. Antibody production following exposure to antigen.

ond booster shot of antigen many months later returns the antibody concentration to the level of the secondary response. This is the reason why booster shots are given to protect against certain pathogens that one might encounter throughout life, such as *Clostridium tetani* (the cause of tetanus). In addition to memory B cells, memory T cells also contribute to immunological memory.

Some people are born without the ability to produce protective antibodies. Because they are unable to produce antibodies, they have no gamma globulins in their blood. This abnormality is called *agammaglobulinemia*. These persons are very susceptible to infections by even the least virulent microorganisms in their environment. One treatment for agammaglobulinemia that is often successful consists of a bone marrow transplant, which involves the transfer of precursor white blood cells from a closely related person. Some of these cells become lymphocytes. These lymphocytes may be implanted in the lymph nodes and become immunocompetent, *i.e.,* capable of being stimulated by antigens to produce antibodies.

Persons who produce an insufficient amount of antibodies are said to have *hypogammaglobulinemia*. Their resistance to infection is lower than normal, so they usually do not recover from infectious diseases as readily as most other persons.

Some patients are immunosuppressed (unable to make antibodies) following the administration of immunosuppressive drugs or agents, such as the antilymphocytic serum given before organ transplant surgery. AIDS patients are infected with a virus (human immunodeficiency virus; HIV) which destroys the helper T cells (T_H cells) that are required in the processing of T-dependent antigens and are also involved in cell-mediated immune responses. These patients usually succumb to secondary infections to which they have little resistance.

The Immune System

The primary functions of the immune system are to (1) differentiate between "self" and "non-self" ("foreign"), and (2) destroy that which is "non-self." The immune system involves very complex interactions between many different types of cells and cellular secretions. Although it encompasses the whole body, the lymphatic system is the site and source of most immune activity. The cells involved in the immune responses originate in bone marrow, from which most blood cells develop (Fig. 11-11). Three lines of lymphocytes—B cells, T cells, and NK (natural killer) cells—are derived from lymphoid stem cells of bone marrow.

T CELLS
About half of the stem cells migrate to the thymus gland where they differentiate into T lymphocytes or T cells ("T" for thymus), including helper T cells (T_H cells), suppressor T cells (T_S cells), cytotoxic T cells (T_C cells), and delayed hypersensitivity T cells (T_D cells). Thymus processing of T cells begins shortly before birth. T cells are small lymphocytes found in the blood, lymph, and lymphoid tissues. About

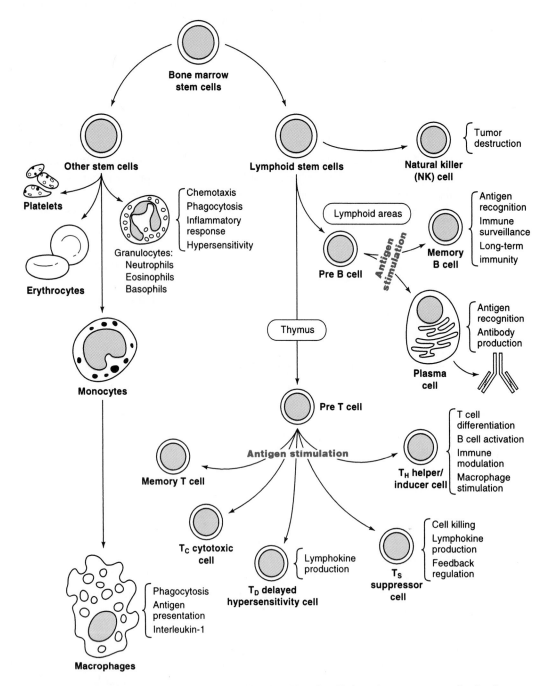

FIGURE 11-11. Differentiation of blood and lymph cells from bone marrow cells; development of the "soldiers" that participate in nonspecific and specific defense mechanisms.

70 to 80% of the lymphocytes in peripheral blood are T cells. T cells do not actually produce humoral antibodies, but they aid in the control of antibody production and are involved in cell-mediated immune responses (*e.g.,* tissue transplant rejection; cellular immunity to mycobacteria, fungi, and viruses; cytotoxicity of virus-infected cells and tumor cells).

B CELLS

Other lymphocytic stem cells differentiate in the liver and intestinal lymphoid areas into B lymphocytes or B cells, originally named for the bursa lymphoid area in birds, but now the "B" is generally assumed to stand for bone marrow. About 10 to 15% of the lymphocytes in peripheral blood are B cells. B cells migrate to lymphoid tissues where they produce antibodies that circulate through lymph and blood to protect the individual (humoral immunity). Thousands of B cells exist in the body, even though these cells live only about 1 to 2 weeks. When stimulated by an antigen, each B cell is capable of producing hundreds of specific antibodies per second. Cells of the immune system are shown in Figure 11-11.

Immune Responses

Many cells, cellular secretions, and reactions are involved in immune responses. The specific humoral response always depends on the presence of specific antibodies for each antigen or antigenic determinant. Some chemicals and complex chemical reactions are nonspecific and yet they depend on antigen-antibody complexes for activation (the complement system, for example). An antigen-antibody complex (or immune complex) is formed when a specific antibody combines with the antigenic determinant that stimulated the production of that antibody. Although some types of cell-mediated responses occur in the presence of antibodies, others do not involve antigen-antibody complexes.

HUMORAL IMMUNITY

Humoral immunity (or antibody-mediated immunity) involves the production of antibodies, as opposed to cell-mediated immunity (discussed in the next section) which does not involve antibody production. For antibodies to be produced within the body, a complex series of events must occur, some of which are not completely understood. It is known that macrophages, T cells, and B cells often are involved in a cooperative effort. (The processing of antigens within the body is actually far more complex than the abbreviated explanation that follows.)

The majority of antigens are referred to as T-dependent antigens because T cells (specifically helper T cells; also called T_H cells) are involved in their processing; in other words, the processing is dependent upon T cells. T-dependent antigens are usually complex proteins containing large numbers of different antigenic determinants with little repetition among themselves. T-dependent antigens are processed in the following manner: When bacteria invade, they are first ingested and digested by macrophages. The macrophages then display various bacterial antigenic determinants on their surface, where they are attached to cell membrane proteins called major histocompatibility complex (MHC) markers. At this point, the macrophages are referred to as *antigen-presenting cells*. T_H cells coming into contact with these antigenic determinants on the surface of the antigen-presenting cells are sensitized or primed by the antigenic determinants to send out chemical signals (cytokines) which serve to stimulate the production of antibodies by B cells. Note that the T_H cells assist in the production of antibodies, but do not manufacture antibodies themselves.

When the chemical signal reaches a B cell that is capable of recognizing that particular signal, the B cell will proliferate (*i.e.*, it will produce a clone of identical B cells). It is thought that IgD molecules (discussed later) on the surface of the B cell enable it to recognize the chemical signal. Some of the members of the newly formed clone mature into antibody-producing plasma cells. Antibodies are expelled in gunfire fashion for several days until the plasma cell dies. Each plasma cell makes only one type of antibody—the one that will bind with the antigenic determinant that activated the B cell and stimulated production of that antibody. Members of the clone that do not become plasma cells, and some primed T cells, remain in the body as memory cells, able to respond very quickly should the antigen appear again at a later date.

T_H cells induce B cells to produce antibody, whereas suppressor T cells (T_S cells) inhibit antibody production. In this way, the level of antibodies is neither excessive nor insufficient if the control mechanisms are working properly. Because the level of antibodies is regulated by T_H and T_S cells, they are referred to as regulatory T cells. Acting together, they ensure that the immune response is effective but not destructive.

Some antigens, called T-independent antigens, do not require antigen-presenting cells or T_H cells in their processing. T-independent antigens are large polymeric molecules (usually polysaccharides) containing repeating antigenic determinants. Processing of T-independent antigens is initiated when an appropriate B cell makes physical contact with the antigenic determinant. The activated B cell next produces a clone of identical B cells. Some of the members of the newly formed clone mature into antibody-producing plasma cells whereas others become memory cells.

When an antibody combines with an antigen, an antigen-antibody complex (or immune complex) is formed. Antigen-antibody complexes are capable of activating the complement cascade (via the classical pathway) resulting in, among other effects, the activation of leukocytes, lysis of bacterial cells, and increased phagocytosis as a result of opsonization. Thus, acute extracellular bacterial infections are controlled almost entirely by antibody-mediated immunity (AMI).

CELL-MEDIATED IMMUNITY

Antibodies are unable to enter cells, including cells containing intracellular pathogens. Fortunately, there is an arm of the immune system capable of controlling chronic infections by intracellular parasites (bacteria, protozoa, fungi, viruses). It is called *cell-mediated immunity* (CMI)—a complex system of interactions between many types of cells and cellular secretions (cytokines). (Only a brief overview of CMI can be provided here.) Included among the various cells that participate in CMI are macrophages, T_H cells, T_C cells, T_D cells, natural killer (NK) cells, killer (K) cells, and granulocytes. Although CMI does not involve the production of antibodies, antibodies produced during humoral immunity may play a minor role in cell-mediated responses. A typical cell-mediated cytotoxic response would involve the following steps:

> Step 1: A macrophage engulfs and partially digests a pathogen. Fragments (antigenic determinants) of the pathogen are then displayed on the surface of the macrophage (*i.e.*, the macrophage acts as an antigen-presenting cell).
>
> Step 2: A T_H cell binds to one of the antigenic determinants being displayed on the macrophage surface. The T_H cell produces *lymphokines* (cytokines produced by lymphocytes) which reach an effector cell of the immune system (*e.g.*, a T_C cell, NK cell, or K cell).
>
> Step 3: The effector cell binds to a target cell (*i.e.*, a pathogen-infected host cell displaying the same antigenic determinant on its surface).
>
> Step 4. Vesicular contents of the effector cell are discharged. These include perforin and other proteins/enzymes, which literally punch holes in the target cell membrane. Other cytokines released by effector cells are tumor necrosis factor, lymphotoxin, and NK *cytotoxic* factor.
>
> Step 5. Toxins produced by the effector cells enter the target cell, causing disruption of DNA and organelles. The target cell dies.

Both humoral and cell-mediated immune responses play a role in the body's defense against viral infections. In cytolytic viral infections (*e.g.*, herpes infections), the viruses can be neutralized and destroyed by antibodies and the complement system when they move in body fluids from a lysed cell to an intact cell. When the virus is established within body cells, the cell-mediated immune response can destroy the virus-infected cells, preventing viral multiplication. If the virus is not completely destroyed, however, it may become latent in nerve ganglion cells, as in herpes infections.

T_C cells, NK cells, and K cells kill infected host cells when pathogens are established inside the cells. Thus, infected liver cells are destroyed in hepatitis infections during the body's battle against the disease. The "AIDS virus" (HIV) that targets T_H cells is particularly destructive because it destroys the very cells that would have helped fight the infection. The lack of T_H cells impairs both humoral and cell-mediated immunity, making AIDS patients very susceptible to many opportunistic infections and malignancies.

Another type of T cell (T_D cell) is involved in delayed hypersensitivity reactions. This response is discussed later in this chapter.

NK AND K CELLS

Both NK and K cells are in a subpopulation of lymphocytes called large granular lymphocytes. Although they morphologically resemble lymphocytes, NK and K cells lack typical T- or B cell surface molecules ("markers"). They also differ from T- and B cells in other ways. For example, they do not proliferate in response to antigen and appear not to be involved in antigen-specific recognition. As their names imply, NK and K cells kill target cells, including foreign cells, virus-infected cells, and tumor cells. NK cells are able to attach directly to cell surfaces, whereas K cells attach to antibody-coated cells. K cells have receptors on their surface for the F_C region of IgG antibody molecules (F_C regions and IgG antibodies are discussed in a following section), enabling them to attach to antibody-coated target cells; this is known as antibody-dependent cellular cytotoxicity. The apparent specificity of K cells is dependent upon the presence of IgG antibody on the surface of target cells. Although firm evidence is lacking for an "immune surveillance" system within our bodies that monitors for and destroys malignant cells, both NK and K cells may participate in such a system.

Antibody Structure and Function

Antibodies belong to a class of glycoproteins called immunoglobulins. All antibodies are immunoglobulins, but not all immunoglobulins are antibodies. Antibodies are produced by plasma cells in response to stimulation of B cells by foreign antigens. Antibodies found in the blood are called humoral or circulating antibodies. Those that provide protection against infectious diseases are called protective antibodies.

The basic structure of an immunoglobulin molecule resembles the letter Y (Fig. 11-12). It consists of two identical light polypeptide chains, two identical heavy polypeptide chains, two antigen-binding sites, and an F_C region. In this basic form, the molecule is referred to as a monomer. The light chains contain fewer amino acids than the heavy chains. The chains are connected to each other by disulfide (–S–S–) bonds. The monomer is bivalent in the sense that it has two sites (antigen-binding sites) that can bind specifically to the antigenic determinant that stimulated production of that antibody. The F_C region enables the molecule to bind to cells that possess surface receptors that recognize the F_C region.

Studies of the gamma globulin component of human blood have revealed that five classes (or isotypes) of immunoglobulins exist; they have been designated IgA, IgD, IgE, IgG, and IgM. Each may consist of several subclasses. The functions of each of these classes are listed in Table 11-4.

Immunoglobulin A (IgA) represents about 15 to 20% of the immunoglobulins in human serum. About 80% of the IgA molecules are monomers, weighing about 150,000 daltons. Most of the remaining 20% exists in the form of dimers—two monomers held together by a protein chain called a J-chain ("J" for joining)—weighing about 370,000 daltons. IgA is the predominant immunoglobulin in saliva, tears, seminal fluid, colostrum, breast milk, and mucous secretions of the lungs and gastrointestinal tract. Secretory IgA (sIgA) is a dimer that contains another protein called the secretory component. The secretory component apparently facilitates the transport of sIgA into secretions. IgA molecules serve to protect the

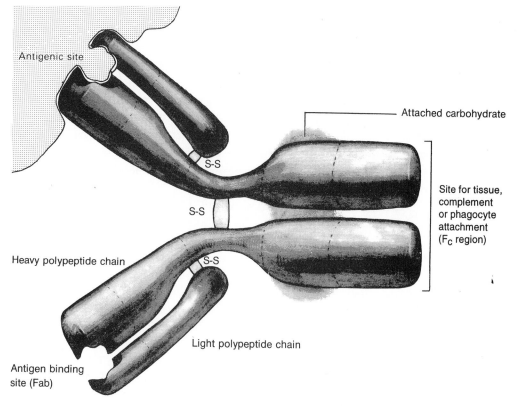

FIGURE 11-12. Basic structure of immunoglobulin IgG.

TABLE 11-4. Immunoglobulin Classes and Functions			
Ig Class	**Molecular Weight**	**% in Serum (Approx.)**	**Functions**
IgA	160,000 to 385,000	15–20	Protects the mucous membranes and internal cavities against infection; found as secretory antibodies in tears, saliva, colostrum, and other secretions
IgD	184,000	<1%	Fetal antigen receptor; controls antigen stimulation of B-cells; found in blood and on lymphocytes
IgE	188,000	<1%	Found on surfaces of basophils and mast cells; causes allergies, drug sensitivity, anaphylaxis, and immediate hypersensitivity; combats parasitic diseases
IgG	146,000 to 170,000	70–75	Protects against disease; attaches to phagocytes and tissues fixes complement; crosses placental barrier; causes certain immunological diseases; found in blood and lymph
IgM	970,000	10	Protects against early infection; bactericidal to gram-negative bacteria; fixes complement; found in blood and lymph

external openings and mucous membranes from the attachment, colonization, and invasion of pathogens. The IgA in colostrum and breast milk helps protect nursing newborns. In the intestine, IgA attaches to viruses, bacteria, and protozoal parasites, such as *Entamoeba histolytica,* and prevents the pathogens from adhering to mucosal surfaces, thus preventing invasion. In addition to monomeric and dimeric IgA, there are also trimeric IgA molecules.

Immunoglobulin D (IgD) is a monomeric molecule that makes up less than 1% of serum immunoglobulin, but is found in large quantities on the surface of B cells. Its function is unknown, but it is possible that the IgD molecules on the B cell's surface serve as antigen receptors and determine which specific antigen that particular B cell is able to respond to.

Immunoglobulin E (IgE) is also called P-K antibody (in honor of the two scientists, Prausnitz and Küstner, who first identified it). IgE is scarce in serum, but is found in large quantities on the surface of basophils and mast cells, where the monomeric molecules adhere by their F_C regions. Basophils are granulocytes that circulate in the blood. Mast cells are very similar to basophils morphologically, but they are found in tissues—especially tissues that surround the eyes, nose, respiratory tract, and gastrointestinal tract. The cytoplasmic granules of basophils and mast cells contain histamine, serotonin, and other chemical mediators that are responsible for allergic manifestations. Thus, the IgE molecules on the surface of basophils and mast cells play a major role in allergic responses (described later).

Immunoglobulin G (IgG) is the most abundant immunoglobulin type in serum, accounting for 70 to 75% of the total immunoglobulin pool. This monomeric molecule, the smallest of the immunoglobulins, is the only class of immunoglobulin that can cross the placenta. Maternal antibodies that cross the placenta help protect the newborn during its first months of life. Antigen-bound IgG can bind to and activate complement, a process known as complement fixation. IgG molecules can bind to a wide range of cellular receptors to promote phagocytosis and antibody-dependent cytotoxicity. As a result of "memory cells," high levels of IgG are produced very rapidly during the secondary response to antigens (described earlier).

Immunoglobulin M (IgM), the largest of the immunoglobulins (weighing about 1,000,000 daltons), accounts for approximately 10% of the serum immunoglobulin pool. It is a pentamer, consisting of five monomers held together by a J-chain. Because a pentamer has 10 antigen-binding sites, IgM can potentially bind to 10 identical antigenic determinants. IgM antibodies are the first antibodies formed in the primary response to antigens (including pathogens), although IgG antibodies become the most prevalent class. IgM is the most efficient complement-fixing immunoglobulin.

MONOCLONAL ANTIBODIES

Purified antibodies that are directed against specific antigens have been produced in laboratories by an innovative technique in which a single plasma cell that produces only one specific type of antibody is fused with a rapidly dividing tumor cell. The new long-lived, antibody-producing cell is called a *hybridoma*. These hybridomas are capable of producing large amounts of specific antibodies called *monoclonal antibodies*. The first monoclonal antibodies were produced in 1975 and, since then, many uses have been found for them. They are commonly used in immunodiagnostic procedures (IDPs)—immunological procedures used in laboratories to diagnose diseases. The first diagnostic kit containing monoclonal antibodies was approved for use in the United States in 1981. Many other monoclonal antibody-based IDPs have been developed over the past 20 years. Monoclonal antibodies are also being evaluated for possible use in fighting diseases, killing tumor cells, boosting the immune system, and preventing organ rejection.

Hypersensitivity

Hypersensitivity occurs when the defensive immune responses have gone awry; *i.e.,* by being overly sensitive. Sometimes, instead of protecting a person, immune responses irritate and damage certain cells in the body. This can be compared to the person who builds a fire in the living room to warm the house which results in burning the house down.

There are several different types of hypersensitivity reactions; some involve various antibodies and others do not. All depend on the presence of antigen and T cells sensitized to that antigen. Hypersensitivity reactions are divided into two general categories, immediate and delayed, depending on the nature of the immune reaction and the time required for an observable reaction to occur (Fig. 11-13).

An *immediate hypersensitivity reaction* occurs from within a few minutes to 24 hours. There are three categories of immediate reactions, referred to as Type I, Type II, and Type III hypersensitivity reactions. *Type I hypersensitivity reactions* (also known as anaphylactic reactions) include classic allergic responses such as hay fever symptoms, asthma, hives, and the gastrointestinal symptoms that result from food allergies; allergic responses to insect stings and drugs; and anaphylactic shock. These reactions all involve IgE antibodies and the release of chemical mediators (especially histamine) from mast cells and basophils.

Type II hypersensitivity reactions are cytotoxic reactions, including the cytotoxic reactions seen in blood transfusions, Rh incompatibility reactions, and myasthenia gravis, involving IgG or IgM antibodies and complement. A typical Type II hypersensitivity reaction might follow this sequence: (1) A particular drug binds to the surface of a cell;

Type	IMMEDIATE			DELAYED
	I	**II**	**III**	**IV**
	Allergy, Anaphylaxis	Cytolysis, Blood type reactions	Auto-immunity	TB skin test, Transplantation rejection

FIGURE 11-13. Types of hypersensitivity reactions.

(2) Anti-drug antibodies then bind to the drug; (3) This initiates complement activation on the cell surface; (5) The complement cascade leads to lysis of the cell.

Type III hypersensitivity reactions are immune complex reactions such as those that occur in serum sickness and certain autoimmune diseases (*e.g.,* systemic lupus erythematosus [SLE], rheumatoid arthritis), involving IgG or IgM antibodies, complement, and neutrophils. Serum sickness is a cross-reacting antibody immune reaction in which antibodies formed to globular proteins in horse serum (used for antivenom treatments) may also bind with similar proteins in the patient's blood. The formation of these immune complexes (antigen + antibody + complement) causes the symptoms of fever, rash, kidney malfunction, and joint lesions of serum sickness. Certain complications (sequelae) of untreated or inadequately treated strep throat and other *Streptococcus pyogenes* infections are the result of Type III hypersensitivity reactions. IgG and IgM antibodies produced in response to *S. pyogenes* infection may bind with streptococcal antigens (*e.g.,* M-protein). The resultant immune complexes become deposited in heart tissue, joints, or the glomeruli of the kidney, causing rheumatic fever, arthritis, and glomerulonephritis.

Type IV hypersensitivity reactions are referred to as *delayed-type hypersensitivity* (DTH); such reactions are usually observed after 24 hours. These reactions are part of cell-mediated immunity (CMI); they occur in tuberculin and fungal skin tests, contact dermatitis, and transplantation rejection. DTH is the prime mode of defense against intracellular bacteria and fungi. DTH involves a variety of cell types, including macrophages, T_C cells, T_D cells, NK cells, and K cells, but antibodies do not play a major role.

A classic example of a DTH reaction is a positive TB skin test. Purified protein derivative (PPD), a protein derived from *Mycobacterium tuberculosis,* is injected intradermally into a person. If an "immunological memory" of that particular protein exists in that person's body, a DTH reaction will occur, producing the typical swelling and redness associated with a positive test result. The following events occur to produce the positive reaction: (1) Within 2 to 3 hours after injection of the PPD, there is an influx of polymorphonuclear cells (PMNs) into the site; (2) This is followed by an influx of lymphocytes and macrophages, while the PMNs disperse; (3) Within 12 to 18 hours, the area becomes red (*erythematous*) and swollen (*edematous*); (4) The *erythema* and *edema* reach maximum intensity between 24 and 48 hours; (5) With time, as the swelling and redness disappear, the lymphocytes and macrophages disperse.

A positive TB skin test result may indicate any of four possibilities: (1) that the person was infected with *M. tuberculosis* at some time in the past, but the organisms were killed by the host defense mechanisms; (2) that the person harbors live *M. tuberculosis* organisms, but does not actually have tuberculosis; (3) that the person has active tuberculosis; (4) that the person had received BCG vaccine at some time in the past. Many countries (not including the United States) routinely immunize their citizens against tuberculosis using BCG vaccine. Although this vaccine is only about 50% effective in preventing tuberculosis, it does cause individuals to have positive TB skin test results for variable time periods following immunization.

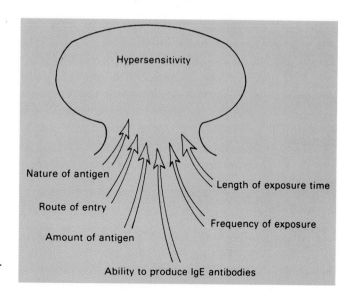

FIGURE 11-14. Factors in the development of hypersensitivity.

A reaction that is similar to the positive TB skin test occurs in contact dermatitis (contact hypersensitivity), following contact with certain metals, the catechols of poison ivy, cosmetics, and topical medications. The rejection of transplanted tissues containing foreign histological (tissue) antigens appears to occur in a similar manner, except that lymphokines and antibodies cause the rejection of the transplant.

THE ALLERGIC RESPONSE

Type I immediate hypersensitivity is probably the most commonly observed type of hypersensitivity because over one-half of the American population is allergic to something. People who are prone to allergies (*atopic* individuals) produce IgE (sometimes called reagin) antibodies when they are exposed to *allergens* (antigens that cause allergic reactions). The IgE molecules bind to the surface of basophils and mast cells by their F_C regions. The type and severity of an allergic reaction depend on a combination of factors, including the nature of the antigen, the amount of antigen entering the body, the route by which it enters, the length of time between exposures to the antigen, the person's ability to produce IgE antibodies, and the site of IgE attachment (Fig. 11-14).

Examples of Allergens	
Animal dander	Insect venom
Drugs (*e.g.,* penicillin)	Latex
Foods (*e.g.,* peanuts, shellfish, dairy products)	Mold spores
House dust (dust-mite feces)	Pollens

The allergic reaction results from the presence of IgE antibodies bound to basophils in the blood or to mast cells in connective tissues, produced following the person's first exposure to the allergen. When the allergen binds to cell-bound IgE during a subsequent exposure to the allergen, the sensitized cells respond by degranulation—the discharge and outpouring of irritating and damaging substances (chemical mediators) from the cytoplasmic granules (Figures 11-15, 11-16). These mediators of the allergic responses include histamine, prostaglandins, serotonin, bradykinin, slow-reacting substance of anaphylaxis (SRS-A), leukotrienes, and chemicals that attract eosinophils (eosinophilotactic agents).

LOCALIZED ANAPHYLAXIS

Type I hypersensitivity reactions (*anaphylactic reactions*) may be localized or systemic. Localized reactions usually involve mast cell degranulation, whereas systemic reactions usually involve basophil degranulation. Hay fever, asthma, and hives are examples of localized anaphylaxis. The symptoms depend on how the allergen enters the body and the sites of IgE attachment. If the allergen (*e.g.,* pollens, dust, fungal spores) is inhaled and deposits on the mucous membranes of the respiratory tract, the IgE antibodies that are produced attach to the mast cells in that area. Subsequent exposure to those inhaled allergens allows them to bind to the attached IgE, causing mast cell degranulation. The released histamine initiates the classic symptoms of hay fever. Antihistamines function by binding to and, thus, blocking the sites where histamine binds. Antihistamines are not as effective in treating asthma, however, because the mediators of this lower respiratory allergy include chemical mediators in addition to histamine. Allergens (*e.g.,* food and drugs) entering through the digestive tract can also sensitize the host, and subsequent exposure may result in the symptoms of food allergies (hives, vomiting, and diarrhea).

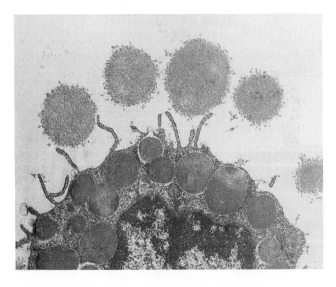

FIGURE 11-15. Transmission electron micrograph showing rat mast cell degranulation. (TEM by P. Engelkirk)

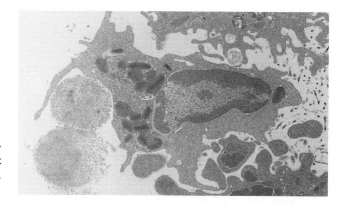

FIGURE 11-16. Transmission electron micrograph showing phagocytosis of rat mast cell granules by a rat eosinophil. (TEM by P. Engelkirk)

SYSTEMIC ANAPHYLAXIS

Systemic *anaphylaxis* results from the release of chemical mediators from basophils in the bloodstream. It occurs throughout the body and, thus, tends to be a more serious condition than localized anaphylaxis. It may lead to a severe, potentially fatal, condition known as *anaphylactic shock*. Most often, the allergens involved in systemic anaphylaxis are drugs or insect venom to which the host has been sensitized. Penicillin is an example of a hapten—a substance which must first bind to a host blood protein (a carrier protein) before IgE antibodies are produced. The IgE antibodies then bind to circulating basophils. Subsequent injections of penicillin into the sensitized host may cause degranulation of the basophils and release of large amounts of histamine and other chemical mediators into the circulatory system.

The shock reaction usually occurs immediately (within 20 minutes) after re-exposure to the allergen. The first symptoms are flushing of the skin with itching, headache, facial swelling, and difficulty in breathing; this is followed by falling blood pressure, nausea, vomiting, abdominal cramps, and urination (caused by smooth muscle contractions). In many cases, acute respiratory distress, unconsciousness, and death may follow shortly. Swift treatment with epinephrine (adrenaline) and antihistamine usually stops the reaction.

Healthcare professionals must take particular care to ask patients if they have any allergies or sensitivities before administering drugs. In particular, those people with allergies to penicillin and other drugs and to insect stings should wear Medic-Alert tags so that they do not receive improper treatment during a medical crisis.

LATEX ALLERGY

In 1997, the National Institute for Occupational Safety and Health (NIOSH) issued a warning that "workers exposed to latex gloves and other products containing natural rubber latex may develop allergic reactions, such as skin rashes; hives; nasal, eye, or sinus symptoms; asthma; and (rarely) shock." A year later, it was estimated that as many as 17% of healthcare workers develop latex allergy, primarily as a result of wearing latex gloves. Latex can trigger any of three types of reactions: (1) irritant contact dermatitis (not a true allergy because the immune system is not involved); (2) allergic contact dermatitis (a type of delayed hypersensitivity or Type IV allergy; this is the most common type of reaction); (3) immediate type hypersensitivity (a systemic Type I, IgE-mediated reaction; can be very serious). Once a person has become sensitized to latex, a reaction may occur even when the individual is not actually wearing latex gloves. Cornstarch is the powder most commonly used in latex gloves, and inhalation of allergen-laden cornstarch particles is sufficient to cause allergic symptoms. Latex-sensitive employees should avoid latex-containing items, but this is difficult to do in a hospital environment. It has been estimated that more than 20,000 medical products contain latex. Alternatives to powdered latex gloves are powder-free latex gloves and gloves made of materials other than latex.

SKIN TESTING AND ALLERGY SHOTS

Anaphylactic reactions can be prevented by avoiding known allergens. In some cases, skin tests are used to identify the offending allergens. Then, desensitization may be accomplished by injecting small doses of allergen, repeatedly, several days apart. This treatment may be effective by causing the production of increased amounts of circulating IgG antibody instead of IgE. In theory, the IgG should bind with the allergen and block its attachment to the basophil- and/or mast cell-bound IgE. Such circulating IgG molecules, produced in response to allergy shots, are called *blocking antibodies.* This preventive measure usually works better for inhaled allergens than for allergens that are ingested or injected.

AUTOIMMUNE DISEASES

An *autoimmune disease* results when a person's immune system no longer recognizes certain body tissues as "self" and attempts to destroy those tissues as being "non-self" or "foreign." This may occur with certain tissues that are not exposed to the immune system during fetal development so that they are not recognized as self. Such tissues may include the lens of the eye, the brain and spinal cord, and sperm. Subsequent exposure to this tissue (by surgery or injury) may allow antibodies (IgG or IgM) to be formed, which together with complement could cause destruction of these tissues, resulting in blindness, allergic encephalitis, or sterility.

It is believed that certain drugs and viruses may alter the antigens on host cells, thus inducing the formation of autoantibodies or sensitized T cells to react against these altered tissue cells. Examples of autoimmune diseases include autoimmune hemolytic anemia (which destroys erythrocytes), rheumatoid arthritis (which affects joints); systemic lupus erythematosus (SLE; which affects kidneys, lung, skin, and brain); Grave's disease and Hashimoto's thyroiditis (which affect the thyroid); myasthenia gravis and multiple sclerosis (which affect muscle); and Addison's disease (which affects adrenal glands). Specific autoimmune diseases are the result of Types II, III, or IV hypersensitivity reactions.

Immune Deficiency

If a person's immune system is functioning properly, that person is said to be immunocompetent. If a person's immune system is not functioning properly, that person is said to be immunosuppressed, immunodepressed, or immunocompromised. The most common cause of immune deficiency worldwide is malnutrition. In addition, there are acquired and inherited immunodeficiencies.

Acquired immunodeficiencies may be caused by drugs (*e.g.*, cancer chemotherapeutic agents and drugs given to transplant patients), irradiation, or certain infectious diseases (*e.g.*, HIV infection). HIV infection leads to a decrease in T_H cells, which, in turn, prevents the production of antibodies against T-dependent antigens and, consequently, the inability to fight off certain pathogens. These pathogens overwhelm the patient's host defenses, eventually causing death. AIDS patients usually die from a variety of devastating infectious diseases, including viral, bacterial, fungal, and parasitic diseases.

Inherited immunodeficiency diseases can be the result of deficiencies in antibody production, complement activity, phagocytic function, or NK cell function. Two inherited immunodeficiency diseases have already been mentioned: agammaglobulinemia and hypogammaglobulinemia. Others include severe combined immune deficiency (SCID), DiGeorge syndrome, Wiskott-Aldrich syndrome, chronic granulomatous disease, and Chediak-Higashi syndrome. Bone marrow transplantation and gene therapy may be of value in treating certain immunodeficiency diseases.

Immunodiagnostic Procedures

Historically, the length of time it takes to get lab results has been the most common criticism of the clinical microbiology laboratory. Sometimes days or even weeks are necessary to isolate pathogens from clinical specimens, to get them growing in pure culture and large numbers, and to perform the tests necessary to identify them. With certain infectious diseases, it is impossible to isolate the pathogens, either because they are obligate intracellular pathogens or are extremely fastidious.

One solution to these problems has been the development of *immunodiagnostic procedures* (IDPs)—laboratory procedures that help to diagnose infectious diseases by detecting antigens or antibodies in clinical specimens. The results of such procedures are often available on the same day that the clinical specimen is collected from the patient. Immunodiagnostic procedures performed on serum specimens are sometimes referred to as *serologic procedures* and are performed in a section of the clinical microbiology laboratory known as the *Serology Section*.

Some IDPs are designed to detect antigens, whereas others detect antibodies (Fig. 11-17). Detection of antigens in a clinical specimen is an indication that a particular pathogen is present in the patient, thus, providing direct evidence that the

Immunodiagnostic Procedures

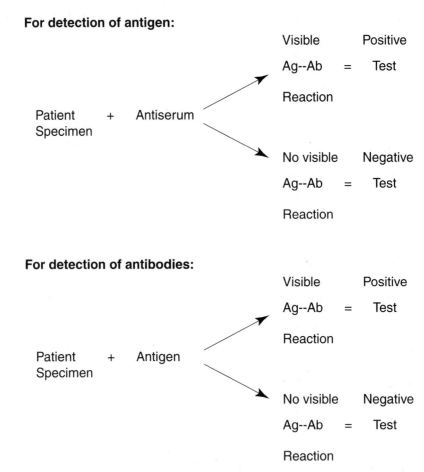

For detection of antigen:

For detection of antibodies:

FIGURE 11-17. Immunodiagnostic procedures (IDPs). Depending upon the type of IDP being performed, the visible antigen-antibody (Ag-Ab) reaction might be agglutination (clumping) of cells or latex particles, formation of a precipitin line or band, fluorescence, or production of a color (as in enzyme immunoassays).

patient is infected with that pathogen. Detection of antibodies directed against a particular pathogen is indirect evidence of infection with that pathogen. There are three possible explanations for the presence of antibodies to a particular pathogen: (1) present infection (*i.e.*, the person is currently infected with the pathogen); (2) past infection (*i.e.*, the person was infected with the pathogen in the past, and antibodies are still present in the person's body); (3) vaccination (*i.e.*, the antibodies are the result of the person having been vaccinated against that particular pathogen at some time in the past). Because several explanations are possible for the presence of antibodies in a clinical specimen, the presence of antigens provides the best proof of current infection. Unfortunately, antigen detection procedures are not available for many infectious diseases. Another problem with antibody detection procedures is that it takes a person about 10 to 14 days to produce antibodies; thus, even if the person is infected with a particular pathogen, antibodies will not be detectable for about 2 weeks.

Two ways to increase the value of antibody detection procedures to diagnose present infection are (1) to specifically test for IgM antibodies and (2) to use paired sera. Because IgM antibodies are the first antibodies to be produced during the initial exposure to an antigen (the primary response) and are relatively short-lived, the presence of IgM antibodies directed against a particular pathogen is evidence that the pathogen is currently infecting the individual. To test paired sera, one serum specimen (called the acute serum) is collected during the acute stage of the disease and one (called the convalescent serum) is collected 2 weeks later. A significant rise in antibody titer (concentration) between the acute and convalescent sera is evidence that the patient was actively producing antibodies against that pathogen during the 2-week period and, therefore, that that pathogen is the cause of the patient's current infection.

The reagents used to detect either antigens or antibodies are purchased by laboratories from commercial companies. The reagent used to detect antigens contains antibodies, and is called an *antiserum*. An antiserum is usually prepared by inoculating a laboratory animal with the pathogen (usually dead pathogens are used), and then collecting blood from the animal several weeks later. The blood is allowed to clot and the serum is drawn off. The reagent used to detect antibodies contains antigens. This is usually a suspension of dead pathogens.

A variety of different laboratory tests have been designed so that a visible reaction will be observed if an antigen-antibody reaction takes place. Such tests, which include *agglutination* (involving the clumping of particles such as red blood cells or latex beads), *precipitin* (involving the production of a precipitate), immunofluorescence procedures and enzyme-linked immunosorbent assays (ELISAs), are represented diagrammatically in Table 11-5. In the Blood Bank, agglutination tests are used to learn a person's blood type, which is determined by the types of antigens that are present on that person's red blood cells. An example of a precipitin procedure is shown in Figure 11-18.

For detection of antigen, the clinical specimen is mixed with a particular antiserum. A visible reaction indicates that the antigen is present in the clinical specimen and that the test is considered to be positive. If the visible reaction is not observed, then the antigen was not present in the specimen and the test is considered to be negative. Example: A drop of cerebrospinal fluid (CSF) from a patient with meningitis is mixed with a drop of antiserum containing antibodies against *Haemophilus influenzae*. A visible antigen-antibody reaction is evidence that the patient's CSF contained *H. influenzae* antigens, and the patient is diagnosed as having meningitis due to *H. influenzae*.

TABLE 11-5. Immunodiagnostic Procedures for Detection of Antibodies in a Patient's Serum

Reaction In Vitro	REAGENTS			RESULTS	
	Antigen	Antibody	Other	+	−
Agglutination	Red blood cells or bacteria	Patient's serum		Clumping	No clumping
Precipitin	Toxins, hormones, proteins	Patient's serum	Agar or solution	Precipitate	No precipitate
Lysis by Complement	Cells, bacteria	Patient's serum	Complement	Lysis	No Lysis
Fluorescent Antibody Technique	Pathogen	Patient's serum	Fluorescein-tagged rabbit antiserum	Fluorescent pathogen	No fluorescence
Opsonization	Bacteria	Patient's serum	Phagocytes	More bacteria in phagocytes	Fewer bacteria in phagocytes

Reaction in Vitro	REAGENTS			RESULTS	
	Antigen	Antibody	Other	+	−
Capsular Swelling (Quellung Reaction)	Encapsulated bacteria	Patient's serum		Calsule appears to swell	No appearance of swelling
Immobilization	Motile bacteria	Patient's serum		No motility	Motility
Radio-immunoassay	Patient's serum	Radioactively-tagged rabbit antiserum		High radioactivity	Low or no radioactivity
Enzyme-linked assay	Test microbe	Patient's serum	Enzyme linked antibody +Substrate	Color change	No color change

For detection of antibodies, the clinical specimen is mixed with a suspension of a particular antigen. A visible reaction indicates that antibodies against that pathogen are present in the clinical specimen and the test is considered to be positive. If the visible reaction is not observed, then antibodies against that pathogen are not present in the specimen and the test is considered to be negative. Example: A drop of serum from a patient suspected of having Lyme disease is mixed with a suspension of *Borrelia burgdorferi* (the etiologic agent of Lyme disease). A visible antigen-antibody reaction is evidence that the patient's serum contained antibodies against *B. burgdorferi,* and the patient is diagnosed as having Lyme disease.

The Quellung reaction, another type of IDP, is performed in the following manner: A drop of spinal fluid or sputum containing Gram-positive cocci (or a drop of a suspension of Gram-positive cocci prepared from a colony) is placed on a glass microscope slide. A loopful of *Streptococcus pneumoniae* antiserum is then added and mixed well. Next, a drop of methylene blue dye is added and mixed well. A glass cover slip is placed over the mixture and, after 10 minutes, the mixture is examined microscopically, using the oil immersion lens and reduced light. If the cocci are *S. pneumoniae*, antibodies will have bound to the capsule, making the capsule appear swollen (Fig. 11-19); actually, the bound antibodies change the refractive index, making the capsule more visible, but it is not really swollen. This represents a positive Quellung reaction and the organism can now be identified as *S. pneumoniae*. Using a variety of different commercially available antisera, the specific capsular serotype of this particular *S. pneumoniae* can also be determined.

Skin testing is another type of IDP, but one that is performed *in vivo* (in the patient) rather than *in vitro* (in the laboratory). In skin testing, antigens are injected within or beneath the skin (intradermally or subcutaneously, respectively). Examples of such tests are the tuberculosis skin test (previously described), the Schick test, the Dick test, and the Schultz-Charlton reaction.

The Schick test is used to determine an individual's susceptibility to diphtheria toxin. A small amount of toxin is injected intradermally into one forearm, and an equal amount of heat-inactivated toxin is similarly injected into the other forearm. If the individual possesses antitoxins (antibodies that neutralize the toxin), indicating that the person is protected from the toxin, there will be no reaction at either injection site. Individuals lacking antitoxins and, thus, lacking protection, will have

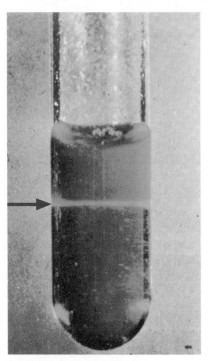

FIGURE 11-18. Precipitin tube test. An antigen-containing specimen is overlaid onto agar containing an antiserum. At a critical point of interface, where the antigen and antibody concentrations are optimal, a visible precipitate (arrow) forms. This precipitin band is composed of antigen-antibody complexes.

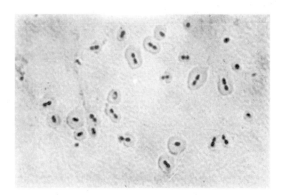

FIGURE 11-19. A positive Quellung reaction. The capsules of *Streptococcus pneumoniae* appear swollen as a result of a change in refractive index caused by the binding of antibodies to capsular polysaccharide (see text for details).

a positive reaction (redness) at the site of toxin injection, but a negative reaction at the other injection site.

The Dick test is similar to the Schick test, but is used to determine an individual's susceptibility to the erythrogenic toxin produced by some strains of *Streptococcus pyogenes*. Erythrogenic toxin is responsible for the rash and other manifestations of scarlet fever. An individual possessing protective antibodies will have a negative reaction at the injection site. An individual lacking protective antibodies will have a positive reaction.

The Schultz-Charlton reaction can be used to diagnose scarlet fever. An antitoxin against erythrogenic toxin is injected intradermally at a rash site. If the rash blanches (clears) at that site, the diagnosis of scarlet fever is made. If the rash does not blanch, the rash is due to some other cause.

■ REVIEW OF KEY POINTS

- Certain human host defense mechanisms are classified as nonspecific, whereas others are classified as specific. Nonspecific host defense mechanisms serve to protect the body from all types of foreign substances or pathogens. Specific host defense mechanisms are directed against specific foreign substances or pathogens that have entered the body.
- Another way to categorize human host defense mechanisms is to divide them into first, second, and third lines of defense. The first and second lines of defense are nonspecific, whereas the third line of defense is specific.
- The first line of defense includes innate resistance; physical barriers such as intact skin and mucous membranes; chemical, physiological and temperature barriers; microbial antagonism by indigenous microflora; as well as nutritional status and state of health.
- The second line of defense includes nonspecific cellular and chemical responses such as inflammation, fever, interferon production, activation of the complement cascade, iron balance, cellular secretions, activation of blood proteins,

chemotaxis, phagocytosis, neutralization of toxins, and the cleanup and repair of damaged areas.

- The complement system involves about 30 different blood proteins that interact in a step-wise manner known as the complement cascade. Complement activation by immune complexes or other mechanisms aids in the initiation and amplification of inflammation, attraction and activation of leukocytes, lysis of bacteria and other foreign cells, and enhanced phagocytosis (opsonization).

- Prostaglandins are biologically reactive lipids that aid in platelet aggregation, immune response, inflammation, increased capillary permeability, pain production, diarrhea, autoimmune responses, and fever production.

- Phagocytes rid the body of unwanted or harmful substances, such as dead cells, unused cellular secretions, dust, debris, and microbes. Following their attraction to a site by chemotactic substances, they attach to, surround, ingest, and digest the material.

- Indications of inflammation include redness, heat, edema, pain, often pus formation, and sometimes loss of function of the inflamed area. The purposes of the inflammatory response are to localize an infection, prevent the spread of microbial invaders, neutralize toxins, and aid in the repair of damaged tissue.

- The third line of defense against pathogens is the specific immune response, which usually involves the production of antibodies which will recognize, bind to, and inactivate or destroy specific microorganisms or their toxins. Also, there are protective cell-mediated immune responses in which antibodies play only minor roles, if any. Cell-mediated immune responses involve a variety of cell types, including macrophages and various types of lymphocytes.

- Immunity to an infectious disease may be innate or acquired. If acquired, the immunity may have been acquired actively (where antibodies were actively produced by the person) or passively (where the person received antibodies that were produced by others). Active acquired immunity may occur naturally or artificially. Likewise, passive acquired immunity may occur naturally or artificially.

- The production of antibodies in response to a pathogen that has entered the body is an example of natural, active acquired immunity. The production of antibodies in response to a vaccine is an example of artificial, active acquired immunity. A fetus receiving antibodies that were produced by its mother is an example of natural, passive acquired immunity. A soldier receiving antibodies contained in a shot of gamma globulin is an example of artificial, passive acquired immunity.

- Immunology is the scientific study of immune responses, including active and passive acquired immunity to infectious agents, antibody production, cell-mediated immune responses, allergic responses, other types of hypersensitivity reactions, autoimmunity, and immunodiagnostic procedures.

- The amount and type of antibodies produced by a given antigenic stimulation depend on the nature of the antigen, the site of antigenic stimulus, the amount of antigen, and the number of times the person is exposed to the antigen.

- The immune system involves complex interactions between different types of cells and cellular secretions, occurring mostly in the lymphatic system. Cells involved in the immune responses originate in bone marrow (from which most blood cells develop); they include B cells (antibody producers), T cells (helper T cells, suppressor T cells, cytotoxic T cells, and delayed hypersensitivity T cells), and natural killer (NK) cells.

- Antibodies are in a class of proteins known as immunoglobulins. There are five types of glycoprotein antibodies, designated as IgA, IgD, IgE, IgG, and IgM. IgD, IgE, and IgG are monomers, consisting of two light chains and two heavy chains. Although IgA may be a monomer, a dimer, or a trimer, the dimer form is most common. IgM is a pentamer.

- Hypersensitivity reactions may be immediate or delayed, depending on the nature of the immune reaction and the time required for an observable reaction. Type I hypersensitivity reactions (anaphylactic reactions) include the classic allergic responses of hay fever, asthma and hives, resulting from allergies to pollen, mold spores, animal dander, foods, insect venom, drugs, and other allergens. Hypersensitivity reactions range from relatively mild, localized reactions to very severe, systemic reactions (*e.g.*, anaphylactic shock).

- Type II hypersensitivity reactions are cytotoxic reactions. An example of a Type II reaction is the massive destruction of red blood cells that occurs when a person receives a unit of incompatible blood (*e.g.*, if a Type A person receives a unit of Type B blood). Type III reactions are immune complex reactions, resulting from the deposition of immune complexes beneath various membranes in the body (*e.g.*, glomerulonephritis). Type IV reactions are delayed-type hypersensitivity (cell-mediated) reactions, such as those that occur in positive TB and fungal skin tests, contact dermatitis, and transplant rejection. Autoimmune diseases may be Type II, Type III, or Type IV hypersensitivity reactions.

- Immunodiagnostic procedures (IDPs) are laboratory tests of value in diagnosing infectious diseases by detecting either antigens or antibodies in clinical specimens. IDPs performed on serum specimens are called serologic procedures.

Problems and Questions

1. What factors contribute to natural resistance to disease?
2. List some nonspecific defenses of the skin, respiratory system, digestive system, and urogenital tract.
3. List some of the blood proteins that aid in the destruction of invading microorganisms.
4. Describe the process of phagocytosis.
5. What are the four main symptoms of inflammation?
6. What causes the four main symptoms of inflammation?
7. What is the immune response? Is it always a protective mechanism?
8. What types of substances are effective antigens?
9. Where in the body are antibodies formed?
10. What steps are involved in the production of antibodies?
11. What is the anamnestic response?
12. What is the main difference between active and passive acquired immunity?
13. How are antibodies transferred in passive immunity?
14. List five classes of immunoglobulins. Where are each of these found?
15. What are the main differences among the four types of hypersensitivity?
16. What is an allergen? Cite some examples.
17. Give an example of a delayed hypersensitivity reaction.
18. Describe an example of an autoimmune disease.
19. List four types of immunodiagnostic procedures.
20. List four *in vivo* antigen-antibody tests and when they are used.

Self Test

After you have read Chapter 11, reviewed the chapter outline, examined the objectives, studied the new terms, and answered the problems and questions above, complete the following self test.

MATCHING EXERCISES

Complete each statement from the list of words provided.

Resistance Against Pathogens

bile	lysozyme	innate
interferon	specific	indigenous
species	digestive enzymes	microflora
complement	nonspecific	

1. Complement, interferon, and phagocytes are some of the body's _____ defenses.
2. _____ is secreted by the liver, stored in the gallbladder, and released into the small intestine, where it lowers the surface tension of particles, such as bacteria, to make them more digestible.

3. An enzyme found in nasal secretions that lyses the cell walls of certain bacteria is _____.

4. When antibodies are formed that bind with the specific antigens that caused their formation, the body is activating a _____ host defense mechanism against foreign substances.

5. The usually harmless microorganisms that reside on the skin and in the mucous membranes of many body systems are the _____.

6. _____ is a protein that is secreted by cells infected by viruses and serves to prevent the virus from multiplying in surrounding cells.

7. A certain species of animals resistant to a disease found in other species is said to have _____ resistance.

8. Lysosomes contain many _____, which aid in the destruction of phagocytized pathogens.

9. The natural, inherited defense mechanisms that give certain individuals or species some resistance to certain diseases are part of _____ resistance.

10. A complex group of proteins found in blood plasma that aids in the inactivation and destruction of bacteria is the _____ system.

Immunology

immunology	anamnestic response	antibodies
agammaglobulinemia	hypogammaglobu-	IgE
antitoxins	linemia	primary response
IgM	immunoglobulins	antigen
allergen	anaphylactic response	IgG
complement	IgA	serum
immunocompetent	plasma	

1. When blood clots, the remaining liquid is called _____.

2. When a person is not able to produce antibodies, he has an abnormality known as _____.

3. The smallest but most abundant of the antibodies found in the serum is _____.

4. Any foreign material that can stimulate the production of antibodies is an _____.

5. When a person's immune system is functioning properly, the person is said to be _____.

6. _____ is the antibody type that is produced in response to allergens, and binds to basophils and mast cells.

7. An antigen to which some people become allergic is called an _____.

8. The antibodies that are formed in response to toxin or toxoid antigens are _____.

9. When small lymphocytes are initially stimulated by antigens (such as vaccines) to develop into plasma cells to produce antibodies, this process is known as the _____.

10. The liquid portion of unclotted blood is called _____.

11. The largest serum antibody and the first to be formed in the primary response against pathogens is _____.

12. If an individual produces less than the normal amount of antibodies, the abnormality is called _____.

13. A group of serum proteins that are activated by antigen-antibody complexes is _____.

14. When memory lymphocytes are restimulated by antigens to produce the same antibodies again (as when a booster vaccine is given), the _____ has occurred.

15. The secretory antibody found in tears, saliva, and colostrum is _____.

16. The glycoprotein molecules that are produced by B cells in response to the presence of antigens are called _____.

17. A Type I hypersensitivity reaction is a/an _____.

Immune Response

allergy	delayed	natural active acquired
artificial active acquired	autoimmune	immunity
immunity	artificial passive	natural passive acquired
immediate	acquired immunity	immunity

1. Hay fever, asthma, and anaphylactic shock are all examples of _____ hypersensitivity.

2. The production of antibodies during the course of an infectious disease is known as _____.

3. A positive TB skin test is an example of _____ hypersensitivity.

4. Production of antibodies in response to a vaccine is an example of _____.

5. The rejection of transplanted tissues or organs, such as a kidney transplant, is an example of _____ hypersensitivity.

6. Giving gamma globulin or immune serum globulin to an individual is an example of _____.

7. When people's bodies produce antibodies against their own tissues and destroy their own organs, such as the heart, kidney, or thyroid, we say that they have a/an _____ disease.

8. A newborn having temporary maternal antibodies to protect him/her from disease is an example of _____.

Immunodiagnostic Procedures (IDPs)

antigens	agglutination	antitoxins
antibodies	antiserum	erythrocytes

1. Immunodiagnostic procedures detect either _____ or _____ in clinical specimens.

2. _____ are used as the reagent in IDPs that detect antigens.

3. A serum containing specific antibodies is called a/an _____.
4. _____ directed against toxins are called _____.
5. Immunodiagnostic procedures that detect _____ are more useful for diagnosis of a patient's present illness than IDPs that detect _____.
6 _____ are used as the reagent in IDPs that detect antibodies.
7. The type of IDP used to determine a persons blood type is called a/an _____ procedure.
8. A person's blood type is determined by learning the type of _____ that are present on that person's _____.

TRUE OR FALSE (T OR F)

___ 1. Phagocytes easily ingest and digest all types of bacteria.
___ 2. The symptoms of inflammation are heat, swelling, redness, pain, and sometimes loss of function.
___ 3. One purpose of the inflammatory response is to destroy bacteria.
___ 4. The main function of the ciliated epithelial cells that line the lower respiratory tract is to engulf and digest bacteria.
___ 5. There are indigenous microflora in the healthy bladder.
___ 6. Periodic flushing of urine through the urethra helps prevent pathogens from invading the urinary bladder.
___ 7. Interferon defends the body by lysing the cell wall of gram-negative bacteria.
___ 8. Certain types of leukocytes are the most efficient phagocytes in the body.
___ 9. The reticuloendothelial system is a system of tissues, including those of the spleen, liver, lymph nodes, blood vessels, and intestines.
___ 10. Either proteins or large polysaccharides may serve as antigens.
___ 11. Proteins are more effective antigens than polysaccharides.
___ 12. An effective vaccine must always contain live pathogens.
___ 13. A good vaccine stimulates the body to produce protective antibodies.
___ 14. An attenuated vaccine contains live, weakened pathogens that are no longer capable of causing disease, but are capable of stimulating the production of protective antibodies against that pathogen.
___ 15. Colostrum is a good source of antibodies to temporarily protect newborns.
___ 16. Passive immunity is always temporary.
___ 17. The TB skin test reaction is an example of immediate hypersensitivity.
___ 18. Rheumatic fever can be a complication of β-hemolytic streptococcal infections.
___ 19. *In vitro* refers to tests being performed in the living animal.
___ 20. Agglutination procedures are used to determine a person's blood type.
___ 21. A series of *in vitro* antigen-antibody tests on a sick person that shows an increasing antibody titer indicates that the patient's body is fighting the infection.
___ 22. Antitoxins are often used in the treatment of diphtheria, botulism, and tetanus.

MULTIPLE CHOICE

1. The body's fine line of defense against pathogens is
   a. antibody molecules
   b. phagocytes
   c. unbroken skin
   d. complement
   e. T cells

2. The migration of leukocytes in response to chemical agents is called
   a. phagocytosis
   b. leukocytosis
   c. leukopenia
   d. chemotaxis
   e. leukocidin

3. Interferon, an antiviral substance, has the following properties:
   a. it is species-specific
   b. it is not viral-specific
   c. it prevents viruses from multiplying
   d. all of the above
   e. none of the above

4. The reticuloendothelial system (RES)
   a. contains macrophages
   b. includes Kupffer cells, dust cells, and microglia
   c. is responsible for removing bacteria and other particles from circulating body fluids
   d. all of the above
   e. none of the above

5. The second line of defense against pathogens includes
   a. phagocytosis
   b. inflammation
   c. complement
   d. all of the above
   e. none of the above

6. Which of the following is not one of the four steps in phagocytosis?
   a. cell division
   b. digestion
   c. attachment
   d. chemotaxis
   e. ingestion

7. Which of the following is not one of the four major signs of inflammation?
   a. pain
   b. heat
   c. redness
   d. pus formation
   e. swelling

8. A substance that causes fever is said to be
   a. pyogenic
   b. pyocyanin
   c. an exudate
   d. pyrogenic
   e. an exotoxin

9. Cell-mediated immunity does *not* always involve
   a. a variety of cell types
   b. antibody production
   c. a delayed reaction
   d. cytokines

10. Antibodies are secreted by
    a. T_H cells
    b. macrophages
    c. basophils
    d. plasma cells
    e. T_C cells

11. Humoral immunity involves all of the following except
    a. macrophages
    b. B cells
    c. antibodies
    d. plasma cells
    e. antigens

12. Immunity that develops as a result of actual infection is
    a. natural passive acquired immunity
    b. artificial active acquired immunity
    c. natural active acquired immunity
    d. artificial passive acquired immunity

13. Natural passive acquired immunity would result from
    a. a gamma globulin injection
    b. a vaccine
    c. ingestion of colostrum
    d. having the measles

14. The vaccines used to protect people from diphtheria and tetanus are
    a. subunit vaccines
    b. inactivated vaccines
    c. attenuated vaccines
    d. toxoids
    e. antitoxins

15. The Schultz-Charlton test involves the injection of
    a. antitoxin into a test animal
    b. antitoxin into a patient
    c. toxin into a test animal
    d. toxin into a patient

CHAPTER 12

Major Infectious Diseases of Humans

SKIN INFECTIONS
Healthy Skin
Opportunists
Major Microbial Opportunists
  Gram-Positive Cocci
  Gram-Positive Bacilli
  Gram-Negative Bacilli
Viral Infections of the Skin
  Chickenpox and Shingles
  Measles, Hard Measles,
    Rubeola
  German Measles, Rubella
  Warts
Bacterial Infections of the Skin
  Impetigo, Impetigo of the
    Newborn, Scalded Skin
    Syndrome
  Scarlet Fever (Scarlatina),
    Erysipelas
  Folliculitis (Hair Follicle Infections),
    Furuncles (Boils), Carbuncles,
    Styes
  Acne
  Anthrax, Woolsorter's Disease
  Leprosy (Hansen Disease)
Fungal Infections of the Skin
  Dermatophytosis, Tinea Infections,
    Ringworm, Dermatomycosis
Burn and Wound Infections
  Burns
  Wound and Surgical Infections
  Infections Associated with Bites

EYE INFECTIONS
Viral Infections of the Eye
Bacterial Infections of the Eye
  Bacterial Conjunctivitis, Pink Eye
  Chlamydial Conjunctivitis, Inclusion
    Conjunctivitis, Paratrachoma
  Trachoma, Chlamydia
    Keratoconjunctivitis
  Gonococcal Conjunctivitis,

  Gonorrheal Ophthalmia
    Neonatorum

INFECTIOUS DISEASES
OF THE MOUTH
Indigenous Microflora of the Oral
  Cavity
Bacterial Infections of the Oral
  Cavity
  Acute Necrotizing Ulcerative
    Gingivitis (ANUG), Vincent's
    Angina, Trench Mouth

EAR INFECTIONS
Viral and Bacterial Ear Infections
  Otitis Media, Middle Ear Infection
  Otitis Externa, External Otitis, Ear
    Canal Infections, Swimmer's Ear

INFECTIOUS DISEASES OF THE
RESPIRATORY SYSTEM
Nonspecific Respiratory
  Infections
  Pneumonia
Viral Respiratory Infections
  Common Cold, Acute Viral Rhinitis,
    Acute Coryza
  Hantavirus Pulmonary Syndrome
    (HPS)
  Influenza, Flu
Specific Bacterial Respiratory
  Infections
  Diphtheria
  Legionellosis, Legionnaires'
    Disease, Pontiac Fever
  Mycoplasmal Pneumonia, Primary
    Atypical Pneumonia
  Streptococcal Pharyngitis, Strep
    Throat
  Tuberculosis, TB
  Whooping Cough, Pertussis
Fungal Respiratory Infections
  Cryptococcosis
  Histoplasmosis

  Pneumocystis carinii Pneumonia
    (PCP), Interstitial Plasma-Cell
    Pneumonia

INFECTIOUS DISEASES OF THE
GASTROINTESTINAL (GI) TRACT
Viral Gastrointestinal Infections
  Viral Gastroenteritis, Viral Enteritis,
    Viral Diarrhea
  Viral Hepatitis
Bacterial Gastrointestinal
  Infections
  Bacterial Gastritis and Ulcers
  Campylobacter
    Gastroenteritis
  Cholera
  Enterovirulent Escherichia coli
  Enterohemorrhagic E. Coli
    (EHEC) Diarrhea
  Enterotoxigenic E. Coli (ETEC)
    Diarrhea, Traveler's
    Diarrhea
  Typhoid Fever, Enteric Fever
  Shigellosis, Bacillary Dysentery
Bacterial Foodborne Intoxications,
  Foodborne Infections, Food
  Poisoning
  Botulism
  Clostridium perfringens Food
    Poisoning
  Staphylococcal Food Poisoning
Protozoal Gastrointestinal
  Diseases
  Primary Amebiasis, Amebic
    Dysentery
  Cryptosporidiosis
  Giardiasis

INFECTIOUS DISEASES OF THE
GENITOURINARY (GU) TRACT
Sexually Transmitted Diseases
  Urinary Tract Infections, Urethritis,
    Cystitis, Ureteritis, Prostatitis

Viral Genitourinary Tract Infections
Anogential Herpes Viral Infections
Genital Warts, Genital Papillomatosis, Condyloma Acuminatum
Bacterial Genitourinary Tract Infections
Genital Chlamydial Infections, Genital Chlamydiasis
Gonorrhea
Syphilis
Protozoal Genitourinary Tract Infections
Trichomoniasis
Other Sexually Transmitted Diseases
INFECTIOUS DISEASES OF THE CIRCULATORY SYSTEM
Viral Lymphatic and Cardiovascular Infections
Acquired Immunodeficiency Syndrome (AIDS)

Colorado Tick Fever
Infectious Mononucleosis, "Mono," "Kissing Disease"
Mumps, Infectious Parotitis
Rickettsial and Ehrlichial Cardiovascular Infections
Rocky Mountain Spotted Fever, Tickborne Typhus Fever
Endemic Typhus Fever, Murine Typhus Fever, Fleaborne Typhus
Epidemic Typhus Fever, Louseborne Typhus
Ehrlichiosis
Other Bacterial Cardiovascular Infections
Infective Endocarditis
Lyme Disease, Lyme Borreliosis
Plague, Black Death, Bubonic Plague, Pneumonic Plague, Septicemic Plague
Tularemia, Rabbit Fever
Protozoal Cardiovascular Infections

Babesiosis
Malaria
Toxoplasmosis
African Trypanosomiasis, African Sleeping Sickness
American Trypanosomiasis, Chagas' Disease
INFECTIOUS DISEASES OF THE NERVOUS SYSTEM
Viral Nervous System Infections
Poliomyelitis, Infantile Paralysis
Rabies
Bacterial Nervous System Infections
Botulism
Tetanus, Lockjaw
Protozoal Nervous System Infections
African Trypanosomiasis
Primary Amebic Meningoencephalitis, Naegleriasis

OBJECTIVES

After studying this chapter, you should be able to:

- *Name the major organs that might become infected in each body system*
- *List the most common members of the indigenous microflora usually found in the various body systems*
- *Outline the etiologic agent, reservoir, mode of transmission, pathogenesis, treatment, and control measures for the major infectious diseases of each body system*
- *For each body system, list some examples of diseases that are caused by bacteria, viruses, fungi, and protozoa.*

NEW TERMS

Arbovirus
Botulinal toxin
Cervicitis
Choleragin
Conjunctivitis
Cystitis
Cytotoxins
Dermatophytes

Encephalitis
Encephalomyelitis
Endocarditis
Epididymitis
Gingivitis
Hepatomegaly
Immunocompetent person
Immunosuppressed person

Keratitis
Keratoconjunctivitis
Lymphadenitis
Lymphadenopathy
Lymphangitis
Lymphocytosis
Malaise
Mastitis

Meninges (sing. *meninx*)
Meningitis
Meningoencephalitis
Myelitis
Myocarditis
Necrotoxin
Nephritis
Oncogenic
Oophoritis

Orchitis
Parotitis
Pericarditis
Periodontitis
Proctitis
Prophylaxis
Prostatitis
Pyelonephritis
Salpingitis

Sebum
Septicemia
Splenomegaly
Tetanospasmin
Tinea infections
Toxemia
Ureteritis
Urethritis

In previous chapters you have studied the disease process, how pathogens cause disease, modes of transmission, how the body attempts to defend itself, and the chemotherapeutic agents and vaccines used to cure and prevent certain infectious diseases. This chapter summarizes the major infectious diseases of the skin, eyes, ears, oral cavity, respiratory system, gastrointestinal (GI) tract, genitourinary (GU) tract, cardiovascular system, and nervous system. Some infections involve several body systems simultaneously or involve the movement of the pathogen from one area of the body to another. The source of the pathogen may be an opportunistic member of the indigenous microflora, but usually the causative agent is transmitted to the recipient from a reservoir of infection (see Chapter 9).

The early signs and symptoms of many diseases are flu-like; usually slight fever, headache, fatigue, *malaise* (a feeling of bodily discomfort), gastrointestinal upset, or sneezing and coughing develops in the patient. Only specific signs and symptoms characteristic of the major diseases are mentioned here.

SKIN INFECTIONS

Healthy Skin

As seen in Figure 12–1, the structure of skin is not simple. Healthy, intact skin serves as a formidable, protective barrier for the underlying tissues. Vast numbers of microbes survive on and within the epidermal layers, in pores, and in hair follicles (see Table 8–1). Microbial growth here is controlled by the (1) amount of moisture present, (2) pH, (3) temperature, (4) salinity of perspiration, (5) chemical wastes such as urea and fatty acids, and (6) other microbes present that secrete fatty acids and antimicrobic substances. Proper hygienic cleanliness and washing serves to flush away dead epithelial cells, many transient and resident microbes, and the odorous organic materials present in perspiration, *sebum* (sebaceous gland secretions), and microbial secretions.

Once the skin barrier is broken (*e.g.*, by wounds, surgery, or burns), the opportunists may infect underlying tissues, invade capillaries and lymph, and be carried by blood, lymph, or phagocytes to many regions of the body.

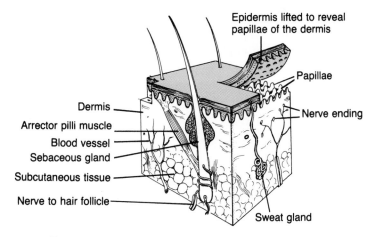

Epidermis lifted to reveal
papillae of the dermis

Papillae

Dermis

Nerve ending

Arrector pilli muscle

Blood vessel

Sebaceous gland

Subcutaneous tissue

Nerve to hair follicle

Sweat gland

FIGURE 12-1. Cross-section of the skin.

Opportunists

Most microbes that colonize the skin are harmless, but when the ecological balance of the skin environment changes chemically, physically, or microbiologically, several genera of aerobes and anaerobes may cause infections (Fig. 12–2).

Major Microbial Opportunists

GRAM-POSITIVE COCCI

Staphylococcus epidermidis, Staphylococcus aureus, Micrococcus species, and *Streptococcus* species are facultative anaerobes that may invade through breaks in skin and cause local, deep, or systemic infections. Many of these anaerobes produce invasive enzymes and damaging exotoxins capable of causing serious diseases, such as toxic shock syndrome (TSS), which is primarily caused by *S. aureus,* but can also be caused by *Streptococcus pyogenes.* TSS is a nationally notifiable disease in the United States. A total of 128 U.S. cases of TSS were reported to the CDC during 1998.

GRAM-POSITIVE BACILLI

These pleomorphic rods, frequently referred to as diphtheroids, include *Corynebacterium, Brevibacterium,* and *Propionibacterium* species, which frequently cause hair follicle and sweat gland infections.

Nationally notifiable diseases are infectious diseases that must be reported by U.S. physicians and health-care facilities to the Centers for Disease Control and Prevention (CDC), where data regarding these diseases are maintained and published periodically in *Morbidity and Mortality Weekly Report* (*MMWR*). Currently, 52 infectious diseases are considered to be nationally notifiable diseases. Most of them are described in this chapter, as are certain infectious diseases not currently considered to be nationally notifiable. Totals cited in this chapter for the years 1996 and 1997 are final figures, but the 1998 totals are considered "provisional," reflecting the number of cases reported to the CDC as of December 31, 1998.

GRAM-NEGATIVE BACILLI

In moist areas, armpits, perineum, and between toes, *Pseudomonas* and some enteric rods may be found. Microbes can grow profusely on the organic compounds in perspiration and sebum, producing malodorous fatty acids. Many deodorants contain antimicrobial agents that inhibit growth of these bacteria.

Viral Infections of the Skin

CHICKENPOX AND SHINGLES

Characteristics. Chickenpox (also known as varicella) is a respiratory infection and generalized viremia with local vesicular lesions on the skin of the face, thorax, and back that become encrusted. Vesicles also form in mucous membranes. A rash usually appears first on the trunk and later on the face, neck, arms and legs. Although usually a mild, self-limiting disease, it can be severely damaging to a fetus. Reye's (pronounced "rize") syndrome (a severe encephalomyelitis with liver damage) may follow clinical chickenpox if aspirin is given to children under 16 years of age. Secondary bacterial infections (*e.g.*, pneumonia, otitis media, bacteremia) frequently occur. Shingles (also

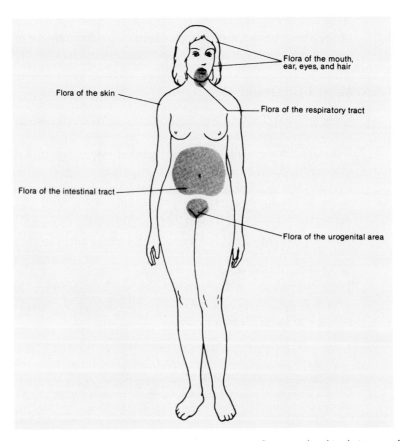

FIGURE 12-2. Areas where most of the indigenous microflora reside: skin, hair, mouth, ears, eyes, throat, nose, intestinal tract, and genitourinary tract.

known as herpes zoster) is a reactivation of varicella provirus in adults; it is an inflammation of sensory ganglia of cutaneous sensory nerves, producing fluid-filled blisters and pain. Shingles may occur at any age, but is most common after age 50.

Pathogen. Varicella-zoster (VZ) virus; a herpes virus.

Reservoir. Humans.

Transmission. Person-to-person by direct contact, droplet or airborne spread of vesicle fluid or secretions of the respiratory tract.

Incubation Period. 14 to 21 days; contagious from 1 to 2 days before onset of rash until all crops of vesicles have crusted.

Epidemiology. Chickenpox is the leading cause of vaccine-preventable death in the United States. Each year during the period 1990–1994, chickenpox was the underlying cause of death in an average of 43 children under 15 years of age. During the period 1988–1995, up to 10,000 children were hospitalized each year for chickenpox or its complications. A total of 98,727 U.S. cases of chickenpox were reported to the CDC during 1997.

Prevention and Control. A live, attenuated varicella virus vaccine (VARIVAX®) was licensed in the United States in 1995; susceptible children may receive the vaccine at any time after the first birthday. Isolation for 1 week after eruption of vesicles or until vesicles become dry. Airborne and contact precautions for hospitalized patients (see Chapter 10).

Treatment. Mild cases require treatment of symptoms only; wet compresses help soothe the itching. Varicella-zoster immune globulin (VZIG) is available for those with impaired immune system and is effective if given within 96 hours after exposure. Antiviral chemotherapy (acyclovir for severe cases of varicella; valacyclovir, famciclovir or acyclovir for herpes zoster) may be effective.

MEASLES, HARD MEASLES, RUBEOLA

Characteristics. Highly communicable viral disease with fever, cough, light sensitivity, Koplik spots in mouth, red blotchy skin rash. The rash starts around the ears and on the face and neck; in severe cases, it spreads over the trunk, arms and legs. Complications include bronchitis, pneumonia, otitis media, and encephalitis. Rarely, autoimmune subacute sclerosing panencephalitis (SSPE) may follow a latent period.

Pathogen. Measles (rubeola) virus; a paramyxovirus.

Reservoir. Humans.

Transmission. Airborne by droplet spread or direct contact with nasal or throat secretions.

Incubation Period. 7 to 14 days; contagious from 2 to 4 days before rash until 2 to 5 days after the onset of rash.

Epidemiology. Eighty-nine U.S. cases were reported to the CDC in 1998, the lowest number of measles cases reported for a single year since measles became a nationally reportable disease in 1912.

Prevention and Control. Live, attenuated measles vaccine is administered as part of the measles-mumps-rubella (MMR) vaccine; the initial dose should be administered between 12 and 15 months of age. Vaccination produces immunity in over 95% of recipients. Airborne precautions for hospitalized patients.

Treatment. Bed rest, fluids, preventive nursing care to prevent complications and secondary infections. Acetaminophen or ibuprofen may be given to reduce fever.

GERMAN MEASLES, RUBELLA

Characteristics. Fine, pinkish, flat rash begins 1 or 2 days after the onset of symptoms; starts on the face and neck and spreads to the trunk, arms, and legs. German measles is a milder disease than is hard measles, with fewer complications. During first trimester of pregnancy, it may cause congenital rubella syndrome in fetus.

Pathogen. Rubella virus; a togavirus (Fig. 12-3).

Reservoir. Humans.

Transmission. Droplet spread or direct contact with nasopharyngeal secretions of infected people.

Incubation Period. 14 to 21 days; contagious from shortly before the onset of symptoms until the rash disappears; infected newborns may be infective for many months.

Epidemiology. Worldwide occurrence; 345 U.S. cases were reported to the CDC in 1998.

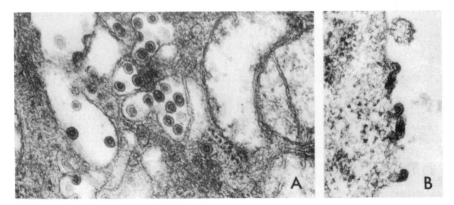

FIGURE 12-3. Development of rubella virus in the surface and cytoplasmic membranes of infected cell cultures. (A) Viral particles budding from cytoplasmic membranes into vacuoles and cytoplasm. Numerous mature virions are present within vacuoles (original magnification, ×60,000). (B) Viral particles budding from the surface of an infected cell (original magnification, ×60,000). (Oshiro LS et al: J Gen Virol 5:205)

Prevention and Control. Live, attenuated rubella vaccine is administered as part of the measles-mumps-rubella (MMR) vaccine; the initial dose should be administered between 12 and 15 months of age. Vaccine is recommended for all susceptible premarital and postpartum women. Droplet Precautions for hospitalized patients (see Chapter 10).

Treatment. A mild disease that seldom requires treatment.

WARTS

Characteristics. Many varieties including common warts (verrucae vulgaris), venereal warts (condyloma acuminatum, see Fig. 12–19), and plantar warts. Most warts are harmless, but some can become cancerous.

Pathogens. About 60 to 70 different types of human papillomaviruses (of the papovavirus group of DNA viruses).

Reservoir. Humans.

Transmission. Transmission usually by direct contact. Genital warts are sexually transmitted. Easily spread from one area of the body to another, but most are not very contagious from person to person (genital warts are an exception).

Incubation Period. Range is 1 to 20 months, but usually 2 to 3 months.

Epidemiology. Worldwide occurrence; not a nationally notifiable disease in the United States.

Prevention and Control. Avoid direct contact with another person's warts. Condom use may reduce the transmission of venereal warts.

Treatment. May disappear without treatment. Salicylic acid plasters, freezing with liquid nitrogen, 5-fluorouracil, or interferon may be required, depending upon type, location, and severity.

Bacterial Infections of the Skin

IMPETIGO, IMPETIGO OF THE NEWBORN, SCALDED SKIN SYNDROME

Characteristics. Small pus-filled blisters (pustules), then scabby, crusted eruptions (Fig. 12–4), itching, peeling of skin.

Pathogens. Staphylococcus aureus or *Streptococcus pyogenes* or both. *S. aureus* is responsible for impetigo of the newborn (impetigo neonatorum) and staphylococcal scalded skin syndrome (SSSS), both of which may occur as epidemics in hospital nurseries.

Pathogenicity. Spreads through skin by producing hyaluronidase. SSSS is produced by strains of *S. aureus* that produce epidermolytic toxin, which causes the top layer of skin (epidermis) to split from the rest of the skin.

Reservoir. Humans.

Transmission. Twenty to thirty percent of the general population are nasal carriers

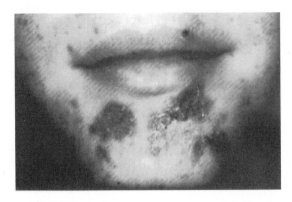

FIGURE 12-4. Impetigo.

of *S. aureus.* Transmission is through direct contact with a person having a purulent lesion or a carrier. In hospitals, it is spread by hands of health-care workers.

Incubation Period. Variable, but commonly 4 to 10 days.

Prevention and Control. Handwashing. Healthcare workers having pustules, boils, abscesses, and other minor lesions should not be permitted to work in the nursery. Antibiotic ointments for nasal carriers. Use of antiseptic solutions during epidemics. Contact precautions for hospitalized patients.

Diagnosis. The infecting strain must be isolated on culture media and identified using biochemical- or enzyme-based tests. Staphylococci are catalase-positive, Gram-positive cocci in clusters. The coagulase test differentiates *S. aureus* (coagulase-positive) from other species of *Staphylococcus* (coagulase-negative). Susceptibility testing must be performed because many strains of *S. aureus* have become multiply drug-resistant.

Treatment. Such topical antiseptics and antibiotics as penicillinase-resistant penicillins, cephalosporin, or clindamycin; vancomycin for infections with methicillin-resistant strains of *S. aureus* (MRSA).

SCARLET FEVER (SCARLATINA), ERYSIPELAS
Characteristics. Scarlet fever: widespread, pink-red rash, most obvious on the abdomen, sides of the chest, and in skinfolds. Erysipelas: hot, red eruptions (St. Anthony's fire).

Pathogen. *Streptococcus pyogenes;* also known as group A, β-hemolytic streptococcus and "Strep A."

Pathogenicity. Scarlet fever is caused by erythrogenic toxin, produced by some strains of *S. pyogenes.* It can be a complication (sequela) of strep throat (or strep pharyngitis). Some strains of *S. pyogenes* are the "flesh-eating bacteria" (see Insight Box) and some strains produce a toxin that causes toxic shock syndrome (TSS), although the majority of TSS cases are caused by *S. aureus.*

Reservoir. Humans.

Transmission. Person-to-person via large respiratory droplets or direct contact with patients or carriers.

Incubation Period. Usually 1 to 3 days.

Prevention and Control. Contact and droplet precautions for hospitalized patients; isolation until after 24 hours of antibiotic therapy.

Diagnosis. The infecting strain must be isolated on culture media and identified using biochemical- or enzyme-based tests. Streptococci are catalase-negative, Gram-positive cocci, usually in chains. *S. pyogenes* is β-hemolytic. The A-disk (bacitracin sensitivity) test differentiates between *S. pyogenes* (bacitracin-sensitive) and other β-hemolytic streptococci (bacitracin-resistant). Currently, because *S. pyogenes* has not yet developed resistance to penicillin, susceptibility testing is not routinely performed. Some strains have become resistant to other antimicrobial agents, however.

Treatment. Penicillin, amoxicillin, erythromycin, cephalosporins, clindamycin.

FOLLICULITIS (HAIR FOLLICLE INFECTIONS), FURUNCLES (BOILS), CARBUNCLES, STYES

Characteristics. An infected hair follicle may progress to a large, tender, swollen pustule (boil) which, in turn, may progress to a furuncle, involving subcutaneous tissue. A stye (or sty) is an infected sebaceous gland at the edge of the eyelid or under it.

Pathogen. *Staphylococcus aureus* (Gram-positive cocci in clusters).

Reservoir. Humans (rarely animals).

Transmission. Auto-infection occurs in carriers; contact with someone who has a purulent lesion or is a carrier.

Incubation Period. Variable, but usually 4 to 10 days.

Prevention and Control. Good personal hygiene, especially handwashing. Use of liquid antibacterial soap. Avoid common use of toilet articles.

Diagnosis. See previous section on "Impetigo, Impetigo of the Newborn, Scalded Skin Syndrome."

Treatment. Culture and susceptibility testing should be performed. Topical bacitracin or polymyxin B. Oral antibiotics for boils near the nose to prevent spread to the brain. People with recurring boils may require oral antibiotics for months or years.

ACNE

Characteristics. A common condition in which pores become clogged with dried sebum, flaked skin, and bacteria; leads to the formation of blackheads and whiteheads (collectively known as acne pimples) and inflamed, infected abscesses; more common among teenagers.

Pathogens. *Propionibacterium acnes* and other *Propionibacterium* spp. (all are anaerobic, Gram-positive bacilli).

INSIGHT
"Microbes In The News"—"Flesh-Eating" Bacteria

"Flesh-eating" bacteria, a term coined by the press, received worldwide attention in the news media, especially tabloid newspapers, following a small number of cases of infection with these organisms in an English town in 1994. However, medical scientists had been aware of such bacteria earlier.

The so-called "flesh-eating" bacteria are especially invasive strains of *Streptococcus pyogenes,* a Gram-positive coccus also known as group A, beta-hemolytic streptococcus. This is the bacterium that causes strep throat and its various sequelae (complications), such as scarlet fever, rheumatic fever, and glomerulonephritis.

The "flesh-eating" strains of *S. pyogenes* produce proteases, enzymes capable of destroying proteins and enabling the bacteria to invade human epithelial cells. They also produce a toxin (pyogenic toxin) that causes a type of toxic shock syndrome. Such strains rely on genetic information obtained from a type of bacterial virus (bacteriophage) that infects them.

Scientists have been aware of invasive group A streptococcal infections for many years, but the number of cases has increased significantly in the past few years. According to the Centers for Disease Control and Prevention (CDC), an estimated 10,000 to 15,000 cases of invasive strep infections occur in the United States per year, causing an estimated 2000 to 3000 deaths. Although this is a large number of infections and deaths, it does not constitute an epidemic. About 5 to 10% of the cases involve necrotizing fasciitis (the proper name for the "flesh-eating" cases). As one newspaper article stated, "the news coverage is more widespread than the bacteria."

Necrotizing fasciitis and toxic shock syndrome can occur following entry of the invasive strains into cuts, bruises, and even chickenpox lesions, and often begins as a mild skin lesion. It can quickly spread, damaging nearby tissue and causing gangrene. Distant abscesses, pneumonia, multi-organ failure, shock, and death may occur. People with strep throat should be particularly careful not to contaminate any external lesions they might have with their own saliva. Laboratory workers and other healthcare personnel should also be very careful when handling throat swabs, throat culture plates, and when working with *S. pyogenes* isolates. They should protect any open skin lesions on their hands by wearing gloves, avoid creation of aerosols when working with *S. pyogenes* cultures, and carefully sterilize all streptococcal-contaminated tissue, culture media, glassware, and any eating utensils used by patients with streptococcal infections.

Reservoir. Humans.

Prevention and Control. Cleanliness, good personal hygiene.

Treatment. Depends upon severity. Topical drugs, such as clindamycin, erythromycin, tretinoin (retinoic acid), benzoyl peroxide, and sulfur resorcinol. Oral antibiotics, such as tetracycline, minocycline, erythromycin, doxycycline.

ANTHRAX, WOOLSORTER'S DISEASE

Characteristics. Anthrax can affect the skin (cutaneous anthrax), the lungs (inhalation or pulmonary anthrax), or the GI tract, depending upon the portal of entry of the etiologic agent. In cutaneous anthrax, depressed blackened lesions called eschars occur, caused by a necrotoxin (a substance that kills cells). Inhalation and gastrointestinal anthrax are often fatal, but cutaneous anthrax usually is not.

Pathogen. Bacillus anthracis, a spore-forming, Gram-positive bacillus.

Pathogenicity. Endospore germinates; bacteria produces necrotoxins.

Reservoir. Spores are present in soil, animal hair, wool, hides, and products made from them.

Transmission. Entry of endospores through breaks in skin, inhalation of spores, or ingestion of bacteria in contaminated meat.

Incubation Period. A few hours to 7 days; most often, cases occur within 48 hours of exposure; transmission from person to person is very rare.

Epidemiology. No U.S. cases were reported to the CDC in 1998.

Prevention and Control. A vaccine is available for people at high risk (*e.g.*, veterinarians, laboratory workers, employees of textile mills, and military personnel) and animals at risk. Avoid contact with infected animals and soils contaminated by spores from infected animals. Thoroughly wash, disinfect, or sterilize hair, wool, and other animal products.

Diagnosis. Isolation of *B. anthracis* from blood, lesions, or discharges and identification using biochemical- or enzyme-based tests. Use paired sera to demonstrate a rise in antibody titer between acute and convalescent serum specimens.

Treatment. Standard precautions for hospitalized patients (see Chapter 10). Skin infections are treated with penicillin injections or oral tetracycline or erythromycin. Lung infections require IV penicillin.

LEPROSY (HANSEN DISEASE)

Characteristics. Neural, tuberculoid form: lesions on skin and peripheral nerves, loss of sensation. Cutaneous, lepromatous form: progressive disfiguring nodules in skin; invades throughout body.

Pathogen. *Mycobacterium leprae,* an acid-fast bacillus, cultured only in laboratory animals (nine-banded armadillos, mouse footpads).

Reservoirs. Humans; armadillos in Texas and Louisiana have a naturally occurring disease that is identical to experimental leprosy in this animal, suggesting that transmission from armadillos to humans is possible.

Transmission. The exact mode of transmission has not been clearly established. The organisms may gain entrance through the respiratory tract or broken skin. Does not appear to be easily transmitted from person to person. Prolonged, close contact with an infected individual appears to be necessary. The tuberculoid form of leprosy isn't contagious.

Incubation Period. Nine months to twenty years, with an average of about 4 years for tuberculoid leprosy and 8 years for lepromatous leprosy.

Epidemiology. The worldwide prevalence of leprosy was estimated to be about 2.5 million in 1994; 102 U.S. cases were reported to the CDC in 1998. Most U.S. cases involve people who emigrated from developing countries.

Prevention and Control. Early treatment can prevent or correct most major deformities. Evidence suggests that infectiveness is usually lost within 3 months of continuous and regular treatment with dapsone or clofazimine, or within 3 days of treatment with rifampin. Standard precautions for hospitalized patients.

Diagnosis. *M. leprae* cannot be grown on artificial media. Demonstration of acid-fast bacilli in skin smears or skin biopsies.

Fungal Infections of the Skin

Many fungi cause skin lesions and may enter the body through the skin, such as *Sporothrix* and *Candida,* but the most common are the *dermatophytes,* the fungi that cause superficial fungal infections of the skin, hair, and nails (called ringworm or *tinea infections*). The cell-mediated immune responses to fungi result in inflammation and limitation of the spread of the fungi.

DERMATOPHYTOSIS, TINEA INFECTIONS, RINGWORM, DERMATOMYCOSIS

Characteristics. The dermatophytoses are named in accordance with the site of infection. Fungal lesions of the scalp (tinea capitis), beard area (tinea barbae), groin area (tinea cruris or jock itch), trunk of the body (tinea corporis), foot (tinea pedis or athlete's foot), and nails (tinea unguium or onychomycosis). Some fungal infections cause only limited irritation, scaling, and redness. Others cause itching, swelling, blisters, and severe scaling.

Pathogens. Various species of filamentous fungi, including *Microsporum, Epidermophyton,* and *Trichophyton* spp.

Reservoirs. Humans, animals, and soil.

Transmission. Direct or indirect contact with lesions of humans or animals, contaminated floors, shower stalls, locker room benches, barber clippers, combs, hairbrushes, clothing. Spores enter through breaks in skin and moist areas, and germinate into filamentous growths.

Incubation Period. 4 to 14 days.

Prevention and Control. Maintain good hygienic practices. Keep susceptible areas of body clean and dry. Use fungicidal foot powders and fungicidal agents in shower stalls, on locker room floors and benches. Standard precautions for hospitalized patients.

Diagnosis. Microscopic examination of potassium hydroxide (KOH) preparations of skin scrapings can reveal the presence of fungal hyphae. Dermatophytes can be cultured on various media including Sabouraud dextrose agar. Molds are identified by a combination of macroscopic and microscopic observations.

Treatment. Keep infected areas clean and dry. Topical application of antifungal powders and creams containing miconazole (Lotrimin®), clotrimazole, econazole, ketoconazole. Oral griseofulvin or ketoconazole may be required for more severe or stubborn infections.

Burn and Wound Infections

When the protective skin barrier is broken by burns, wounds, or surgical procedures, opportunistic indigenous microflora and environmental bacteria can invade and cause local or deep-tissue infections. The bacteria may also become systemic and produce exotoxins, causing severe damage to the individual.

BURNS

Many patients die after being burned severely due to loss of body fluids and the presence of toxic microbial invaders, including *S. aureus, S. pyogenes, Pseudomonas aeruginosa,* and many fungi. These organisms usually grow aerobically in the burned area. They can produce many exotoxins, such as cytotoxins and necrotoxins (which damage cells and tissue), and neurotoxins (which affect the nervous system), which may ultimately cause the death of the patient.

Burn victims must be treated in an aseptic environment in which healthcare personnel are gloved, gowned, and masked. The burns may be left open to the air to speed healing or covered with artificial skin or film to reduce contamination and loss of fluids. Some antimicrobial topical agents (*e.g.,* silver sulfadiazine) may be used to reduce the possibility of infections during the prolonged healing period.

WOUND AND SURGICAL INFECTIONS

Most gunshot, stab, puncture, bite, and abrasive wounds are contaminated during the wounding process. Microbes introduced are frequently anaerobes from dust or dirt and indigenous microflora that grow rapidly deep within the wound; other facultative anaerobes flourish on the surface of the wound, secreting enzymes which enable them to invade the blood and lymph. Bacterial and fungal spores may also be introduced, causing local and deep-seated infections. Any traumatic injury must be opened and thoroughly cleansed to remove debris and inhibit the growth of microbes; then it is usually covered lightly to prevent further contamination.

Staphylococcus and *Streptococcus* species are the most frequent bacteria that cause focal infections. They may invade by secreting hyaluronidase and other spreading factors to cause severe toxic systemic infections of many sites, including the brain, spinal cord, and bone. *P. aeruginosa* from soil and feces is notorious for causing deep, antibiotic-resistant infections. This Gram-negative bacillus produces protease enzymes that enable it to move through tissues and an exotoxin that inhibits protein synthesis. Antimicrobial susceptibility tests are essential to ensure that the appropriate drug is administered.

Anaerobic *Clostridium* species are of grave concern in puncture wounds. Endospores of *Clostridium tetani* (see Fig. 12–28) may be introduced from soil or fecal contamination. When these spores germinate into vegetative bacteria, neurotoxins are produced that cause involuntary muscle spasms and respiratory failure. The availability of vaccines and antitoxins has greatly reduced deaths from tetanus in developed countries.

Clostridium perfringens, the major etiologic agent of gas gangrene, is not only of concern in accidental wounds but in surgical sites as well. After the contaminating spores germinate, the vegetative pathogens produce many invasive enzymes and exotoxins, resulting in gas-filled necrotic areas. Rapid and extensive destruction of muscle tissue may necessitate amputation of an infected extremity.

Fungal infections, such as sporotrichosis, are caused by the introduction of soil fungi and spores that invade cutaneous and lymphatic tissues through wounds that are sometimes as small as a thorn prick. This slowly progressing disease is usually localized and self-limiting, but severe cases may be treated with oral potassium iodide, amphotericin B, or itraconazole.

INFECTIONS ASSOCIATED WITH BITES

Human and animal bites introduce oral microflora into the wound, and such wounds frequently become infected with these organisms. Laboratory request slips accompanying specimens collected from infected bite wounds must clearly state that the specimen is from a bite wound, as well as the type of bite (*i.e.,* human, cat, dog, etc.). Because specific pathogens are frequently associated with certain animals, alerted laboratory personnel will then process the specimens in a manner that will enable isolation of those particular pathogens. Infections with *Pasteurella multocida* and *P. haemolytica* (Gram-negative bacilli), for example, are frequently associated with cat and dog bites. Other diseases associated with animal bites include cat-scratch disease (caused by *Bartonella henselae,* a Gram-negative bacillus), infections due to *Capnocytophaga* spp. (Gram-negative bacilli), plague (caused by *Yersinia pestis,* a Gram-negative bacterium), rabies (a viral disease), rat-bite fever (caused by *Spirillum minor,* a spirochete, or *Streptobacillus moniliformis,* a Gram-negative bacillus), tetanus (caused by a neurotoxin produced by *Clostridium tetani,* an anaerobic, spore-forming, Gram-positive bacillus), and tularemia (caused by *Francisella tularensis,* a Gram-negative bacillus).

Many other pathogens are introduced through the skin to the circulatory or nervous systems by bites of arthropod vectors (ticks, mites, fleas, lice, biting flies, and mosquitoes). These diseases include encephalitis, Colorado tick fever, Rocky Mountain spotted fever, typhus, plague, tularemia, malaria, and Lyme disease; they are discussed elsewhere in this chapter and in Appendix C.

Treating a bite wound immediately is the most valuable way to prevent these diseases. The area must be cleaned thoroughly with hot, soapy water. Following cleaning and disinfection, the wound should be left open to the air to prevent microbial growth. Bite wounds should not be closed tightly with stitches and bandages because this procedure would encourage the growth of anaerobic bacteria. Vaccines are available for prevention of rabies and tetanus.

EYE INFECTIONS

The eye consists of tissues similar to and contiguous with the skin. The anatomy and structure of the eye are illustrated in Figure 12–5. The external surface of the eye is lubricated, cleansed, and protected by tears, mucus, and sebum. Thus, continual production of tears and the presence of lysozyme and other antimicrobial

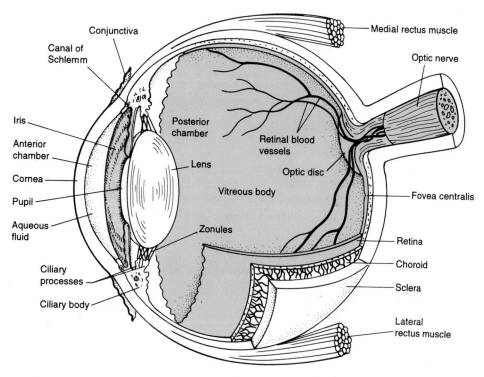

FIGURE 12-5. Anatomy of the eye.

substances found in tears greatly reduce the numbers of indigenous microflora found on the eye surfaces.

Infections of the eye caused by bacteria (including chlamydias) and viruses should be differentiated from allergic manifestations and irritation by microscopic examination of the *exudate* (oozing pus), culture of pathogens, or immunodiagnostic procedures (*e.g.*, immunofluorescent procedures and enzyme-linked immunosorbent assays [ELISAs]). *Conjunctivitis* is an infection or inflammation of the conjunctiva—the thin, tough lining that covers the inner wall of the eyelid and the sclera (the white of the eye). *Keratitis* is an infection or inflammation of the cornea—the domed covering over the iris and lens. *Keratoconjunctivitis* involves both the cornea and conjunctiva.

Viral Infections of the Eye

Adenoviruses, enteroviruses, and herpes simplex viruses can cause conjunctivitis, keratitis, and/or keratoconjunctivitis. People with viral infections should wash their hands thoroughly before inserting or removing contact lenses or otherwise touching their eyes. Antiviral agents such as trifluridine, vidarabine, or idoxuridine may be prescribed for herpes simplex infections, either as an ointment or eye drops. Contact precautions for hospitalized patients should be employed.

Bacterial Infections of the Eye

BACTERIAL CONJUNCTIVITIS, PINK EYE

Characteristics. Irritation, reddening of conjunctiva, edema of eyelids, mucopurulent discharge, sensitivity to light; highly contagious.

Pathogens. *Haemophilus influenzae* biogroup *aegyptius* and *Streptococcus pneumoniae* are the most common causes, although many different bacteria can cause pink eye.

Reservoir. Humans.

Transmission. Human to human via contact with eye and respiratory discharges; contaminated fingers, clothing, eye makeup, eye medications, ophthalmic instruments, and contact lens-wetting and lens-cleaning agents.

Incubation Period. Usually 1 to 3 days, depending on causative agent.

Prevention and Control. Personal hygiene; standard precautions for hospitalized patients, sterilization of fomites.

Treatment. Ophthalmic tetracycline, erythromycin, gentamicin, or sulfonamide, depending on the susceptibility of the pathogen.

CHLAMYDIAL CONJUNCTIVITIS, INCLUSION CONJUNCTIVITIS, PARATRACHOMA

Characteristics. In neonates, acute conjunctivitis with mucopurulent discharge; may result in mild scarring of conjunctivae and cornea; may be concurrent with chlamydial nasopharyngitis or pneumonia; in adults, may be concurrent with nongonococcal urethritis or cervicitis.

Pathogens. Certain serotypes (serovars) of *Chlamydia trachomatis.*

Reservoir. Humans.

Transmission. Contact with genital discharges of infected people; contaminated fingers to eye; infection in newborns via infected birth canal; nonchlorinated swimming pools.

Incubation Period. Usually 5 to 12 days in newborns; 6 to 19 days for adults.

Diagnosis. Chlamydias do not grow on artificial media; diagnosis by cell culture and/or immunodiagnostic procedures.

Prevention and Control. Identification and treatment of chlamydial genital infections of expectant parents; standard precautions for hospitalized patients; use of condoms; chlorination of swimming pools; povidone-iodine, tetracycline, or erythromycin ophthalmic ointment or drops in eyes of newborns. Chlamydias are not susceptible to penicillin.

Treatment. Tetracycline, erythromycin, ofloxacin, or azithromycin.

TRACHOMA, CHLAMYDIA KERATOCONJUNCTIVITIS

Characteristics. Highly contagious, acute or chronic conjunctival inflammation, resulting in scarring of cornea and conjunctiva, deformation of eyelids, and blindness.

Pathogens. Certain serotypes (serovars) of *Chlamydia trachomatis*.

Reservoir. Humans.

Transmission. Direct contact with infectious ocular or nasal secretions or contaminated articles; also spread by flies.

Epidemiology. Trachoma is most common in poverty-stricken areas of the hot, dry Mediterranean countries and the Far East. It is the leading cause of blindness in the world. Trachoma occurs only rarely in the United States; it is not a nationally notifiable disease in the United States.

Incubation Period. 5 to 12 days.

Diagnosis. Microscopic observation of intracellular chlamydial elementary bodies in epithelial cells of Giemsa-stained conjunctival scrapings or by an immunofluorescent procedure; alternatively, the chlamydias can be isolated from specimens using cell culture techniques.

Prevention and Control. Personal hygiene; improved sanitation and living conditions; avoid common use of toilet articles and towels; standard precautions for hospitalized patients.

Treatment. Topical ophthalmic tetracycline or erythromycin, oral sulfonamides, tetracyclines, erythromycin or azithromycin.

GONOCOCCAL CONJUNCTIVITIS, GONORRHEAL OPHTHALMIA NEONATORUM

Characteristics. Acute redness and swelling of conjunctiva, purulent discharge (Fig. 12–6); corneal ulcers, perforation, and blindness, if untreated.

Pathogen. Neisseria gonorrhoeae; Gram-negative, kidney bean-shaped diplococci; also known as gonococci or GC.

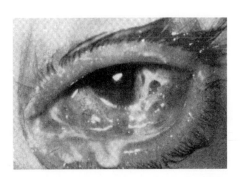

FIGURE 12-6. Purulent conjunctivitis caused by *Neisseria gonorrhoeae.*

Reservoir. Humans (the maternal cervix).

Transmission. Contact with the infected birth canal during delivery; adult infection can result from finger-to-eye contact with infectious genital secretions.

Incubation Period. Usually 1 to 5 days.

Diagnosis. Microscopic observation of Gram-negative cocci in smears of purulent material; isolation of *N. gonorrhoeae* on culture media (chocolate agar or modified chocolate agar, such as Thayer-Martin agar, Martin-Lewis agar, New York City agar, or Transgrow®).

Prevention and Control. Culturing for *N. gonorrhoeae* during the prenatal period; instill 1% silver nitrate solution in eyes of neonates; ophthalmic erythromycin or tetracycline ointments may be used to prevent chlamydial, as well as gonococcal eye infections.

Treatment. Ceftriaxone, cefixime, ciprofloxacin, or ofloxacin for penicillinase-producing *N. gonorrhoeae* (PPNG); penicillin G for penicillin-sensitive strains; mother and infant should also be treated for chlamydial infection. Standard precautions for hospitalized patients.

INFECTIOUS DISEASES OF THE MOUTH

The oral cavity is a complex ecosystem suitable for growth and interrelationships of many types of microorganisms, including a variety of Gram-positive and Gram-negative bacteria. The actual indigenous microflora of the mouth varies greatly from person to person.

When the anatomy of the mouth (Fig. 12–7) is studied, it becomes obvious that there are many areas where bacteria can attach, colonize, and proliferate, even in the presence of the normal defenses. In the healthy mouth, saliva secreted by salivary and mucous glands helps control the growth of opportunistic oral flora. Saliva contains enzymes (including lysozyme), immunoglobulins (IgA), and buffers to control the near-neutral pH and continually flushes microbes and food particles through the mouth. Other antimicrobial secretions and phagocytes are found in the mucus that coats the oral surfaces. The hard, complex, calcium tooth enamel, bathed in protective saliva, usually resists damage by oral microbes; however, if the ecological balance is upset or is not properly maintained, oral disease may result.

Indigenous Microflora of the Oral Cavity

The indigenous microflora of the mouth has been shown recently to include about 300 identified species of bacteria, both aerobes and anaerobes. Many additional, as-yet-unidentified bacteria also live there. Some members of the oral microflora are beneficial; they produce secretions that are antagonistic to other bacteria. Although several species of *Streptococcus* (*S. salivarius, S. mitis, S. sanguis,* and *S. mutans*) and *Actinomyces* species often interact to protect the oral surfaces, in other circumstances, they are involved in oral disease.

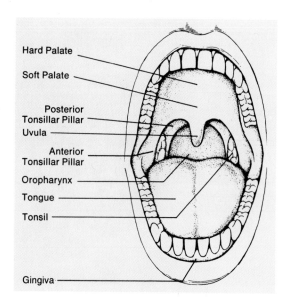

FIGURE 12-7. Anatomy of the mouth.

Bacterial Infections of the Oral Cavity

The anaerobic environment produced by oxidation-reduction reactions of the oral flora allow certain genera of anaerobic bacteria (*e.g.*, *Bacteroides, Porphyromonas, Fusobacterium, Prevotella, Actinomyces,* and *Treponema* spp.) to become involved in the production of oral diseases. The coating that forms on unclean teeth, called dental plaque, is a coaggregation of bacteria and their products. Many of these microorganisms produce a slime layer or glycocalyx that enables them to attach firmly and cause damage to the tooth enamel. Certain carbohydrates, especially sucrose, are metabolized by streptococci (especially *S. mutans,* Fig. 12–8), lactobacilli, and *Actinomyces* spp., producing lactic acid, which rapidly dissolves the tooth enamel. When plaque remains on teeth for more than 72 hours, it hardens into tartar or calculus, which cannot be completely removed by brushing and flossing.

The progressive microbial activities involving formation of dental plaque, dental caries (tooth decay), *gingivitis* (infection of the gingiva or gums), and *periodontitis* (a more severe infection, involving tooth and bone loss) result from a unique microbial population, reduced host defenses, improper diet, and poor dental hygiene. These diseases are the consequence of at least four microbial activities, including (1) formation of dextran (a polysaccharide) from sugars by streptococci, (2) acid production by lactobacilli, (3) deposition of calculus by *Actinomyces,* and (4) secretion of inflammatory substances (endotoxin) by *Bacteroides* species. This combination of circumstances damages the teeth, soft tissues (gingiva), alveolar bone, and the periodontal fibers attaching teeth to bone. Diseases such as gingivitis, periodontitis, and trench mouth (described below) are collectively known as periodontal diseases.

Periodontal diseases can be prevented by maintaining good health, proper oral hygiene (brushing, tartar-control toothpaste, flossing), an adequate diet without

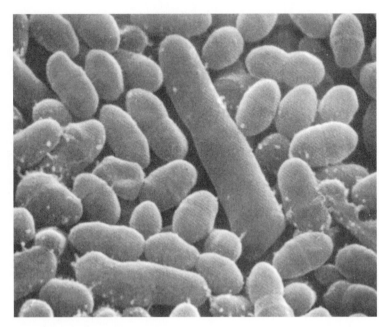

FIGURE 12-8. Electron micrograph of *Streptococcus mutans* in dental plaque. (SEM by Joan Foster)

sugars, and regular fluoride treatments to help control the microbial population and to prevent destructive bacterial interactions.

Severe gingivitis and periodontitis require professional care by a specially-trained dentist called a periodontist. Using techniques known as scaling and planing, periodontists remove tartar that has accumulated on tooth surfaces up to one-fifth of an inch below the gum line—areas where brushing and flossing cannot reach. Following dental surgery, periodontists often prescribe a chlorhexidine mouth rinse as a temporary substitute for brushing and flossing.

ACUTE NECROTIZING ULCERATIVE GINGIVITIS (ANUG), VINCENT'S ANGINA, TRENCH MOUTH

Characteristics. The term trench mouth comes to us from World War I, where soldiers fighting in trenches developed the infection. It is usually the result of a combination of poor oral hygiene, physical or emotional stress, and poor diet. It involves painful, bleeding gums, erosion of gum tissue, and swollen lymph nodes under the jaw.

Pathogens. Trench mouth is a noncontagious, synergistic infection involving two or more microorganisms. The most commonly involved bacteria are *Prevotella intermedia* (an anaerobic, Gram-negative bacillus) and an as-yet-unnamed intermediate sized spirochete. Other bacteria may also participate, such as *Bacteroides* and *Fusobacterium* spp. (anaerobic, Gram-negative bacilli), *Actinomyces* spp. (anaerobic, Gram-positive bacilli), and *Borrelia* and *Treponema* spp. (anaerobic, Gram variable, spirochetes).

Prevention and Control. As is true for other periodontal diseases, trench mouth can be prevented by good oral hygiene. Standard precautions for hospitalized patients.

Treatment. Periodontal treatment, including a thorough cleaning, removal of tartar and dead gum tissue, 1.5% hydrogen peroxide or chlorhexidine rinses, and sometimes antibiotics.

EAR INFECTIONS

When the anatomy of the ear (Fig. 12–9) is studied, one observes that there are only three pathways for pathogens to enter: (1) through the eustachian (auditory) tube, from the throat and nasopharynx; (2) from the external ear; and (3) through the blood or lymph. Usually, bacteria are trapped in the middle ear when a bacterial infection in the throat and nasopharynx causes the eustachian tube to close. The result is an anaerobic condition in the middle ear, allowing anaerobes and facultative anaerobes to grow and cause pressure on the tympanic membrane (eardrum). Swollen lymphoid (adenoid) tissues, viral infections, and allergies may also close the eustachian tube, especially in young children.

Viral and Bacterial Ear Infections

OTITIS MEDIA, MIDDLE EAR INFECTION

Characteristics. Often develops as a complication of the common cold. Persistent and severe earache; temporary hearing loss; pressure in middle ear; bulging of the

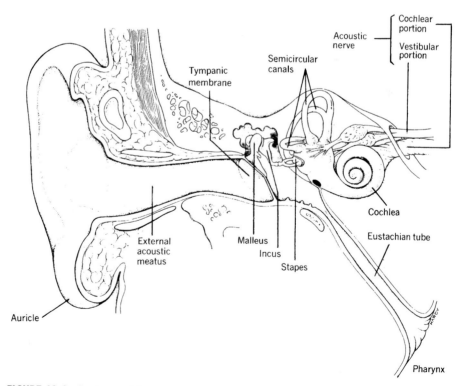

FIGURE 12-9. Anatomy of the ear.

eardrum (tympanic membrane); nausea, vomiting, diarrhea, and fever in young children; may lead to rupture of the eardrum, bloody discharge and then pus from the ear. Severe complications, including bone infection, permanent hearing loss, and meningitis, may occur. Most common in young children, particularly those between 3 months to 3 years of age.

Pathogens. The three most common bacterial causes of otitis media are *Streptococcus pneumoniae, Haemophilus influenzae,* and *Moraxella catarrhalis.* Other bacterial causes include *Streptococcus pyogenes* and *Staphylococcus aureus.* Viruses include measles virus, parainfluenza virus, and respiratory syncytial virus (RSV).

Diagnosis. If there is a discharge from the ear, a sample should be sent to the microbiology laboratory for culture and susceptibility testing (C&S). Beta-lactamase testing should be performed on isolates of *H. influenzae* and *S. pneumoniae.*

Treatment. Oral antibiotics, such as amoxicillin or penicillin. Many strains of *H. influenzae* and *S. pneumoniae* have become resistant to ampicillin and penicillin, respectively. If the eardrum is bulging, the doctor may perform a myringotomy; an opening is made through the ear drum to allow fluid to drain from the middle ear.

OTITIS EXTERNA, EXTERNAL OTITIS, EAR CANAL INFECTIONS, SWIMMER'S EAR

Characteristics. Infection of ear canal with itching, pain, a malodorous discharge, tenderness, redness, swelling, impaired hearing; most common during the summer swimming season; trapped water in the external ear canal can lead to wet, softened skin, which is more easily infected by bacteria or fungi.

Pathogens. Escherichia coli, Pseudomonas aeruginosa, Proteus vulgaris, Staphylococcus aureus; rarely by a fungus, such as *Aspergillus.*

Reservoirs. Contaminated swimming pool water; sometimes indigenous microflora; articles inserted in ear canal for cleaning out debris and wax.

Prevention and Control. Prevent contaminated water from entering and being trapped by wax in external ear canal. Attempts to clean the ear canal with cotton swabs can push wax and debris toward the ear drum, where it can accumulate.

Diagnosis. Material from the infected ear canal should be sent to the microbiology laboratory for C&S. Most strains of *Pseudomonas aeruginosa* are multiply drug-resistant.

Treatment. Removal of wax and debris from the ear canal by a physician; antibiotic ear drops; analgesics may be necessary to relieve pain.

INFECTIOUS DISEASES OF THE RESPIRATORY SYSTEM

To simplify discussion of the functions, defenses, and diseases of the respiratory system (Fig. 12–10), the respiratory system is often separated into the upper respiratory

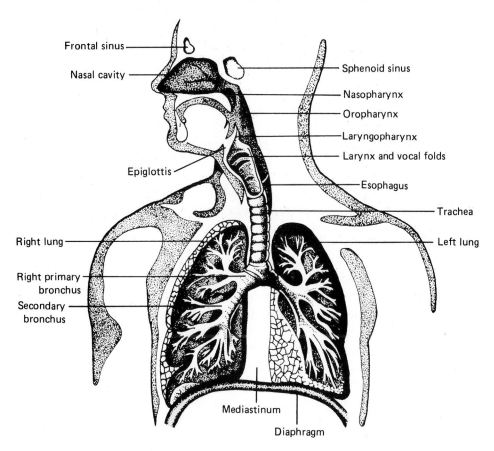

FIGURE 12-10. The respiratory system.

tract (URT) consisting of the nose and pharynx (throat), and the lower respiratory tract (LRT), including the larynx, trachea, bronchial tubes, and alveoli. The most common diseases are those of the URT (*e.g.,* colds and sore throats), which may predispose the patient to more serious infections, such as sinusitis, otitis media, bronchitis, and pneumonia.

Typical indigenous microflora found in the mucosa of the nose and throat include bacterial species of *Streptococcus, Staphylococcus, Haemophilus, Corynebacterium, Neisseria, Bacteroides, Branhamella, Fusobacterium,* and *Actinomyces.* Many of these microorganisms may cause opportunistic diseases of the respiratory tract. Respiratory tract infections are the most common cause of death from infectious diseases.

Nonspecific Respiratory Infections

PNEUMONIA

Characteristics. An acute nonspecific infection of the small air sacs (alveoli) and tissues of the lung, with fever, productive cough (meaning that sputum is coughed up), acute chest pain, chills, and shortness of breath; diagnosed by abnormal chest sounds and chest radiographs. Pneumonia is often a secondary infection that follows a primary viral respiratory infection.

Pathogens. May be caused by Gram-positive or Gram-negative bacteria, mycoplasmas, viruses, fungi, or protozoa. Community-acquired bacterial pneumonia is most frequently caused by *Streptococcus pneumoniae* (pneumococcal pneumonia). Other bacterial pathogens include *Haemophilus influenzae, Staphylococcus aureus, Klebsiella pneumoniae* and occasionally other Gram-negative bacilli and anaerobic members of the oral flora. Atypical pathogens include *Legionella* (legionellosis), *Mycoplasma pneumoniae* (mycoplasmal pneumonia; primary atypical pneumonia), *Chlamydia pneumoniae* (chlamydial pneumonia). Psittacosis (ornithosis; parrot fever), a type of pneumonia caused by *Chlamydia psittaci,* is normally acquired by inhalation of respiratory secretions and desiccated droppings of infected birds (*e.g.,* parrots, parakeets). Fungi such as *Histoplasma capsulatum* (histoplasmosis), *Coccidioides immitis* (coccidioidomycosis), *Candida albicans* (candidiasis), *Cryptococcus neoformans* (cryptococcosis), *Blastomyces* (blastomycosis), *Aspergillus* (aspergillosis) and *Pneumocystis carinii* (previously considered to be a protozoan) may be etiologic agents of pneumonia, especially in immunocompromised individuals. Even various species of bread molds can cause pneumonia in immunosuppressed patients; a condition known as mucormycosis or zygomycosis. Viral pneumonia may be caused by adenoviruses, respiratory syncytial virus (RSV), parainfluenza viruses, cytomegalovirus, measles virus, chickenpox virus, and other viruses.

Hospital-acquired bacterial pneumonia is most often caused by Gram-negative bacilli, especially *Klebsiella, Enterobacter, Serratia,* and *Acinetobacter* species. *Pseudomonas aeruginosa* and *S. aureus* are also frequent causes of nosocomial pneumonias. Pneumonia is the most common fatal infection acquired in hospitals.

Reservoir. Humans.

Transmission. Droplet inhalation, direct oral contact. Contact with contaminated hands and fomites.

Incubation Period. Depends upon the pathogen involved.

Epidemiology. About 2 million people develop pneumonia in the United States per year. Of those, approximately 40,000 to 70,000 die. In developing countries, pneumonia and dehydration from severe diarrhea are the leading causes of death.

Prevention and Control. A pneumococcal vaccine is available for high-risk individuals (*e.g.,* the elderly, people with asplenia, sickle cell disease, HIV infection). Avoid crowding. Handwashing. Standard precautions for hospitalized patients.

Diagnosis. Sputum specimens should be sent to the microbiology laboratory for C&S.

Treatment. Antibiotic therapy depends on the causal agent identified in sputum or other LRT specimens.

Viral Respiratory Infections

COMMON COLD, ACUTE VIRAL RHINITIS, ACUTE CORYZA

Characteristics. A viral infection of the lining of the nose, sinuses, throat, and large airways. Produces coryza (profuse discharge from nostrils), sneezing, runny eyes, sore throat, chilliness, malaise (tiredness); may be accompanied by laryngitis and/or bronchitis; secondary bacterial infections, including sinusitis and otitis media may follow.

Pathogens. Many different viruses cause colds, including over 100 serotypes of rhinovirus, adenoviruses (Fig. 12–11), coronavirus, RSV, influenza viruses, parainfluenza viruses, etc.

Reservoir. Humans.

Transmission. Respiratory secretions by way of hands and fomites; direct contact with or inhalation of airborne droplets.

Incubation Period. Between 12 hours and 5 days; usually about 48 hours.

Period of Communicability. From 24 hours before onset of symptoms to 5 days after onset.

Epidemiology. Most common in fall, winter, and spring; not a nationally notifiable disease in the United States.

Prevention and Control. Frequent handwashing. Sanitary disposal of oral and nasal discharges. Disinfect eating and drinking utensils. Avoid crowding and contact with infected individuals. An oral, live adenovirus vaccine has proven effective in military recruits.

Treatment. Patient should stay warm and comfortable and drink plenty of fluids. Aspirin, acetaminophen, ibuprofen, nasal decongestants, antihistamines, steam inhalation, and cough suppressants may provide temporary relief. Antibiotics are ineffective against viruses.

HANTAVIRUS PULMONARY SYNDROME (HPS)

Characteristics. Acute viral disease characterized by fever, myalgias (muscular pain), GI complaints, cough, difficulty in breathing, and hypotension (decreased blood pressure).

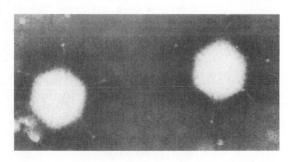

FIGURE 12–11. Electron micrograph of purified type 5 adenovirus particles embedded in sodium silicotungstate (original magnification, ×350,000). (Valentine RC, Pereira HG: J Mol Biol 13:13, 1965)

Pathogens. At least four different hantaviruses (Sin Nombre, Bayou, Black Creek Canal, New York) have caused HPS in the United States.

Reservoir. At least 15 different species of rodents in the United States, including the deer mouse and cotton rat.

Transmission. Inhalation of infected, aerosolized rodent feces, urine, and saliva.

Incubation Period. 8 to 21 days.

Epidemiology. The Sin Nombre hantavirus was the cause of the "mystery illness" that occurred in the Four Corners area of New Mexico and Arizona in 1993. Between 1993 and 1997, 176 U.S. cases were identified in 28 states. The overall case fatality rate was 44.3%. Nineteen cases of HPS were reported to the CDC in 1998, but HPS is not notifiable in all states.

Prevention and Control. Avoid rodent-infested areas. Sweeping of dried rodent excretions can produce aerosols. Standard precautions for hospitalized patients.

Diagnosis. Immunodiagnostic procedures.

Treatment. Ribavirin (an antiviral agent).

INFLUENZA, FLU

Characteristics. A specific, acute, viral respiratory infection with fever, chills, headache, aches and pains throughout the body (most pronounced in the back and legs), sore throat, cough, nasal drainage; sometimes causing bronchitis, pneumonia, and death in severe cases; nausea, vomiting, and diarrhea may occur, particularly in children. Although the term "stomach flu" is often heard, influenza viruses rarely cause gastrointestinal symptoms. "Stomach flu" (also known as the "24-hour flu") is caused by different viruses.

Pathogens. Influenza viruses, types A, B, and C, are RNA viruses in the Orthomyxovirus family. Influenza A viruses cause severe symptoms and are associated with pandemics and widespread epidemics. Influenza B viruses cause less severe disease and more localized outbreaks. Influenza C viruses do not cause epidemics or significant disease.

Reservoir. Humans are the primary reservoir; pigs and ducks also serve as reservoirs. Because pig cells have receptors for both avian and human strains of influenza virus, pigs serve as "mixing bowls," resulting in new viruses containing RNA segments from both avian and human strains.

Transmission. Airborne spread; direct contact.

Incubation Period. 1 to 3 days.

Period of Communicability. About 3 to 7 days from onset of symptoms.

Epidemiology. The Spanish flu epidemic (also known as the swine flu epidemic) of 1918–1919 killed 20 to 40 million people worldwide, including about 675,000

Americans. Other devastating pandemics occurred in 1889, 1957, and 1968. The 1957 Asian flu and 1968 Hong Kong flu pandemics killed about 70,000 and 34,000 Americans, respectively. Flu epidemics occur in the United States almost every year, affecting 10 to 20% of the population. In the winter of 1997, a serious "bird flu" or "chicken flu" epidemic was prevented by the slaughter of about 1.5 million chickens and ducks in markets in Hong Kong and surrounding territories. Influenza in not a nationally notifiable disease in the United States.

Prevention and Control. Yearly immunization; prophylactic amantadine or rimantadine may be used against type A for high-risk patients. Good personal hygiene and avoiding crowds during epidemics may prevent infection. Droplet precautions for hospitalized patients.

Diagnosis. Isolation of influenza virus from pharyngeal or nasal secretions, using cell culture techniques; immunodiagnostic procedures; demonstration of a rise in antibody titer between acute and convalescent sera.

Treatment. Bed rest and plenty of liquids. Avoid exertion. Rimantadine and amantadine interfere with the replication cycle of type A (but not type B) influenza viruses; either can cause adverse reactions. Aspirin should not be given to children because of the danger of Reye's (pronounced "rize") syndrome, involving the central nervous system (CNS) and liver; this is true for any viral infection in children.

Specific Bacterial Respiratory Infections

DIPHTHERIA

Characteristics. An acute, contagious bacterial disease, primarily involving tonsils, pharynx, larynx, and nose; a cytotoxin causes the formation of a tough, adherent, grayish pseudomembrane, which may cause difficulty in breathing; sore throat; swollen and tender cervical lymph nodes; swelling of the neck; CNS and heart may be affected; sometimes fatal; there is also a cutaneous form of diphtheria, which is more common in the tropics.

Pathogen. *Corynebacterium diphtheriae,* pleomorphic, Gram-positive bacilli that form characteristic V-, L-, and Y-shaped arrangements of bacilli. Only strains infected with a particular corynebacteriophage (called β-phage) are toxigenic (toxin producing); the exotoxin (diphtheria toxin) is coded for by a bacteriophage gene known as the Tox gene. It is the toxin that causes the heart and nerve damage.

Reservoir. Humans.

Transmission. Airborne droplets, direct contact, contaminated fomites, raw milk.

Incubation Period. Usually 2 to 5 days.

Epidemiology. At one time, diphtheria was a major killer of children in the United States. However, as a result of widespread vaccination with diphtheria toxoid (an altered form of diphtheria toxin), only one U.S. case was reported to the CDC in 1998. Unfortunately, diphtheria continues to be a major killer of children in developing countries, where epidemics are occurring.

Prevention and Control. Vaccination with diphtheria toxoid, usually as diphtheria-tetanus-pertussis (DTP) vaccine and diphtheria-tetanus (DT) booster vaccines at appropriate intervals; patient isolation, quarantine, and disinfection of all fomites. Contact precautions for hospitalized patients with cutaneous diphtheria. Droplet precautions for hospitalized patients with pharyngeal diphtheria.

Diagnosis. A nasopharyngeal swab and a throat swab, preferably containing a sample of the pseudomembrane, should be sent to the microbiology laboratory for culture. Special media called Loeffler serum medium and cystine-tellurite or Tinsdale medium are used for culture and identification of *C. diphtheriae*. Toxigenicity is determined using laboratory animals (rabbits or guinea pigs).

Treatment. Administer antitoxin (horse serum containing antibodies against diphtheria toxin) in known and strongly suspected cases, with erythromycin or penicillin G in confirmed cases. Treat carriers with erythromycin or penicillin G.

LEGIONELLOSIS, LEGIONNAIRES' DISEASE, PONTIAC FEVER

Characteristics. An acute bacterial pneumonia with headache, high fever, dry cough followed by a productive cough, chills, shortness of breath, diarrhea, pleural and abdominal pain; there is about a 20% fatality rate; Pontiac fever is not associated with pneumonia or death.

Pathogens. *Legionella pneumophila,* a Gram-negative bacillus; additional *Legionella* spp. can also cause the disease.

Reservoir. Environmental water sources; ponds, lakes, creeks, hot-water and air-conditioning systems, cooling towers, shower heads, whirlpool tubs, humidifiers, tap water and water distillation systems, perhaps soil and dust; aerosols have been produced by vegetable misting devices in supermarkets.

Transmission. Airborne from water and perhaps dust; probably not person-to-person.

Incubation Period. 2 to 10 days for Legionnaire's disease; most often 24 to 48 hours for Pontiac fever.

Epidemiology. Legionnaire's disease was first recognized as a disease following an outbreak in a Philadelphia hotel in 1976, but evidence exists that prior epidemics and deaths were due to *Legionella* spp. Epidemics continue to occur, often associated with hotels, cruise ships, hospitals, and supermarkets. It usually affects the elderly, people with preexisting respiratory disease, heavy smokers and heavy drinkers. A total of 1327 U.S. cases were reported to the CDC in 1998.

Prevention and Control. Tap water should not be used in respiratory therapy equipment. Decontamination of implicated sources by superchlorination or superheating. Standard precautions for hospitalized patients.

Diagnosis. Sputum, blood, and urine specimens should be sent to the microbiology laboratory for C&S. *Legionella* spp. stain poorly and require cysteine and other nutrients to grow. The recommended culture medium is buffered charcoal yeast extract agar. Immunodiagnostic procedures are available for detection of *Legionella* organisms or antigen in aspirates, tissue specimens, urine and body fluids.

Treatment. Erythromycin, clarithromycin, azithromycin, a fluoroquinolone, with or without rifampin.

MYCOPLASMAL PNEUMONIA, PRIMARY ATYPICAL PNEUMONIA

Characteristics. Gradual onset with headache, fatigue (malaise), dry cough, sore throat, and less often, chest discomfort and rash. Scant sputum at first, which may increase later. Illness may last from a few days to a month or more. Most common in people 5 to 35 years of age. Pneumonias produced by mycoplasmas and chlamydias are the most common types of atypical pneumonias (*i.e.*, pneumonias that are caused by organisms other than those that are the typical causes of pneumonia).

Pathogen. Mycoplasma pneumoniae; tiny, Gram-negative bacteria, lacking cell walls.

Reservoir. Humans.

Transmission. Droplet inhalation; direct contact with an infected person or articles contaminated with nasal secretions or sputum from an ill, coughing patient.

Incubation Period. 6 to 32 days.

Prevention and Control. Avoidance of crowded living and sleeping quarters; proper disposal of tissues and other soiled articles; handwashing; droplet precautions for hospitalized patients.

Diagnosis. Demonstration of a rise in antibody titer (concentration) between acute and convalescent sera (see Chapter 11). On artificial media, *M. pneumoniae* produces tiny "fried egg" colonies, having a dense central area and a less dense periphery.

Treatment. Although mycoplasmal pneumonia can be severe, most cases are mild and most people recover without treatment. Erythromycin, tetracycline, clarithromycin or azithromycin for severe cases.

STREPTOCOCCAL PHARYNGITIS, STREP THROAT

Characteristics. An acute bacterial infection of the throat with sore throat, chills, fever, headache; beefy red throat; white patches of pus on pharyngeal epithelium; enlarged tonsils; enlarged and tender cervical lymph nodes. The infection may spread to the middle ear, sinuses, or the organs of hearing. Untreated strep throat can lead to complications (sequelae) such as scarlet fever (due to erythrogenic toxin), rheumatic fever and glomerulonephritis. The latter two conditions result from the deposition of immune complexes beneath heart and kidney tissue, respectively.

Some strains produce a pyrogenic exotoxin that causes toxic shock syndrome and some strains (the so-called esh-eating bacteria) can cause necrotizing fasciitis.

Pathogen. *Streptococcus pyogenes;* β-hemolytic, catalase-negative, Gram-positive cocci in chains (Fig. 12—12); also known as Group A streptococcus or Strep A.

Reservoir. Humans.

Transmission. Human-to-human by direct contact, usually hands; aerosol droplets; secretions from patients, and nasal carriers; contaminated dust, lint, handkerchiefs; contaminated milk and milk products have been associated with foodborne outbreaks of streptococcal pharyngitis.

Incubation Period. 1 to 3 days.

Period of Communicability. Usually 10 to 21 days if untreated, but can be longer; 24 hours or less if adequately treated with an antibiotic.

Epidemiology. Over 200,000 cases per year in the United States, mostly among children (3 to 15 years); not a nationally noti able disease in the United States.

Prevention and Control. Throat cultures followed by antibiotic treatment; personal hygiene and cleanliness; no protective immunity following recovery; no effective vaccines; standard precautions for hospitalized patients.

Diagnosis. The sole purpose of a routine throat culture is to determine if a patient does or does not have strep throat. If β-hemolytic streptococci are isolated, they are tested to determine if they are group A streptococci. Rapid strep tests (based on detection of antigen) can be performed on throat swabs, but if the test is negative, a more traditional test (such as a throat culture and bacitracin susceptibility) should be performed.

Treatment. Penicillin G or penicillin V (amoxicillin); clindamycin, erythromycin or a cephalosporin may be used for patients allergic to penicillins.

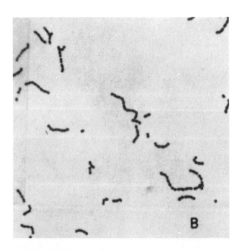

FIGURE 12-12. Chain formation characteristic of streptococci in liquid media. Smear made from 24-hr culture in serum broth (original magnification, ×1000).

TUBERCULOSIS, TB

Characteristics. An acute or chronic mycobacterial infection of the pulmonary tract; productive cough; cold sweating at night; shortness of breath; may invade lymph nodes to cause systemic disease; tuberculosis may affect many areas of the body, including the kidney, urinary bladder, bones, reproductive organs, joints, skin, bowel, adrenal glands, and brain.

Pathogens. Primarily *Mycobacterium tuberculosis* (a slow-growing, acid-fast, Gram-positive bacillus); occasionally other *Mycobacterium* spp. (*e.g.*, *M. africanum* or *M. bovis* from cattle).

Reservoir. Primarily humans; rarely, primates, cattle, other mammals.

Transmission. Airborne droplets; prolonged direct contact with infected individuals. Bovine tuberculosis may result from exposure to infected cattle or ingestion of contaminated milk.

Incubation Period. Four to twelve weeks from infection until pulmonary lesions and positive TB skin test; symptoms of tuberculosis usually occur within 6 to 12 months after infection, but may not occur for a year or more.

Period of Communicability. For as long as viable *M. tuberculosis* organisms are being discharged in the infected person s sputum. Effective chemotherapy usually eliminates communicability within a few weeks.

Epidemiology. Worldwide, tuberculosis is a more frequent cause of death than any other pathogen. Approximately one-third of the world s population is infected with *M. tuberculosis.* The World Health Organization (WHO) estimated that in 1996 there were 8 million new cases of tuberculosis and 3 million deaths from the disease. The total number of infected individuals was estimated to be 1.7 billion. A resurgence of tuberculosis in the United States occurred during the late 1980s and early 1990s, primarily as a result of the HIV/AIDS epidemic and the emergence of multidrug-resistant strains of *M. tuberculosis.* During 1998, a total of 14,756 new U.S. cases of tuberculosis were reported to the CDC. Other *Mycobacterium* spp. also commonly cause infections in AIDS patients.

Prevention and Control. Annual skin testing for individuals at high risk of infection (*e.g.*, pulmonary ward personnel). Preventive treatment with isoniazid (INH) for individuals who convert to being skin test-positive. Isolation and treatment of infected individuals. Airborne precautions for hospitalized patients. Tuberculin testing of humans and cattle; chest radiograph and prompt treatment of TB skin test-positive people. A vaccine (BCG vaccine) is available for individuals at high risk of infection, but it is not routinely used in the United States. Individuals receiving BCG vaccine will have a positive TB skin test for variable periods of time following vaccination.

Diagnosis. Demonstration of acid-fast bacilli (AFB) in sputum specimens provides a rapid, presumptive diagnosis of tuberculosis. Isolation of *M. tuberculosis* on

L wenstein-Jensen or Middlebrook culture media takes about 3 to 6 weeks, due to the organism s long generation time (about 18 to 24 hours). A variety of more rapid techniques are available for isolation and identi cation of *M. tuberculosis,* including automated and semi-automated instruments, DNA probes, polymerase chain reaction, and gas-liquid chromatography. Because many strains of *M. tuberculosis* are multidrug-resistant, susceptibility testing should be done as soon as possible. Infected patients show a positive delayed hypersensitivity skin test (the Mantoux pu-ri ed protein derivative [PPD] tuberculin skin test) and pulmonary tubercles may be seen on chest radiographs. A positive skin test may mean past infection, present infection, or BCG vaccination.

Treatment. Because multidrug-resistant strains of *M. tuberculosis* are increasingly common, treatment consists of a combination of antimicrobial drugs including isoniazid (INH), rifampin (RIF), pyrazinamide (PZA), and ethambutol (EMB) or streptomycin (SM). When drug susceptibility results become available, a speci c drug regimen can be selected. Sputum conversion usually occurs within 4 to 8 weeks of appropriate chemotherapy.

WHOOPING COUGH, PERTUSSIS

Characteristics. A highly contagious, acute bacterial childhood (usually) infection. The rst stage (the prodromal or catarrhal stage) of the disease involves mild, cold-like symptoms. The second stage (the paroxysmal stage) produces severe, uncontrollable coughing ts. The coughing often ends in a prolonged, high-pitched, deeply indrawn breath (the whoop, from which whooping cough gets its name). The coughing ts produce a clear, tenacious mucus, and vomiting; they may be so severe as to cause lung rupture, bleeding in the eyes and brain, rectal prolapse, or hernia. The third stage (the recovery or convalescent stage) usually begins within 4 weeks of onset. Parapertussis is a similar but milder disease.

Pathogens. Pertussis is caused by *Bordetella pertussis,* a small, encapsulated, non-motile, Gram-negative coccobacillus that produces endotoxin and exotoxins. Parapertussis is caused by *B. parapertussis.* A related organism, *B. bronchiseptica,* causes respiratory infections in animals, including kennel cough in dogs.

Reservoir. Humans.

Transmission. Airborne via droplets produced by coughing.

Incubation Period. 6 to 20 days.

Epidemiology. Worldwide occurrence; 6279 U.S. cases were reported to the CDC in 1998.

Prevention and Control. Vaccination of all young children with DTP (D for diphtheria, T for tetanus, P for pertussis) containing diphtheria and tetanus toxoids and *B. pertussis* antigens (an acellular vaccine). Droplet precautions for hospitalized patients.

Diagnosis. Nasopharyngeal aspirates or swabs should be sent to the microbiology laboratory. Special media, such as Bordet-Gengou agar (a potato-based medium) or Regan-Lowe agar (a charcoal/horse blood medium) are used to isolate *B. pertussis*. Nucleic acid and immunodiagnostic procedures are also available.

Treatment. Erythromycin, doxycycline, or azithromycin. During treatment, mucus may be suctioned out of the patient s throat and it may be necessary to insert a tube into the windpipe (trachea) to deliver oxygen directly to the lungs.

Fungal Respiratory Infections

CRYPTOCOCCOSIS

Characteristics. A deep mycosis usually presenting as a meningitis, although infection of the lungs, kidneys, prostate, skin and bone also occur; a common infection in AIDS patients.

Pathogens. Two subspecies of *Cryptococcus neoformans;* an encapsulated yeast.

Reservoir. Pigeon nests, pigeon droppings, other bird droppings, soil.

Transmission. By inhalation.

Incubation Period. Unknown; pulmonary infection may precede brain infection by months or years.

Epidemiology. Not a nationally noti able disease in the United States, where an estimated 300 cases occur each year.

Prevention and Control. Chemical decontamination of bird droppings; avoid inhalation of dried bird droppings; standard precautions for hospitalized patients.

Diagnosis. Cryptococcal meningitis can be presumptively diagnosed by observing encapsulated budding yeasts in an India Ink preparation (see App. D) of spinal uid. Culture and identi cation using biochemical tests are required for de nitive diagnosis. Immunodiagnostic procedures are available.

Treatment. Antifungal agents: amphotericin B, with or without ucytosine, followed by uconazole.

HISTOPLASMOSIS

Characteristics. A systemic mycosis of varying severity, ranging from asymptomatic to acute to chronic; the primary lesion is usually in the lungs. The acute disease involves malaise, fever, chills, headache, myalgia, chest pains, and a nonproductive cough (no sputum).

Pathogens. Two variants of *Histoplasma capsulatum,* a dimorphic fungus that grows as a mold in soil and as a yeast in animal and human hosts.

Reservoir. Soil containing bird droppings, especially chicken droppings; bat droppings in caves.

Transmission. Inhalation of conidia (asexual spores).

Incubation Period. 3 to 17 days, commonly 10 days.

Epidemiology. Although it is the most common systemic fungal disease in the U.S., histoplasmosis is not a nationally noti able disease, occurring primarily in the Ohio, Mississippi, and Missouri River valleys.

Prevention and Control. Minimize exposure to dust in a contaminated environment, especially around chicken coops and in bat caves; standard precautions for hospitalized patients.

Diagnosis. Culture and identi cation by biochemical tests. Produces mold colonies when incubated at room temperature and yeast colonies when incubated at body temperature. Conversion from the mold form to the yeast form can sometimes be accomplished in the laboratory.

Treatment. Antifungal agents: itraconazole or amphotericin B.

PNEUMOCYSTIS CARINII PNEUMONIA (PCP), INTERSTITIAL PLASMA-CELL PNEUMONIA

Characteristics. An acute to subacute pulmonary disease found in malnourished, chronically ill children; premature infants; and immunosuppressed patients (patients whose immune systems are not functioning properly) such as AIDS patients. A common, contributory cause of death in AIDS patients. *Pneumocystis* causes an asymptomatic infection in immunocompetent people (people whose immune systems are functioning properly). Patients have fever, dif culty in breathing, rapid breathing, dry cough, and cyanosis; pulmonary in ltration of alveoli with frothy exudate; usually fatal in untreated patients.

Pathogen. *Pneumocystis carinii* (Fig. 12—13); has both protozoal and fungal properties; was classi ed as a protozoan for many years; currently classi ed as a fungus.

Reservoir. Humans.

Transmission. Mode of transmission not known. Perhaps direct contact; transfer of pulmonary secretions from infected to susceptible persons; perhaps airborne.

Incubation Period. Not known.

Prevention and Control. Prophylaxis of immunosuppressed patients with trimethoprim-sulfamethoxazole (TMP-SMX); careful disinfection of respiratory therapy equipment; standard precautions for hospitalized patients; avoid placing a patient with PCP in the same room with an immunocompromised patient.

Diagnosis. Demonstration of *Pneumocystis* in material from bronchial brushings, open lung biopsy, lung aspirates, or smears of tracheobronchial mucus by various staining and immunodiagnostic procedures. *Pneumocystis carinii* cannot be cultured.

Treatment. TMP-SMX or pentamidine.

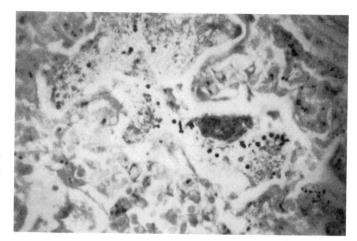

FIGURE 12-13. *Pneumocystis carinii* appears as dark oval bodies when stained with a special silver stain. Foamy material is also seen throughout the several alveoli in this section of the lung (original magnification, ×250).

INFECTIOUS DISEASES OF THE GASTROINTESTINAL (GI) TRACT

The digestive tract consists of a long tube with many expanded areas designed for digestion of food, absorption of nutrients, and elimination of undigested materials (Fig. 12—14). Transient and resident microbes continuously enter and leave the GI tract. Most of the microorganisms ingested with food are destroyed in the stomach and duodenum by the low pH (gastric contents have a pH of approximately 1.5) and are inhibited from growing in the lower intestines by the resident micro ora (microbial antagonism). They are then ushed from the colon during defecation, along with large numbers of indigenous microbes.

The largest number of resident microorganisms is found in the lower small intestine and colon. The availability of nutrients, moisture, and a constant temperature of 37¡C allows an estimated 400 to 500 species to live in the colon. Because it is anaerobic in the colon, the organisms that live there are obligate anaerobes, aerotolerant anaerobes, or facultative anaerobes. The most common obligate anaerobes are species of *Bacteroides* (Gram-negative bacilli); *Bi dobacterium, Eubacterium* and *Clostridium* (all Gram-positive bacilli); *Peptostreptococcus* (Gram-positive cocci); and *Veillonella* (Gram-negative cocci). Facultative anaerobes are less abundant than obligate anaerobes, but are better understood because they are easier to isolate and study in the laboratory. Included in this group are many genera and species of the family *Enterobacteriaceae,* including the coliforms that serve as indicators of fecal contamination of water supplies (Chapter 10). Also known as enteric bacilli, the indigenous members of the family *Enterobacteriaceae* include *Citrobacter, Enterobacter, Escherichia, Klebsiella,* and *Proteus* spp.

Many viruses, most of which are harmless, are also found in fecal material; however, some can cause gastroenteritis under certain conditions. Some may be ingested with contaminated food, such as hepatitis A virus in fecally contaminated raw oysters.

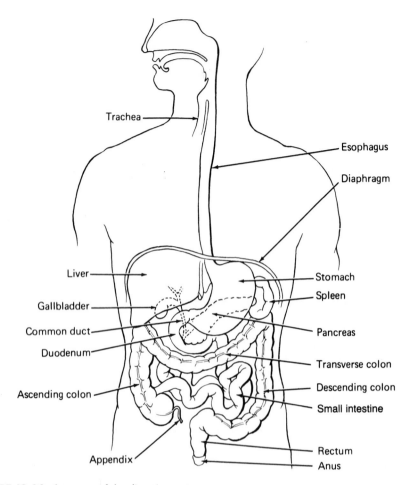

FIGURE 12-14. Anatomy of the digestive system.

Viral Gastrointestinal Infections

VIRAL GASTROENTERITIS, VIRAL ENTERITIS, VIRAL DIARRHEA

Viral gastroenteritis may be an acute or epidemic illness in infants, children, and adults. The most common viruses infecting children in their rst years of life are rotaviruses, enteric adenoviruses, caliciviruses, astroviruses, and perhaps some Norwalk viruses. Those infecting children and adults include certain Norwalk viruses and rotaviruses. Although sometimes referred to as stomach u or 24-hour u, keep in mind that u is an abbreviated form of in uenza, which is a respiratory disease.

Symptoms include nausea, vomiting, diarrhea, abdominal pain, myalgia, headache, malaise, and low-grade fever. Although most often a self-limiting disease lasting 24 to 48 hours, viral gastroenteritis (especially due to rotaviruses) can be fatal in infants and young children. In developing countries, rotavirus infections are responsible for over 800,000 diarrheal deaths per year.

Humans are the sole reservoir and transmission is most often by the fecal-oral route. Airborne transmission and contact with contaminated fomites may cause hospital epidemics. Oral rehydration therapy is often necessary. Standard precautions are practiced for most hospitalized patients with viral gastroenteritis; however, contact precautions should be practiced for diapered or incontinent patients with rotavirus infections.

VIRAL HEPATITIS

Hepatitis, or in ammation of the liver, can have many causes, including alcohol, drugs, and viruses. Viral hepatitis refers to hepatitis caused by any one of about a dozen different viruses, including hepatitis A virus (HAV), hepatitis B virus (HBV), hepatitis C virus (HCV), hepatitis D virus (HDV) and hepatitis E virus (HEV), hepatitis G virus (HGV), hepatitis GB virus A (HGBV-A), hepatitis GB virus B (HGBV-B), and hepatitis GB virus C (HGBV-C). Hepatitis can also occur as a result of viral diseases such as infectious mononucleosis, yellow fever, and cytomegalovirus infection. See Table 12—1 for information about viral types, sources, modes of transmission, and types of disease.

Within the United States, acute hepatitis is most commonly caused by HAV (about 47% of cases), HBV (about 34%), or HCV (about 17%). In 1998, the number of cases of hepatitis A, hepatitis B, and hepatitis C reported to the CDC were 22,028; 8651; and 4840, respectively. The number of actual cases is thought to be much higher. An estimated 16,000 people die each year of chronic liver disease associated with persistent hepatitis virus infection.

Vaccines are available for HAV and HBV. The HAV vaccine, which contains inactivated virus grown in cell culture, is recommended for people at increased risk of acquiring hepatitis A (including military personnel and others traveling to regions where HAV is endemic, homosexual and bisexual males, and users of illicit drugs). The HBV vaccine is a subunit vaccine, produced by the yeast *Saccharomyces cerevisiae* by genetic engineering. It is recommended for persons at high risk of acquiring HBV infection, such as infants born to HBV antigen-positive mothers, household contacts of HBV carriers, homosexual and bisexual males, and users of illicit drugs. It is required for healthcare workers exposed to blood.

In addition to vaccination against HBV, healthcare personnel practice Standard Precautions (Chapter 10), which are designed to reduce the risk of transmission of bloodborne and other pathogens in hospitals. Hepatitis B immune globulin can be given to unvaccinated people who have been exposed to HBV, perhaps by accidental needlestick injury.

Bacterial Gastrointestinal Infections

BACTERIAL GASTRITIS AND ULCERS

Characteristics. Infection with *Helicobacter pylori* can cause chronic bacterial gastritis and duodenal ulcers. Gastritis is suspected when a person has upper abdominal pain with nausea or heartburn. People with duodenal ulcers may experience gnawing, burning, aching, mild to moderate pain just below the breastbone, an

TABLE 12-1. Hepatitis Viruses

Name of Virus	Type of Virus	Mode(s) of Transmission	Type of Disease
HAV	a nonenveloped, linear ssRNA picornavirus (a type of enterovirus)	Predominantly fecal-oral; waterborne and foodborne epidemics are common, especially in developing countries; consumption of contaminated raw shellfish is a common cause	Acute, self-limited disease; referred to as hepatitis A and infectious hepatitis; most hepatitis A infections cause no symptoms and go unrecognized
HBV	an enveloped, circular dsDNA hepadnavirus	Sexual intercourse, mother-to-infant, direct parenteral exposure (e.g., via contaminated blood or blood products, needles and syringes shared by injecting drug users); can be transmitted by healthy carriers	Persistent infection, chronic hepatitis, hepatocellular carcinoma; referred to as serum hepatitis
HCV (previously called Non A-Non B hepatitis virus)	an enveloped, linear ssRNA flavivirus	Predominantly direct parenteral exposure; HCV causes at least 80% of hepatitis cases arising from blood transfusions; commonly transmitted by infecting drug users who share needles and syringes	Persistent infection, chronic hepatitis, hepatocellular carcinoma; HCV is the leading cause of chronic hepatitis in the United States.
HDV (also known as Delta virus)	an enveloped, circular ssRNA viral satellite (a defective RNA virus)	Parenteral; commonly transmitted by infecting drug users who share needles and syringes; coinfection with HBV is necessary	Persistent infection, chronic hepatitis
HEV	a nonenveloped, linear ssRNA calicivirus	Predominantly fecal-oral	Acute, self-limited disease
HGV	a linear ssRNA flavivirus	Parenteral	Can cause chronic hepatitis
HGB viruses (the 3 GB viruses were initially isolated from the blood of a Chicago surgeon having the initials G.B.)	RNA viruses (perhaps flaviviruses)	Unknown	Unknown

empty feeling, and hunger. The pain usually occurs when the stomach is empty. Drinking milk, eating, or taking antacids generally relieves the pain, but it usually returns 2 or 3 hours later. Gastric ulcers and gastric adenocarcinoma are also epidemiologically associated with *H. pylori* infection. Gastric ulcers can cause swelling of the tissues leading into the small intestine, which prevents food from easily passing out of the stomach. This, in turn, can cause pain, bloating, nausea, or vomiting after eating. Gastric ulcers and duodenal ulcers are types of peptic ulcers. Complications of peptic ulcers include penetration, perforation, bleeding, and obstruction.

Pathogen. *Helicobacter pylori* is a curved, microaerophilic, capnophilic, Gram-negative bacillus that is found on the mucus-secreting epithelial cells of the stomach. No other bacteria are known to grow in the extremely acidic stomach.

Reservoir. Humans are the only known reservoir.

Transmission. Probably ingestion; presumed to be either oral-to-oral or fecal-to-oral.

Diagnosis. Diagnostic techniques include staining and culturing of gastric and duodenal biopsy specimens, the urea breath test, the NH_4 excretion test, DNA probes, and immunodiagnostic procedures. In the urea breath test, the patient ingests radioactively labeled urea and his or her breath is analyzed 60 minutes later for radioactively labeled CO_2. The enzyme urease, produced by *H. pylori,* splits the urea into ammonia and CO_2; hence, the presence of CO_2 indicates the presence of *H. pylori.* In the NH_4 excretion test, the patient consumes urea containing radioactively labeled nitrogen. The ammonia produced in the stomach by *H. pylori* is absorbed into the blood, excreted in the urine, and the amount of radioactively labeled NH_4 in the urine is measured.

Treatment. Treatment for asymptomatic infection is controversial. However, symptomatic *H. pylori* infection can be treated with a combination of tetracycline, metronidazole, and bismuth subsalicylate (*e.g.,* Pepto-Bismol°).

CAMPYLOBACTER GASTROENTERITIS

Characteristics. An acute bacterial enteric disease ranging from asymptomatic to severe, with diarrhea, nausea, vomiting, fever, malaise, abdominal pain; usually self-limiting, lasting 2 to 5 days. Stools may contain gross or occult blood, mucus, and WBCs.

Pathogens. *Campylobacter jejuni* and less commonly, *C. coli;* curved, S-shaped, or spiral-shaped Gram-negative bacilli; often having a gull-winged morphology following cell division (two curved bacilli not yet separated from each other); microaerophilic and capnophilic; optimal growth temperature of 42°C.

Reservoirs. Animals, including poultry, cattle, sheep, swine, rodents, birds, kittens, puppies, and other pets. Most raw poultry is contaminated with *C. jejuni,* necessitating proper methods of cleaning and disinfecting in the kitchen (see Chapter 9).

Transmission. Ingestion of contaminated food (*e.g.*, chicken, pork), raw milk, water; contact with infected pets, farm animals, contaminated cutting boards.

Incubation Period. 1 to 10 days, usually 2 to 5 days.

Prevention and Control. Irradiating foods; thorough cooking of foods, especially poultry; pasteurization of milk; chlorinating or boiling water supplies; handwashing after animal contact and food preparation; standard precautions for hospitalized patients.

Diagnosis. Recovery of *Campylobacter* from stool specimens, using selective medium (Campy blood agar, containing several antimicrobial agents to suppress growth of other bacteria), a Campy gas mixture (5% O_2, 10% CO_2, 85% N_2), and 42°C incubation.

Treatment. Replacement of fluids and electrolytes; a fluoroquinolone or erythromycin may be of value early in the illness or to eliminate the carrier state.

CHOLERA

Characteristics. An acute, bacterial, diarrheal disease with profuse watery stools, occasional vomiting, and rapid dehydration; if untreated, circulatory collapse, renal failure, and death may occur. Over 50% of untreated people with severe cholera die.

Pathogens. Certain biotypes of *Vibrio cholerae* serogroup 01; curved (comma-shaped), Gram-negative bacilli that secrete an *enterotoxin* (a toxin that adversely affects cells in the intestinal tract) called *choleragin*. Other *Vibrio* spp. (*V. parahemolyticus, V. vulnificus*) also cause diarrheal diseases. Vibrios are halophilic (salt-loving) and are, thus, found in marine environments.

Reservoirs. Humans and aquatic reservoirs (copepods and other zooplankton).

Transmission. Fecal-oral route; contact with feces or vomitus of infected people; ingestion of fecally contaminated water and foods (especially raw or undercooked shellfish and other seafood); flies.

Incubation Period. A few hours to 5 days; usually 2 to 3 days.

Epidemiology. Occurs worldwide, with periodic epidemics and pandemics. The most recent western Hemisphere cholera pandemic started in Peru in 1991; by 1994, more than 950,000 cases had been reported in 21 countries in the Western Hemisphere. Only 12 U.S. cases were reported to the CDC in 1998. Most U.S. cases involve the ingestion of raw or undercooked seafood (*e.g.*, oysters) from the coastal waters of Louisiana and Texas.

Prevention and Control. Purification of water supplies and proper sewage disposal. Boiling of water supplies. Avoiding inadequately cooked fish and shellfish. Although a vaccine is available, it provides only partial protection. Fly control. Standard precautions for hospitalized patients. Handwashing. Disinfection of feces, vomitus, and articles used by patients. Prophylactic tetracycline or doxycycline for contacts.

Diagnosis. Rectal swabs or stool specimens should be inoculated onto thiosulfate-citrate-bile-sucrose (TCBS) agar; different *Vibrio* spp. produce different reactions on this medium. Biochemical tests are used to identify the various species. Biotyping is accomplished using commercially available antisera.

Treatment. Prompt replacement of lost body fluids, salts, and minerals is essential. IV fluids if necessary. Early treatment with tetracycline is effective.

ENTEROVIRULENT *ESCHERICHIA COLI*

The strains and serotypes of *E. coli* that reside in the human colon as part of the indigenous microflora are opportunistic pathogens. Under usual circumstances, they cause no harm and are, in fact, somewhat beneficial. However, if these organisms gain access to certain extracolonic sites (*e.g.,* urinary bladder, bloodstream, a wound), they can cause serious infections. There are other strains and serotypes of *E. coli* that are not indigenous microflora of the human colon; some are collectively referred to as enterovirulent *E. coli* (see Insight Box). Two general types, the enterohemorrhagic *E. coli* and the enterotoxigenic *E. coli,* are discussed below.

ENTEROHEMORRHAGIC *E. COLI* (EHEC) DIARRHEA

Characteristics. Hemorrhagic, watery diarrhea; abdominal cramping. Usually there is no fever or only a slight fever. About 5% of infected people (especially children under age 5 and the elderly) develop hemolytic-uremic syndrome (HUS), with anemia, low platelet count, and kidney failure.

Pathogens. *E. coli* O157:H7 (possessing a cell wall antigen designated "O157" and a flagellar antigen designated "H7") is the most commonly involved EHEC serotype; others include O26:H11, O111:H8, and O104:H21; these are Gram-negative bacilli that produce potent cytotoxins called Shiga-like toxins (because of their close resemblance to certain toxins produced by *Shigella dysenteriae*).

Reservoirs. Cattle; also infected humans.

Transmission. Fecal-oral route; inadequately cooked, fecally contaminated beef; unpasteurized milk; person-to-person; fecally contaminated water.

Incubation Period. 3 to 8 days, usually 3 to 4 days.

Epidemiology. The first recognized outbreak of diarrhea due to enterohemorrhagic *E. coli* (O157:H7) occurred in 1982, involving contaminated hamburger meat (*i.e.,* hamburger meat contaminated with cattle feces). Since then, there have been several well-publicized epidemics involving the same serotype. Not all of the outbreaks have involved meat; some have involved unpasteurized milk and apple juice, lettuce, and other raw vegetables. The CDC have estimated that *E. coli* O157:H7 infection accounts for a minimum of 20,000 cases of illness and 250 deaths in the United States per year. A total of 2741 U.S. cases were reported to the CDC in 1996.

Prevention and Control. Enforcement of measures to minimize contamination of meat by animal intestinal contents during the slaughtering process. Pasteurization.

INSIGHT
"Microbes In The News"—*E. coli* 0157:H7

Escherichia coli is a Gram-negative bacillus that lives in the intestinal tract of virtually everyone. As long as these "garden variety" strains of *E. coli* remain in the intestinal tract, they cause us no harm and, in fact, even do some good by producing certain vitamins. However, because they are opportunistic pathogens, they do possess the ability to cause harm when they gain access to certain parts of the body (*e.g.,* urinary bladder, bloodstream, wounds) where they do not belong. Intestinal strains of *E. coli* are the primary cause of urinary tract infections, septicemia, and nosocomial infections.

But, there are other strains (or serotypes) of *E. coli* that are not members of our indigenous flora and are enteric pathogens whenever they are ingested. These are referred to as *enterovirulent E. coli,* and there are at least five known types.

1. *Enterotoxigenic E. coli* (ETEC) are a leading cause of infant diarrhea and mortality in developing countries and the leading cause of traveler's diarrhea. These strains produce enterotoxins that cause a profuse, watery diarrhea with no blood or mucus, abdominal cramping, vomiting, and dehydration. ETEC possess fimbriae (pili) which enable the bacteria to adhere to the intestinal wall.
2. *Enteroinvasive E. coli* (EIEC) are able to invade and proliferate within intestinal cells, much like *Shigella,* killing the infected cells. EIEC cause dysentery with blood and mucus present in the patient's stool specimens.

3. *Enteropathogenic E. coli* (EPEC) primarily causes infant diarrhea, mucus in stools, fever and dehydration. These strains destroy the microvilli of the small intestine.
4. The so-called *enteroaggregative E. coli* (EAEC) strains are less well understood, but they are known to cause infant diarrhea in developing countries.
5. *Enterohemorrhagic E. coli* (EHEC), also known as *verocytotoxic E. coli* (VTEC), has been the subject of media attention and congressional inquiry. The serotype known as O157:H7 has caused a number of epidemics that have been associated with eating undercooked, contaminated hamburger meat.. The first hamburger-associated epidemic occurred in 1982. EHEC strains cause hemorrhagic (bloody) diarrhea and a severe, sometimes fatal, urinary tract condition known as hemolytic uremic syndrome (HUS) in children. Although serotype O157:H7 is the most frequently encountered EHEC serotype, others are known. Hamburger meat has not been the only source of EHEC strains; other sources have included mayonnaise, unpasteurized apple juice, fermented hard salami, raw vegetables (*e.g.,* lettuce and radish sprouts), well water, lake water, and even a municipal water system. Diarrhea due to EHEC strains is more common in the United States than diarrhea due to *Shigella.*

Adequate cooking of meat (especially ground beef), to 160°C, until no longer pink. Proper handwashing practices. Proper cleaning and disinfection of kitchen utensils, cutting boards, countertops, and sponges. Proper methods of sewage disposal and water treatment, including swimming pools. Standard precautions for hospitalized cases; contact precautions for diapered or incontinent patients.

Diagnosis. E. coli O157:H7 infection should be suspected in any patient with bloody diarrhea. Stool specimens should be inoculated onto sorbitol-MacConkey (SMAC) agar. Colorless, sorbitol-negative colonies should then be assayed for

O157 antigen using commercially available antiserum. Other immunodiagnostic procedures are available.

Treatment. Fluid and electrolyte replacement; the role of antimicrobial agents in the treatment of *E. coli* O157:H7 infections is uncertain; the use of antimobility agents is contraindicated.

ENTEROTOXIGENIC *E. COLI* (ETEC) DIARRHEA, TRAVELER'S DIARRHEA

Characteristics. Profuse, watery diarrhea with or without mucus or blood, vomiting, abdominal cramping; dehydration and low-grade fever may occur.

Pathogens. Many different serotypes of enterotoxigenic *E. coli* that produce either a heat-labile toxin, a heat-stable-toxin, or both toxins.

Reservoir. Infected humans.

Transmission. Fecal-oral route; ingestion of fecally contaminated food or water.

Incubation Period. 10 to 72 hours.

Epidemiology. Enterotoxigenic strains of *E. coli* are the most common cause of traveler's diarrhea worldwide, and a common cause of diarrheal disease in young children in developing countries.

Prevention and Control. While traveling, patronize restaurants with a reputation for safety and don't eat food or drink beverages sold by street vendors. Eat only cooked foods and peeled fruits and vegetables. Avoid salads and uncooked vegetables. Drink boiled water. Prophylactic use of bismuth subsalicylate (Pepto-Bismol®). Prophylactic use of antimicrobial agents is controversial. Proper sewage disposal and water treatment. Handwashing. Standard precautions for hospitalized patients.

Diagnosis. Isolation of the organism from stool specimens, followed by demonstration of enterotoxin production, DNA probe techniques, or immunodiagnostic procedures.

Treatment. Fluid and electrolyte replacement therapy. Bland diet. The use of antimicrobial agents is controversial, except in severe cases, where loperamide (Imodium®) (not in children) and trimethoprim-sulfamethoxazole or tetracycline may be of value. Many strains of ETEC are multidrug-resistant.

SALMONELLOSIS

Characteristics. Gastroenteritis with sudden onset of headache, abdominal pain, diarrhea, nausea, and sometimes vomiting. Dehydration may be severe. May develop into septicemia or localized infection in any tissue of the body.

Pathogens. Gastrointestinal salmonellosis is caused by members of the family *Enterobacteriaceae* that some microbiologists call *Salmonella typhimurium* and *Salmonella enteritidis* (of which there are over 2000 serotypes) and other microbiologists call *Salmonella* serotype typhimurium and *Salmonella* serotype enteritidis. They are Gram-negative bacilli that invade intestinal cells, release endotoxin, and produce cytotoxins and enterotoxins. About 200 of the *Salmonella* enteritidis serotypes cause gastrointestinal salmonellosis in the United States.

Reservoirs. A wide range of domestic and wild animals, including poultry, swine, cattle, rodents, reptiles (*e.g.*, pet iguanas and turtles), pet chicks, dogs and cats; also humans (*e.g.*, patients, carriers).

Transmission. Ingestion of contaminated food (*e.g.*, eggs, unpasteurized milk, meat, poultry, raw fruits and vegetables); fecal-oral transmission from person to person; food handlers; contaminated water supplies.

Incubation Period. 6 to 72 hours, usually 12 to 36 hours.

Epidemiology. 45,471 U.S. cases of salmonellosis were reported to the CDC in 1996.

Prevention and Control. Education of food handlers. Handwashing after playing with animals and handling foods. Pasteurization. Thorough cooking of eggs, meat, poultry. Thorough cleaning and disinfection of food preparation areas. Standard precautions for hospitalized patients. Water purification. Effective sewage disposal. Newspaper articles published before the Thanksgiving holiday often contain safety precautions regarding the handling and cooking of turkeys.

Diagnosis. Stool specimens should be submitted to the microbiology laboratory for C&S. *Salmonella* spp. are non-lactose-fermenters and, thus, produce colorless colonies on MacConkey agar. Biochemical tests are used for identification and commercially available antisera are used for serotyping.

Treatment. Fluid and electrolyte replacement therapy. In severe cases, treatment with cefotaxime, ceftriaxone, or a fluoroquinolone may be necessary. Many strains of *Salmonella* are multidrug-resistant.

TYPHOID FEVER, ENTERIC FEVER

Characteristics. A systemic bacterial disease with fever, severe headache, malaise, anorexia, a rash on the trunk in about 25% of patients, nonproductive cough, and constipation. Bacteremia; pneumonia; gall bladder, liver, bone infection; endocarditis; meningitis, and other complications may occur. About 10% of untreated patients die.

Pathogens. *Salmonella typhi* (the typhoid bacillus); Gram-negative bacilli that release endotoxin and produce exotoxins. A similar, but less severe, infection is caused by *S. paratyphi*.

Reservoirs. Humans for typhoid and paratyphoid; rarely, domestic animals for paratyphoid. Some people become carriers following infection, shedding the pathogens in their feces or urine.

Transmission. Fecal-oral route; food or water contaminated by feces or urine of patients or carriers; oysters harvested from fecally contaminated waters; fecally contaminated fruits and raw vegetables; from feces to food by flies.

Incubation Period. 3 days to 3 months, usually 1 to 3 weeks.

Epidemiology. Worldwide, an estimated 17 million cases per year with approximately 600,000 deaths. A total of 324 U.S. cases were reported to the CDC in 1998. Mary Mallon ("Typhoid Mary") was a particularly infamous carrier. She worked as a cook in New York State during the early 1900s, working in several private homes and a hospital. She was responsible for a number of outbreaks of typhoid fever and three deaths ("Everywhere that Mary went, typhoid fever was sure to follow."). Because she refused to have a gall bladder operation to eliminate the pathogens and refused to change her profession, she was mandatorily and permanently confined to a hospital to prevent her from causing further cases. Eventually, she was allowed to assist with hospital chores. Mary Mallon died in 1938 at the age of 70. There are currently an estimated 2000 typhoid carriers in the United States, most with chronic gallbladder disease.

Prevention and Control. Handwashing. Control of flies, including fly-proof latrines. Water purification. Proper disposal of sewage. Careful food preparation. Boil shellfish for at least 10 minutes. A vaccine is available for individuals at high risk of infection (*e.g.,* travelers to endemic areas). Pasteurization of milk. Standard precautions for hospitalized patients.

Diagnosis. Isolation of *S. typhi* from blood, urine, feces, or bone marrow. Identification by biochemical tests. Immunodiagnostic procedures are available.

Treatment. Fluid and electrolyte replacement therapy; use of a fluoroquinolone or ceftriaxone. Some strains are multidrug-resistant. With prompt antibiotic treatment, only about 1% of patients die. Carriers can also be treated with antibiotics.

SHIGELLOSIS, BACILLARY DYSENTERY

Characteristics. An acute bacterial infection of the lining of the small and large intestine; diarrhea with blood, mucus, and pus; nausea, vomiting, cramps, fever, as many as 20 bowel movements a day; sometimes *toxemia* (toxins in the blood) and convulsions (in children); other serious complications (*e.g.,* hemolytic uremic syndrome) may occur.

Pathogens. Shigella dysenteriae, S. flexneri, S. boydii, and *S. soneii;* nonmotile, Gram-negative bacilli; members of the family *Enterobacteriaceae;* plasmid associated with toxin production and virulence; relatively few (10 to 100) organisms are required to cause disease.

Reservoir. Humans.

Transmission. Direct or indirect fecal-oral transmission from patients or carriers; fecally contaminated hands and fingernails; fecally contaminated food, milk, drinking water; flies can transfer organisms from latrines to food.

Incubation Period. Usually 1 to 3 days.

Epidemiology. Worldwide, shigellosis is estimated to cause approximately 600,000 deaths per year, with about two thirds of the cases and most of the deaths occurring in children under 10 years of age; 25,978 U.S. cases were reported to the CDC in 1996.

Prevention and Control. Handwashing; construction and maintenance of fly-proof latrines; proper sewage disposal and water purification procedures; care in food preparation and handling; infected individuals should neither prepare nor serve food; standard precautions for hospitalized patients; contact precautions for diapered or incontinent patients.

Diagnosis. Presence of leukocytes in stool specimens. Immediate inoculation of GN (for Gram-negative) enrichment broth and solid media (such as MacConkey, xylose-lysine-deoxycholate [XLD], and Hektoen enteric [HE] agar) with fresh feces or rectal swab. *Shigella* spp. produce colorless colonies on MacConkey agar because they are non-lactose-fermenters. Identification by culture, biochemical, and immunodiagnostic procedures.

Treatment. Illness is usually self-limited. Fluid and electrolyte replacement therapy. A fluoroquinolone is the drug of choice. Many strains are multidrug-resistant. Antimotility agents are contraindicated; they may prolong illness.

Bacterial Foodborne Intoxications, Foodborne Infections, Food Poisoning

The term "food poisoning" is broad and may include diseases resulting from the ingestion of chemical contaminants as well as bacteria or bacterial toxins, phycotoxins, mycotoxins, viruses, or protozoa. In this section, only diseases resulting from the ingestion of bacteria or their toxins are described. Technically, diseases resulting from the ingestion of toxin-producing bacteria are called "infections," whereas diseases resulting from the ingestion of preformed bacterial toxins are called "intoxications." The distinction is based upon where the toxin is actually produced—in the body (*in vivo*) or in the food (*in vitro*). Incubation time (the time that elapses between ingestion and onset of symptoms) may be influenced by a number of factors, including (1) whether toxin-producing bacteria are ingested (in which case, it will take additional time for the organisms to multiply and produce toxin *in vivo*) or a preformed toxin is ingested, (2) the number of organisms ingested, and (3) the amount of a preformed toxin that is ingested. Table 12–2 contains a brief synopsis of foodborne diseases, some of which are described more fully elsewhere in this chapter.

BOTULISM
Characteristics. Botulism (a neuromuscular disease involving a flaccid type of paralysis) is the most severe form of food poisoning, often resulting in death. There are actually three types of botulism: (1) classical foodborne botulism, (2) infant

TABLE 12-2. Etiologic Agents of Food Poisoning

Category of Pathogen	Pathogen	Causes an Infection (Toxin Produced In Vivo)/Incubation Time	Causes an Intoxication (Toxin Produced In Vitro)/Incubation Time
Algae	Phycotoxins produced by various species of marine algae		Yes/variable, depending upon the species involved
Bacteria	*Bacillus cereus*		Yes/1 to 6 hrs for vomiting toxin; 6 to 24 hrs for diarrheal toxin
	Campylobacter spp.	Yes/usually 2 to 5 days	
	Clostridium botulinum	Yes, in wound and infant botulism/variable	Yes, in classical foodborne-botulism/ usually 12 to 36 hours
	Clostridium perfringens	Yes, causes an infection if vegetative bacteria or spores are ingested/ usually 10 to 12 hrs	Yes, causes an intoxication if toxins are ingested
	Certain serotypes of *Escherichia coli*	Yes/usually 3 to 4 days for enterohemorrhagic strains; usually 6 to 48 hrs for enterotoxigenic strains	
	Listeria monocytogenes	Yes/unknown for diarrheal disease; 2 to 6 wks for invasive disease	
	Salmonella spp.	Yes/usually 6 to 48 hrs for diarrheal disease; usually 7 to 14 days for typhoid fever	
	Shigella	Yes/usually 2 to 4 days	
	Staphylococcus aureus		Yes/usually 2 to 4 hrs
	Vibrio spp.	Yes, some species/usually 1 to 5 days for *V. cholerae*	Yes, some species/4 to 30 hrs for *V. parahemolyticus*
Fungi	Mycotoxins produced by various fungi, including *Aspergillus* spp., rusts, smuts, and certain mushrooms		Yes/variable, depending upon the species involved
Protozoa	*Cryptosporidium parvum*	Yes; pathogenic mechanism unknown/2 to 28 days; median 7 days	
	Cyclospora cayetanensis	Yes; pathogenic mechanism unknown/1–11 days; median 7 days	

(continued)

Category of Pathogen	Pathogen	Causes an Infection (Toxin Produced In Vivo)/Incubation Time	Causes an Intoxication (Toxin Produced In Vitro)/Incubation Time
	Giardia lamblia	Yes; pathogenic mechanism unknown/3 to 25 days; median 7 days	
Viruses	Hepatitis A virus	Yes; pathogenesis does not involve toxin production/15 to 50 days; median 28 days production	

botulism, and (3) wound botulism. Classical foodborne botulism results from the ingestion of food (often home-canned fruits or vegetables) containing botulinum toxin (a potent neurotoxin); thus, in the case of classical foodborne botulism, the exotoxin is produced *in vitro*. Infant botulism results from ingestion of *Clostridium botulinum* spores (most often in honey), germination of the spores in the infant's intestinal tract, and production of botulinal toxin *in vivo*. Wound botulism is similar to gas gangrene, in that clostridial spores enter a wound, germinate, and the toxin is produced *in vivo*. Botulinum toxin may cause nerve damage, visual difficulty, respiratory failure, flaccid paralysis of voluntary muscles, brain damage, coma, and death within a week, if untreated. Respiratory failure is the usual cause of death.

Pathogen. Clostridium botulinum, a spore-forming, Gram-positive, anaerobic bacillus that produces botulinum toxin, one of the most potent toxins known.

Reservoirs. Dust, soil, foods contaminated with dirt, honey, corn syrup, inadequately heated home-canned foods, neutral pH foods, lightly cured foods.

Transmission. Ingestion of foods in which *C. botulinum* has produced botulinum toxin (classical foodborne botulism); ingestion of foods containing *C. botulinum* (infant botulism); entry of *C. botulinum* spores into wounds (wound botulism); botulism has also occurred among IV drug abusers.

Incubation Period. Neurologic symptoms of foodborne botulism usually occur with 12 to 36 hours of ingestion of the toxin-containing food.

Epidemiology. 119 U.S. cases of botulism (all types) were reported to the CDC in 1996; of those, 25 were foodborne botulism and 80 were infant botulism; most patients with infant botulism have been between 2 weeks and 1 year of age.

Prevention and Control. Careful washing, canning, processing, and cooking of food. Before eating, home-canned vegetables must be boiled (with stirring) for at least 10 minutes to destroy botulinal toxins. Botulinum toxin may be present in non-swollen cans and in foods lacking any sort of "off-odor." Thorough cleaning of wounds. Never feed raw honey to infants or add it to their milk or foods.

Diagnosis. Botulism is diagnosed by demonstrating botulinum toxin in the patient's serum or gastric aspirate or in the incriminated food; or by culture of *C. botulinum* from a gastric aspirate or stool or a wound culture in the case of wound botulism or feces in the case of infant botulism.

Treatment. Intravenous administration of botulinum antitoxin (horse serum). Debridement of wound and penicillin in cases of wound botulism. Standard precautions for hospitalized patients.

CLOSTRIDIUM PERFRINGENS FOOD POISONING

Characteristics. A gastrointestinal toxemia with colic, diarrhea, nausea, rarely vomiting and fever; usually a mild disease lasting 1 day or less; rarely fatal in healthy people.

Pathogen. *Clostridium perfringens,* a Gram-positive, spore-forming, enterotoxin-producing, anaerobic bacillus.

Reservoirs. Spores in soil, GI tract of humans and animals (cattle, swine, poultry, fish).

Transmission. Ingestion of food (usually meat and gravies) contaminated by dirt or feces, kept at moderate temperatures allowing bacterial growth and exotoxin production.

Incubation Period. 6 to 24 hours, usually 10 to 12 hours.

Prevention and Control. Cook food well, especially meat dishes; hot dishes should be served while still hot; proper preparation of meat and poultry; handwashing.

Diagnosis. Demonstration of *C. perfringens* in food or patient's stool or detection of enterotoxin in the patient's stool.

Treatment. None required; standard precautions for hospitalized patients.

STAPHYLOCOCCAL FOOD POISONING

Characteristics. A gastroenteritis intoxication with an abrupt and often violent onset, with severe nausea, cramps, and vomiting; often with diarrhea and sometimes with below normal temperature and decreased blood pressure; rarely fatal.

Pathogens. Enterotoxin-producing strains of *Staphylococcus aureus* growing in foods.

Reservoirs. Humans (skin, abscesses, nasal secretions); occasionally cows with infected udders, dogs and fowl.

Transmission. Ingestion of *S. aureus*-contaminated foods containing staphylococcal enterotoxin (a type of heat-stable exotoxin); foods prepared from contaminated milk or milk products (*e.g.,* cheese); typically contaminated foods include custard, cream-filled pastries, salad dressings, sandwiches, processed meats and fish.

Incubation Period. 30 min to 8 hours, usually 2 to 4 hours.

Prevention and Control. Sanitation and cleanliness in the kitchen; handwashing; keep perishable foods either hot or cold; temporarily exclude people with boils and abscesses from preparing or handling food.

Diagnosis. In an outbreak, recovery of staphylococci from or detection of entero-toxin in an epidemiologically implicated food. Isolation of large numbers of enterotoxin-producing staphylococci from stool or vomitus. Phage typing can be performed to determine if the staphylococci recovered from the food are the same phage type as those isolated from the patient.

Treatment. Fluid replacement if needed; standard precautions for hospitalized patients.

Protozoal Gastrointestinal Diseases

PRIMARY AMEBIASIS, AMEBIC DYSENTERY

Characteristics. A protozoal intestinal disease, which may be asymptomatic, mild, or severe; often with dysentery, fever, chills, bloody or mucoid diarrhea or consti-pation, and colitis. Amebae may invade mucous membranes of the colon, forming abscesses. Amebae may be also be disseminated via the bloodstream to extrain-testinal sites, leading to abscesses of the liver, lung, brain, and other organs.

Pathogen. *Entamoeba histolytica;* an ameba in the subphylum Sarcodina; occurs in two stages: the cyst stage (the dormant, infective stage) and the motile, metaboli-cally active, reproducing trophozoite stage (the ameba). Diarrheal specimens con-tain cysts and trophozoites.

Reservoir. Symptomatic or asymptomatic humans.

Transmission. Ingestion of fecally contaminated food or water; flies on food; soiled hands of infected food handlers; oral-anal sexual contact.

Incubation Period. A few days to several months, commonly 2 to 4 weeks.

Prevention and Control. Sanitary sewage disposal; protection and treatment of water supplies; personal hygiene and handwashing before preparing and eating food; stopping the practice of fertilizing crops with human feces; control flies around foods; standard precautions for hospitalized patients.

Diagnosis. Microscopic observation of *E. histolytica* trophozoites and/or cysts in stained smears of fecal specimens. Physical features of *E. histolytica* trophozoites and cysts enable differentiation from most other pathogenic and nonpathogenic amebae found in stool specimens.

Treatment. Metronidazole (Flagyl®) or tinidazole for symptomatic cases; fluid and electrolyte replacement; iodoquinol or paromomycin for asymptomatic carriers.

CRYPTOSPORIDIOSIS

Characteristics. A parasitic infection that may be asymptomatic or may cause diarrhea, cramping, and abdominal pain; may be prolonged, fulminant, and fatal in immunosuppressed patients.

Pathogen. *Cryptosporidium parvum*, a sporozoan parasite.

Reservoirs. Humans, cattle and other domestic animals.

Transmission. Fecal-oral transmission; person-to-person, animal-to-person, contaminated water or food.

Incubation Period. 1 to 12 days, with an average of about 7 days.

Epidemiology. A total of 3068 U.S. cases of cryptosporidiosis were reported to the CDC in 1998, but cryptosporidiosis is not notifiable in all states; thus, the actual number of cases that year was probably higher.

Prevention and Control. Thorough handwashing after handling calves and other animals with diarrhea; proper treatment (including filtration) of drinking water; boiling of drinking water, especially by immunosuppressed persons; standard precautions for hospitalized patients.

Diagnosis. Microscopic observation of small (4 to 6 μm), acid-fast oocysts in stained smears of fecal specimens. Immunodiagnostic procedures are available.

Treatment. Paromomycin; fluid and electrolyte replacement.

GIARDIASIS

Characteristics. A protozoal infection of the duodenum (the uppermost portion of the small intestine); may be asymptomatic, mild, or severe; with diarrhea, steatorrhea (loose, pale, malodorous, fatty stools), abdominal cramps, bloating, abdominal gas, fatigue, and possibly weight loss.

Pathogens. *Giardia lamblia* (see Fig. 3–8B and Fig. C-2 in Appendix C), a flagellated protozoan. Trophozoites attach to mucosal membranes. Trophozoites and/or cysts are expelled in feces.

Reservoirs. Humans; possibly beaver and other wild and domestic animals.

Transmission. Fecal-oral route; ingestion of cysts in fecally contaminated water or foods; person-to-person by soiled hands to mouth (as occurs in daycare centers).

Incubation Period. 3 to 25 days or longer, usually 7 to10 days.

Prevention and Control. Proper treatment (including filtration) of public water supplies; routine chlorination does not destroy cysts, especially in cold water; sanitary disposal of feces; boil all emergency water supplies; prevent fecal contamination of toys and others objects in daycare centers; thorough cleaning and disinfection of fecally soiled articles (*e.g.,* washcloths used in daycare centers).

Diagnosis. Microscopic observation of trophozoites (Appendix C) and/or cysts in stained smears of fecal specimens. Immunodiagnostic procedures are available.

Treatment. Metronidazole (Flagyl®), tinidazole, furazolidone, or paromomycin.

INFECTIOUS DISEASES OF THE GENITOURINARY (GU) TRACT

When the anatomy of the male and female urinary and reproductive systems are studied, it is easy to locate many areas where infections may occur (Fig. 12–15). The urinary tract is usually protected from pathogens by the frequent flushing action of urination. The acidity of normal urine also discourages growth of many microorganisms. Indigenous microflora are found at and near the outer opening

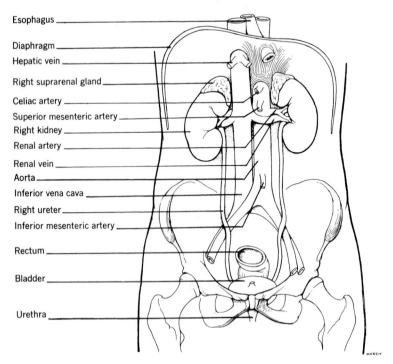

Esophagus

Diaphragm

Hepatic vein

Right suprarenal gland

Celiac artery

Superior mesenteric artery

Right kidney

Renal artery

Renal vein

Aorta

Inferior vena cava

Right ureter

Inferior mesenteric artery

Rectum

Bladder

Urethra

FIGURE 12-15. Urinary tract.

(meatus) of the urethra of both males and females. Inhabitants of the distal urethra include coagulase-negative *Staphylococcus, Streptococcus, Lactobacillus,* diphtheroids, nonpathogenic *Neisseria* and *Mycobacterium,* and *Mycoplasma* spp., as well as some Gram-negative enteric bacteria. Additionally, the female genital area supports the growth of many other microorganisms. In the adult vaginal microflora, there are many species of *Lactobacillus, Staphylococcus, Streptococcus, Enterococcus, Neisseria, Clostridium, Actinomyces, Prevotella,* diphtheroids, enteric bacilli, and *Candida.* The balance among these microbes depends on the estrogen levels and pH of the site. Should any of these or other microorganisms invade further into the GU tract, a variety of nonspecific infections may occur. The rest of the reproductive systems of both sexes should be free of microbial life (Fig. 12–16).

Sexually Transmitted Diseases

The term sexually transmitted disease (STD), formerly called venereal disease (VD), includes any of the infections transmitted by sexual activities. They are diseases of not only the reproductive and urinary tracts, but also of the skin, mucous membranes, blood, lymphatic and digestive systems, and many other body areas. Epidemic STDs of the 1990s, such as acquired immunodeficiency syndrome (AIDS), chlamydial and herpes infections, gonorrhea, and syphilis are expected to continue into the 21st century. Because "the AIDS virus" (HIV) primarily causes damage to helper T cells and, thus, inhibits antibody production, it is discussed later with diseases of the circulatory system. Previously discussed diseases such as hepatitis B, amebiasis, and giardiasis can also be considered sexually transmitted diseases, as can many other diseases.

URINARY TRACT INFECTIONS, URETHRITIS, CYSTITIS, URETERITIS, PROSTATITIS

Inflammatory infections of the urinary tract include infections of the urethra (*urethritis*), urinary bladder (*cystitis*), ureters (*ureteritis*), prostate (*prostatitis*), and kidney (*nephritis* or *pyelonephritis*). These infections may be caused by any of a variety of microorganisms, introduced by poor personal hygiene, sexual intercourse, the insertion of catheters, etc. The most common cause of cystitis and pyelonephritis is *E. coli;* other common causes of cystitis are species of *Klebsiella, Proteus, Enterococcus, Staphylococcus,* and *Pseudomonas.* The most common cause of urethritis is *N. gonorrhoeae* (gonorrhea is discussed in a following section), but nonspecific or nongonococcal urethritis (NGU) is frequently caused by species of *Chlamydia, Ureaplasma,* and *Mycoplasma,* usually transmitted sexually.

Viral Genitourinary Tract Infections

ANOGENITAL HERPES VIRAL INFECTIONS

Characteristics. In general, herpes simplex infections are characterized by a localized primary lesion, latency, and a tendency for localized recurrence. In women, the principal sites of primary anogenital herpes virus infection are the cervix and

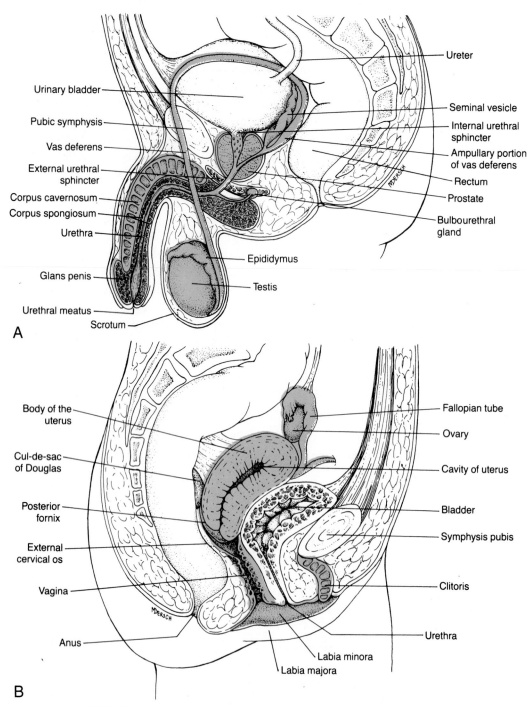

FIGURE 12-16. Reproductive systems. (*A*) Male; (*B*) Female.

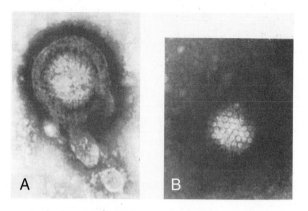

FIGURE 12-17. Herpes simplex virus particles embedded in phosphotungstate. (*A*) Enveloped virion showing the thick envelope surrounding the nucleocapsid. (*B*) Naked viral particle; the structure of the capsomeres is plainly visible (original magnification, ×200,000). (Watson DH, et al: Virology 19:250, 1963)

vulva, with recurrent disease affecting the vulva, perineal skin, legs and buttocks. In men, lesions appear on the penis, and in the anus and rectum of persons engaging in anal sex. The initial symptoms are usually itching, tingling, and soreness, followed by a small patch of redness, and then a group of small, painful blisters. The blisters break and fuse to form painful, circular sores, which become crusted after a few days. The sores heal in about 10 days but may leave scars. The initial outbreak is more painful, prolonged, and widespread than subsequent outbreaks and may be associated with fever.

Pathogens. Usually herpes simplex virus, type 2 (HSV 2); occasionally HSV 1 (Fig. 12–17).

Reservoir. Humans.

Transmission. Direct sexual contact; oral-genital, oral-anal, or anal-genital contact during presence of lesions (Fig. 12–18); mother-to-fetus or mother-to-neonate transmission occurs during pregnancy and birth.

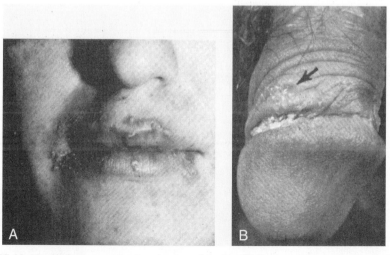

FIGURE 12-18. (*A*) Recurrent herpes simplex of the lip. (*B*) Recurrent herpes progenitalis.

Incubation Period. 2 to 12 days.

Prevention and Control. Refrain from intercourse with person with herpetic lesions; use of latex condoms; cesarean delivery before membranes rupture; acyclovir may be used prophylactically to reduce the incidence of recurrences and of herpes infections in immunosuppressed patients.

Diagnosis. Observation of characteristic cytologic changes in tissue scrapings or biopsy specimens; immunodiagnostic procedures.

Treatment. Treatment does not cure genital herpes, but antiviral agents such as acyclovir, famciclovir or valacyclovir may shorten the outbreak.

GENITAL WARTS, GENITAL PAPILLOMATOSIS, CONDYLOMA ACUMINATUM

Characteristics. Genital warts start as tiny, soft, moist, pink or red swellings, which grow rapidly and may develop stalks. Their rough surfaces give them the appearance of small cauliflowers. Multiple warts often grow in the same area, most often on the penis in males and the vulva, vaginal wall, cervix, and skin surrounding the vaginal area in females. Genital warts also develop around the anus and in the rectum in males or females who engage in anal sex (Fig. 12–19).

Pathogens. Human papillomaviruses (HPV) of the papovavirus group of DNA

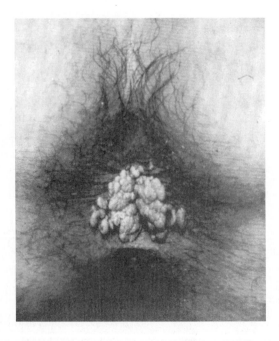

FIGURE 12-19. Anogenital warts (condylomata acuminata or moist warts).

viruses (human wart viruses); HPV genotypes 16 and 18 have been associated with cervical cancer.

Reservoir. Humans.

Transmission. Direct contact, usually sexual; through breaks in skin or mucous membranes; from mother to neonate during birth.

Incubation Period. 1 to 20 months, usually 2 to 3 months.

Prevention and Control. Avoid direct contact with another person's warts; use of latex condoms; cesarean section may be considered if genital papillomatosis is very extensive.

Treatment. External genital warts may be removed by laser, cryotherapy (freezing) or surgery. Various chemical treatments are available, such as podophyllin in tincture of benzoin, trichloroacetic acid, 5-fluorouracil (5-FU), or interferon alpha-2b.

Bacterial Genitourinary Tract Infections

GENITAL CHLAMYDIAL INFECTIONS, GENITAL CHLAMYDIASIS

Characteristics. The most frequent cause of nongonococcal urethritis (NGU), causing mucopurulent urethral discharge, urethral itching, and burning on urination; may also cause epididymitis, infertility, and proctitis in men. Most commonly causes endocervical and urethral infections, salpingitis, infertility, and chronic pelvic pain in women. Infection during pregnancy may result in premature rupture of membranes and preterm delivery, as well as conjunctivitis and pneumonia in neonates. May be concurrent with gonorrhea.

Pathogens. Certain serotypes of *Chlamydia trachomatis;* tiny, obligately intracellular, Gram-negative bacteria. Less common causes of NGU are *Ureaplasma ureolyticum* (closely related to mycoplasmas), herpes simplex viruses, and *Trichomonas vaginalis.*

Reservoir. Humans.

Transmission. Direct sexual contact or mother-to-neonate during birth.

Incubation Period. 7 to 14 days or longer.

Epidemiology. Chlamydial infections were the most common notifiable infectious diseases in the United States in 1998 (593,097 cases reported to the CDC that year). The number of U.S. chlamydial infections reported to CDC that year exceeded the number of reported U.S. gonorrhea cases by about 248,000.

Prevention and Control. Personal cleanliness, sex education, use of latex condoms; prophylactic treatment of sexual contacts.

Diagnosis. Identification of *C. trachomatis* by cell culture, staining, and immunodiagnostic procedures.

Treatment. Azithromycin or doxycycline; erythromycin for neonates and pregnant women.

GONORRHEA

Characteristics. Gonorrhea can be manifested in a multitude of ways, some of which involve the GU tract (described below) and some of which do not (*e.g.,* conjunctivitis [Fig. 12–6], rash, pharyngitis, proctitis, arthritis). Gonococcal infections of the GU tract include urethritis and epididymitis in males and cervicitis, Bartholinitis (inflammation of the Bartholin's ducts), pelvic inflammatory disease (PID), salpingitis (inflammation of the Fallopian tubes), endometritis, and vulvovaginitis in females. Urethral discharge and painful urination are common in infected males, usually starting 2 to 7 days after infection. Infected women may be asymptomatic for weeks or months, during which time severe damage to the reproductive system may occur.

Pathogen. *Neisseria gonorrhoeae,* also known as gonococcus or GC; Gram-negative diplococci; some strains (called penicillinase-producing *N. gonorrhoeae,* or PPNG) possess plasmids containing the gene for penicillinase production; some strains are multiply drug resistant.

Reservoir. Humans.

Transmission. Direct mucous membrane-to-mucous membrane contact, usually sexual contact; adult-to-child (may indicate sexual abuse); mother-to-neonate during birth.

Incubation Period. Usually 2 to 7 days, but sometimes longer.

Epidemiology. 345,087 U.S. cases were reported to the CDC during 1998.

Prevention and Control. Avoid intercourse with infected people; use of latex condoms; vaginal and cervical cultures for pregnant women; standard precautions for hospitalized patients.

Diagnosis. Typical appearance of Gram-stained urethral discharge from male patients, with numerous white blood cells and numerous intra-and extracellular Gram-negative diplococci. Culture on chocolate agar or a modified chocolate agar (such as Thayer-Martin medium, Martin-Lewis medium, New York City agar, or Transgrow®). Beta-lactamase testing of isolates, followed by antimicrobial susceptibility testing if beta-lactamase-positive. Isolates are identified using biochemical tests. Immunodiagnostic procedures are available.

Treatment. Ceftriaxone, cefixime, ciprofloxacin or ofloxacin; patients with gonorrhea should be treated presumptively for co-infection with *C. trachomatis* with azithromycin or doxycycline.

SYPHILIS

Characteristics. A treponemal disease that occurs in three stages: a painless, primary lesion known as a chancre (Fig. 12–20A); a secondary skin rash (especially on the palms and soles) about 4 to 6 weeks later, with fever and mucous membrane lesions (Fig. 12–20B); a long latent period (which can be as long as 5 to 20 years); and then a tertiary stage with damage to the central nervous system (CNS), cardiovascular system, visceral organs, bones, sense organs, and other sites. Usually, the damage to the CNS or heart is not reversible.

Pathogen. *Treponema pallidum;* a Gram-variable, tightly coiled spirochete that is too thin to be seen with brightfield microscopy.

Reservoir. Humans.

Transmission. Direct contact with lesions, body secretions, mucous membranes, blood, semen, saliva, and vaginal discharges of infected people, usually during sexual contact; blood transfusions; transplacentally from mother to fetus.

Incubation Period. 10 days to 3 months, usually 3 weeks.

Epidemiology. During 1986 to 1990, an epidemic of syphilis occurred throughout the United States but, since then, syphilis rates have declined each year. In 1998, 7183 U.S. cases of primary/secondary syphilis and 399 U.S. cases of congenital syphilis were reported to the CDC.

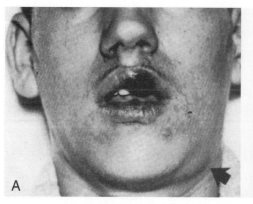 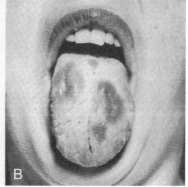

FIGURE 12-20. Syphilis. (A) Primary syphilis, showing a chancre of the lip with unilateral adenopathy (arrow). (B) Secondary syphilis, with mucous patches on the tongue.

Prevention and Control. Sex education; avoid sexual contact with infected persons; use of latex condoms; serologic tests for high-risk persons and pregnant women; standard precautions for hospitalized patients.

Diagnosis. Primary syphilis can be diagnosed by darkfield microscopy of material scraped from the margin of chancres. Many immunodiagnostic procedures are available, such as the RPR, VDRL, and FTA-Abs tests for detecting antibodies in serum or spinal fluid specimens and fluorescent antibody procedures for detecting antigen in material obtained from lesions or lymph nodes.

Treatment. Penicillin G (benzathine penicillin); tetracycline, doxycycline, or ceftriaxone for penicillin-sensitive persons.

Protozoal Genitourinary Tract Infections

TRICHOMONIASIS

Characteristics. A sexually transmitted protozoal disease causing vaginitis in women, with a profuse, thin, foamy, malodorous, greenish-yellowish discharge; may cause urethritis or cystitis; often asymptomatic; rarely symptomatic in men, but may cause prostatitis, urethritis, or infection of the seminal vesicles.

Pathogen. Trichomonas vaginalis; a flagellate (see Fig. C-2 in App. C).

Reservoir. Humans.

Transmission. Direct contact with vaginal and urethral discharges of infected people during sexual intercourse.

Incubation Period. 4 to 20 days, average 7 days; asymptomatic carriers may harbor *T. vaginalis* for years.

Epidemiology. Not a nationally notifiable disease in the United States. It has been estimated that approximately one-third of the U.S. cases of vaginitis are caused by *T. vaginalis* (another third by *Candida albicans* and another third by bacteria).

Prevention and Control. Avoid sexual relations with infected persons; use of latex condoms; concurrent treatment of sexual partners of infected individuals.

Diagnosis. Vaginitis due to *T. vaginalis* can be diagnosed by performing a wet mount examination (see App. D) of vaginal discharge material and observing the motile trophozoites. Culture procedures are also available. Sometimes *T. vaginalis* trophozoites are seen in urine and Papanicolaou (Pap) smears.

Treatment. Metronidazole (Flagyl®) or tinidazole.

Other Sexually Transmitted Diseases

Many other pathogens may be sexually transmitted. Three, seen more often in parts of the world other than in the United States, are chancroid, granuloma inguinale, and lymphogranuloma venereum (LGV). Chancroid, caused by the Gram-negative bacterium *Haemophilus ducreyi,* can be treated with azithromycin or ceftriaxone. Granuloma inguinale, a chronic infection caused by a Gram-negative bacterium named *Calymmatobacterium granulomatis* (*Donovania granulomatis*), is treatable with trimethoprim-sulfamethoxazole. LGV is a chlamydial infection involving the lymph nodes, rectum, and reproductive tract. It is caused by certain serotypes of *Chlamydia trachomatis* (see Fig. 3–4) and treated with tetracycline or erythromycin. It should be noted that many sexually transmitted diseases are transmitted simultaneously. A total of 386 U.S. cases of chancroid were reported to the CDC during 1996; neither granuloma inguinale nor LGV are notifiable diseases in the United States.

INFECTIOUS DISEASES OF THE CIRCULATORY SYSTEM

The circulatory system, consisting of the cardiovascular system and lymphatic system (Fig. 12–21), carries blood and lymph throughout the body. Included in these fluids are many cells: erythrocytes, leukocytes (including lymphocytes), and platelets (thrombocytes). Most leukocytes function to protect the body from pathogens by phagocytosis or antibody production (Chapter 11).

Normally, the blood is sterile; it contains no resident microflora. Transient *bacteremia* (the temporary presence of bacteria in the blood) often results from dental extractions, wounds, bites, and damage to the intestinal, respiratory, or reproductive tract mucosa. However, when pathogenic organisms are capable of resisting or overwhelming the phagocytes and other body defenses—or when the individual is immunosuppressed or is otherwise more susceptible than normal—a systemic disease called *septicemia* may occur. A patient with septicemia experiences chills, fever, and prostration (extreme exhaustion), and has bacteria and/or their toxins in his or her bloodstream. Lymph occasionally picks up microorganisms from the intestine, lungs, and other areas, but these transient organisms are usually quickly engulfed by phagocytic cells in the liver and lymph nodes.

Viruses often invade the circulatory system and damage certain "target" cells, an example being HIV, which destroys helper T cells, causing the immune system to malfunction. Some cardiovascular diseases are the result of *toxemia,* the presence of bacterial toxins in the bloodstream. Toxic shock syndrome, scarlet fever, Lyme disease, and septic shock are examples of such diseases. Certain fungal and protozoal cardiovascular diseases may cause damage by actually clogging the capillaries in various regions of the body.

Viral Lymphatic and Cardiovascular Infections

Nonspecific diseases of the lymphatic system include *lymphadenitis* (inflamed and swollen lymph nodes), *lymphadenopathy* (diseased lymph nodes), and *lymphangitis* (inflamed lymphatic vessels). Nonspecific diseases of the cardiovascular system include *endocarditis* (inflammation of the endocardium—the endothelial membrane

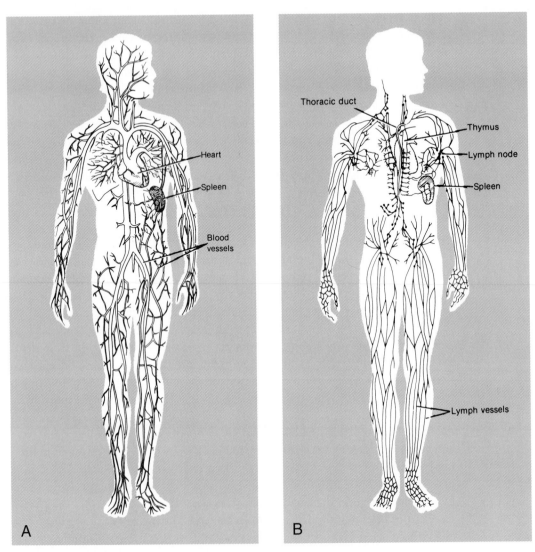

FIGURE 12-21. (A) The cardiovascular system. (B) The lymphatic system.

that lines the cavities of the heart), *myocarditis* (inflammation of the myocardium—the muscular walls of the heart), and *pericarditis* (inflammation of the pericardium—the membranous sac around the heart). Any of these conditions can be the result of infectious diseases.

ACQUIRED IMMUNODEFICIENCY SYNDROME (AIDS)

Characteristics. Signs and symptoms of acute HIV infection (*i.e.*, infection with "the AIDS virus") usually occur within days to weeks after initial exposure and last from a few days to more than 10 weeks (usually less than 14 days). Unfortunately, this syndrome is often undiagnosed or misdiagnosed because HIV antibodies are not usually detected during this early phase of infection. Signs and symptoms of acute HIV infection include fever, rash, headache, lymphadenopathy, pharyngitis, myalgia, arthralgia (joint pain), aseptic meningitis, retro-orbital pain, weight loss, depression, GI distress, night sweats, and oral or genital ulcers.

Acquired immunodeficiency syndrome (AIDS) is a severe, life-threatening syndrome which represents the late clinical stage of infection with HIV. Invasion and destruction of helper T cells (Chapter 11; Fig. 12–22) leads to suppression of the patient's immune system (immunosuppression). Because the immune system of HIV-infected people is unable to produce antibodies in response to T-dependent antigens (Chapter 11), secondary infections caused by viruses (*e.g.,* Cytomegalovirus, herpes simplex), protozoa (*e.g., Cryptosporidium, Toxoplasma*), bacteria (*e.g.,* mycobacteria), and/or fungi (*e.g., Candida, Cryptococcus, Pneumocystis*) become systemic and cause death of the patient. AIDS patients die as a result of overwhelming infections caused by a variety of pathogens, often opportunistic pathogens. Kaposi's sarcoma (a previously rare type of cancer) is a frequent complication of AIDS; it is thought to be caused by a type of herpes virus. Previously considered to be a universally fatal disease, certain combinations of drugs, referred to as "cocktails," are

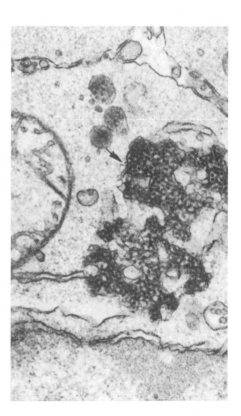

FIGURE 12-22. A large inclusion (arrow) within the cytoplasm of a peripheral blood mononuclear cell from a patient with AIDS (original magnification, ×40,000).

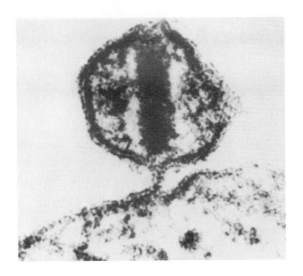

FIGURE 12-23. Extracellular HIV retrovirus particle from a lymphocyte of a patient with ARC. Note the dense cylindrical core.

extending the life of some AIDS patients. However, most (80 to 90%) AIDS patients have died within 3 to 5 years after the diagnosis of AIDS is made.

Pathogens. Human immunodeficiency virus (HIV) (Fig. 12–23); two types have been identified—HIV-1 (more common in the United States) and HIV-2; in a family of viruses known as retroviruses. Most likely, HIV-1 first invades dendritic cells in the genital and oral mucosa; these cells then fuse with CD4+ lymphocytes (helper T cells) and spread to deeper tissues. HIV-1 can be cultured from plasma about 5 days following infection.

Reservoir. Humans.

Transmission. Direct sexual contact, homosexual or heterosexual; sharing of contaminated needles and syringes by IV drug abusers; transfusion of contaminated blood and blood products; transplacental transfer from mother-to-child; breast feeding by HIV-infected mothers; needlestick, scalpel, broken glass injury.

Incubation Period. Variable; less than 1 year to 10 years or longer; generally 1 to 3 months; about half of infected adults will have developed AIDS within 10 years of infection.

Epidemiology. Pandemic. As of February 1998, it was estimated that more than 30 million people were infected with HIV worldwide, with as many as 16,000 individuals becoming infected per day in 1997. Approximately 90% of all people with HIV infection live in developing countries (about two-thirds of them live in sub-Saharan Africa). During 1997, about 5.8 million people became infected with HIV and approximately 2.3 million people died as a result of AIDS. The death rate in south and southeast Asia is 9 times higher, and in Africa it is 60 times higher, than the death rate in North America. As of September 1996, a total of 566,000 U.S. AIDS cases had been reported to the CDC; new U.S. cases reported to the CDC during 1997 and 1998 totaled 57,953 and 46,311, respectively (see Fig. 9–11).

Prevention and Control. Education; avoid sexual contact with high-risk persons (*e.g.,* prostitutes and other people having multiple sex partners, IV drug abusers); use of latex condoms; standard precautions for hospitalized patients; healthcare workers should exercise extreme caution with used needles, scalpels, and broken glass.

Diagnosis. Acute HIV-1 infection cannot be diagnosed with standard antibody detection tests; such tests first become positive approximately 22 to 27 days after acute infection. Antigen detection tests that detect the p24 antigen of HIV-1 and tests that detect viral RNA can be used to diagnose acute infection, however.

Treatment. The routine use of anti-retroviral drugs and prophylactic drugs to prevent opportunistic infections may prolong the survival of HIV-infected people for many years. The two main categories of anti-retroviral drugs are reverse transcriptase inhibitors and protease inhibitors. Use of other antimicrobial agents to treat secondary infections.

COLORADO TICK FEVER

Characteristics. One of 10 known tickborne infectious diseases in the United States (see Shaded Box). An acute viral infection with fever, headache, fatigue, aching, and occasionally rash, encephalitis, myocarditis, and tendency to bleed.

Pathogen. An unclassified arthropodborne virus.

Reservoirs. Tick-infested small mammals, such as ground squirrel, porcupine, chipmunk, mice.

Tickborne Diseases of the United States

Babesiosis (a protozoal disease)
Colorado tick fever (a viral disease)
Human granulocytic ehrlichiosis (a bacterial disease)
Human monocytic ehrlichiosis (a bacterial disease)
Lyme disease (a bacterial disease)
Powassan virus encephalitis (a viral disease)
Rocky Mountain spotted fever (a bacterial disease)
Tickborne relapsing fever (a bacterial disease)
Tularemia (a bacterial disease)
Q fever (a bacterial disease)
(Note: tick paralysis also occurs in the United States, but it is not an infectious disease.)

Transmission. Bite of virus-infected tick.

Incubation Period. Usually 4 to 5 days.

Epidemiology. Endemic in the mountainous regions above 5000 feet elevation of western United States and Canada. Occurs most frequently in hikers, fishermen, and those who work in endemic areas. Not a nationally notifiable disease in the United States.

Prevention and Control. Area-wide application of acaricides and control of tick habitats (*e.g.*, leaf litter and brush). Avoid tick-infested areas and animals, whenever possible. Wear light-colored clothing, long pants, and long-sleeved shirts in tick-infested woods or fields. Use tick repellents—diethyltoluamide (DEET) on skin and permethrin on pant legs and sleeves. For persons exposed to tick-infested habitats, prompt careful inspection for and removal of crawling or attached ticks. Remove ticks carefully, to include the head and mouth parts. Avoid crushing ticks during removal. Standard precautions for hospitalized patients.

Treatment. None, except removal of tick, rest and nutritious diet.

INFECTIOUS MONONUCLEOSIS, "MONO," "KISSING DISEASE"

Characteristics. An acute viral disease; may be asymptomatic or may be characterized by fever, sore throat, lymphadenopathy (especially posterior cervical lymph nodes), *splenomegaly* (enlarged spleen), and fatigue; usually a self-limited disease of one to several weeks duration; rarely fatal.

Pathogen. Epstein-Barr virus (EBV), a human herpes virus that is known to be *oncogenic* (cancer causing); invades B lymphocytes.

Reservoir. Humans.

Transmission. Person-to-person, by direct contact with saliva; kissing facilitates spreading among adolescents; can be transmitted via blood transfusion.

Incubation Period. 4 to 6 weeks.

Prevention and Control. Avoid salivary contamination from infected individuals, common drinking containers; don't kiss infected individuals; disinfect articles contaminated with nose and throat discharges; handwashing; standard precautions for hospitalized patients .

Diagnosis. Immunodiagnostic procedures.

Treatment. None, except rest and nutritious diet; avoid heavy lifting and contact sports to prevent rupturing the spleen; avoid use of aspirin in children to prevent Reye's syndrome.

MUMPS, INFECTIOUS PAROTITIS

Characteristics. An acute viral infection characterized by fever, swelling and tenderness of the salivary glands; complications can include *orchitis* (inflammation of the testes), *oophoritis* (inflammation of the ovaries), meningitis, encephalitis, deafness, pancreatitis, arthritis, mastitis, nephritis, thyroiditis, and pericarditis.

Pathogen. Mumps virus, a paramyxovirus.

Reservoir. Humans.

Transmission. Droplet spread and direct contact with the saliva of an infected person.

Incubation Period. 12 to 25 days, commonly 18 days.

Epidemiology. 606 U.S. cases were reported to the CDC in 1998.

Prevention and Control. Attenuated MMR (measles-mumps-rubella) vaccine for children and nonimmune adults; droplet precautions for hospitalized patients.

Treatment. No specific chemotherapy.

Rickettsial and Ehrlichial Cardiovascular Infections

ROCKY MOUNTAIN SPOTTED FEVER, TICKBORNE TYPHUS FEVER

Characteristics. A tickborne rickettsial disease characterized by sudden onset of moderate to high fever, extreme exhaustion (prostration), muscle pain, severe headache, chills, conjunctival infection, and maculopapular rash on extremities on about the third day, which spreads to the palms, soles, and much of the body; in about 4 days, small purplish areas (petechiae) develop as a result of bleeding in the skin; although death is uncommon, it can occur.

Pathogen. *Rickettsia rickettsii;* a Gram-negative bacterium; an obligate intracellular pathogen that invades endothelial cells (cells that line blood vessels).

Reservoir. Infected ticks on dogs, rodents, and other animals.

Transmission. Bite of infected tick.

Incubation Period. 3 to 14 days after tick bite.

Epidemiology. Occurs in the Western Hemisphere, including all parts of the United States, especially the Atlantic seaboard. A total of 332 U.S. cases were reported to the CDC in 1998.

Prevention and Control. See previous section on prevention and control of Colorado tick fever. Check dogs for presence of ticks and carefully remove any crawling or attached ticks. Standard precautions for hospitalized patients.

Diagnosis. Immunodiagnostic procedures.

Treatment. Tick removal; IV fluids may be needed; tetracycline or chloramphenicol.

ENDEMIC TYPHUS FEVER, MURINE TYPHUS FEVER, FLEABORNE TYPHUS

Characteristics. An acute febrile disease (similar to, but milder than, epidemic typhus, which is described next) with shaking chills, headache, fever, and a faint, pink rash.

Pathogen. *Rickettsia typhi;* a Gram-negative bacterium; an obligate intracellular pathogen.

Reservoir. Infected rat fleas.

Transmission. Rat → flea → human; rat fleas defecate the rickettsiae into bite site and other fresh skin wounds.

Incubation Period. 1 to 2 weeks, commonly 12 days.

Epidemiology. Worldwide occurrence, but rare in the United States (fewer than 80 cases reported annually); not a nationally notifiable disease in the United States.

Prevention and Control. Apply insecticide to rat-infested areas, then use rodent control measures; avoid contact with rats; standard precautions for hospitalized patients.

Treatment. Tetracycline or chloramphenicol.

EPIDEMIC TYPHUS FEVER, LOUSEBORNE TYPHUS

Characteristics. An acute rickettsial disease, often with sudden onset of headache, chills, prostration, fever and general pains. A rash appears on the fifth or sixth day, initially on the upper trunk, followed by spread to the entire body, but usually not to the face, palms or soles. May be fatal if untreated.

Pathogen. *Rickettsia prowazekii;* a Gram-negative bacterium; an obligate intracellular.

Reservoir. Humans.

Transmission. Humans to body lice (*Pediculus humanus;* Fig. 12–24) to humans; infected lice defecate while feeding and the rickettsiae in the feces are rubbed into the bite wound or other superficial abrasions.

Incubation Period. 1 to 2 weeks, commonly 12 days.

Epidemiology. Occurs in colder climates, where people may live under unhygienic conditions and are louse-infested; in World War I, body lice were referred to as "cooties" by soldiers; not a nationally notifiable disease in the United States.

Prevention and Control. Use of insecticide to kill body lice; improve living conditions and personal hygiene; standard precautions for hospitalized patients.

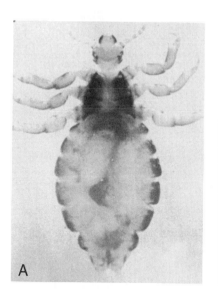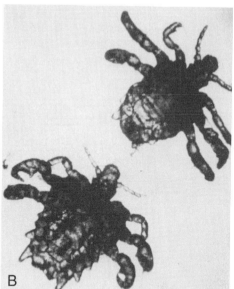

FIGURE 12-24. (A) *Pediculus humanus,* the louse that is the vector for epidemic typhus. (B) Pubic lice, *Phthirus pubis.*

Treatment. Tetracycline or chloramphenicol.

EHRLICHIOSIS

Characteristics. An acute, febrile illness ranging from asymptomatic to mild to severe and life-threatening. Patients usually present with acute influenza-like illness with fever, headache, and generalized malaise. Reminiscent of Rocky Mountain spotted fever, without the rash. The estimated fatality rate is about 5%.

Pathogens. Ehrlichia spp.; Gram-negative coccobacilli; closely related to rickettsiae; obligate intraleukocytic pathogens. *Ehrlichia chaffeenis* invades human monocytes, causing human monocytic ehrlichiosis (HME). Another species (similar to or identical to *E. equi*) invades human granulocytes, causing human granulocytic ehrlichiosis (HGE). *E. canis, E. equi,* and *E. phagocytophila* cause ehrlichiosis in dogs, horses, and ruminants.

Reservoir. Unknown.

Transmission. Tick bite.

Incubation Period. 7 to 21 days.

Epidemiology. The first human U.S. case of ehrlichiosis (a person with HME) occurred in 1991. As of June 1998, approximately 500 ehrlichiosis cases had been confirmed by the CDC. Cases of HME are more common than HGE cases. Most HME cases have occurred in the southeast and mid-Atlantic states, whereas most HGE cases have occurred in states with high rates of Lyme disease (particularly Connecticut,

Minnesota, New York, and Wisconsin). In these states, the tick that transmits the HGE agent is the same tick that transmits *Borrelia burgdorferi,* the etiologic agent of Lyme disease. The two different types of ehrlichiosis seem to be transmitted by different species of ticks. Not a nationally notifiable disease in the United States.

Prevention and Control. See previous section on prevention and control of Colorado tick fever. Standard precautions for hospitalized patients.

Diagnosis. Immunodiagnostic procedures and nucleic acid assays.

Treatment. Tetracycline or doxycycline; rifampin in pregnant women.

Other Bacterial Cardiovascular Infections

INFECTIVE ENDOCARDITIS

Infective (or infectious) endocarditis is caused by a pathogen, usually a bacterium or a fungus. It is characterized by the presence of vegetations (bacteria and blood clots) on or within the endocardium, most commonly involving a heart valve. Abnormal or damaged valves are most susceptible to infection, although valves can become contaminated during open heart surgery. The vegetations can break loose and be transported to vital organs, where they can block arterial blood flow. Obviously, such obstructions are very serious, possibly leading to strokes, heart attacks, and death.

The two most common types of infective endocarditis are acute bacterial endocarditis and subacute bacterial endocarditis. Acute bacterial endocarditis is usually due to colonization of heart valves by virulent bacteria such as *Staphylococcus aureus* (the most common cause), *Streptococcus pneumoniae, Neisseria gonorrhoeae, Streptococcus pyogenes,* and *Enterococcus faecalis.* In subacute bacterial endocarditis (SBE), heart valves are infected by less virulent organisms such as alpha-hemolytic streptococci or oral origin (viridans) streptococci, *Staphylococcus epidermidis, Enterococcus* spp., and *Haemophilus* spp. Fungal endocarditis is rare, but cases of *Candida* and *Aspergillus* endocarditis do occur.

Oral streptococci can enter the bloodstream following minor or major dental procedures, oral surgery, and aggressive tooth brushing. Phlebotomy procedures and insertion of IV lines sometimes force organisms from the skin into the bloodstream. Injecting drug users are at high risk of developing infective endocarditis as a result of contaminated needles, syringes, and drug solutions.

Blood cultures are required for diagnosis of infective endocarditis. Treatment will depend upon the specific pathogen involved and the antimicrobial susceptibility results.

LYME DISEASE, LYME BORRELIOSIS

Characteristics. A tickborne disease characterized by three stages: 1) an early, distinctive, red skin lesion (usually at the site of the tick bite), expanding to a diameter of 6 inches (15 cm), often with a central clearing; 2) early systemic manifestations which may include fatigue, chills, fever, headache, stiff neck, muscle pain, joint aches, with or without lymphadenopathy; and 3) neurologic abnormalities (*e.g.,* aseptic meningitis, facial paralysis, myelitis, and encephalitis) and cardiac abnormalities (*e.g.,* arrhythmias, pericarditis) several weeks or months after the initial symptoms appear.

Pathogen. *Borrelia burgdorferi;* a Gram-negative, loosely coiled spirochete.

Reservoirs. Ticks, rodents (especially deer mice), and mammals (especially deer).

Transmission. Tickborne.

Incubation Period. Distinctive skin lesion usually appears 3 to 33 days after tick bite.

Epidemiology. The first U.S. cases occurred in 1975 in Lyme, Connecticut. Since then, Lyme disease has been reported in 45 states (mainly the mid-Atlantic, Northeast, and North Central states) and it occurs in many other areas of the world. A total of 14,646 U.S. cases were reported to the CDC in 1998. Lyme disease is the most common vector-borne disease in the United States.

Prevention and Control. See previous section on prevention and control of Colorado tick fever. Vaccines for dogs and humans are available. Standard precautions for hospitalized patients.

Diagnosis. Observation of the characteristic targetlike skin lesion plus immunodiagnostic procedures. *B. burgdorferi* can be grown in the laboratory on a special medium (Barbour-Stoenner-Kelley [BSK] medium at 33° C).

Treatment. Doxycycline or amoxicillin or cefuroxime axetil.

PLAGUE, BLACK DEATH, BUBONIC PLAGUE, PNEUMONIC PLAGUE, SEPTICEMIC PLAGUE

Characteristics. An acute, often severe zoonosis, involving rodents and their fleas. Initial signs and symptoms may include fever, chills, malaise, myalgia, nausea, prostration, sore throat and headache. Bubonic plague is named for the swollen, inflamed, and tender lymph nodes (buboes) that develop, usually lymph nodes receiving drainage from the site of the flea bite. In about 90% of cases, the inguinal (groin area) lymph nodes are involved. Pneumonic plague, which is highly communicable, involves the lungs; it can result in localized outbreaks or devastating epidemics. Septicemic plague, septic shock, meningitis, and death may occur. During the middle ages, plague was referred to as the "black death" because of the darkened, bruised appearance of the corpses. The blackened skin and foul smell were the result of cell necrosis and hemorrhaging into the skin.

Pathogen. *Yersinia pestis;* a nonmotile, bipolar-staining, Gram-negative coccobacillus; sometimes referred to as the plague bacillus.

Reservoirs. Wild rodents (especially ground squirrels in the United States) and their fleas; rarely, rabbits, wild carnivores and domestic cats.

Transmission. Rodent → flea → human. Also, handling of tissues of infected rodents, rabbits, and other animals, and droplet transmission from person-to-person (in pneumonic plague).

Incubation Period. 1 to 7 days; 2 to 4 days in pneumonic plague.

Epidemiology. Plague probably dates back a thousand or more years B.C. In the last 2000 years, the disease has killed millions of people, perhaps hundreds of millions. Huge plague epidemics occurred in Asia and Europe, including the European plague epidemic of 1348–1350 which killed about 44% of the population (40 million people out of 90 million). The last major plague epidemic in Europe occurred in 1721. Plague still occurs, but the availability of insecticides and antibiotics have greatly reduced the incidence of this dreadful disease. Only 8 U.S. cases were reported to the CDC in 1998.

Prevention and Control. Certain occupations and lifestyles (*e.g.,* hunting, trapping, camping) carry an increased risk of exposure. Avoid contact with rodents and their fleas. Use of insecticides and insect repellents. A vaccine is available for high-risk persons. Standard precautions for hospitalized patients with bubonic plague, but strict isolation and droplet precautions for patients with pneumonic plague until at least 72 hours following initiation of effective therapy.

Diagnosis. Observation of typical appearance (bipolar-staining bacilli that resemble safety pins) in Gram-stained or Wright-Giemsa-stained sputum, CSF, or material aspirated from a bubo. Culture, biochemical tests, immunodiagnostic tests, and antimicrobial susceptibility testing.

Treatment. Streptomycin with or without a tetracycline.

TULAREMIA, RABBIT FEVER

Characteristics. An acute zoonosis with a variety of clinical manifestations, depending upon portal of entry into the body. Most often presents as a skin ulcer and regional lymphadenitis. Ingestion results in pharyngitis, abdominal pain, diarrhea, and vomiting. Inhalation results in pneumonia and septicemia, with a 30 to 60% fatality rate.

Pathogen. Francisella tularensis; a small, pleomorphic, Gram-negative coccobacillus; some strains are more virulent than others.

Reservoir. Wild animals, especially rabbits, muskrats, beavers; some domestic animals; hard ticks.

Transmission. Tick bite; ingestion of contaminated meat or drinking water; entry of organisms into wound while skinning infected animals; inhalation of dust; animal bites.

Incubation Period. Variable; 1 to 14 days, usually 3 to 5 days.

Epidemiology. Ninety-six U.S. cases were reported to the CDC in 1994; since then, tularemia has not been a nationally notifiable disease in the United States.

Prevention and Control. See previous section on prevention and control of Colorado tick fever. Use impervious gloves when handling and skinning animals, especially rabbits; cook wild meat well. Vaccinate high-risk persons. In the laboratory, wear gloves, use a biological safety cabinet, and prevent production of aerosols. Standard precautions for hospitalized patients.

Diagnosis. Culture, biochemical tests, and immunodiagnostic procedures.

Treatment. Streptomycin, gentamicin, or a tetracycline.

Protozoal Cardiovascular Infections

BABESIOSIS

Characteristics. A protozoal disease that may include fever, chills, myalgia, fatigue, jaundice, and anemia; potentially severe, sometimes fatal, especially in splenectomized people; patients may be simultaneously infected with Lyme disease, which is transmitted by the same species of ticks.

Pathogens. *Babesia microti* and other *Babesia* spp., including *B. divergens* in Europe; intraerythrocytic sporozoan parasites.

Reservoirs. Rodents for *B. microti;* cattle for *B. divergens.*

Transmission. Tick bite; rarely by blood transfusion.

Incubation Period. 1 week to 12 months.

Epidemiology. Occurs in Europe, Mexico, and the United States; not a nationally notifiable disease in the United States; most U.S. cases occur in New York and New England.

Prevention and Control. See previous section on prevention and control of Colorado tick fever. Standard precautions for hospitalized patients.

Diagnosis. Observation and identification of the parasites within red blood cells in Giemsa-stained blood smears; differentiation from malarial parasites is necessary; immunodiagnostic procedures are also available.

Treatment. Clindamycin plus quinine.

MALARIA

Characteristics. A systemic sporozoan infection with fever, chills, sweating, headache, nausea; the cycle of chills, fever and sweating is repeated daily, every other day, or every third day, depending upon the particular pathogen; in addition to these symptoms, falciparum malaria may be accompanied by cough, diarrhea, respiratory distress, shock, renal and liver failure, pulmonary and cerebral edema, coma and death.

Pathogens. Four different species of *Plasmodium* cause human malaria: *P. vivax* (the most common species), *P. falciparum* (the most deadly), *P. malariae,* and *P.*

ovale. These sporozoan protozoa have a complex life cycle (see Appendix C), involving a female *Anopheles* mosquito, the liver and erythrocytes of an infected human, and many life cycle stages.

Reservoir. Humans.

Transmission. Injection of sporozoites into the human blood stream while an infected female *Anopheles* mosquito takes a blood meal; also by blood transfusion or contaminated needles and syringes

Incubation Period. The time between injection of sporozoites to the appearance of clinical symptoms depends upon the *Plasmodium* species: 7 to 14 days for *P. falciparum*, 8 to 14 days for *P. vivax* and *P. ovale*, 7 to 30 days for *P. malariae;* can be much longer, however (8 to 10 months or even longer).

Epidemiology. A major health problem in many tropical and subtropical countries, with an estimated 300 to 500 million cases and 1.5 to 2.7 million deaths annually. About 90% of all malaria cases occur in Africa, where approximately 1 million children die from malaria each year. A total of 1361 U.S. cases were reported to the CDC in 1998, mostly imported cases; a few non-imported, mosquito-transmitted cases of malaria occur in the United States each year.

Prevention and Control. Mosquito control measures, such as filling and draining areas of impounded water, larvicides, and insecticides; protective measures (spraying, window screens, bed nets, insect repellents) to minimize mosquito bites; chemoprophylaxis using chloroquine phosphate (in areas where malarial parasites are chloroquine-sensitive) or mefloquine, doxycycline, or primaquine (in areas where malarial parasites are chloroquine-resistant). Standard precautions for hospitalized patients.

Diagnosis. Observation and identification of a *Plasmodium* sp. in Giemsa-stained blood smears; mixed infections (infection with more than one *Plasmodium* sp.) occur in certain geographic areas.

Treatment. Various drugs, depending upon species, susceptibility of the parasite, and route of administration (oral or parenteral).

TOXOPLASMOSIS

Characteristics. A systemic sporozoan disease which, in immunocompetent people, may be asymptomatic or may resemble infectious mononucleosis. Serious disease, even death, may occur in immunocompromised persons, involving the CNS, lungs, muscles, and heart. Cerebral toxoplasmosis is common in AIDS patients. Infection during early pregnancy may lead to fetal infection, causing death of the fetus or serious birth defects.

Pathogen. Toxoplasma gondii; an intracellular sporozoan.

Reservoirs. Definitive hosts include cats and other felines that usually acquire infection by eating infected rodents or birds. Intermediate hosts include rodents, birds, sheep, goats, swine, and cattle.

Transmission. Humans usually become infected by eating infected raw or undercooked meat (usually pork or mutton), containing the cyst form of the pathogen, or ingesting oocysts shed in the feces of infected cats. Oocysts may be present in food or water contaminated by feline feces. Children may inhale or ingest oocysts from sandboxes containing cat feces. Infection can also be acquired transplacentally, by blood transfusion, or by organ transplantation.

Incubation Period. 5 to 20 days, depending upon mode of transmission.

Prevention and Control. Cook meats thoroughly. Wear gloves when cleaning cat litter box or gardening. Dispose of cat feces daily (before the parasite becomes infective) in toilet or by burying. Pregnant women should avoid cat litter boxes, cat feces-contaminated soil, and eating rare or raw meat. Wash hands after handling cats. Standard precautions for hospitalized patients.

Treatment. Pyrimethamine and sulfadiazine with folinic acid.

AFRICAN TRYPANOSOMIASIS, AFRICAN SLEEPING SICKNESS

Characteristics. A systemic disease caused by hemoflagellates (flagellated protozoa in the blood stream). Early stages include a painful chancre at site of a tsetse fly bite, fever, intense headache, insomnia, lymphadenitis, anemia, edema and rash. Later stages include body wasting, falling asleep ("sleeping sickness"), coma, and death if untreated.

Pathogens. Two different subspecies of *Trypanosoma brucei* cause African trypanosomiasis. *T. brucei gambiense* causes most cases of "sleeping sickness"; the disease may last several years. *T. brucei rhodesiense* causes a more rapidly fatal form of African trypanosomiasis; usually lethal within weeks or months without treatment.

Reservoirs. Humans, wild animals, cattle.

Transmission. To humans by bite of infected tsetse fly (*Glossina*).

Incubation Period. 3 days to a few weeks for *T. brucei rhodesiense;* several months or years for *T. brucei gambiense.*

Prevention and Control. Avoid tsetse fly bites; treatment of infected humans.

Diagnosis. Observation of the trypomastigote form of the parasite in blood, lymph or CSF. See Fig. C-2 in App. C. Immunodiagnostic procedures.

Treatment. Suramin, eflornithine, or pentamidine isethionate.

AMERICAN TRYPANOSOMIASIS, CHAGAS' DISEASE

Characteristics. An acute disease in children, with an inflammatory response at the site of the reduviid bug bite, fever, malaise, lymphadenopathy, *hepatomegaly* (enlarged liver), splenomegaly. May be asymptomatic. Chronic irreversible complications include heart damage, arrhythmias, enlarged esophagus (megaesophagus), and enlarged colon (megacolon).

Pathogen. *Trypanosoma cruzi;* occurs as a hemoflagellate (trypomastigote form) and as an intracellular parasite (amastigote form) without a flagellum.

Reservoirs. Humans and over 150 different species of domestic and wild animals, including dogs, cats, rodents, carnivores and primates.

Transmission. The vectors in Chagas' disease are insects known as reduviid bugs (also called triatome bugs, kissing bugs, assassin bugs, cone-nosed bugs, etc.). The bugs become infected when they take blood meals from infected animals. The bugs defecate as they take a blood meal or feed at the corner of a sleeping person's eye. The feces, containing the parasite, are then rubbed into the bite wound or into the eye. Transmission by blood transfusion also occurs.

Incubation Period. 5 to 14 days if transmission is by reduviid bug bite; 30 to 40 days if by blood transfusion.

Epidemiology. Primarily a disease of South America, Central America, and Mexico, where an estimated 16 to 18 million people are infected. Chagas' disease kills approximately 40,000 people each year. A few cases have occurred in the United States (by bug bite or blood transfusion). As increasing numbers of infected people enter the United States from endemic areas, there is a growing concern about the safety of the blood supply. Currently, donor blood is not routinely screened in the United States for the presence of *T. cruzi*. Chagas' disease is not a nationally notifiable disease in the United States.

Prevention and Control. Elimination of the reduviid bug vectors from thatched roofs and animal burrows beneath homes and use of bed nets in endemic areas. Screening of blood and organ donors from endemic areas.

Diagnosis. Observation of trypomastigotes in blood or amastigotes in tissue or lymph node biopsies. Immunodiagnostic procedures. Xenodiagnosis is performed in endemic countries. In this procedure, sterile (non-infected) reduviid bugs are allowed to take blood meals from persons suspected of having Chagas' disease (the bite is painless). The bugs are then taken to a laboratory, where their feces are periodically checked for the presence of the parasite.

Treatment. Nifurtimox or benznidazole.

INFECTIOUS DISEASES OF THE NERVOUS SYSTEM

The nervous system is composed of the central nervous system (CNS) and the peripheral nervous system. The CNS (Fig. 12–25) consists of the brain, the spinal cord, and the three membranes (called the *meninges*) that cover the brain and spinal cord. The CNS is well protected and remarkably resistant to infection; it is encased in bone, bathed and cushioned in cerebrospinal fluid (CSF), and nourished by capillaries. These capillaries make up the blood-brain barrier, supplying nutrients but not allowing larger particles, such as macromolecules (*e.g.*, antibodies and most antibiotics), cells of the immune system, and microorganisms to pass from the blood into the brain. The peripheral nervous system consists of nerves that branch from the brain and spinal cord.

There are no indigenous microflora of the nervous system. Microbes must gain access to the CNS through trauma (fracture or medical procedure), via the blood and lymph to the CSF, or along the peripheral nerves.

An infection of the meninges is called *meningitis; encephalitis* is an infection of

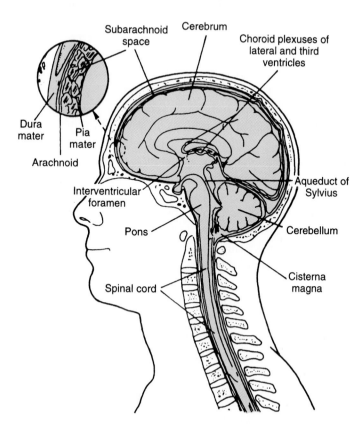

FIGURE 12-25. Central nervous system, illustrating the brain, spinal cord, and meninges.

the brain itself; and *myelitis* is an infection of the spinal cord. Several viruses may produce viral or aseptic meningitis and encephalitis. The most common are enteroviruses, coxsackieviruses, echoviruses, and mumps viruses; others include arboviruses, polioviruses, adenoviruses, measles, herpes, and varicella viruses. Arboviruses (arthropod-borne viruses) are introduced by mosquito vectors and cause several forms of viral encephalitis (*e.g.,* eastern and western equine encephalitis, St. Louis encephalitis, California encephalitis).

Traditionally, the three major causes of bacterial meningitis have been *Haemophilus influenzae* (the primary cause in children), *Neisseria meningitidis* (the primary cause in adolescents), and *Streptococcus pneumoniae* (the primary cause in the elderly) (see Fig. 9–6). However, vaccination of children with the Hib vaccine has drastically reduced the incidence of *H. influenzae* meningitis in children. The major causes of bacterial meningitis in neonates are *Streptococcus agalactiae* (Group B, β-hemolytic streptococci), *E. coli* and other members of the family *Enterobacteriaceae,* and *Listeria monocytogenes.* Less common causes of bacterial meningitis are *Staphylococcus aureus, Pseudomonas aeruginosa, Salmonella,* and *Klebsiella.* Free-living amebae that may cause *meningoencephalitis* (an infection of both the brain and the meninges) are in the genera *Naegleria* and *Acanthamoeba.* Other protozoa that may invade the meninges are *Toxoplasma* and *Trypanosoma.* Occasionally, fungal pathogens, especially *Cryptococcus neoformans* (an encapsulated yeast), cause meningitis.

Early symptoms of bacterial meningitis include fever, headache, stiff neck, sore throat, and vomiting. Then neurological symptoms of dizziness, convulsions, minor paralysis, and coma occur; death may result within a few hours. Meningitis is a medical emergency and steps must be taken immediately to determine the cause. Diagnosis is usually made by a combination of patient symptoms, physical examination, and Gram stain and culture of the cerebrospinal fluid. Treatment of viral meningitis is supportive. The use of specific antibacterial or antifungal agents will depend upon the particular bacterial or fungal pathogen involved and the results of antimicrobial susceptibility testing.

Several CNS diseases are caused by toxins. Examples of bacterial neurotoxins are botulinal toxin (the exotoxin that causes botulism), and tetanospasmin (the cause of tetanus). Diseases caused by fungal toxins (mycotoxins) include ergot from grain molds and mushroom poisoning. *Gonyaulax,* an alga found in algal "blooms," produces neurotoxins, which may concentrate in bivalve shell fish and cause paralytic symptoms following ingestion of the contaminated shell fish. A variety of other algae also produce neurotoxins (refer back to Chapter 3).

Viral Nervous System Infections

POLIOMYELITIS, INFANTILE PARALYSIS
Characteristics. In most patients, a minor illness with fever, malaise, headache, nausea and vomiting. In about 1% of patients, the disease progresses to severe mus-

cle pain, stiffness of the neck and back, with or without flaccid paralysis. Major illness is more likely to occur in older children and adults.

Pathogens. Poliovirus types 1, 2, and 3; small RNA enteroviruses (Fig. 12–26).

Reservoir. Humans.

Transmission. Person-to-person, primarily via the fecal-oral route; also by throat secretions.

Incubation Period. 3 to 35 days, usually 7 to 14 days for paralytic cases; the maximum extent of paralysis is usually reached within 3 to 4 days after onset of symptoms.

Epidemiology. Although once a major health problem in the United States, vaccines became available in the 1950s. Only 1 U.S. case was reported to the CDC in 1998. The World Health Organization (WHO) is attempting to eradicate polio worldwide by the year 2000.

Prevention and Control. Immunize all infants and children with a series of Salk injectable, inactivated polio vaccine (IPV) or Sabin oral, attenuated poliovirus vaccine (OPV); standard precautions for hospitalized patients; adequate sewage and water treatment precautions.

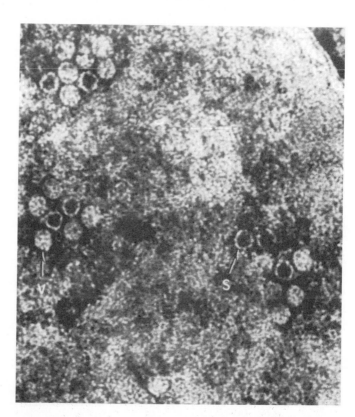

FIGURE 12-26. Development of poliovirus particles in pieces of cytoplasmic matrix of artificially disrupted cells. Particles in various stages of assembly from empty shells (s) to complete virions (v) can be seen (original magnification, ×200,000). (Horne RW, Nagington J: J Mol Biol 1:333, 1959. Copyright by Academic Press, Inc., Ltd.)

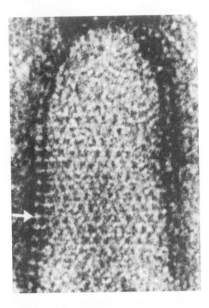

FIGURE 12-27. Intact rabies virus particle negatively stained with phosphotungstate. On the left (arrow) are well-resolved surface projections, 6–7 nm long (original magnification, ×400,000). (Hummeler K, et al: J Virol 1:152, 1967)

Diagnosis. Isolation of polio virus from stool samples, CSF, or oropharyngeal secretions using cell culture techniques; immunodiagnostic procedures.

Treatment. Respiratory assistance and physical therapy as needed.

RABIES

Characteristics. A usually fatal, acute viral *encephalomyelitis* (inflammation of the brain and spinal cord) of mammals, with mental depression, restlessness, headache, fever, malaise, paralysis (which usually starts in the lower legs and moves upward through the body), salivation, spasms of throat muscles induced by a slight breeze or drinking water, convulsions, and death caused by respiratory failure.

Pathogen. Rabies virus (Fig. 12–27), a rhabdovirus; a large, complex, enveloped RNA virus.

Reservoirs. Many wild and domestic mammals, including dogs, foxes, coyotes, wolves, jackals, skunks, raccoons, mongooses, and bats.

Transmission. Usually, the bite of a rabid animal introduces virus-laden saliva; airborne transmission from bats in caves; person-to-person by saliva is theoretically possible, but has never been documented.

Incubation Period. Usually 3 to 8 weeks, but can be shorter or longer.

Epidemiology. Rabies is endemic in every country of the world except Antarctica and in every state except Hawaii. An estimated 30,000 to 50,000 people die of rabies annually throughout the world. A total of 7084 animal cases were reported to the CDC in 1998, but no human cases were reported that year.

Prevention and Control. Vaccinate all pets; avoid all sick, aggressive, or friendly wild animals; prophylactically vaccinate high-risk persons (*e.g.,* veterinarians, veterinary technicians, spelunkers); standard precautions for hospitalized patients; immunization of contacts having open wounds or mucous membrane exposure to the patient's saliva.

Diagnosis. Virus isolation using cell culture techniques or immunodiagnostic procedures; observation of Negri bodies in animal brain tissue.

Treatment. Prompt and proper treatment of bite wounds; administration of vaccine or human rabies immune globulin (HRIG) prior to development of symptoms.

Bacterial Nervous System Infections

BOTULISM (DESCRIBED PREVIOUSLY UNDER FOOD POISONING)

TETANUS, LOCKJAW

Characteristics. An acute neuromuscular disease induced by a bacterial exotoxin (tetanospasmin), with painful muscular contractions, primarily of the masseter (the muscle that closes the jaw) and neck muscles; spasms, rigid paralysis, respiratory failure, and death may result.

Pathogen. Clostridium tetani (Fig. 12–28); a motile, Gram-positive, anaerobic, spore-forming bacillus that produces a potent neurotoxin called tetanospasmin.

Reservoir. Soil contaminated with human, horse, or other animal feces; *C. tetani* is a member of the indigenous intestinal flora of humans and animals.

Transmission. Spores of *C. tetani* are introduced into a puncture wound, burn, or needle stick by contamination with soil, dust, or feces. Under anaerobic conditions in the wound, spores germinate into vegetative *C. tetani* cells which produce the exotoxin in vivo.

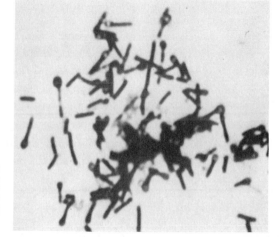

FIGURE 12-28. Cells of *Clostridium tetani* after 24 hours on a cooked meat—glucose medium. Note spherical, terminal endospores (original magnification, ×4,500).

Incubation Period. Usually 3 to 21 (average 10) days, although it can be shorter or longer.

Epidemiology. Thirty-five U.S. cases were reported to the CDC in 1998.

Prevention and Control. Active immunization of all children with DTP, a combined diphtheria-tetanus-pertussis vaccine, containing tetanus toxoid. Boosters of DT should be given at 10-year intervals and after a wound or accident with no booster for 5 years. Non-immunized persons may be given human tetanus immune globulin (TIG) following a puncture wound.

Diagnosis. Attempts to isolate *C. tetani* from wounds or demonstrate antibody production are rarely successful.

Treatment. Dirty wounds should be cleaned and left open, when feasible, to inhibit the anaerobic growth; administer penicillin to destroy wound organisms, antitoxin (TIG) to neutralize the exotoxin, and vaccinate with tetanus toxoid.

Protozoal Nervous System Infections

AFRICAN TRYPANOSOMIASIS (PREVIOUSLY DESCRIBED UNDER PROTOZOAL CARDIOVASCULAR INFECTIONS)

PRIMARY AMEBIC MENINGOENCEPHALITIS, NAEGLERIASIS

Characteristics. A usually fatal infection of the brain and meninges with sore throat, severe frontal headache, hallucinations, nausea, vomiting, high fever, and stiff neck; death usually occurs within 10 days.

Pathogen. *Naegleria fowleri;* a free-living ameba found in soil and fresh water, especially stagnant water.

Reservoir. Aquatic and soil habitats; stagnant, contaminated pond or lake water; "the old swimming hole."

Transmission. Infection is usually acquired by exposure of the nasal passages to contaminated fresh water, most often by diving, water skiing, or swimming; the amebae colonize the nasal tissues and then invade the brain and meninges via the olfactory nerves.

Incubation Period. 3 to 7 days.

Prevention and Control. Educate the public to the dangers of swimming, diving, water skiing in ameba-contaminated water.

Diagnosis. Amebae may be observed in wet mounts of unstained CSF specimens, but they are difficult to see unless the light intensity is reduced. The CSF will contain leukocytes and, unless the amebae are moving, they will be difficult to distinguish from leukocytes. The amebae do not stain with the Gram-staining procedure. Although the amebae can be cultured, the culture procedure is usually not performed in most clinical microbiology laboratories. Sadly, most cases of P.A.M. have been diagnosed by post-mortem examination of brain tissue obtained at autopsy.

Treatment. Amphotericin B.

Problems and Questions

1. Name six factors that affect the number of microbes on the skin. How may opportunists invade through the skin?
2. Which bacteria cause most burn, wound, and bite infections? What types of toxins do they produce?
3. How do oral bacteria interact to cause dental caries? How can you prevent dental caries?
4. By what pathways may microbes invade to cause ear infections? Which bacteria are usually found in ear infections?
5. Why is pneumonia called a nonspecific disease? Name several pathogens that might cause pneumonia.
6. Differentiate between foodborne infections and foodborne intoxications. Cite some examples of each.
7. Which protozoa cause gastrointestinal infections? How do they get there?
8. How are urinary tract infections transmitted? Why might *E. coli* be found more often in female cystitis than in male cystitis?
9. How may sexually transmitted infections be prevented? List genital infections that may cause congenital and neonatal infections. How could transmission to fetus and newborn be prevented?
10. Differentiate between typhus fever and typhoid fever. How are they transmitted and controlled?
11. By what means might children be exposed to plague, tularemia, toxoplasmosis, and rabies? After exposure, how should the children be treated to prevent or reduce the symptoms of these diseases?
12. List several anatomical sites, systems, or organs where there are no indigenous microflora. Why might this be so?
13. What are the usual symptoms of infectious diseases of the nervous system? Define meningitis, encephalitis, myelitis, meningoencephalitis, and encephalomyelitis.
14. List several diseases caused by neurotoxins. Cite several nonspecific diseases discussed in this chapter.

Self Test

After you have read Chapter 12, reviewed the study outline, examined the objectives, studied the new terms, and answered the problems and questions above, complete the following self test.

MATCHING EXERCISES

Match each of the diseases under Column I with the appropriate item in Column II. An answer from Column II may be used once or more than once.

Diseases: Types of Etiologic Agents

Column I
- __ 1. Gonorrhea
- __ 2. Trachoma
- __ 3. Tuberculosis
- __ 4. Histoplasmosis
- __ 5. Amebiasis
- __ 6. Candidiasis
- __ 7. Syphilis
- __ 8. AIDS
- __ 9. Common cold
- __ 10. Strep throat
- __ 11. Legionellosis
- __ 12. Rocky Mountain spotted fever
- __ 13. Warts
- __ 14. Athlete's foot
- __ 15. Botulism
- __ 16. Whooping cough
- __ 17. Measles
- __ 18. Mumps
- __ 19. Trichomoniasis
- __ 20. Hepatitis

Column II
- __ a. Virus
- __ b. Bacterium
- __ c. Fungus
- __ d. Protozoan

Diseases: Etiologic Agents

Column I
- __ 1. Primary atypical pneumonia
- __ 2. Pneumococcal pneumonia
- __ 3. Legionellosis
- __ 4. Histoplasmosis
- __ 5. Whooping cough
- __ 6. Diphtheria
- __ 7. Strep throat
- __ 8. Acute rhinitis
- __ 9. Otitis media
- __ 10. Thrush
- __ 11. Boils, carbuncles
- __ 12. Colds

Column II
- __ a. *Bordetella pertussis*
- __ b. *Streptococcus pneumoniae*
- __ c. Influenza virus
- __ d. *Mycoplasma pneumoniae*
- __ e. *Legionella pneumophila*
- __ f. *Histoplasma capsulatum*
- __ g. *Coccidioides immitis*
- __ h. *Candida albicans*
- __ i. *Staphylococcus aureus*
- __ j. Rhinovirus
- __ k. *Corynebacterium diphtheriae*
- __ l. *Streptococcus pyogenes*
- __ m. *Haemophilus influenzae*
- __ n. *Staphylococcus epidermidis*
- __ o. *Streptococcus agalactiae*
- __ p. Mumps virus

TRUE AND FALSE (T OR F)

___ 1. Chickenpox can be diagnosed by finding Guarnieri bodies as cytoplasmic inclusions in infected cells.

___ 2. Reye's syndrome is severe encephalomyelitis and severe liver involvement in adults following a viral infection treated with aspirin.

___ 3. Measles is a mild childhood disease easily controlled by vaccination.

___ 4. Koplik's spots are small bluish-yellow spots that occur in the mouth on the buccal mucosa of rubella patients.

___ 5. Infection of a fetus with rubella during the first trimester can cause severe birth defects.

___ 6. Patients most susceptible to *P. aeruginosa* infections are hospitalized burn patients.

___ 7. Trachoma is the leading cause of blindness worldwide, especially prevalent in areas where poor hygienic practices exist.

___ 8. Lymphogranuloma venereum (LGV) is a syndrome with initial symptoms of conjunctivitis.

___ 9. The causative agent of the most common form of bacterial pneumonia is *Streptococcus pneumoniae.*

___ 10. The majority of respiratory tract infections are not bacterial but viral.

___ 11. A vaccine against colds is possible because of the immunological similarity of the group of viruses that causes colds.

___ 12. *Corynebacterium diphtheriae* excretes a powerful endotoxin that results in the formation of a pseudomembrane that can block respiratory passages.

___ 13. Control of diphtheria is based entirely on the mass immunization of children.

___ 14. Complications following some streptococcal infections include rheumatic fever and acute glomerulonephritis.

___ 15. A positive tuberculin skin test is interpreted as indicating an active infection.

___ 16. Typhoid fever is the most serious type of salmonellosis.

___ 17. Urinary tract infections are almost always caused by strict pathogens.

___ 18. Catheterization is not performed routinely for the collection of urine samples for microbiological examinations.

___ 19. *Trichomonas vaginalis* appears to be spread as a sexually transmitted disease.

___ 20. AIDS is caused by a virus that specifically destroys certain lymphocytes.

___ 21. Mumps virus infects the sublingual salivary glands.

___ 22. Rocky Mountain spotted fever (RMSF) is caused by a rickettsial organism, transmitted by body lice.

___ 23. Pneumonic plague is almost always fatal and is very contagious.

___ 24. An infection or inflammation of the membranes enclosing the brain and spinal cord is called meningitis.

___ 25. Antibiotics are useful in the treatment of rabies.

MULTIPLE CHOICE

1. Reye's syndrome is sometimes associated with
   a. influenza complications
   b. chickenpox
   c. the use of aspirin to treat fever in viral infections of children
   d. all of the above

2. The eye is generally protected from damage by
   a. sebum lubricating the eyeball
   b. tears washing the eye
   c. mucous secretions that moisten the eye
   d. all of the above

3. *Chlamydia trachomatis* causes
   a. skin infections
   b. secondary infections of burns
   c. cold sores
   d. eye disease

4. *Bacteroides* spp. are of concern in oral disease because they
   a. cause gum infections and loosen teeth
   b. are anaerobic
   c. survive in the gingival sulcus
   d. produce endotoxins
   e. all of the above

5. Saliva functions in oral health by
   a. rinsing teeth and buffering acids
   b. providing IgA to destroy oral flora
   c. providing proper anaerobic conditions
   d. all of the above

6. Otitis media is frequently caused by
   a. bacteria of the indigenous microflora of the throat
   b. *Streptococcus pneumoniae*
   c. *Haemophilus influenzae*
   d. all of the above

7. The most common infectious disease of the upper respiratory tract is
   a. pneumonia
   b. strep throat
   c. otitis media
   d. common cold
   e. influenza

8. Infectious hepatitis is caused by
   a. HAV
   b. HBV
   c. HCV
   d. HDV
   e. HEV

9. An infection of the bladder is known as
   a. urethritis
   b. ureteritis
   c. cystitis
   d. nephritis

10. The most common cause of pharyngitis is
    a. viruses
    b. *Streptococcus pyogenes*
    c. anaerobic bacteria
    d. *Streptococcus pneumoniae*

11. Enterotoxins are exotoxins that affect the
    a. respiratory tract
    b. central nervous system
    c. cardiovascular system
    d. gastrointestinal tract
    e. genitourinary tract

12. Which of the following statements about rabies is true?
    a. Once a person is bitten by a rabid animal, it is impossible to prevent rabies in that person.
    b. Human rabies is very common in the United States.
    c. A vaccine is available to protect high-risk individuals from developing rabies.
    d. Animal rabies is uncommon in the United States.

13. Which of the following organisms is the most common cause of urethritis?
    a. *Escherichia coli*
    b. *Neisseria gonorrhoeae*
    c. *Chlamydia trachomatis*
    d. *Candida albicans*

14. Which of the following organisms is the most common cause of cystitis?
    a. *Escherichia coli*
    b. *Neisseria gonorrhoeae*
    c. *Chlamydia trachomatis*
    d. *Candida albicans*

15. The most common sexually transmitted disease in the United States is caused by
    a. *Escherichia coli*
    b. *Neisseria gonorrhoeae*
    c. *Chlamydia trachomatis*
    d. *Candida albicans*
    e. *Trichomonas vaginalis*

Appendices

APPENDIX A

Taxonomic Categories of Selected Medically Important Bacteria

Taxonomic Category	Selected Medically Important Genera
Spirochetes	*Borrelia, Leptospira, Treponema*
Aerobic/microaerophilic, motile, helical/vibrioid Gram-negative bacteria	*Campylobacter, Helicobacter, Spirillum*
Gram-negative aerobic/microaerophilic bacilli and cocci	*Alcaligenes, Bordetella, Brucella, Francisella, Legionella, Moraxella, Neisseria, Pseudomonas*
Facultatively anaerobic Gram-negative bacilli	*Citrobacter, Edwardsiella, Enterobacter, Escherichia, Gardnerella, Haemophilus, Klebsiella, Morganella, Pasteurella, Proteus, Providencia, Salmonella, Serratia, Shigella, Vibrio, Yersinia*
Anaerobic Gram-negative bacilli	*Bacteroides, Fusobacterium, Porphyromonas, Prevotella*
Anaerobic Gram-negative cocci	*Veillonella*
Rickettsias and chlamydias	*Rickettsia, Chlamydia*
Gram-positive cocci	*Peptostreptococcus, Staphylococcus, Streptococcus*
Endospore-forming Gram-positive bacilli and cocci	*Bacillus, Clostridium*
Regular, nonsporeforming Gram-positive bacilli	*Lactobacillus, Listeria*
Irregular, nonsporeforming Gram-positive bacilli	*Actinomyces, Arachnia, Bifidobacterium, Corynebacterium, Eubacterium, Mobiluncus, Propionibacterium*
Mycobacteria	*Mycobacterium*
Nocardioform actinomycetes	*Nocardia*
Mycoplasmas (or Mollicutes): cell wall-less bacteria	*Mycoplasma, Ureaplasma*

Source: Bergey's Manual of Determinative Bacteriology, 9th Ed. Holt JG et al. Williams & Wilkins, Baltimore, 1994.

Appendices

APPENDIX B

Compendium of Important Bacterial Pathogens of Humans

Bacillus anthracis (Buh-sil'-us an'-thray-sis). An aerobic, sporeforming, Gram-positive bacillus; the etiologic agent of anthrax in humans, cattle, swine, sheep, rabbits, guinea pigs, and mice; a cutaneous, respiratory, or gastrointestinal disease, depending upon the portal of entry

Bacteroides (Bak-ter-oy'-dez) species. Anaerobic, Gram-negative bacilli; common members of the indigenous microflora of the oral cavity, gastrointestinal tract, and vagina; opportunistic pathogens that cause a variety of infections, including appendicitis, peritonitis, abscesses, and post-surgical wound infections

Bordetella pertussis (Bor-duh-tel'-uh per-tus'-sis). A fastidious, Gram-negative coccobacillus; the etiologic agent of whooping cough, which is also called "pertussis"

Borrelia burgdorferi (Boh-ree'-lee-uh burg-door'-fur-eye). A Gram-negative, loosely-coiled spirochete; the etiologic agent of Lyme disease; transmitted from infected deer and mice to humans by tick bite

Campylobacter jejuni (Kam'-pih-low-bak'-ter juh-ju'-nee). A curved, Gram-negative bacillus, having a characteristic corkscrew-like motility; a common cause of gastroenteritis with malaise, myalgia, arthralgia, headache, and cramping abdominal pain

Chlamydia (Kluh-mid'-ee-uh) species. Gram-negative bacilli that are obligate intracellular pathogens; unable to grow on artificial media; etiologic agents of non-gonococcal urethritis (NGU), trachoma, inclusion conjunctivitis,

lymphogranuloma venereum, pneumonia, and psittacosis (ornithosis)

Clostridium botulinum (Klos-trid'-ee-um bot-yu-ly'-num). An anaerobic, spore-forming, Gram-positive bacillus; common in soil; produces a neurotoxin called botulin, which causes botulism, a very serious and sometimes fatal type of food poisoning

Clostridium difficile (Klos-trid'-ee-um dif'-fuh-seal). An anaerobic, sporeforming, Gram-positive bacillus can colonize the intestinal tract, where overgrowth (superinfection) commonly occurs following ingestion of oral antibiotics; this organism produces two toxins—an enterotoxin that causes antibiotic-associated diarrhea (AAD), and a cytotoxin that causes pseudomembranous colitis (PMC); a common cause of nosocomial infections

Clostridium perfringens (Klos-trid'-ee-um purr-frin-'jens). An anaerobic, spore-forming, Gram-positive bacillus; common in feces and soil; the most common cause of gas gangrene (myonecrosis); produces an enterotoxin that produces relatively mild food poisoning

Clostridium tetani (Klos-trid'-ee-um tet'-an-eye). An anaerobic, spore-forming, Gram-positive bacillus; common in soil; produces a neurotoxin called tetanospasmin, which causes tetanus

Corynebacterium diphtheriae (Kuh'-ry-nee-bak-teer'-ee-um dif-thee'-ree-ee). A pleomorphic, Gram-positive bacillus; virulent strains are lysogenic; they cause diphtheria and produce a powerful exotoxin that causes tissue degeneration

Enterococcus (En-ter-oh-kok'-us) species. Gram-positive cocci; common members of the indigenous microflora of the gastrointestinal tract; opportunistic pathogens; a fairly common cause of cystitis and nosocomial infections; some strains, called vancomycin-resistant enterococci (VRE), are multiply drug-resistant

Escherichia coli (Esh-er-ick'-ee-uh koh'-ly). A member of the family *Enterobacteriaceae;* a Gram-negative bacillus; a facultative anaerobe; a very common member of the indigenous microflora of the colon; an opportunistic pathogen; the most common cause of septicemia and urinary tract and nosocomial infections; some serotypes (the enterovirulent *E. coli*) are always pathogens

Francisella tularensis (Fran'-suh-sel-luh tool-uh-ren'-sis). A Gram-negative bacillus; the etiologic agent of tularemia; may enter the body by inhalation, ingestion, tick bite, or penetration of broken or unbroken skin; frequently follows contact with infected animals (*e.g.,* rabbits)

Fusobacterium (Few'-zoh-bak-teer'-ee-um) species. Anaerobic, Gram-negative bacilli; common members of the indigenous microflora of the oral cavity, gastrointestinal tract, and vagina; opportunistic pathogens that cause a variety of infections, including oral and respiratory infections

Haemophilus influenzae (He-mof'-uh-lus in-flu-en'-zee). A fastidious, Gram-negative bacillus; a facultative anaerobe; encapsulated; found in low numbers as indigenous microflora of the upper respiratory tract; an opportunistic pathogen; a cause of bacterial meningitis; causes about one-third of ear infections; causes respiratory infections, but is *not* the cause of influenza (caused by influenza viruses); some strains are ampicillin-resistant

Helicobacter pylori (Hee'-luh-ko-bak-ter py-lor'-ee). A curved, Gram-negative bacillus; capable of colonizing the stomach; a common cause of ulcers

Klebsiella pneumoniae (Kleb-see-el'-uh new-moh'-nee-ee). A member of the family *Enterobacteriaceae;* a Gram-negative bacillus; a facultative anaerobe; a common member of the indigenous microflora of the colon; an opportunistic pathogen; a fairly common cause of pneumonia and cystitis

Lactobacillus (Lak-toh-buh-sil'-us) species. Gram-positive bacilli; some species are found in foods (*e.g.,* yogurt, cheese); other species are common members of the indigenous microflora of

the vagina and gastrointestinal tract; rarely pathogenic

Legionella pneumophila (Lee-juh-nel'-luh new-mah'-fill-uh). An aerobic, Gram-negative bacillus; common in soil and water; the etiologic agent of a pneumonia (legionellosis); can contaminate water tanks and pipes; has caused epidemics in hotels, hospitals, and cruise ships

Listeria monocytogenes (Lis-teer'-ee-uh mon-oh-sigh-toj'-uh-nees). A Gram-positive bacillus; the etiologic agent of listeriosis; can cause meningitis, encephalitis, septicemia, endocarditis, abortion, and abscesses; enters the body via ingestion of contaminated foods (*e.g.,* cheese)

Mycobacterium leprae (My'-koh-bak-teer'-ee-um lep'-ree). An aerobic, acid-fast, Gram-positive bacillus; referred to as the leprosy bacillus or Hansen's bacillus; the etiologic agent of leprosy (Hansen's disease); transmitted from person to person; has been found in wild armadillos, which are now used as laboratory animals to propagate this organism

Mycobacterium tuberculosis (My'-koh-bak-teer'-ee-um tu-ber'-kyu-loh'-sis). An acid-fast, Gram-positive bacillus; causes tuberculosis; many strains are multiply drug resistant

Mycoplasma pneumoniae (My'-koh-plaz-muh new-moh'-nee-ee). A small, pleomorphic, Gram-negative bacillus; lacks a cell wall; the etiologic agent of atypical pneumonia

Neisseria gonorrhoeae (Ny-see'-ree-uh gon-or-ree'-ee). Also known as gonococci; a fastidious, Gram-negative diplococcus; microaerophilic and capnophilic; always a pathogen; causes gonorrhea; many strains are penicillin-resistant

Neisseria meningitidis (Ny-see'-ree-uh men-in-jih'-tid-is). Also known as meningococci; an aerobic, Gram-negative diplococcus; found as indigenous microflora of the upper respiratory tract of some people ("carriers"); a common cause of bacterial meningitis; also causes respiratory infections

Nocardia (No-kar'-dee-uh) species. Aerobic, acid-fast, Gram-positive bacilli; the etiologic agent of nocardiosis (a respiratory disease) and mycetoma (a tumor-like disease, most often involving the feet)

Peptostreptococcus (Pep'-toh-strep-toh-kok'-us) species. Anaerobic Gram-positive cocci; common members of the indigenous microflora of the gastrointestinal tract, vagina, and oral cavity;

opportunistic pathogens that cause a variety of infections, including abscesses, oral infections, and appendicitis

Porphyromonas (Porf'-uh-row-mow'-nus) species. Anaerobic, Gram-negative bacilli; common members of the indigenous microflora of the oral cavity and gastrointestinal tract; opportunistic pathogens that cause a variety of infections, including abscesses, oral infections, and bite wound infections

Prevotella (Pree'-voh-tel'-luh) species. Anaerobic Gram-negative bacilli; common members of the indigenous microflora of the vagina and gastrointestinal tract; opportunistic pathogens that cause a variety of infections, including abscesses

Proteus (Pro'-tee-us) species. Members of the family *Enterobacteriaceae;* Gram-negative bacilli; facultative anaerobes; common members of the indigenous microflora of the colon; opportunistic pathogens; a fairly common cause of cystitis

Pseudomonas aeruginosa (Su-doh-moh'-nas air-uj-in-oh'-suh). An aerobic, Gram-negative bacillus; produces a characteristic blue-green pigment (pyocyanin); has a characteristic fruity odor; causes burn wound, ear, urinary tract, and respiratory infections; one of the major causes of nosocomial infections; most strains are multiply drug resistant and resistant to some disinfectants

Rickettsia (Rih-ket'-see-uh) species. Gram-negative bacilli that are obligate intracellular pathogens; unable to grow on artificial media; the etiologic agents of typhus and typhus-like diseases (*e.g.,* Rocky Mountain spotted fever); rickettsial diseases are transmitted by arthropods (ticks, fleas, mites, lice)

Salmonella (Sal'-moh-nel'-uh) species. Members of the family *Enterobacteriaceae;* Gram-negative bacilli; facultative anaerobes; a fairly common cause of food poisoning, especially cases caused by contaminated poultry; *Salmonella typhi* is the etiologic agent of typhoid fever

Shigella (She-gel'-uh) species. Members of the family *Enterobacteriaceae;* Gram-negative bacilli; facultative anaerobes; a major cause of gastroenteritis and childhood mortality in the developing nations of the world

Staphylococcus aureus (Staf'-ih-low-kok'-us aw'-ree-us). Frequently referred to as "staph," as in staph infection; a Gram-positive coccus in clusters; a facultative anaerobe; found in low numbers on skin; the nasal passages of some people (carriers) are colonized with *S. aureus;* an oppor-

tunistic pathogen; a major cause of skin, soft tissue, respiratory, bone, joint, and endovascular infections; a very common cause of nosocomial infections; produces many toxins, including cytotoxins, exfoliative toxins, and leukocidin; some strains produce a toxin that causes toxic shock syndrome; some strains (those that produce enterotoxin) cause food poisoning; produces various enzymes, including protease, lipase, and hyaluronidase that destroy tissue, coagulase that causes clot formation, and staphylokinase that dissolves clots; some strains (the MRSA) are multiply drug resistant

Streptococcus agalactiae (Strep-toh-kok'-us ay-guh-lak'-tee-ee). Also known as group B streptococcus; a beta-hemolytic, Gram-positive coccus; often colonizes the vagina; a frequent cause of neonatal meningitis

Streptococcus pneumoniae (Strep-toh-kok'-us new-moh'-nee-ee). Also known as pneumococci; an alpha-hemolytic, Gram-positive diplococcus; a facultative anaerobe; found in low numbers as indigenous microflora of the upper respiratory tract; an opportunistic pathogen; the most common cause of bacterial pneumonia; a common cause of bacterial meningitis; causes about one-third of ear infections; some strains are penicillin-resistant

Streptococcus pyogenes (Strep-toh-kok'-us py-oj'-uh-nees). Also known as group A streptococcus; a beta-hemolytic, Gram-positive coccus in chains; a facultative anaerobe; infrequently found in low numbers as indigenous microflora of the upper respiratory tract; an opportunistic pathogen; the cause of "strep throat"; causes skin and wound infections; some strains (those that produce erythrogenic toxin) are capable of causing scarlet fever; some strains are referred to as "flesh-eating bacteria"; some strains produce a toxin that causes toxic shock syndrome

Treponema pallidum (Trep-oh-nee'-muh pal'-luh-dum). A very thin, tightly coiled spirochete; the etiologic agent of syphilis

Vibrio cholerae (Vib'-ree-oh khol'-er-ee). An aerobic, curved (comma-shaped), Gram-negative bacillus; halophilic; lives in salt water; the etiologic agent of cholera

Yersinia pestis (Yer-sin'-ee-uh pes'-tis). A Gram-negative bacillus; the etiologic agent of plague in humans, rodents, and other mammals; transmitted from rat to rat and rat to human by the rat flea

Appendices

APPENDIX C
Parasitology

INTRODUCTION

Parasitism is a symbiotic relationship that is of benefit to one party or symbiont (the parasite) and usually detrimental to the other party (the host). This does not mean that the parasite necessarily causes disease in the host, although disease does occur in certain parasitic relationships. In virtually all parasitic relationships, the parasite deprives the host of nutrients.

Parasites are defined as organisms that live *on* or *in* other living organisms (hosts) at whose expense they gain some advantage. There are many types of plant parasites and many types of animal parasites; this discussion will be limited to animal parasites.

Parasites that live on the outside of the host's body are referred to as *ectoparasites,* whereas those that live inside are called *endoparasites.* Arthropods such as mites, ticks, and lice are examples of ectoparasites. Parasitic protozoa and helminths are examples of endoparasites.

The life cycle of a particular parasite may involve one or more hosts. If more than one host is involved, the *definitive host* is defined as the host that harbors the adult or sexual stage of the parasite or the sexual phase of the life cycle. The *intermediate host* is the host that harbors the larval or asexual stage of the parasite or the asexual phase of the life cycle. Parasite life cycles range from simple to complex. There are one-host parasites, two-host parasites, and three-host parasites. Knowing the life cycle of a particular parasite enables public health workers and clinicians to control and diagnose the infection. The rather complex life cycle of malarial parasites is depicted in Figure C-1.

A *facultative parasite* is an organism that can be parasitic, but does not have to live as a parasite. It is capable of living an independent life (apart from a host). The free-living amebae that can cause keratitis and meningoencephalitis are examples of facultative parasites. An *obligate parasite,* on the other hand, has no choice. To survive, it must be a parasite. Most parasites that infect humans are obligate parasites.

Parasitology is the study of parasites and a *parasitologist* is someone who studies parasites. If you were to take an upper division or graduate-level parasitology

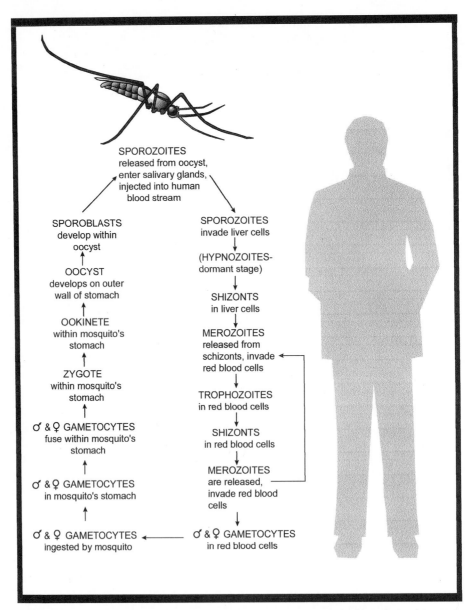

FIGURE C-1. Life cycle of malarial parasites. Malarial parasites have a complex life cycle, involving many different life cycle stages. Humans become infected when an infected, female *Anopheles* mosquito injects sporozoites while taking a blood meal. The sporozoites enter the human blood stream, are transported to the liver, and invade liver cells (hepatocytes), where schizonts (collections of merozoites) develop. Merozoites are released when the schizont ruptures. Each merozoite can invade another liver cell, leading to schizont development, and the release of more merozoites. With *P. vivax* and *P. ovale,* dormant forms (hypnozoites) may remain in hepatocytes, causing relapses months to years later. Eventually, merozoites enter the peripheral blood stream where they invade erythrocytes. Within an erythrocyte, the merozoite transforms into a trophozoite. The trophozoite may mature into any of three life cycle stages: a schizont, a male gametocyte, or a female gametocyte.

course, it would be divided into three areas of study: the study of parasitic protozoa, the study of helminths, and the study of arthropods.

PARASITIC PROTOZOA

Not all protozoa are parasitic. For example, many of the pond water protozoa studied in introductory biology and microbiology courses are not parasites. Some protozoa are facultative parasites, capable of a free-living existence, only becoming parasites when they accidentally gain entrance to the body. *Acanthamoeba* spp. and *Naegleria fowleri* are examples of facultative parasites. These free-living amebae normally reside in soil or water, but can cause serious diseases when they gain entrance to the eyes or central nervous system. Many protozoan parasites of humans are obligate parasites.

Protozoa are in the Kingdom Protista. Most are unicellular, but some (like *Volvox* species; a pond water protozoan) are multicellular (colonial) protozoa. Protozoa are classified taxonomically by their mode of locomotion. Sarcodina (amebae) move by means of pseudopodia ("false feet"). Mastigophora (flagellates) move by means of flagella (see Figure C-2). Ciliata or Ciliophora (ciliates) move by means of cilia. Sporozoa have no pseudopodia, flagella, or cilia, and, therefore, do not move.

Because protozoa are tiny, protozoan infections are most often diagnosed by microscopic examination of body fluids, tissue specimens, or feces. Peripheral blood smears are usually stained with Giemsa stain, whereas fecal specimens are stained with trichrome, iron-hematoxylin, or acid-fast stains. Most parasitic protozoal infections are diagnosed by observing either *trophozoites* or *cysts* in the specimen. The trophozoite is the motile, feeding, dividing stage in a protozoan's life cycle, whereas the cyst is the dormant stage (in some ways, cysts are much like bacterial spores). Table C-1 describes how some important protozoal infections of humans are acquired and diagnosed.

FIGURE C-I. (continued) The diagnosis of malaria is made by microscopic examination of a Giemsa-stained blood smear and observation of parasite-infected red blood cells. When the schizonts rupture, merozoites are released and they invade other erythrocytes. In order for the parasite life cycle to continue, at least one male and one female gametocyte must be ingested by a female *Anopheles* mosquito while taking a blood meal from the infected person. Within the mosquito's stomach, the female gametocyte matures into a female gamete, and the male gametocyte produces several male gametes. A male gamete fuses with a female gamete, producing a zygote. Because the sexual phase of the life cycle occurs in the mosquito, the mosquito is considered to be the definitive host. The zygote matures into a motile form called an ookinete. The ookinete escapes from the stomach by squeezing between cells in the stomach wall, and encysts on the outer wall of the mosquito's stomach, becoming an oocyst. Within the oocyst, sporoblasts develop and mature into sporozoites. When the oocyst bursts open, the sporozoites are released, some of which enter the mosquito's salivary glands. The portion of the life cycle that occurs within the mosquito takes 8 to 35 days, depending upon the particular *Plasmodium* species and temperature.

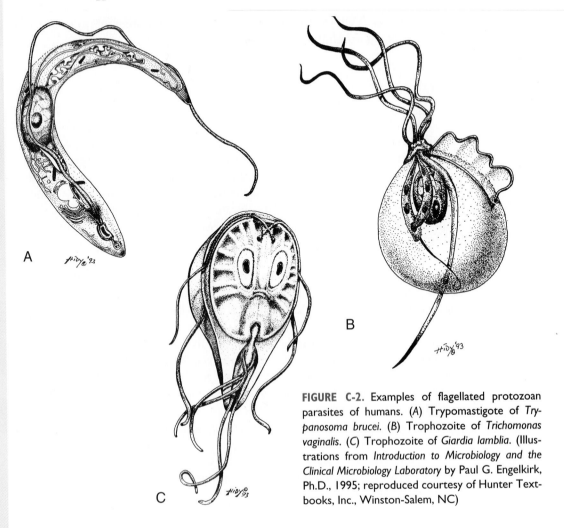

A

B

C

FIGURE C-2. Examples of flagellated protozoan parasites of humans. (A) Trypomastigote of *Trypanosoma brucei*. (B) Trophozoite of *Trichomonas vaginalis*. (C) Trophozoite of *Giardia lamblia*. (Illustrations from *Introduction to Microbiology and the Clinical Microbiology Laboratory* by Paul G. Engelkirk, Ph.D., 1995; reproduced courtesy of Hunter Textbooks, Inc., Winston-Salem, NC)

HELMINTHS

The word *helminth* means parasitic worm. Although helminths are not microorganisms, the various procedures used to diagnose helminth infections are performed in the clinical microbiology laboratory. These procedures often involve the observation of microscopic stages in the life cycles of these parasites. Helminths infect humans, other animals, and plants, but we will limit our discussion to helminth infections of humans.

Helminths are multicellular, eucaryotic organisms in the Kingdom Animalia. The two major divisions of helminths are roundworms (*Nematoda* or *nematodes*) and flatworms (*Platyhelminthes*). The flatworms are further divided into tapeworms (*Cestodes*) and flukes (*Trematodes*). Table C-2 lists some of the more common helminths that cause human infection. Helminths are always endoparasites.

(text continues on page 460)

TABLE C-1. Protozoan Parasites That Infect Humans

Phylum/ Subphylum	Parasite	Disease	How Acquired	How Diagnosed
Sarcodina	*Entamoeba histolytica*	amebiasis; amebic dysentery; extra-intestinal amebic abscesses	usually, ingestion of cysts in contaminated food or water	observation of cysts and / or trophozoites (amebae) in fecal specimens
	Naegleria fowleri	primary amebic meningoencephalitis (PAM)	usually, diving into contaminated pond water (the "old swimming hole")	observation of trophozoites in CSF or cysts in autopsy tissue
Mastigophora	*Giardia lamblia*	giardiasis (a diarrheal disease)	usually ingestion of cysts in contaminated water or food	observation of cysts and / or trophozoites in fecal specimens
	Trichomonas vaginalis	trichomoniasis; causes about one-third of cases of vaginitis	direct contact; trichomoniasis is a sexually transmitted disease (STD)	observation of trophozoites in saline wet mounts of vaginal or urethral discharge material or prostatic secretions
	Leishmania spp.	leishmaniasis	injection of the parasite when a *Phlebotomus* sand fly takes a blood meal	observation of parasite in aspirates or biopsy specimens
	subspecies of *Trypanosoma brucei*	African trypanosomiasis (African sleeping sickness)	injection of the parasite when a Tsetse fly takes a blood meal	observation of trypomastigotes in blood or CSF specimens or lymph node aspirates
	Trypanosoma cruzi	American trypanosomiasis (Chagas' disease)	parasites in the feces of a Reduviid bug get rubbed into bug bite wound	observation of trypomastigotes in blood or amastigotes in biopsy specimens
Ciliata (Ciliophora)	*Balantidium coli*	balantidiasis (a dysenteric disease)	ingestion of cysts in contaminated water	observation of cysts and / or trophozoites in fecal specimens
Sporozoa	*Cryptosporidium parvum*	cryptosporidiosis (a diarrheal disease)	ingestion of oocysts in contaminated water	observation of oocysts in fecal specimens
	Cyclospora cayetanensis	cyclosporiasis (a diarrheal disease)	ingestion of oocysts in contaminated food or water	observation of oocysts in fecal specimens

(continued)

Phylum/ Subphylum	Parasite	Disease	How Acquired	How Diagnosed
	Plasmodium spp.	malaria	injection of sporozoites when a female *Anopheles* mosquito takes a blood meal	observation of trophozoites, schizonts, and / or gametocytes in blood specimens
	Toxoplasma gondii	toxoplasmosis	ingestion of oocysts from cat feces or cysts in contaminated meat	immunodiagnostic procedures

TABLE C-2. Helminths That Commonly Infect Humans

General Category	Scientific Name	Common Name	Disease
Intestinal nematodes	*Ascaris lumbricoides*	Large intestinal roundworm of humans	Ascariasis or *Ascaris* infection (adult worms in small intestine, although they can migrate to other parts of the body)
	Enterobius vermicularis	Pinworm	Pinworm infection or enterobiasis (adult worms in cecum)
	Necator americanus	New world hookworm	Hookworm infection (adult worms in small intestine)
	Strongyloides stercoralis	Threadworm	Strongyloidiasis (adult worms in small intestine)
	Trichuris trichiura	Whipworm	Trichuriasis or whipworm infection (adult worms in colon)
Blood and tissue nematodes	*Dracunculus medinensis*	Guinea worm	Dracunculiasis or guinea worm infection (adult worms in subcutaneous tissue)
	Trichinella spiralis	NA	Trichinosis (adult worms in lining of small intestine)
Filarial nematodes	*Brugia malayi* and *Wuchereria bancrofti*	NA	Filariasis (advanced stages are called elephantiasis) (adult worms in lymph nodes; prelarval stages called microfilariae are present in blood)
	Loa loa	Eye worm	Loiasis (adult worms in subcutaneous tissue; can migrate beneath conjunctiva of eye; microfilariae are present in blood)

General Category	Scientific Name	Common Name	Disease
	Onchocerca volvulus	Blinding worm	Onchocerciasis or "river blindness" (adult worms live in subcutaneous nodules; microfilariae in skin and eyes)
Intestinal cestodes	*Diphyllobothrium latum*	Fish tapeworm	Fish tapeworm infection or diphyllobothriasis (adult tapeworm in small intestine)
	Dipylidium caninum	Dog tapeworm	Dog tapeworm infection (adult tapeworm in small intestine)
	Hymenolepis diminuta	Rat tapeworm	Rat tapeworm infection (adult tapeworm in small intestine)
	Hymenolepis nana	Dwarf tapeworm	Dwarf tapeworm infection (adult tapeworm in small intestine)
	Taenia saginata	Beef tapeworm	Beef tapeworm infection (adult tapeworm in small intestine)
	Taenia solium	Pork tapeworm	Pork tapeworm infection (adult tapeworm in small intestine) and / or cysticercosis (larvae in various tissues)
Tissue cestode	*Echinococcus granulosus*	NA	Hydatidosis or hydatid disease (larvae, called hydatid cysts, in organs or tissues)
Intestinal trematode	*Fasciolopsis buski*	Intestinal fluke	Fasciolopsiasis (adult worms in intestine)
Liver trematodes	*Clonorchis sinensis*	Chinese or oriental liver fluke	Clonorchiasis (adult worms in bile ducts)
	Fascioloa hepatica	Liver fluke	Fascioliasis (adult worms in bile ducts)
Lung trematode	*Paragonimus westermani*	Lung fluke	Paragonimiasis (adult worms in lungs)
Blood trematodes	*Schistosoma haematobium*	Blood fluke	Schistosomiasis (adult worms in blood vessels, especially vessels that surround the urinary bladder)
	Schistosoma japonicum	Blood fluke	Schistosomiasis (adult worms in blood vessels, especially vessels that surround the small intestine)
	Schistosoma mansoni	Blood fluke	Schistosomiasis (adult worms in blood vessels, especially vessels that surround the small intestine)

The typical helminth life cycle includes three major stages—the *egg,* the *larva,* and the *adult.* Adults produce eggs, from which larvae emerge, and the larvae mature into adult worms. The host that harbors the larval stage is called the *intermediate host,* whereas the host that harbors the adult worm is called the *definitive host.* Sometimes helminths have more than one intermediate host or more than one definitive host. The fish tapeworm, for example, is what's known as a 3-host parasite, having one definitive host and two intermediate hosts in its life cycle (see Figure C-3). Dogs, cats, or humans can serve as definitive hosts for the dog tapeworm. Table C-3 contains information concerning intermediate and definitive hosts of various helminths.

Helminth infections are primarily acquired by ingesting the larval stage, although some larvae are injected into the body via the bite of infected insects, and others enter the body by penetrating skin. Table C-4 explains how various helminth infections are acquired and how they are diagnosed in the clinical parasitology laboratory.

Life cycle of *Diphyllobothrium latum* (a 3-host parasite)

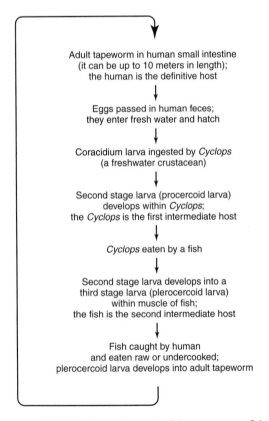

Adult tapeworm in human small intestine
(it can be up to 10 meters in length);
the human is the definitive host

↓

Eggs passed in human feces;
they enter fresh water and hatch

↓

Coracidium larva ingested by *Cyclops*
(a freshwater crustacean)

↓

Second stage larva (procercoid larva)
develops within *Cyclops*;
the *Cyclops* is the first intermediate host

↓

Cyclops eaten by a fish

↓

Second stage larva develops into a
third stage larva (plerocercoid larva)
within muscle of fish;
the fish is the second intermediate host

↓

Fish caught by human
and eaten raw or undercooked;
plerocercoid larva develops into adult tapeworm

FIGURE C-3. Life cycle of *Diphyllobothrium latum,* the fish tapeworm, a 3-host parasite. In this life cycle, the human is the definitive host, the *Cyclops* is the first intermediate host, and the fish is the second intermediate host.

TABLE C-3. Intermediate and Definitve Hosts of Common Helminths

Helminth	Intermediate Host(s)	Definitive Host(s)
Ascaris lumbricoides	None	Human
Enterobius vermicularis	None	Human
Necator americanus	None	Human
Strongyloides stercoralis	None	Human
Trichuris trichiura	None	Human
Dracunculus medinensis	Cyclops (a fresh water crustacean)	Human
Trichinella spiralis	Pig, bear, walrus, human (a dead-end host), etc. (contain both the adult and larval stages)	Pig, bear, walrus, human, etc. (contain both the adult and larval stages)
Brugia malayi and *Wuchereria bancrofti*	Various species of mosquitoes	Human
Loa loa	Chrysops (mango fly)	Human
Onchocerca volvulus	Simulium (black fly)	Human
Diphyllobothrium latum	Cylcops (first intermediate host) and fresh water fish (second intermediate host)	Human
Dipylidium caninum	Flea	Dog, cat, human
Hymenolepis diminuta	Beetle	Rat, mouse, human
Hymenolepis nana	None	Human
Taenia saginata	Cow	Human
Taenia solium	Pig	Human
Echinococcus granulosus	Sheep, human (dead-end host)	Dog
Fasciolopsis buski	Fresh water snails	Human, dog, pig, rabbit
Clonorchis sinensis	Fresh water snails (first intermediate hosts), fresh water fish (second intermediate hosts)	Human, dog, cat, other fish-eating mammals
Fasciola hepatica	Fresh water snails	Human, cow, sheep
Paragonimus westermani	Fresh water snails (first intermediate hosts), crabs or crayfish (second intermediate hosts)	Human, dog, cat, carnivores
Schistosoma spp.	Fresh water snails	Human

Many helminth diseases are endemic in the United States, although people living in the United States do not have the variety or magnitude of helminth infections that occur in other parts of the world. Table C-5 provides estimates of the total number of people, worldwide, who are infected with the most common helminths, while Table C-6 lists the most common helminth infections in the United States.

(text continues on page 465)

TABLE C-4. How Helminth Infections are Acquired and Diagnosed

Helminth	How Infection is Acquired	How Infection is Diagnosed
Ascaris lumbricoides	Ingestion of eggs	Observation of eggs in stool specimens
Enterobius vermicularis	Ingestion of eggs	Observation of eggs in "Scotch tape preps"
Necator americanus	Pentration of skin by infective larvae	Observation of eggs in stool specimens
Strongyloides stercoralis	Penetration of skin by infective larvae	Observation of larvae in duodenal aspirates or stool specimens
Trichuris trichiura	Ingestion of eggs	Observation of eggs in stool specimens
Dracunculus medinensis	Ingestion of infected *Cyclops* in fresh water	Observation of adult worm beneath the skin or emerging from a blister (usually on the ankle or foot)
Trichinella spiralis	Ingestion of pork or bear meat containing larvae	Usually not diagnosed or an incidental finding at autopsy
Brugia malayi and *Wuchereria bancrofti*	Injection of infective larvae by mosquito	Observation of microfilariae in stained blood specimens
Loa loa	Injection of infective larvae by *Chrysops* (mango fly)	Observation of adult worm beneath the skin or in the conjunctiva of the eye; less often, by observing microfilariae in stained blood specimens
Onchocerca volvulus	Injection of infective larvae by *Simulium* (black fly)	Observing microfilariae in "skin snips"
Diphyllobothrium latum	Ingestion of fresh water fish containing second-stage larvae	Observatino of worm segments (proglottids) or eggs in stool specimens
Dipylidium caninum	Ingestion of infected flea	Observation of proglottids or egg packets in stool specimens
Hymenolepis diminuta	Ingestion of infected beetle	Observation of eggs in stool specimens
Hymenolepis nana	Person-to-person (fecal-oral) transmission	Observation of eggs or proglottids (rarely) in stool specimens
Taenia saginata	Ingestion of infected beef	Observation of proglottids or eggs in stool specimens
Taenia solium	Ingestion of infected pork	Observation of proglottids or eggs in stool specimens; cysticercosis may be diagnosed by CT scans, MRI techniques, x-ray, or immunodiagnostic procedures
Echinococcus granulosus	Ingestion of eggs	Observation of cysts by CT scans, MRI techniques, or x-ray; immunodiagnostic procedures
Fasciolopsis buski	Ingestion of raw or undercooked plants (water caltrops, water chestnuts, water bamboo) on which metacercariae are encysted	Observation of eggs in stool specimens
Clonorchis sinensis	Ingestion of infected freshwater fish	Observation of eggs in stool specimens

Helminth	How Infection Is Acquired	How Infection Is Diagnosed
Fasciola hepatica	Ingestion of raw or undercooked aquatic vegetation (e.g., watercress) on which metacercariae are encysted	Observation of eggs in stool specimens
Paragonimus westermani	Ingestion of infected crabs or crayfish	Observation of eggs in sputum or stool specimens
Schistosoma spp.	Penetration of skin by cercariae present in fresh water	Observation of eggs in urine (*S. haematobium*) or stool specimens (*S. japonicum* and *S. mansoni*)

TABLE C-5. Estimated Worldwide Prevalence for the Major Helminth Infections

Helminth Infection	Worldwide Prevalence
Ascariasis	1,000,000,000
Hookworm infection	900,000,000
Trichuriasis	750,000,000
Enterobiasis	400,000,000
Schistosomiasis	200,000,000
Filariasis	90,000,000
Strongyloidiasis	80,000,000
Taeniasis	70,000,000
Clonorchiasis/Opisthorchiasis	18,000,000
Fascioliasis	17,000,000
Trichinosis	11,000,000
Diphyllobothriasis	9,000,000
Paragonimiasis	6,000,000
Dracunculiasis	3,000,000

Reference: *Hopkins DR: Homing in on helminths. Am J Trop Med Hyg 46:626–634, 1992.*

TABLE C-6. The Most Common Helminth Infections in the United States

Helminth	Percentage of Stools That Were Positive
Nematodes	
Hookworm	1.5
Trichuris trichiura	1.2
Ascaris lumbricoides	0.8
Strongyloides stercoralis	0.4
Enterobius vermicularis	0.4 (Also identified in 1094 of 9597 "Scotch tape preps;" the most common helminth infection in the U.S.)
Cestodes	
Hymenolepis nana	0.4
Taenia sp.	<0.1
Taenia saginata	<0.1
Hymenolepis diminuta	<0.1
Diphyllobothrium latum	<0.1
Dipylidium caninum	<0.1
Trematodes	
Clonorchis/Opisthorchis	0.6
Schistosoma mansoni	<0.1
Fasciola hepatica	<0.1
Paragonimus sp.	<0.1

Reference: *Kappus KD, et al.: Intestinal parasitism in the United States: update on a continuing problem. Am J Trop Med Hyg 50:705–713, 1994*
Note: Parasites were found in 20% of the 216,275 stool specimens examined by state diagnostic laboratories in the United States in 1987 (the last year such a survey was performed). Not all parasites listed are endemic in the Unites States; some infections were imported. In 1987, hookworms were the most frequently identified helminths in stool specimens. In all likelihood, most of these infections were acquired outside the United States. Thus, physicians should be aware of the possibility of hookworm infections and infections with other exotic helminths (e.g., *Clonorchis, Opisthorchis,* and *Schistosoma*) in the increasing numbers of immigrants and travelers from countries where these parasites are highly endemic.

TABLE C-7. Parasitic Diseases Ranked Among the Top 21 Fatal Diseases Worldwide by the World Health Organization (1990)

Parasitic Disease	Annual Deaths
Malaria	1–2 million
Schistosomiasis	200,000
Amebiasis	40,000–110,000
Hookworm infection	50,000–60,000
African trypanosomiasis	20,000
Ascariasis	20,000

A 1990 World Health Organization listing of estimated annual deaths ranked six parasitic diseases, including three helminths, among the top 21 fatal diseases worldwide. Those parasitic diseases and the estimated number of annual deaths are shown in Table C-7.

EVASION OF HOST DEFENSE MECHANISMS

To survive within hosts, parasites have developed rather ingenious methods of evading nonspecific and specific defense mechanisms. Within the body, schistosomes coat themselves with host proteins to cover their surface antigens. This camouflage prevents the immune system from recognizing the parasites as being foreign. The trypanosomes that cause African trypanosomiasis are able to vary their surface antigens (a process called antigenic variation). Just about the time that the host has produced antibodies against the parasite's surface antigens, the parasite changes its surface antigens so that the antibodies have nothing to attach to. *Toxoplasma gondii, Trypanosoma cruzi*, and *Leishmania* parasites have protective mechanisms that enable them to live within phagocytic white blood cells. Many parasites evade the immune system by suppressing humoral and/or cell-mediated immune responses (*i.e.*, they cause immunosuppression).

ARTHROPODS

There are many different classes of arthropods, but only three are studied in a parasitology course: *insects* (Class Insecta), *arachnids* (Class Arachnida), and certain

---

TABLE C-8. Ways in Which Arthropods May Be Involved in Human Diseases

Type of Involvement	Example(s)
The arthropod may actually be the *cause* of the disease.	Scabies, a disease in which microscopic mites live in subcutaneous tunnels and cause intense itching
The arthropod may serve as the *intermediate host* in the life cycle of a parasite.	Flea in the life cycle of the dog tapeworm. Beetle in the life cycle of the rat tapeworm. *Cyclops* sp. in life cycle of the fish tapeworm. Tsetse fly in the life cycle of African trypanosomiasis.
	Simulium black fly in the life cycle of onchocerciasis. Mosquito in the transmission of filariasis.
The arthropod may serve as the *definitive host* in the life cycle of a parasite.	Female *Anopheles* mosquito in the life cycle of malarial parasites
The arthropod may serve as a *vector* in the transmission of an infectious disease.	Oriental rat flea in the transmission of plague. Tick in the transmission of Rocky Mountain spotted fever and Lyme disease. Louse in the transmission of epidemic typhus.

TABLE C-9. Arthropods That Serve as Biological Vectors of Human Diseases

Vectors	Disease(s)
Black flies (*Simulium* spp.)	Onchocerciasis ("river blindness") (H)
Cyclops spp.	Fish tapeworm infection (H), guinea worm infection (H)
Fleas	Dog tapeworm infection (H), endemic typhus (B), murine typhus (B), plague (B)
Lice	Epidemic relapsing fever (B), epidemic typhus (B), trench fever (B)
Mites	Rickettsial pox (B), scrub typhus (B)
Mosquitoes	Dengue fever (V), filariasis ("elephantiasis") (H), malaria (P), viral encephalitis (V), yellow fever (V)
Reduviid bugs	American trypanosomiasis (Chagas' disease) (P)
Sand flies (*Phlebotomus* spp.)	Leishmaniasis (P)
Ticks	Babesiosis (P), Colorado tick fever (V), ehrlichiosis (B), Lyme disease (B), relapsing fever (B), Rocky Mountain spotted fever (B), tularemia (B)
Tsetse flies (*Glossina* spp.)	African trypanosomiasis (P)

Key: (B) = a bacterial disease (P) = a protozoal disease
     (H) = a helminth disease (V) = a viral disease

crustaceans (Class Crustacea). The insects that are studied include lice, fleas, flies, mosquitoes, and reduviid bugs. Arachnids include mites and ticks. Crustaceans include crabs, crayfish, and certain *Cyclops* species. Arthropods may be involved in human diseases in any of four ways (shown in Table C-8).

Arthropods may serve as mechanical or biological vectors in the transmission of certain infectious diseases. *Mechanical vectors* merely pick up the parasite at Point A and drop it off at Point B. For example, a house fly could pick up parasite cysts on the sticky hairs of its legs while walking around on animal feces in a meadow. The fly might then fly through an open kitchen window and drop off the parasite cysts while walking on a pie cooling on the counter. A *biological vector*, on the other hand, is an arthropod in whose body the pathogen multiplies or matures (or both). Many arthropod vectors of human diseases are biological vectors (see Table C-9). A particular arthropod may serve as both a host and a biological vector.

APPENDIX D

Microbiology Laboratory Procedures

Although clinical microbiology laboratory procedures should be performed by laboratory personnel, it is sometimes necessary (*e.g.,* nights and weekends in small hospitals) for nurses and other healthcare personnel to perform relatively simple procedures such as inoculation of culture media, Gram staining of clinical specimens, wet mounts, potassium hydroxide preparations (KOH preps), India ink preps, and "Scotch tape" preps. In accordance with the Clinical Laboratory Improvement Act of 1988 (CLIA '88), the minimum qualifications for individuals performing various laboratory tests are determined by the complexity of the laboratory procedures being performed. It is important to note that virtually all clinical microbiology laboratory procedures are designated as being of "moderate-" or "high complexity." Persons planning to perform any clinical laboratory procedures should read and comply with CLIA '88 (and its amendments).

INOCULATION AND INCUBATION OF CULTURE MEDIA

Because the ideal growth temperature of most human pathogens is 37°C, it is important that clinical specimens be inoculated to culture media and incubated as soon as possible following collection of the specimens. When in doubt as to which types(s) of media to inoculate, inoculate a blood agar plate and a chocolate agar plate. The proper way to inoculate the surface of a plate of solid medium is depicted in Figure D-1, and the results of proper inoculation are shown in Figure D-2. Always wear disposable latex gloves when working with clinical specimens.

Step 1: Inoculate approximately one-third of the surface of the medium by *gently* streaking the swab in a zig-zag manner over the surface of the medium. Do not press firmly, as this will tear the medium.
Step 2: Use a sterile inoculating loop to streak the next third of the agar surface.
Step 3: Use the inoculating loop to streak the final third of the agar surface.
Step 4: Place the inoculated plate into a 37°C incubator.

Sometimes a physician may wish to examine a Gram-stained smear of a specimen when no laboratory personnel are available to prepare the smear. To prepare a Gram-stained smear, follow the instructions below.

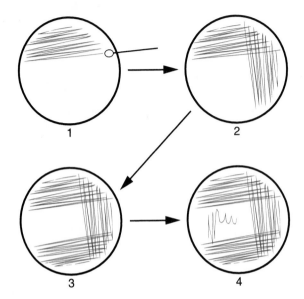

FIGURE D-1. Proper technique for inoculating the surface of agar medium to obtain well-isolated colonies.

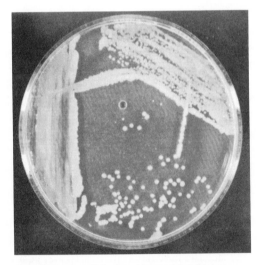

FIGURE D-2. Culture plate showing well-isolated bacterial colonies.

PREPARATION OF A SMEAR FOR GRAM STAINING

Step 1: If the specimen is a swab, prepare a smear by rolling the swab over the center portion of a glass microscope slide. If the specimen is a liquid, dip a sterile swab or sterile inoculating loop into the specimen and then spread a drop of the specimen over the center portion of a glass microscope slide. Let the smear air dry before proceeding.

Step 2: Fix the smear by flooding it with absolute (100%) methanol, leaving the methanol in place for 30 seconds, and then tipping the slide to let the methanol run off.

Step 3: Hold the slide under a stream of gently flowing water to remove the methanol. The smear is now ready for Gram staining.

(**Note:** If absolute methanol is not available, the smear may be fixed by passing it swiftly through a Bunsen burner flame a couple times.)

Steps in Gram Staining

Step 1: Place the smear on a staining rack over a sink. Flood the fixed smear with crystal violet (the primary stain). Allow the crystal violet to remain in contact with the smear for 30 seconds. It is during this step that all bacteria become blue.

Step 2: Wash off the excess crystal violet using a stream of gently flowing water. Too strong a current will wash the smear off the slide.

Step 3: Flood the smear with Gram's iodine (the mordant). Allow the Gram's iodine to remain in contact with the smear for 30 seconds. The Gram's iodine becomes chemically bonded to the crystal violet. All bacteria remain blue.

Step 4: Wash off the excess Gram's iodine using a stream of gently flowing water.

Step 5: Hold the slide at an angle over the sink. Drip the decolorizing agent (usually 95% ethanol) over the smear until no further blue color washes off the smear. This usually only takes a few drops and a few seconds. If performed correctly, Gram-negative bacteria become decolorized (go back to being colorless), while Gram-positive bacteria remain blue.

Step 6: Immediately stop the decolorization process by placing the slide under a stream of gently flowing water. Allowing the alcohol to remain on the smear too long will cause decolorization of Gram-positive bacteria.

Step 7: Flood the smear with safranin (called the counterstain or secondary stain). Allow the safranin to remain in contact with the smear for 1 minute. Safranin is a bright red dye. It is during this step that Gram-negative bacteria become red.

Step 8: Wash off the excess safranin with a stream of gently flowing water.

Step 9: Using bibulous paper or a clean paper towel, gently blot the Gram-stained smear to remove any remaining water. Do not rub. Examine the smear with a microscope. If performed correctly, Gram-positive bacteria will be blue-to-purple and Gram-negative bacteria will be pink-to-red. (Note: Cell wall structure determines whether a bacterial cell is Gram-positive or Gram-negative at the conclusion of the Gram-staining procedure. The thicker the layer of peptidoglycan, the more difficult it is to remove the crystal violet from the cell. Also, the decolorization step dissolves the lipid in the cell walls of Gram-negative bacteria, causing the crystal violet to flow out of the cell.) Refer to Color Figures 1 through 10 to observe the Gram-stained appearance of bacteria, polymorphonuclear cells (PMNs), and epithelial cells.

PREPARATION OF A WET MOUNT

Wet mounts are easier and faster to perform than Gram staining. In a wet mount, a portion of the unstained clinical specimen is examined microscopically. One of the most common uses of a wet mount is in the diagnosis of vaginitis (described below), where vaginal discharge material is examined for the presence of yeasts (*Candida albicans*), protozoa (*Trichomonas vaginalis*), or evidence of bacterial infection.

Steps in the Preparation of a Wet Mount

Step 1: Place a small drop of sterile saline on a clean glass microscope slide.

Step 2: Mix some of the vaginal discharge material into the drop of saline.

Step 3: Place a glass cover slip over the preparation. Do not allow the specimen to dry.

Step 4: Examine the preparation under the microscope, using reduced lighting. The presence of very large quantities of yeast cells is suggestive of yeast vaginitis. A presumptive or tentative diagnosis of trichomoniasis can be made if motile trophozoites of *T. vaginalis* are seen (see Fig. C-2 in Appendix C). The presence of clue cells (squamous epithelial cells densely covered with small coccobacilli) is suggestive of bacterial vaginosis (BV). If BV is suspected, a Gram stain of the vaginal fluid should be prepared. In BV, few Gram-positive bacilli (*Lactobacillus* spp.) will be seen, but large numbers of Gram-negative bacilli (*Bacteroides* spp.), Gram-variable coccobacilli (*Gardnerella vaginalis*), and curved Gram-negative bacilli (*Mobiluncus* spp.) will be seen.

KOH PREPARATION

A potassium hydroxide preparation (KOH prep) is often used to determine if fungi are present in a clinical specimen. The most common specimen types upon which a KOH prep is performed are skin scrapings, hair clippings, and fingernail and toenail clippings. The function of the KOH is to dissolve keratin in the specimen, making it easier to see any fungal elements (yeasts or hyphae) that might be present. Thus, the KOH acts as a clearing agent.

Steps in the KOH Preparation

Step 1: Place the specimen (pus, exudate, tissue, skin, hair, nail clipping) into a drop of 10% potassium hydroxide on a clean glass microscope slide. Mix and then wait for 5 to 30 minutes before proceeding. (Note: gentle warming may be required to clear the specimen.) A drop of lactophenol cotton blue (LPCB) stain can then be added to enhance the visibility of fungal elements, but this is optional.

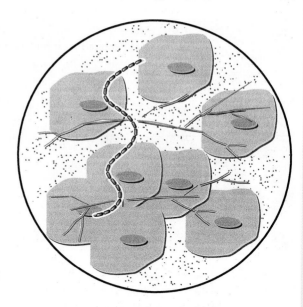

FIGURE D-3. KOH preparation of skin scraping showing fungal hyphae and skin cells.

Step 2: Cover the preparation with a glass coverslip.

Step 3: Examine the preparation under the microscope, with reduced lighting, looking for fungal elements (hyphae, yeast cells, pseudohyphae; Fig. D-3)

INDIA INK PREPARATION

An India ink preparation is of value in the presumptive diagnosis of cryptococcal meningitis, although it is sometimes performed on sputum, blood, or urine.

Steps in the India Ink Preparation

Step 1: Place a drop of spinal fluid on a glass microscope slide.

Step 2: Add a glass coverslip.

Step 3: Place a drop of India ink or nigrosin dye on the glass slide next to the coverslip.

Step 4: Allow the India ink to diffuse under the slide, creating an ink gradient.

Step 5: Examine the preparation microscopically, starting with the X10 objective and then increasing magnification to X400. Examine the region where the ink particle density is neither too light nor too heavy. The presence of budding, encapsulated yeasts is interpreted as a presumptive or tentative diagnosis of cryptococcal meningitis (Fig. D-4). Culture and identification by biochemical tests are necessary for definitive diagnosis.

"SCOTCH TAPE" PREPARATION

Female pinworms (*Enterobius vermicularis*) deposit their eggs on the perianal skin (the skin around the anus). Diagnosis of pinworm infection is made by recovering

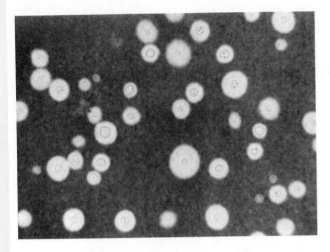

FIGURE D-4. India ink preparation showing encapsulated yeast cells of *Cryptococcus neoformans.*

and observing the eggs, using what is known as the "Scotch tape" preparation (also called the cellophane or cellulose tape preparation). Commercial collection devices are also available for this purpose.

Steps in the "Scotch Tape" Preparation

Step 1: Press the sticky side of a piece of clear cellophane tape or a commercial collection device to the perianal skin early in the morning, before the patient has bathed or defecated.

Step 2: Place the piece of cellophane tape, sticky side down, onto a glass microscope slide or replace the collection device into its transport container, and send it to the microbiology laboratory.

Step 3: Examine the preparation microscopically, using the low power objective. Switch to the high power objective when suspicious objects are observed. Pinworm eggs have a characteristic appearance: football-shaped with one flattened side, smooth shelled, containing a coiled larva, and approximately 20 to 32 × 50 to 60 μm (see Fig. D-5).

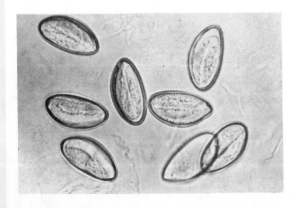

FIGURE D-5. "Scotch tape" preparation showing pinworm (*Enterobius vermicularis*) eggs.

Appendices

APPENDIX E

Useful Conversions

LENGTH CONVERSIONS

To convert inches into centimeters, multiply by 2.54
To convert centimeters into inches, multiply by 0.39
To convert yards into meters, multiply by 0.91
To convert meters into yards, multiply by 1.09

1 mile (mi) = 1.609 kilometers
1 yard (yd) = 0.914 meter
1 foot (ft) = 30.48 centimeters
1 inch (in) = 2.54 centimeters
1 kilometer (km) = 0.62 mile
1 meter (M) = 39.37 inches
1 centimeter (cm) = 0.39 inch
1 millimeter (mm) = 0.039 inch

Note: information about micrometers and nanometers can be found in Figure 1-13 in Chapter 1.

VOLUME CONVERSIONS

To convert gallons into liters, multiply by 3.78
To convert liters into gallons, multipy by 0.26
To convert fluid ounces into milliliters, multiply by 29.6
To convert milliliters into fluid ounces, multiply by 0.034

1 gallon (gal) = 3.785 liters
1 quart (qt) = 0.946 liter
1 pint (pt) = 0.473 liter
1 fluid ounce (fl oz) = 29.573 milliliters
1 liter (L) = 1.057 quarts
1 milliliter (mL) = 0.0338 fluid ounce

WEIGHT CONVERSIONS

To convert ounces into grams, multiply by 28.4
To convert grams into ounces, multiply by 0.035
To convert pounds into kilograms, multiply by 0.45
To convert kilograms into pounds, multiply by 2.2

1 pound (lb) = 0.454 kilogram
1 ounce (oz) = 28.35 grams
1 kilogram (kg) = 2.2 pounds
1 gram (gm) = 0.035 ounce
1 gram = 1,000 milligrams (mg)
1 gram = 1,000,000 micrograms (μg)

TEMPERATURE CONVERSIONS

To convert Celsius (°C) into Fahrenheit (°F), use °F = (°C$\times$1.8) + 32
To convert Fahrenheit (°F) into Celsius (°C), use °C = (°F$-$32) $\times$ 0.556

Reference Temperatures:

	°Celsius	°Fahrenheit
Boiling	100.0	212.0
Body temperature range	42.0	107.6
	41.5	106.7
	41.0	105.8
	40.5	104.9
	40.0	104.0
	39.5	103.1
	39.0	102.2
	38.5	101.3
	38.0	100.4
	37.5	99.5
	37.0	98.6
	36.5	97.7
	36.0	96.8
Room temperature range	25	77
	24	75.2
	23	73.4
	22	71.6
	21	69.8
	20	68
Refrigerature temperature	4	39
Freezing	0	32.0

Appendices

APPENDIX F

Suggested Reading

Alcamo IE: Fundamentals of Microbiology, 5th ed. Menlo Park, Addison Wesley Longman, 1997.

Atlas RM: Microorganisms in our World. St. Louis, Mosby, 1995.

Brock TD: Robert Koch—A Life in Medicine and Bacteriology. Washington, D.C., American Society for Microbiology, 1998.

Brock TD: Milestones in Microbiology. Washington, D.C., American Society for Microbiology, 1998.

de Kruif P: Microbe Hunters. New York, Pocket Books, Inc., 1940.

Dubos R: Pasteur and Modern Science. Washington, D.C., American Society for Microbiology, 1998.

Forbes BA et al: Bailey & Scott's Diagnostic Microbiology, 10th ed. St. Louis, Mosby, 1998.

Baron S et al: Medical Microbiology, 4th ed. Galveston, University of Texas Medical Branch, 1996.

Benenson AS: Control of Communicable Diseases Manual, 16th ed. Washington, D.C., American Public Health Association, 1995.

Berkow R: The Merck Manual of Medical Information. Whitehouse Station, Merck & Company, 1997.

Biddle W: A Field Guide to Germs. New York, Henry Holt and Company, 1995.

Boyd RF: Basic Medical Microbiology, 5th ed. Boston, Little-Brown & Company, 1995.

Clark WR: At War Within—The Double-Edged Sword of Immunity. New York, Oxford University Press, 1995.

Davis BD et al: Microbiology, 4th ed. Philadelphia, JB Lippincott, 1990.

de la Maza LM et al: Color Atlas of Diagnostic Microbiology. St. Louis, Mosby, 1997.

Delost M: Introduction to Diagnostic Microbiology—a Text and Workbook. St. Louis, Mosby, 1996.

Dixon B: Power Unseen—How Microbes Rule the World. New York, W.H. Freeman, 1994.

Engelkirk PG: Introduction to Microbiology and the Clinical Microbiology Laboratory. Winston-Salem, Hunter Textbooks, Inc., 1995.

Garrett L: The Coming Plague. New York, Penguin Books, 1994.

Hart CT, Shears P: Color Atlas of Medical Microbiology. St. Louis, Mosby, 1996.

Howard BJ et al: Clinical and Pathogenic Microbiology, 2nd ed. St. Louis, Mosby, 1994.

Ingraham JL et al: Introduction to Microbiology. Belmont, Wadsworth Publishing Company, 1995.

Isenberg HD et al: Essential Procedures for Clinical Microbiology. Washington, D.C., American Society for Microbiology, 1998.

Jensen MM et al: Microbiology for the Health Sciences, 4th ed. Upper Saddle River, Prentice Hall, 1995.

Koneman EW et al: Color Atlas and Textbook of Diagnostic Microbiology, 5th ed. Philadelphia, Lippincott-Raven, 1997.

Koprowski H, Oldstone MBA: Microbe Hunters Then and Now. Bloomington, Medi-Ed Press, 1996.

Lovelock JE: Gaia—A New Look at Life on Earth. New York, Oxford University Press, 1987.

Mahon CR, Manuselis G, Jr.: Textbook of Diagnostic Microbiology, 2nd ed. Philadelphia, WB Saunders, 1999.

Margulis L, Sagan D: Microcosmos—Four Billion Years of Microbial Evolution. New York, Summit Books, 1986.

McKane L, Kandel J: Microbiology—Essentials and Applications, 2nd ed. New York, McGraw-Hill, 1996.

Morello JA et al: Microbiology in Patient Care, 6th ed. Dubuque, WCB/McGraw-Hill, 1998.

Pelczar MJ et al: Microbiology—Concepts and Applications. New York, McGraw-Hill, 1993.

Prescott L et al: Microbiology, 4th ed. Dubuque, WCB/McGraw-Hill, 1999.

Scheld WM et al: Emerging Infections 1. Washington, D.C., American Society for Microbiology, 1997.

Scheld WM et al: Emerging Infections 2. Washington, D.C., American Society for Microbiology, 1998.

Talaro KP, Talaro A: Foundations in Microbiology, 3rd ed. WCB/McGraw-Hill, 1999.

Tortora GJ et al: Microbiology, 6th ed. Menlo Park, Addison Wesley Longman, 1998.

Turgeon ML: Immunology and Serology in Laboratory Medicine, 2nd ed. St. Louis, Mosby, 1996.

Volk WA et al: Essentials of Medical Microbiology, 5th ed. Philadelphia, Lippincott-Raven, 1996.

Glossary

Abiogenesis (ab′-ee-oh-jen-uh-sis). The theory that life may develop from nonliving matter

Acid (as′-id). A compound that yields a hydrogen ion in a polar solvent such as water; a solution that has a pH lower than 7.0 is said to be *acidic*

Acid-fast stain. A differential staining procedure that differentiates bacteria into those that stain red (acid-fast bacteria) and those that do not stain red (non-acid-fast bacteria); primarily used in the presumptive diagnosis of tuberculosis

Acidophile (uh-sid′-oh-file). An organism that prefers acidic environments; such an organism is said to be *acidophilic*

Acquired immunity. Immunity or resistance acquired at some point in an individual's lifetime

Active acquired immunity. Immunity or resistance acquired as a result of the active production of antibodies

Active carrier. A person who has recovered from an infectious disease, but continues to harbor and transmit the pathogen

Acute disease. A disease having a sudden onset and short duration

Acquired immunodeficiency syndrome (AIDS). A disease characterized by a variety of opportunistic infections and malignancies; caused by human immunodeficiency virus (HIV)

Adenosine triphosphate (uh-den′-oh-seen try-fos′-fate). The major energy-carrying (energy-storing) molecule in a cell

Aerotolerant anaerobe (air-oh-tol′-er-ant an′-air-obe). An organism that can live in the presence of oxygen, but grows best in an anaerobic environment (one that contains no oxygen)

Agammaglobulinemia (ay-gam′-uh-glob′-yu-luh-nee′-me-uh). Absence of, or extremely low levels of, the gamma fraction of serum globulin; sometimes used to denote the absence of immunoglobulins

Agglutination (uh-glue-tuh-nay′-shun). The clumping of particles (such as red blood cells and latex beads) in solution

Agglutination tests. Immunodiagnostic procedures in which antigen-antibody reactions result in the agglutination or clumping of particles

AIDS. See *acquired immunodeficiency syndrome*

Airborne precautions. Specific safety precautions that are practiced in a hospital setting to prevent infections transmitted by the airborne route

Algae (al′-gee), sing. *alga.* Eucaryotic, photosynthetic organisms that range in size from unicellular to multicellular; includes many seaweeds

Algicidal (al′-juh-side-ul) **agent.** A disinfectant that specifically kills algae

Alkaliphile (al′-kuh-luh-file). An organism that prefers alkaline (basic) environments; such an organism is said to be *alkaliphilic*

Allergen (al′-ur-jin). An antigen to which some people become allergic

Allergy (al′-ur-jee). A disease resulting from acquired or induced sensitivity to an allergen

Ameba (uh-me′-bah), pl. *amebae.* A type of protozoan that moves by means of pseudopodia; in the phylum Sarcodina (a subphylum in some classification schemes)

Amino (uh-me′-no) **acids.** The basic units or "building blocks" of proteins

Ammonification (uh-mon′-uh-fuh-kay′-shun). Conversion of nitrogenous compounds (*e.g.,* proteins) into ammonia

Amphitrichous (am-fit′-ri-kus) **bacteria.** Bacteria possessing one flagellum at each end (pole) of the cell

Anabolism (uh-nab′-oh-lizm). That part of metabolism concerned with the building of large compounds from smaller compounds; involves the creation of chemical bonds; requires energy; such chemical reactions are called anabolic or biosynthetic reactions

Anaerobe (an′-air-obe). An organism that does not require oxygen for survival; can exist in the absence of oxygen

Anamnestic (an-am-nes'-tick) **response.** An immune response following exposure to an antigen that the individual is already sensitized to; also known as a *secondary response* or *memory response*

Anaphylactic (an-uh-fuh-lak'-tick) **shock.** Shock following anaphylaxis; may lead to death

Anaphylaxis (an-uh-fuh-lak'-sis). An immediate, severe, sometimes fatal, systemic allergic reaction

Angstrom (ang'-strom). A unit of length, equivalent to 0.1 nanometer; roughly the diameter of an atom

Anion (an'-eye-on). An ion that carries a negative charge

Antagonism (an-tag'-ohn-izm). As used in this book, the killing, injury, or inhibition of one microorganism by products of another

Antibiosis (an'-tee-by-oh'-sis). An association of two organisms which is detrimental to one of them; an antagonistic relationship

Antibiotic (an'-tee-by-ot'-tik). A substance produced by a microorganism that inhibits or destroys other microorganisms

Antibody (an'-tee-bod-ee). A glycoprotein produced by lymphocytes in response to an antigen; often protective

Anticodon (an-tee-ko'-don). The trinucleotide sequence that is complementary to a codon; found on a transfer RNA molecule

Antigen (an'-tuh-jen). A substance, usually foreign, that stimulates the production of antibodies; an *anti*body *gen*erating substance; sometimes called an *immunogen*

Antigenic (an-tuh-jen'-ick) **determinant.** The smallest part of an antigen capable of stimulating the production of antibodies

Antigenic variation. The ability of a microorganism to change its surface antigens

Antimicrobial (an'-tee-my-kro'-be-ul) **agent.** A drug, disinfectant, or other substance that kills microorganisms or suppresses their growth

Antisepsis (an-tee-sep'-sis). Prevention of infection by inhibiting the growth of pathogens

Antiseptic (an-tee-sep'-tick). An agent or substance capable of effecting antisepsis; usually refers to a chemical disinfectant that is safe to use on living tissues

Antiseptic surgery. Surgery performed in a manner to prevent infection by inhibiting the growth of pathogens

Antiseptic techniques. Procedures taken to effect antisepsis; the use of antiseptics

Antiserum (an-tee-see'-rum). A serum containing specific antibodies; also known as an *immune serum*

Antitoxins (an-tee-tok'-sinz). Antibodies produced in response to a toxin; often capable of neutralizing the toxin that stimulated their production

Arbovirus (are'-boh-vy'-rus). A virus that is transmitted by an arthropod; an arthropod-borne virus

Archaebacteria (ark'-ee-back-tier'-ee-uh). "Ancient" bacteria, thought by some scientists to be the earliest types of bacteria

Arthropod (are'-throw-pod). An animal in the phylum Arthropoda; includes insects such as flies, mosquitoes, fleas, and lice, and arachnids such as mites and ticks

Asepsis (a-sep'-sis). A condition in which living pathogens are absent; a state of sterility

Aseptic (ay-sep'-tick) **techniques.** Measures taken to ensure that living pathogens are absent

Asymptomatic (ay'-simp-tow-mat'-ick) **disease.** A disease having no symptoms; also referred to as a *subclinical disease*

Asymptomatic infection. The presence of a pathogen in or on the body, without any symptoms of disease; also referred to as a *subclinical infection*

Atom (at'-um). The smallest particle of matter possessing the properties of an element; composed of protons, neutrons, and electrons

Atomic number. The number of negatively charged electrons in an uncharged atom or the number of protons in its nucleus; the atomic number of an element indicates the element's position in a periodic chart of the elements

Atomic weight. The weight or mass of an atom in relation to the mass of an atom of carbon-12 (^{12}C), which is set equal to 12.000

Atopic (ay-tope'-ick) **person.** Allergic person; one who suffers from allergies

Attenuated (uh-ten'-yu-ay-ted). An adjective meaning weakened, less pathogenic, used to describe certain microorganisms

Attenuated vaccines. Vaccines prepared from attenuated microorganisms

Attenuation (uh-ten-yu-ay-'shun). The process by which microorganisms are attenuated

Autoclave (aw'-toe-klav). An apparatus used for sterilization by steam under pressure

Autogenous (aw-toj′-uh-nus) **vaccine.** A vaccine made from microorganisms or cells obtained from the person's own body

Autoimmune (aw-toh-uh-myun′) **disease.** A disease in which the body produces antibodies directed against its own tissues

Autolysis (aw-tol′-uh-sis). Autodigestion; self digestion

Autotroph (aw′-toe-trof). An organism that uses carbon dioxide as its sole carbon source

Avirulent (ay-veer′-yu-lent). Not virulent; not pathogenic; not capable of causing disease

Axial (ak′-see-ul) **filament.** An organelle of motility possessed by spirochetes

B cell. B lymphocyte; a type of leukocyte that plays an important role in the immune system

Bacillus (bah-sil′-us), pl. *bacilli.* A rod-shaped bacterium; there is also a genus named *Bacillus,* made up of aerobic, gram-positive, spore-forming bacilli

Bacteremia (bak-ter-ee′-me-uh). The presence of bacteria in the bloodstream

Bacteria (back-tier′-ee-uh), sing. *bacterium.* Primitive, unicellular, procaryotic microorganisms

Bactericidal (bak-tear′-eh-sigh′-dull) **agent.** A chemical agent or drug that kills bacteria; a *bactericide*

Bacteriocins (bak-teer′-ee-oh-sinz). Proteins produced by certain bacteria (those possessing bacteriocino-genic plasmids) that can kill other bacteria

Bacteriologist (back′-tier-ee-ol′-oh-jist). One who specializes in the science of bacteriology

Bacteriology (back′-tier-ee-ol′-oh-gee). The study of bacteria

Bacteriophage (back-tier′-ee-oh-faj). A virus that infects a bacterium; also known simply as a *phage*

Bacteriostatic (bak-tear′-ee-oh-stat′-ick) **agent.** A chemical agent or drug that inhibits the growth of bacteria

Bacteriuria (bak-ter-ee′-yu′-ree-uh). The presence of bacteria in the urine

Barophile (bar′-oh-file). An organism that thrives under high environmental pressure; such an organism is said to be *barophilic*

Base. A compound that yields a hydroxyl ion in a polar solvent such as water; a solution that has a pH higher than 7.0 is said to be *basic*

Basophil (bay′-so-fil). A type of granulocyte found in blood; its granules contain acidic substances (*e.g.,* his-tamine) that attract basic dyes

Beta-lysin (β-lysin). An enzyme capable of destroying microorganisms in the bloodstream or within phagocytic cells

Binary (by′-nare-ee) **fission.** A method of reproduction whereby one cell divides to become two cells

Biochemistry (by-oh-kem′-is-tree). The chemistry of living organisms

Biocidal (by-o-sigh′-dull) **agent.** A chemical agent that destroys life, especially microorganisms

Biogenesis (by-oh-gen′-uh-sis). The theory that life originates only from preexisting life and never from nonliving matter

Biological vector. A vector within which a pathogen either multiplies or matures

Bioremediation (by′-oh-ruh-meed′-ee-a-shun). The use of microorganisms to clean up industrial and toxic wastes

Biotechnology (by′-oh-tek-nol′-oh-gee). The use of microorganisms in industry to produce chemicals, an-tibiotics, foods, beverages, and other products.

Biotherapeutic (by′-oh-ther-uh-pu′-tik) **agents.** Microorganisms used to treat various diseases or conditions

Blocking antibodies. IgG antibodies that combine with allergens, thus, preventing them from attaching to IgE antibodies on the surface of basophils and mast cells; produced by the body in response to "allergy shots"

Botulinal (bot′-you-ly-nal) **toxin.** The neurotoxin produced by *Clostridium botulinum;* causes botulism

Candidiasis (kan-duh-dy′-uh-sis). Infection with, or disease caused by, a yeast in the genus *Candida*—usu-ally *Candida albicans;* also known as moniliasis

Capnophile (cap′-no-file). An organism that grows best in the presence of increased concentrations of car-bon dioxide; such an organism is said to be *capnophilic*

Capsid (kap′-syd). The external protein "coat" or covering of a virion

Capsomeres (kap′-so-meers). The protein units that make up the capsid of some virions

Capsule (kap′-sool). An organized layer of glycocalyx, firmly attached to the outer surface of the bacterial cell wall; some yeasts are also encapsulated

Carbohydrates (kar-boh-high′-drates). Organic compounds containing carbon, hydrogen, and oxygen in a ratio of 1:2:1; also known as saccharides

Carrier (keh′-ree-er). An individual with an asymptomatic infection that can be transmitted to other sus-ceptible individuals

Catabolism (kuh-tab′-oh-lizm). That part of metabolism concerned with breaking down large compounds into smaller compounds; involves the breaking of chemical bonds; energy is released; such chemical reactions are called catabolic or degradative reactions

Catalyst (kat'-uh-list). A substance (often an enzyme) that speeds up a chemical reaction, but is not itself consumed or permanently changed in the process

Catalyze (at'-uh-lyz). To act as a catalyst; to speed up a reaction

Cation (kat'-eye-on). An ion that carries a positive charge

Cell (sell). The smallest unit of living structure capable of independent existence

Cell-mediated immunity. A type of immunity involving many different cell types (*e.g.,* macrophages, various lymphocytes), but where antibodies play only a minor role, if any; also known as *delayed hypersensitivity*

Cell membrane (mem'-brain). The protoplasmic boundary of all cells; controls permeability and serves other important functions

Cell theory. The theory stating that all living organisms are composed of cells

Cell wall. The outermost layer of many cells (*e.g.,* algal, bacterial, fungal, and plant cells)

Cellulose (sell'-you-los). A polysaccharide found in the cell walls of algae and plants

Centrioles (sen'-tree-olz). Tubular structures thought to play a role in nuclear division (mitosis) in animal cells and the cells of lower plants

Cervicitis (sir-vuh-sigh'-tis). Inflammation of the neck of the uterus, the cervix uteri

Chemoautotroph (keem'-oh-awe'-toe-trof). An organism that uses chemicals as an energy source and carbon dioxide as a carbon source; a type of autotroph

Chemoheterotroph (keem'-oh-het'-er-oh-trof). An organism that uses chemicals as a source of energy and organic molecules as a source of carbon; a type of heterotroph; sometimes referred to as a *chemoorganotroph*

Chemolithotroph (keem'-oh-lith'-oh-trof). An organism that uses chemicals as a source of energy and inorganic molecules as a source of carbon; a type of lithotroph

Chemostat (keem'-oh-stat). A growth chamber designed to allow input of nutrients and output of cells on a controlled basis

Chemosynthesis (keem'-oh-syn'-thuh-sis). The process of obtaining energy and synthesizing organic compounds from simple inorganic reactions; carried out by some chemoautotrophic bacteria

Chemotactic (keem-oh-tack'-tick) **agents.** Chemical substances that attract leukocytes

Chemotaxis (keem-oh-tack'-sis). The movement of cells in response to a chemical (*e.g.,* the attraction of phagocytes to an area of injury)

Chemotherapeutic (keem'-oh-ther-uh-pyu'-tik) **agent.** Any chemical used to treat any disease.

Chemotherapy (keem'-oh-ther'-uh-pee). The treatment of a disease (including an infectious disease) by means of chemical substances or drugs

Chemotroph (keem'-oh-trof). An organism that uses chemicals as a source of energy

Chitin (ky'-tin). A polysaccharide found in fungal cell walls, but not found in the cell walls of other microorganisms; also found in the exoskeleton of beetles and crabs

Chloroplast (klor'-oh-plast). A membrane-bound organelle found in the cytoplasm of algal and plant cells; a type of *plastid*

Choleragin (kol'er-uh-jen). The enterotoxin produced by *Vibrio cholerae*

Chromatin (kro'-muh-tin). The genetic material of the nucleus; consisting of DNA and associated proteins; during mitotic division, the chromatin condenses and is seen as chromosomes

Chromosome (kro'-mow-soam). A condensed form of chromatin; the location of genes; human diploid cells contain 46 chromosomes (23 pairs); bacterial cells usually contain only one chromosome, which divides to become two just prior to binary fission

Chronic disease. A disease of slow progress and long duration

Ciliates (sil'-ee-itz), sing. *ciliate.* Ciliated protozoa

Ciliophora (sil'-ee-auf'-oh-rah). A phylum of protozoa that includes the ciliates; sometimes referred to as Ciliata

Cilium (sil'-ee-um), pl. *cilia.* A thin, hairlike organelle of motility

Cistron (sis'-tron). The smallest functional unit of heredity; a length of chromosomal DNA associated with a single biochemical function; a gene may consist of one or more cistrons; sometimes used synonymously with gene

Citric (sit'-rik) **acid cycle.** A series of chemical reactions that produces twelve high-energy phosphate bonds; a biochemical pathway; also known as the tricarboxylic acid, TCA cycle, and Krebs cycle

Coagulase (ko-ag'-yu-lace). A bacterial enzyme that causes plasma to clot

Coccobacillus (kok'-ko-buh-sil'-us), pl. *coccobacilli.* A very short bacillus

Coccus (kok'-us), pl. *cocci.* A spherical bacterium

Codon (koh'-don). A sequence of three nucleotides in a strand of mRNA that provides the genetic information (code) for a certain amino acid to be incorporated into a growing protein chain

Coenzyme (koh'-en-zym). A substance that enhances or is necessary for the action of an enzyme; several vitamins are coenzymes; a type of cofactor

Cofactor (koh'-fak'-tor). An atom or molecule essential for the enzymatic action of certain proteins

Colicin (kol'-uh-sin). A type of bacteriocin produced by *Escherichia coli* and other closely related bacteria

Collagen (kol'-luh-jen). The major protein in the white fibers of connective tissue, cartilage, and bone

Collagenase (kol'-uh-juh-nace). A bacterial enzyme that causes the breakdown of collagen

Commensalism (ko-men'-sul-izm). A symbiotic relationship in which one party derives benefit and the other party is unaffected in any way; most members of the indigenous microflora are commensals

Communicable (kuh-myun'-uh-kuh-bul) **disease.** A disease capable of being transmitted person-to-person

Community-acquired infection. Any infection acquired outside of a hospital setting

Competence (kom'-puh-tense). As used in this book, the ability of a bacterial cell to take up free or "naked" DNA, which may lead to transformation; bacteria capable of taking up free DNA are said to be *competent*

Complement (kom'-pluh-ment). A protein complex of 25–30 components (including C1 through C9) found in blood; involved in inflammation, chemotaxis, phagocytosis, and lysis of bacteria

Compound (kom'-pownd). A chemical substance formed by the covalent or electrostatic union of two or more elements

Compound microscope. A microscope containing two or more magnifying lenses

Conidium (ko-nid'-ee-um), pl. *conidia*. An asexual fungal spore

Conjugation (kon-ju-gay'-shun). As used in this book, the union of two bacterial cells, for the purpose of genetic transfer; *not* a reproductive process

Conjunctivitis (kon-junk'-tuh-vi'-tis). Inflammation of the conjunctiva (the mucous membrane that lines the eyelids and covers the anterior portion of the eyeball)

Contact precautions. Specific safety precautions that are practiced in a hospital setting to prevent infections transmitted by contact

Contagious (kon-tay'-jus). Easily transmitted from one person to another, as in *contagious disease*

Contagious disease. A disease easily transmitted person-to-person; a type of communicable disease

Contamination (kon-tam-uh-nay'-shun). As used in this book, a condition indicating the presence of pathogens (which would be referred to as *contaminants*)

Convalescent (kon-vuh-less'-ent) **carrier.** A person who no longer shows the signs of a particular infectious disease, but continues to harbor and transmit the pathogen during the convalescence period

Covalent (koh-vayl'-ent) **bond.** An interatomic bond characterized by the sharing of 2, 4, or 6 electrons; shared electron bonds

Crenated (kree'-nay-ted). Wrinkled, shriveled; *e.g.,* the appearance of red blood cells placed into a hypertonic solution

Crenation (kree-nay'-shun). The process of becoming, or state of being, crenated

Cyanobacteria (sigh'-an-oh-bak-tier'-ee-uh). Photosynthetic bacteria

Cystitis (sis-ty'-tis). Inflammation or infection of the urinary bladder

Cytokines (sigh'-toe-kynz). Soluble protein mediators released by certain cells on contact with antigens; examples include *lymphokines* and *monokines*

Cytology (sigh-tol'-oh-gee). The study of cells

Cytoplasm (sigh'-toe-plazm). That portion of a cell's protoplasm that lies outside the nucleus of the cell

Cytostome (sigh'-toe-stoam). A primitive mouth possessed by some protozoa.

Cytotoxic (sigh-toe-tok'-sik). Detrimental or destructive to cells

Cytotoxins (sigh'-tow-tok'-sinz). Toxic substances that inhibit or destroy cells

Death phase. That part of a bacterial growth curve during which no multiplication occurs and organisms are dying; the fourth and final phase in a bacterial growth curve

Definitive host. A host that harbors the adult or sexual stage of a parasite

Dehydrogenation (dee-hy'-drah-jen-ay'-shun) **reactions.** Chemical reactions in which a pair of hydrogen atoms is removed from a compound, usually by the action of enzymes called dehydrogenases

Dehydrolysis (dee-hy-drol'-uh-sis). A chemical reaction in which two compounds are joined to form a larger compound and water is released in the process; also called dehydration synthesis

Denitrifying (dee'-ni-truh-fy-ing) **bacteria.** Bacteria capable of converting nitrates into nitrogen gas; the process is known as *denitrification*

Deoxyribonucleic (dee-ox′-ee-ry′-bow-new-clay′-ick) **acid (DNA).** A macromolecule containing the genetic code in the form of genes

Dermatophytes (der-mah′-toh-fytes). Fungi that cause superficial mycoses of the skin, hair, and nails; the cause of tinea infections ("ringworm")

Desiccation (des-uh-kay′-shun). The process of being desiccated (thoroughly dried)

Differential (dif-er-en′-shul) **media.** Culture media which enable microbiologists to readily differentiate one organism or group of organisms from another

Dimorphism (dy-more′-fizm). A phenomenon whereby an organism can exist in two shapes or forms; *e.g.*, dimorphic fungi can exist either as yeasts or molds

Dipeptide (dy-pep′-tide). A protein consisting of two amino acids held together by a peptide bond

Diplobacilli (dip′-low-bah-sill′-eye). Bacilli arranged in pairs

Diplococci (dip′-low-kok′-sigh). Cocci arranged in pairs

Disaccharide (die-sack′-uh-ride). A carbohydrate consisting of two monosaccharides; examples include sucrose (table sugar), lactose (milk sugar), and maltose

Disinfect (dis-in-fekt′). To destroy pathogens in or on any substance or to inhibit their growth and vital activity

Disinfectant (dis-in-fek′-tent). A chemical agent used to destroy pathogens or inhibit their growth activity; usually refers to a chemical agent used on nonliving materials

Disinfection (dis-in-fek′-shun). The process of destroying pathogens and their toxins

DNA polymerase (poh-lim′-er-ace). The enzyme necessary for DNA replication

DNA replication (rep-luh-kay′-shun). Production of two DNA molecules (called daughter molecules) from one parent molecule

Droplet precautions. Specific safety precautions that are practiced in a hospital setting to prevent infections transmitted by droplets

Ecology (ee-kol′-oh-jee). The branch of biology concerned with the total complex of interrelationships among living organisms; encompassing the relationships of organisms to each other, to the environment, and to the entire energy balance within a given ecosystem

Ecosystem (ee′-koh-sis-tem). An ecological system that includes all the organisms and the environment within which they occur naturally

Ectoparasite (ek′-toh-par′-uh-site). A parasite that lives on the external surface of its host

Edema (uh-dee′-muh). Swelling due to an accumulation of watery fluid in cells, tissues, or body cavities; swollen areas are described as being *edematous*

Electrolyte (ee-lek′-troh-lite). A substance that decomposes into ions when placed into water

Electron (ee-lek′-tron) **microscope.** A type of microscope that uses electrons as a source of illumination

Electron transport system. A series of biochemical reactions by which energy is transferred stepwise; a major source of energy in some cells

Element (el′-uh-ment). A substance composed of atoms of only one kind

Encephalitis (en-sef-uh-ly′-tis). Inflammation or infection of the brain

Encephalomyelitis (en-sef-uh-low-my′-uh-ly′-tis). Inflammation or infection of the brain and spinal cord

Endemic (en-dem′-ick) **disease.** A disease that is always present in a community or region

Endocarditis (en′-doh-kar-dy′-tis). Inflammation of the endocardium (the innermost lining of the heart)

Endoenzyme (en′-doh-en′-zym). An enzyme produced by a cell that remains within the cell; an intracellular enzyme

Endoparasite (en-doh-par′-uh-site). A parasite that lives within the body of its host

Endoplasmic reticulum (end-oh-plaz′-mick re-tick′-you-lum) **(ER).** A network of membranous tubules and flattened sacs in the cytoplasm of a eucaryotic cell; ER with attached ribosomes is called rough ER (RER) or granular ER; ER having no attached ribosomes is called smooth endoplasmic reticulum (SER)

Endospores (en′-dough-sporz). Thick-walled, resistant bodies formed within bacterial cells for the purpose of survival; a bacterial cell produces only one endospore, and from that endospore emerges (a process known as germination) one bacterial cell; also referred to as *bacterial spores*

Endosymbiont (en′-doh-sym′-be-ont). The party in a symbiotic relationship that lives within the body of the other symbiont

Endotoxin (en-doh-tok′-sin). The lipid portion of the lipopolysaccharide found in the cell walls of gram-negative bacteria; intracellular toxin

Enriched media. Culture media which enable microbiologists to isolate fastidious organisms from samples or specimens and grow them in the laboratory

Enterotoxin (en-ter-oh-tok′-sin). A bacterial exotoxin specific for cells of the intestinal mucosa

Enzyme (en′-zyme). A protein molecule that catalyzes (causes or speeds up) the occurrence of biochemical reactions; remains unchanged in the process; an organic catalyst

Eosinophil (ee-oh-sin′-oh-fil). A type of granulocyte found in blood; its granules contain basic substances (*e.g.*, major basic protein) that attract acidic dyes

Eosinophilia (ee′-oh-sin-oh-fil′-ee-uh). An abnormally high number of eosinophils in the bloodstream

Epidemic (ep-uh-dem′-ick) **disease.** A disease occurring in a higher than usual number of cases in a population during a given time interval

Epidemiology (ep-uh-dee-me-ol′-oh-jee). The study of relationships between the various factors that determine the frequency and distribution of diseases

Epididymitis (ep-uh-did-uh-my′-tis). Inflammation of the epididymis (a tubular structure within the testis)

Episome (ep′-eh-som). An extrachromosomal element (plasmid) that may either integrate into the host bacterium's chromosome or replicate and function stably when physically separated from the chromosome

Epitope (ep′-uh-tope). An antigenic determinant

Erythema (air-uh-thee-muh). Redness of the skin; a reddened area of skin is described as being *erythematous*

Erythrocytes (ee-rith′-roh-sites). Red blood cells

Erythrogenic (ee-rith-roh-jen′-ick) **toxin.** A bacterial toxin that produces redness, usually in the form of a rash

Eucaryotic (you′-kar-ee-ah′-tick) **cells.** Cells containing a true nucleus; organisms having such cells are referred to as *eucaryotes;* can also be spelled eukaryotic

Exfoliative (eks-foh′-lee-uh-tiv) **toxin.** A bacterial toxin that causes shedding or scaling

Exoenzyme (ek-soh-en′-zyme). An enzyme produced by a cell that is released from the cell; an extracellular enzyme

Exotoxin (ek-soh-tok′-sin). A toxin that is released from the cell; an extracellular toxin

Exudate (eks′-yu-date). Any fluid (*e.g.*, pus) that exudes (oozes) from tissue, often as a result of injury, infection, or inflammation

Facultative (fak′-ul-tay-tive) **anaerobe.** An organism that can live either in the presence or absence of oxygen

Facultative intracellular pathogen. An organism that can live either intracellularly or extracellularly

Facultative parasite. An organism that can be a parasite, but is also capable of a free-living existence

Fastidious (fas-tid′-ee-us) **bacterium.** A bacterium that is difficult to isolate from specimens and grow in the laboratory due to its complex nutritional requirements; other types of microbes may also be fastidious

Fatty acid. Any acid derived from fats by hydrolysis (see below)

Fermentation (fer-men-tay′-shun). An anaerobic biochemical pathway in which substances are broken down and energy and reduced compounds are produced; oxygen does not participate in the process

Fermentative pathways. Metabolic pathways in which oxygen does not participate

Fibrinolysin (fy-brin-oh-ly′-sin). See *kinase*

Fibronectin (fi-bro-nek′-tin). A glycoprotein that acts as an adhesive and as a reticuloendothelial mediated host defense mechanism

Fimbriae (fim′-bree-ee), sing. *fimbria.* See pili

Flagellates (flaj′-eh-letz). Flagellated protozoa

Flagellin (flaj′-eh-lin). See *flagella*

Flagella (fluh-jel′-uh), sing. *flagellum.* Whiplike organelles of motility; procaryotic and eucaryotic flagella differ in structure; procaryotic flagella are composed of a protein called *flagellin;* eucaryotic flagella are composed of nine doublet microtubules arranged around two central microtubules (a "9 + 2 arrangement")

Fomites (foh′-mitz). Inanimate objects or substances capable of absorbing and transmitting a pathogen (*e.g.*, clothing, bed linens, towels, eating utensils)

Fungi (fun′-ji), sing. *fungus.* Eucaryotic, non-photosynthetic microorganisms that are saprophytic or parasitic

Fungicidal (fun-juh-sigh′-dull) **agent.** A chemical agent or drug that kills fungi; a *fungicide* or *mycocide*

Gene (jeen). A functional unit of heredity which occupies a specific space (locus) on a chromosome; capable of directing the formation of an enzyme or other protein

Gene product. The substance (usually a protein) that is coded for by a gene

Gene therapy. The insertion of normally functioning genes into a cell to correct problems associated with abnormally functioning genes

Generalized infection. An infection that has spread throughout the body; also known as a *systemic infection*

Generation time. The time required for a cell to split into two cells; also called the *doubling time*

Genetic (juh-net'-ick) **code**. The sequence of nucleotide bases on a DNA molecule that provides the information necessary for cells to produce gene products

Genetic engineering. The insertion of foreign genes into microorganisms to enable the microorganisms to produce specific gene products or to enable them to be used for other purposes

Genetics (juh-net'-iks). The branch of science concerned with heredity

Genotype (jeen'-oh-type). The complete genetic constitution of an individual; *i.e.*, all of that individual's genes

Genus (jee'-nus), pl. *genera*. The first name in binomial nomenclature; contains closely related species

Germ. Slang term for pathogen

Germicidal (jer-muh-sigh'-dull) **agent.** A chemical agent or drug that kills microorganisms; a *germicide*

Gingivitis (jin-juh-vy'-tis). Inflammation or infection of the gingiva (gums)

Glucose (glue'-kohs). A biologically important six-carbon monosaccharide; a hexose; also called dextrose; the product of complete hydrolysis of polysaccharides such as cellulose, starch, and glycogen

Glycocalyx (gly-ko-kay'-licks). Extracellular material which may or may not be firmly attached to the outer surface of the cell wall; capsules and slime layers are examples

Glycogen (gly'-koh-jen). A polysaccharide stored by animal cells as a food reserve; composed of many glucose molecules

Glycolysis (gly-kol'-eh-sis). The anaerobic, energy-producing breakdown of glucose via a series of chemical reactions; a biochemical pathway; also called anaerobic glycolysis

Golgi (goal'-jee) **complex.** A membranous system located within the cytoplasm of a eucaryotic cell; associated with the transport and packaging of secretory proteins; also known as Golgi apparatus or Golgi body

Gram stain. A differential staining procedure named for its developer, Hans Christian Gram, a Danish bacteriologist; differentiates bacteria into those that stain blue-to-purple (gram-positive) and those that stain pink-to-red (gram-negative)

Granulocyte (gran'-yu-loh-site). A granular leukocyte; neutrophils, eosinophils, and basophils are examples

Growth curve. As used in this book, a graphic representation of the change in size of a bacterial population over a period of time; includes a lag phase, a log phase, a stationary phase, and a death phase

Haloduric (hail-oh-dur'-ick). Capable of surviving in a salty environment

Halophile (hail'-oh-file). An organism whose growth is enhanced by a high salt concentration; such an organism is said to be *halophilic*

Hapten (hap'-ten). A small, nonantigenic molecule that becomes antigenic when combined with a large molecule

HBV. Hepatitis B virus; the etiologic agent of "serum hepatitis"

Helminth (hel'-minth). Parasitic worm

Helminthologist (hel'-min-thol'-oh-jist). One who specializes in helminthology

Helminthology (hel'-min-thol'-oh-je). The study of helminths

Hemolysin (he-moll'-uh-sin). A bacterial enzyme capable of lysing erythrocytes, causing release of their hemoglobin

Hemolysis (he-moll'-uh-sis). Destruction of red blood cells (erythrocytes) in such a manner that hemoglobin is liberated into the surrounding environment

Hepatomegaly (hep'-at-oh-meg'-uh-lee). Enlargement of the liver

Heterotroph (het'-er-oh-trof). An organism that uses organic chemicals as a source of carbon; sometimes called an *organotroph*

Histamine (his'-tuh-meen). Potent chemical released from cells (*e.g.*, basophils and mast cells) during some immune reactions; causes constriction of bronchial smooth muscles and vasodilation

Histiocyte (his'-tee-oh-site) or **histocyte** (his-toh-site). A fixed macrophage; *i.e.*, one that remains in tissue and does not "wander"

HIV. Human immunodeficiency virus; the etiologic agent of AIDS

Hospital-acquired infection. See *nosocomial infection*

Host. The organism on or in which a parasite lives

Humoral immunity. A type of immunity where antibodies play a major role

Hyaluronic (high'-uh-lu-ron'-ick) **acid.** A gelatinous, mucopolysaccharide that acts as an intracellular "cement" in body tissue

Hyaluronidase (high'-uh-lu-ron'-uh-dase). A bacterial enzyme that breaks down hyaluronic acid; sometimes called diffusing or spreading factor, because it enables bacteria to invade deeper into tissue

Hybridoma (high-brid-oh'-muh). A tumor produced *in vitro* by fusion of mouse tumor cells and specific-antibody producing cells; used in the production of monoclonal antibodies

Hydrocarbon (high-droh-kar'-bun). An organic compound consisting of only hydrogen and carbon atoms

Hydrogen (high´-droh-jen) **bond.** A bond arising from the sharing of a hydrogen atom; most commonly, where a hydrogen atom links a nitrogen atom to an oxygen atom or to another nitrogen atom

Hydrolysis (hi-drol´-eh-sis). A chemical process whereby a compound is cleaved into two or more simpler compounds with the uptake of the H and OH parts of a water molecule on either side of the chemical bond that is cleaved

Hypersensitivity (high´-per-sen-suh-tiv´-uh-tee). A condition in which there is an exaggerated immune response that causes tissue destruction or inflammation

Hypertonic (hi-per-tahn´-ick) **solution.** A solution having a greater osmotic pressure than cells placed into that solution; a higher concentration of solutes outside the cell

Hyphae (hy´-fee), sing. *hypha.* Long, thin, intertwined, cytoplasmic filaments that make up a mold colony (*mycelium*)

Hypogammaglobulinemia (high´-poh-gam´-uh-glob-yu-luh-nee´-me-uh). Decreased quantity of the gamma fraction of serum globulin, including a decreased quantity of immunoglobulins

Hypotonic (hi-poh-tahn´-ick) **solution.** A solution having a lower osmotic pressure than cells placed into that solution; a lower concentration of solutes outside the cell

Iatrogenic (eye-at-roh-jen´-ick) **infection.** An infection induced by the treatment itself; literally, "physician induced," but could be caused by any healthcare professional

Immune (im-myun´). Free from the possibility of acquiring a particular infectious disease

Immunity (im-myu´-nuh-tee). Being immune or resistant

Immunocompetent (im´-you-no-kom´-puh-tent) **person.** A person who is able to mount a normal immune response

Immunodiagnostic (im´-yu-noh-dy-ag-nos´-tick) **procedures.** Diagnostic test procedures that utilize the principles of immunology; used to detect either antigen or antibody in patients' specimens

Immunogen (ih-myu´-noh-jen). Another name for antigen

Immunoglobulin (im´-yu-noh-glob´-yu-lin). A class of proteins, consisting of two light polypeptide chains and two heavy chains; all antibodies are immunoglobulins, but some immunoglobulins are not antibodies

Immunologist (im-you-nol´-oh-jist). One who specializes in the science of immunology

Immunology (im-you-nol´-oh-je). The study of immunity and the immune system

Immunosuppressed (im´-you-no-sue-pressed) **person.** A person who cannot mount a normal immune response due to suppression or depression of their immune system

In vitro (in vee´-trow). In an artificial environment, as in a laboratory setting; often used in reference to what occurs *outside* an organism

In vivo (in vee´-voh). In a living organism; used in reference to what occurs *within* a living organism

Inactivated vaccines. Vaccines prepared from inactivated (killed) microorganisms

Inclusion bodies. Distinctive structures frequently formed in the nucleus and/or cytoplasm of cells infected with certain viruses

Incubatory (in´-kyu-buh-tor´-ee) **carrier.** A person capable of transmitting a pathogen during the incubation period of a particular infectious disease

Indigenous microflora (in-dij´-uh-nus my-crow-floor-uh). Microorganisms that live on and in the healthy body; also called *indigenous microbiota;* referred to in the past as "normal flora"

Infection (in-fek´-shun). The presence and multiplication of pathogens in or on the body; sometimes used as a synonym for infectious disease

Infectious disease (in-fek´-shus di-zeez´). Any disease caused by a microorganism

Infestation (in-fes-tay´-shun). The presence of ectoparasites (*e.g.,* lice) on the body

Inflammation (in-fluh-may´-shun). A nonspecific pathologic process consisting of a dynamic complex of cytologic and histologic reactions that occur in response to an injury or abnormal stimulation by a physical, chemical, or biologic agent

Inorganic (in-or-gan´-ick) **chemistry.** The science dealing with all types of chemicals except those classified as organic compounds

Inorganic compounds. Chemical compounds in which the atoms or radicals are held together by electrostatic forces rather than by covalent bonds

Interferons (in-ter-fear´-onz). Small, antiviral glycoproteins, produced by cells infected with an animal virus; interferons are cell-specific and species-specific, but not virus-specific

Interleukins (in-ter-lu´-kinz). Lymphokines and polypeptide hormones; interleukin-1 is produced by monocytes; interleukin-2 is produced by lymphocytes

Intermediate host. A host that harbors the larval or asexual stage of a parasite

Ion (eye′-on). A positively or negatively charged atom or group of atoms

Ionic (eye-on′-ick) **bond.** An electrostatic bond

Isotonic (eye-soh-tahn′-ick) **solution.** A solution having the same osmotic pressure as cells placed into that solution; a concentration of solutes outside the cell equal to the concentration of solutes inside the cell

Isotopes (eye′-so-topes). Atoms that are chemically identical, but their nuclei contain different numbers of neutrons; for example, ^{12}C and ^{14}C are isotopes of carbon; ^{14}C has two more neutrons than ^{12}C

Keratitis (ker-uh-tie′-tis). Inflammation of the cornea

Keratoconjunctivitis (ker′-at-oh-kon-junk′-tuh-vi′-tis). Inflammation of the cornea and conjunctiva

Killer cell. A type of cytotoxic T cell involved in cell-mediated immune responses

Kinase (ky′-nace). A bacterial enzyme capable of dissolving clots; also known as *fibrinolysin*

Koch's postulates. A series of scientific steps, proposed by Robert Koch, that must be fulfilled in order to prove that a specific microorganism is the cause of a particular disease

L-forms. Abnormal forms of bacteria that have lost part or all of their rigid cell walls; sometimes the result of exposure of an organism to an antimicrobial agent; also called L-phase variants; the "L" is derived from Lister Institute

Lag phase. That part of a bacterial growth curve during which multiplication of the organisms is very slow or scarcely appreciable; the first phase in a bacterial growth curve

Latent infection. An asymptomatic infection capable of manifesting symptoms under particular circumstances or if activated

Lecithin (less′-uh-thin). A name given to several types of phospholipids that are essential constituents of animal and plant cells

Lecithinase (less′-uh-thuh-nace). A bacterial enzyme capable of breaking down lecithin

Leukocidin (lu-koh-sigh′-din). A bacterial toxin capable of destroying leukocytes

Leukocytes (lu′-koh-sites). White blood cells

Leukocytosis (lu′-koh-sigh-toe′-sis). An increased number of leukocytes in the blood

Leukopenia (lu-koh-pea′-nee-uh). A decreased number of leukocytes in the blood

Lichen (like′-in). An organism composed of a green alga (or a cyanobacterium) and a fungus; an example of a symbiotic relationship known as *mutualism*

Light microscope. A type of microscope that uses visible light as a source of illumination; also called a brightfield microscope

Lipids (lip′-ids). Organic compounds containing carbon, hydrogen, and oxygen that are insoluble in water but soluble in so-called "fat" solvents such as diethyl ether and carbon tetrachloride

Lipopolysaccharide (lip′-oh-pol-ee-sack′-a-ride). A macromolecule of combined lipid and polysaccharide, found in the cell walls of gram-negative bacteria

Lithotroph (lith′-oh-trof). An organism that uses inorganic molecules as a source of carbon

Local infection. An infection that remains localized; that does not spread; also known as a *focal infection*

Logarithmic (log′-uh-rith-mik) **growth phase.** That part of a bacterial growth phase during which maximal multiplication is occurring by geometrical progression; plotting the logarithm (log) of the number of organisms against time produces a straight upward-pointing line; the second phase in a bacterial growth curve; also known as the log phase or exponential growth phase

Logarithmic (log′-uh-ryth-mik) **scale.** A scale (as on graph paper) in which the values of a variable (*e.g.*, number of organisms at a particular point in time) are expressed as logarithms

Lophotrichous (low-fot′-ri-kus) **bacteria.** Bacteria possessing two or more flagella at one or both ends (poles) of the cell

Lymphadenitis (lim′-fad-uh-ny′-tis). Inflammation of a lymph node or lymph nodes

Lymphadenopathy (lim-fad-uh-nop′-uh-thee). A disease process affecting a lymph node or lymph nodes

Lymphangitis (lim-fan-ji′-tis). Inflammation of the lymphatic vessels

Lymphocytosis (lim′-foh-sigh-toe′-sis). An increased number of lymphocytes in the blood

Lymphokines (lim′-foh-kinz). Soluble proteins released by sensitized lymphocytes; examples include chemotactic factors and interleukins; lymphokines represent one category of **cytokines**

Lyophilization (ly-ahf′-eh-leh-zay′-shun). Freeze-drying; a method of preserving microorganisms and foods

Lysogenic (lye-so-jen′-ick) **bacteria.** Pertaining to bacteria in the state of lysogeny

Lysogenic conversion. Alteration of the genetic constitution of a bacterial cell due to the integration of viral genetic material into the host cell genome

Lysogenic cycle. When viral genetic material remains latent or inactive in an infected host cell

Lysogeny (lye-soj′-eh-nee). When viral genetic material is integrated into the genome of the host cell

Lysosomes (lye′-so-somz). Membrane-bound vesicles found in the cytoplasm of eucaryotic cells, containing a variety of digestive enzymes, including lysozyme

Lysozyme (lye′-so-zyme). A digestive enzyme found in lysosomes, tears, and other body fluids; especially destructive to bacterial cell walls

Lytic cycle. When a virus takes over the metabolic "machinery" of the host cell, reproduces itself, and ruptures (lyses) the host cell to allow the newly assembled virions to escape

Macrophage (mak′-roh-faj). A large phagocytic cell that arises from a monocyte

Malaise (muh-laz′). A generalized feeling of discomfort or uneasiness

Mast cell. A tissue cell that closely resembles a basophil

Mastigophora (mas′-ti-gof′-uh-rah). A subphylum of Protozoa in the phylum Sarcomastigophora; the flagellates; considered a phylum in some classification schemes

Mastitis (mass-ty′-tis). Inflammation of the breast

Mechanical vector. A vector that merely transports a pathogen to a susceptible host, but within which the pathogen neither multiplies nor matures

Medical asepsis (ay-sep′-sis). The absence of pathogens in a patient's environment

Medical aseptic (ay-sep′-tick) **technique.** Procedures followed and steps taken to ensure medical asepsis

Meninges (muh-nin′-jez)., sing. *meninx*. As used in this book, the membranes that surround the brain and spinal cord

Meningitis (men-in-ji′-tis). Inflammation or infection of the meninges

Meningoencephalitis (muh-ning′-go-en-sef-uh-ly′-tis). Inflammation or infection of the brain and its surrounding membranes

Mesophile (meez′-oh-file). A microorganism having an optimum growth temperature between 25°C and 40°C; such an organism is said to be *mesophilic*

Mesosome (me′-so-zom). A procaryotic cell organelle (an infolding of the cytoplasmic membrane) possibly involved in cellular respiration

Messenger RNA (mRNA). The type of RNA that contains the exact same genetic information as a single gene on a DNA molecule; also called informational RNA

Metabolism (muh-tab′-oh-lizm). The sum of all the chemical reactions occurring in a cell; consists of *anabolism* and *catabolism*

Metabolite (muh-tab′-oh-lite). Any product of metabolism

Microaerophiles (my-krow-air′-oh-files). Organisms requiring oxygen, but in concentrations lower than the 20–21% found in air; usually around 5%

Microbial (my-krow′-be-ul). Pertaining to microorganisms

Microbial antagonism (an-tag′-un-izm). The killing, injury, or inhibition of one microbe by the substances produced by another

Microbial ecology. Study of the interrelationships among microbes and other microbes, other living organisms, and the non-living environment

Microbicidal (my-krow′-buh-sigh′-dull) **agent.** A chemical or drug that kills microorganisms; a **microbicide**

Microbiologist (my′-crow-by-ol′-oh-jist). One who specializes in the science of microbiology

Microbiology (my′-crow-by-ol′-oh-je). The study of microorganisms

Microbistatic (my-krow′-buh-stat′-ick) **agent.** A chemical agent or drug that inhibits the growth of microorganisms

Micrometer (my′-crow-me-ter). A unit of length, equal to one-millionth of a meter

Microorganisms (my′-crow-or′-gan-izms). Very small organisms; usually microscopic; also called *microbes;* includes algae, bacteria, fungi, protozoa, and viruses

Microscope (my′-crow-skope). An optical instrument that permits one to observe a small object by producing an enlarged image of the object

Microtubules (my-kro′-two-bules). Cylindrical, cytoplasmic tubules found in the cytoskeleton of eucaryotic cells; may be related to the movement of chromosomes during nuclear division

Mitochondria (my-toe-kon′-dree-uh), sing. *mitochondrion*. Eucaryotic organelles involved in cellular respiration for the production of energy; "energy factories" of the cell

Mitosis (my-toe′-sis). A process of cell reproduction consisting of a sequence of modifications of the nucleus that result in the formation of two daughter cells with exactly the same chromosome and DNA content as that of the original cell

Molecule (mol′-ee-kyul). The smallest possible quantity of a di-, tri-, or polyatomic substance that retains the chemical properties of the substance

Monoclonal (mon-oh-klo′-nul) **antibodies.** Antibodies produced by a clone of genetically identical hybrid cells

Monocyte (mon′-oh-site). A relatively large mononuclear leukocyte

Monokines (mon′-oh-kinz). Soluble protein mediators released by sensitized monocytes and macrophages; monokines represent one category of *cytokines*

Monosaccharides (mon-oh-sak′-uh-rides). Carbohydrates that cannot be broken down into any simpler sugar by simple hydrolysis; simple sugars containing 3 to 7 carbon atoms; the basic units of polysaccharides

Monotrichous (mah-not′-ri-kus) **bacteria.** Bacteria possessing only one flagellum

Motile (mow′-till). Possessing the ability to move

Mutagen (myu′-tah-jen). Any agent that can cause a mutation; *e.g.,* radioactive substances, x-rays, or certain chemicals; such an agent is said to be mutagenic

Mutant (myu′-tant). A phenotype in which a mutation is manifested

Mutation (myu-tay′-shun). An inheritable change in the character of a gene; a change in the sequence of base pairs in a DNA molecule

Mutualism (myu′-chew-ul-izm). A symbiotic relationship in which both parties derive benefit

Mycelium (my-see′-lee-um), pl. *mycelia.* A fungal colony; composed of a mass of intertwined hyphae

Mycologist (my-kol′-oh-jist). One who specializes in the science of mycology

Mycology (my-kol′-oh-gee). The study of fungi

Mycosis (my-ko′-sis), pl. *mycoses.* A fungal disease

Mycotoxins (my′-ko-tox-inz). Toxins produced by fungi

Myelitis (my-uh-ly′-tis). Inflammation or infection of the spinal cord

Myocarditis (my′-oh-kar-dy′-tis). Inflammation of the myocardium (the muscular walls of the heart)

Nanometer (nan′-oh-me-ter). A unit of length, equal to one-billionth of a meter

Natural (NK) killer cell. A type of cytotoxic human blood lymphocyte

Necrotoxin (nek′-roh-tok′-sin). An exotoxin that destroys cells

Negative stain. A staining procedure in which unstained objects can be seen against a stained background

Nephritis (nef-ry′-tis). Inflammation of the kidneys

Neurotoxin (new′-roh-tok′-sin). A bacterial toxin that attacks the nervous system

Neutralism (new′-trul-izm). A symbiotic relationship in which organisms occupy the same niche but do not affect one another

Neutrophil (nu′-tro-fil). A type of granulocyte found in blood; its granules contain neutral substances that attract neither acidic nor basic dyes; also called a *polymorphonuclear cell, poly,* or *PMN*

Nitrifying bacteria. Bacteria capable of converting ammonia to nitrites and nitrates; the process is known as *nitrification*

Nitrogen-fixing bacteria. Bacteria capable of converting nitrogen gas to ammonia; the process is known as *nitrogen fixation*

Nonendemic (non′-en-dem′-ick) **disease.** A disease that is not always present (not endemic) in a particular community or region

Nonpathogen (non′-path′-oh-jen). A microorganism that does not cause disease; such an organism is said to be *nonpathogenic*

Nosocomial (nose-oh-koh′-me-ul) **infection.** Any infection acquired while one is hospitalized; a hospital-acquired infection

Nuclear (new′-klee-er) **membrane.** The membrane that surrounds the chromosomes and nucleoplasm of a eucaryotic cell

Nucleic (new-klay′-ick) **acids.** Macromolecules consisting of linear chains of nucleotides; DNA, mRNA, tRNA, and rRNA are examples

Nucleolus (new-klee′-oh-lus). A dense portion of the nucleus, where ribosomal RNA (rRNA) is produced

Nucleoplasm (new′-klee-oh-plazm). That portion of a cell's protoplasm that lies within the nucleus

Nucleotides (new′-klee-oh-tides). The basic units or "building blocks" of nucleic acids, each consisting of a purine or pyrimidine combined with a pentose (ribose or deoxyribose) and a phosphate group

Nucleus (new′-klee-us), pl. *nuclei.* That portion of a eucaryotic cell that contains the nucleoplasm, chromosomes, and nucleoli

Obligate aerobe (air′-obe). An organism that requires 20–21% oxygen (the amount found in the air we breathe) to survive

Obligate anaerobe (an′-air-obe). An organism that cannot survive in oxygen

Obligate intracellular pathogen. A pathogen that must reside within another living cell; examples include viruses, chlamydias, and rickettsias

Obligate parasite. An organism that must be a parasite; incapable of a free-living existence

Octad. A packet of eight cocci.

Oncogenic (ong-koh-jen'-ick). An adjective meaning cancer-causing

Oophoritis (oh-of-or-eye'-tis). Inflammation or infection on an ovary

Opportunistic pathogen (op-poor-tune'-is-tick path'-oh-jen). A microbe with the potential to cause disease, but does not do so under ordinary circumstances; may cause disease in susceptible persons with lowered resistance; also called an opportunist

Opsonins (op'-soh-ninz). Substances (such as antibodies or complement components) that enhance phagocytosis

Opsonization (op'-suh-nuh-zay'-shun). The process by which bacteria are altered to be more readily and more efficiently engulfed by phagocytes; often involves coating the bacteria with antibodies and/or complement components

Orchitis (or-ky'-tis). Inflammation or infection of the testes

Organelles (or'-guh-nelz). General term for the various and diverse structures contained within a cell (*e.g.*, mitochondria, Golgi complex, nucleus, endoplasmic reticulum, and lysosomes)

Organic (or-gan'-ick) **chemistry.** The science concerned with covalently linked atoms, centering around carbon compounds of this type

Organic compounds. Chemical compounds composed of atoms held together by covalent bonds

Osmosis (oz-moh'-sis). The process by which a solvent (*e.g.*, water) moves through a semipermeable membrane from a solution having a lower concentration of solutes (dissolved substances) to a solution having a higher concentration of solutes

Osmotic (oz-maht'-ick) **pressure.** A measure of the tendency for water to move into a solution by osmosis; always a positive value

Oxidation (ok-seh-day'-shun). As used in this book, the loss of one or more electrons, thus, making an atom more electropositive

Oxidative pathways. Metabolic pathways requiring the participation of oxygen

Oxidation-reduction reactions. Any chemical reaction which must, *in toto*, comprise both oxidation and reduction; sometimes referred to as *redox reactions*

Pandemic (pan-dem'-ick) **disease.** A disease in epidemic proportions worldwide

Parasite (par'-uh-sight). An organism that lives on or in another living organism (called the host)

Parasitism (par'-uh-suh-tizm). A symbiotic relationship in which one party (the parasite) derives benefit at the expense of the other party (the host)

Parasitologist (par'-uh-suh-tol'-oh-jist). One who specializes in the science of parasitology

Parasitology (par'-uh-suh-tol'-oh-jee). The study of parasites

Parenteral (puh-ren'-ter-ul) **injection.** Injection of substances directly into the bloodstream

Parotitis (par-oh-ty'-tis). Inflammation of the parotid gland (a salivary gland located near the ear); also known as *parotiditis*

Passive acquired immunity. Immunity or resistance acquired as a result of receipt of antibodies produced by another person or by an animal

Passive carrier. A person who harbors a particular pathogen without ever having had the infectious disease it causes

Pasteurization (pas'-tour-i-zay'-shun). A heating process that kills pathogens in milk, wines, and other beverages

Pathogen (path'-oh-jen). Disease-causing microorganism; such an organism is said to be *pathogenic*

Pathogenicity (path'-oh-juh-nis'-uh-tee). The ability to cause disease

Pathologist (pah-thol'-oh-jist). A physician who is a specialist in pathology

Pathology (pah-thol-oh-gee). The study of disease, especially structural and functional changes that result from disease processes

Pellicle (pel'-uh-kul). As used in this book, a thickened outer membrane possessed by certain protozoa

Peptidoglycan (pep'-tuh-doh-gly'-kan). A complex structure found in the cell walls of eubacteria, consisting of carbohydrates and proteins

Pericarditis (per'-ee-kar-dy'-tis). Inflammation of the pericardium (the membrane or sac around the heart)

Periodontitis (purr'-ee-oh-don-ty'-tis). Inflammation or infection of the *periodontium* (tissues that surround and support the teeth)

Peritrichous (peh-rit'-ri-kus) **bacteria.** Bacteria possessing flagella over their entire surface

Petri (pea′-tree) **dish.** A shallow, circular container made of thin glass or clear plastic, with a loosely fitting, overlapping cover; used in microbiology laboratories for cultivation of microorganisms on solid media

Phagocyte (fag′-oh-site). A cell capable of ingesting bacteria, yeasts, and other particulate matter by phagocytosis; amebae and certain white blood cells are examples of phagocytic cells

Phagocytosis (fag′-oh-sigh-toe′-sis). Ingestion of particulate matter involving the use of pseudopodia to surround the matter

Phagolysosome (fag-oh-ly′-soh-sohm). A membrane-bound vesicle formed by the fusion of a phagosome and a lysosome

Phagosome (fag′-oh-sohm). A membrane-bound vesicle containing an ingested particle (*e.g.*, a bacterium); found in phagocytic cells

Phenol (fee′-nol) **coefficient test.** A laboratory procedure used to determine the effectiveness of a disinfectant as compared to the effectiveness of phenol

Phenotype (fee′-no-type). Manifestation of a genotype; all of the attributes or characteristics of an individual

Photoautotroph (foh′-toe-aw′-toe-trof). An organism that uses light as an energy source and carbon dioxide as a carbon source; a type of autotroph

Photoheterotroph (foh′-toe-het′-er-oh-trof). An organism that uses light as an energy source and organic compounds as a carbon source; a type of heterotroph; sometimes referred to as a *photoorganotroph*

Photolithotroph (foh′-toe-lith′-oh-trof). An organism that uses light as an energy source and inorganic chemicals as a carbon source; a type of lithotroph

Photosynthesis (foe-toe-sin′-thuh-sis). Production of organic substances, using light energy; a cell that produces organic substances in this manner is said to be *photosynthetic*

Phototroph (foh′-toe-trof). An organism that uses light as an energy source

Phycologist (fy-kol′-oh-jist). One who specializes in the science of phycology

Phycology (fy-kol′-oh-gee). The study of algae

Phycotoxins (fy′-ko-tox-inz). Toxins produced by algae

Pili (py′-ly), sing. *pilus.* Hairlike surface projections possessed by some bacteria (called piliated bacteria); most are organelles of attachment; also called *fimbriae;* specialized pili, called *sex pili,* are described below

Pinocytosis (pin′-oh-sigh-toe′-sis). A process resembling phagocytosis, but used to engulf and ingest liquids rather than solid matter

Plasma (plaz′-muh). The liquid portion of circulating blood

Plasma (plaz′-muh) **cell.** An antibody-secreting cell produced from a stimulated B cell

Plasmid (plaz′-mid). An extrachromosomal genetic element; a molecule of DNA that can stably function and replicate while physically separate from the bacterial chromosome

Plasmolysis (plaz-moll′-uh-sis). Cell shrinkage due to a loss of water from the cell's cytoplasm

Plasmoptysis (plaz-mop′-tuh-sis). The escape of cytoplasm from a ruptured cell

Plastids. Membrane-bound organelles containing photosynthetic pigment; the sites of photosynthesis; a *chloroplast* (containing chlorophyll) is a type of plastid

Pleomorphism (plee-oh-more′-fizm). Existing in more than one form; also known as polymorphism; an organism that exhibits pleomorphism is said to be *pleomorphic*

Polymer (pol′-uh-mer). A large molecule consisting of repeated units; nucleic acids, polypeptides, and polysaccharides are examples

Polypeptide (pol-ee-pep′-tide). A protein consisting of more than three amino acids held together by peptide bonds

Polyribosomes (pol-ee-ry′-boh-somz). Two or more ribosomes connected by a molecule of messenger RNA (mRNA)

Polysaccharide (pol-ee-sack′-uh-ride). Carbohydrate consisting of many sugar units; glycogen, cellulose, and starch are examples

Precipitate (pre-sip′-uh-tate). As a noun, a solid material that separates out of a solution or suspension

Precipitin (pre-sip′-uh-tin). A type of precipitate formed by the combination of antigen and specific antibody

Precipitin tests. Immunodiagnostic procedures in which antigen-antibody reactions result in the formation of *precipitin*

Primary disease. The initial disease; may predispose the patient to secondary disease(s); if the primary disease is an infection, it is referred to as a *primary infection*

Prion (pree′-on). An infectious agent consisting of at least one protein, but no demonstrable nucleic acid

Procaryotic (pro'-kar-ee-ah'-tick) **cells.** Cells lacking a true nucleus; organisms having such cells are referred to as *procaryotes;* can also be spelled prokaryotic

Proctitis (prok-ty-tis). Inflammation of the mucous membrane of the rectum

Properdin (pro-pare'-din). A normal serum gamma globulin that participates in the complement pathway

Prophage (pro'-faj). A temperate bacteriophage mutant whose genome does not contain all of the normal components and cannot become fully infective, yet can replicate indefinitely in the bacterial genome; also known as a *defective bacteriophage*

Prophylactic (pro'-fuh-lak'-tick) **agent.** A drug used to prevent a disease

Prophylaxis (pro-fuh-lak'-sis). Prevention of a disease or a process that can lead to a disease; *e.g.,* taking antimalarial medication in a malarious area

Prostaglandins (pros-tuh-glan'-dinz). Physiologically active tissue substances that cause many effects, including vasodilation, vasoconstriction, and stimulation of smooth muscle

Prostatitis (pros-tuh-ty'-tis). Inflammation or infection of the prostate

Prostration (pros-tray'-shun). Marked loss of strength; the patient is lying flat (prostrate)

Proteins (pro'-teens). Macromolecules consisting of two, three, or more amino acids

Protoplasm (pro'-toe-plazm). The semifluid matter within living cells; *cytoplasm* and *nucleoplasm* are two types of protoplasm

Protoplast (pro'-toe-plast). A bacterial cell which has lost its cell wall; the bacterium loses its characteristic shape and becomes round

Protozoa (pro-toe-zoe'-uh), sing. *protozoan.* Eucaryotic microorganisms frequently found in water and soil; some are pathogens; usually unicellular

Protozoologist (pro'-toe-zoe-ol'-oh-jist). One who specializes in protozoology

Protozoology (pro'-toe-zoe-ol'-oh-gee). The study of protozoa

Pseudopodium (su-doe-poh'-dee-um), pl. *pseudopodia.* A temporary extension of protoplasm that is extended by an ameba or white blood cell for locomotion or the engulfment of particulate matter; also called a pseudopod

Psychroduric (sigh-krow-dur'-ick) **organisms.** Organisms able to endure very cold temperatures

Psychrophile (sigh'-krow-file). An organism that grows best at a low temperature (0 to 32°C), with optimum growth occurring at 15 to 20°C; such an organism is said to be *psychrophilic*

Psychrotroph (sigh'-krow-trof). A psychrophile that grows best at refrigerator temperature (4°C); such an organism is said to be *psychrotrophic*

Purine (pure'-een). A molecule found in certain nucleotides and, therefore, in nucleic acids; adenine and guanine are purines found in both DNA and RNA

Pus. A fluid product of inflammation, containing leukocytes and dead cells

Pyelonephritis (py'-uh-low-nef-ry'-tis). Inflammation of certain areas of the kidneys, most often the result of bacterial infection

Pyogenic (py-oh-jen'-ick). Pus-producing; causing the production of pus

Pyrimidine (pi-rim'-uh-deen). A molecule found in certain nucleotides and, therefore, in nucleic acids; thymine and cytosine are pyrimidines found in DNA; cytosine and uracil are pyrimidines found in RNA

Pyrogen (py'-roh-jen). An agent that causes a rise in body temperature; such an agent is said to be *pyrogenic*

Reduction (ree-duk'-shun). As used in this book, the gain of one or more electrons, thus, making an atom more electronegative

Reservoirs (rez'-ev-wars) **of infection.** Living or nonliving material in or on which a pathogen multiplies and/or develops

Resident microflora. Members of the indigenous microflora which are more or less permanent

Resolving power. The ability of the eye or an optical instrument to distinguish detail, such as the separation of closely adjacent objects; also called *resolution*

Reticuloendothelial (ree-tick'-yu-loh-en-doh-thee'-lee-ul) **system (RES).** A collection of phagocytic cells that includes macrophages and cells that line the sinusoids of the spleen, lymph nodes, and bone marrow

Reverse isolation. When a patient is isolated to protect him or her from infection

Ribonucleic (ry-boe-new-klee'-ick) **acid (RNA).** A macromolecule of which there are three main types: messenger RNA (mRNA), ribosomal RNA (rRNA), and transfer RNA (tRNA); found in all cells, but only in certain viruses (RNA viruses)

Ribosomal (rye-boh-so'-mul) **RNA (rRNA).** The type of RNA molecule found in ribosomes

Ribosomes (ry'-boh-soams). Organelles which are the sites of protein synthesis in both procaryotic and eucaryotic cells

RNA polymerase (poh-lim'-er-ace). The enzyme necessary for transcription (see below)

Rough endoplasmic reticulum (RER). See *endoplasmic reticulum*

Salpingitis (sal-pin-jy'-tis). Inflammation of the fallopian tube

Salt. A compound formed by the interaction of an acid and a base; ionizable in water; sodium chloride (NaCl) is an example

Sanitization (san'-uh-tuh-zay'-shun). The process of making something sanitary (healthful)

Saprophyte (sap'-row-fight). An organism that lives on dead or decaying organic matter; such an organism is said to be *saprophytic*

Sarcodina (sar'-ko-dy'-nah). A subphylum of protozoa in the phylum Sarcomastigophora; includes the amebae; considered a phylum in some classification schemes

Sarcomastigophora (sar'-ko-mass-ti-gof'-oh-rah). A phylum of protozoa of the subkingdom Protozoa, characterized by flagella, pseudopodia, or both; contains the subphyla *Sarcodina* and *Mastigophora*

Sebum (see'-bum). The oily secretion produced by sebaceous glands of the skin

Secondary disease. A disease that follows the initial disease; if the secondary disease is an infection, it is referred to as a *secondary infection*

Selective media. Culture media which allow a certain organism or group of organisms to grow while inhibiting growth of all other organisms

Selective permeability. An attribute of membranes whereby only certain substances are able to cross the membranes

Sepsis. The presence of pathogens and/or their toxins in the bloodstream; may lead to *septicemia*

Septic shock. A type of shock resulting from sepsis or septicemia

Septicemia (sep-tuh-see'-me-uh). A serious disease consisting of chills, fever, prostration, and the presence of pathogens and/or their toxins in the blood

Serologic (ser-oh-loj'-ick) **procedures.** Immunodiagnostic test procedures performed using serum

Serology (suh-rol'-oh-jee). That branch of science concerned with serum and serologic procedures

Serum (seer'-um), pl. *sera*. The liquid portion of blood following coagulation (clotting)

Sex pilus. A specialized pilus through which one bacterial cell (the donor cell) transfers genetic material to another bacterial cell (the recipient cell), in a process called *conjugation*

Shock. A sudden, often severe, physical and/or mental disturbance, usually resulting from low blood pressure and a lack of oxygen in organs

Signs of a disease. Abnormalities indicative of disease that are discovered on examination of a patient; objective findings; examples include abnormal lab results; abnormal heart or breath sounds; lumps; abnormalities revealed by x-rays, CAT scans, MRI, EKG, and ultrasound

Simple microscope. A microscope containing only one magnifying lens

Simple stain. A single dye used to stain bacteria or other microbes

Slime layer. A non-organized, non-attached layer of glycocalyx surrounding a bacterial cell

Slime mold. A eucaryotic organism having characteristics of protozoa and fungi; there are two types: cellular and acellular slime molds

Smooth endoplasmic reticulum (SER). See *endoplasmic reticulum*

Solute (sol'-yute). The dissolved substance in a solution; for example, sucrose (table sugar) dissolved in water

Solution (soh-loo'-shun). A homogenous molecular mixture; generally, a substance dissolved in water (an aqueous solution)

Solvent (sol'-vent). A liquid in which another substance dissolves

Source isolation. When a patient is isolated to protect other persons from infection

Species (spe'-shez), pl. *species*. A specific member of a given genus; *e.g., Escherichia coli* is a species in the genus *Escherichia;* the name of a particular species consists of two parts—the generic name ("the first name") and the specific epithet ("the second name"); singular species is abbreviated sp., while plural species is abbreviated spp.

Specific epithet. The second part ("second name") in the name of a species; taken by itself, the specific epithet is not a name

Splenomegaly (splen-oh-meg'-uh-lee). Enlargement of the spleen

Spirochetes (spy'-roh-keets). Spiral-shaped bacteria; *e.g., Treponema pallidum,* the etiologic agent of syphilis

Sporadic (spoh-rad'-ick) **disease.** A disease that occurs occasionally, usually affecting one person; neither endemic nor epidemic

Sporicidal (spor-uh-sigh'-dull) **agent.** A chemical agent that kills spores; a *sporicide*

Sporozoea (spor-oh-zoh′-ee-uh). A large class of protozoa, containing organisms that do not move by cilia, flagella, or pseudopodia; includes the malarial parasites; considered a phylum in some classification schemes

Sporulation (spor′-you-lay′-shun). Production of spores

Standard precautions. Safety precautions taken by healthcare workers to protect themselves and their patients from infection; these precautions are taken for *all* patients and *all* patient specimens (body substances); incorporates safety precautions previously referred to as universal precautions or universal body substance precautions

Staphylococci (staff′′-eh-low-kok′-sigh). Cocci arranged in clusters, such as in the genus *Staphylococcus*

Staphylokinase (staf′-uh-low-ky′-nace). A kinase produced by *Staphylococcus aureus*

Starch. A polysaccharide storage material found in plants

Stationary phase. That part of a bacterial growth phase during which organisms are dying at the same rate at which new organisms are being produced; the third phase in a bacterial growth curve

STD. Sexually transmitted disease

Sterile (stir′-ill). Free of all living microorganisms, including spores

Sterile techniques. Techniques used in an attempt to create an environment that is sterile (devoid of microorganisms)

Sterilization (stir′-uh-luh-zay′-shun). The destruction of *all* microorganisms in or on something (*e.g.,* in water or on surgical instruments)

Streptobacilli (strep′-toh-bah-sill′-eye). Bacilli arranged in chains of varying lengths

Streptococci (strep′-toh-kok′-sigh). Cocci arranged in chains of varying lengths, such as in the genus *Streptococcus*

Streptokinase (strep′-toh-ky′-nace). A kinase produced by streptococci

Subclinical disease. See *asymptomatic disease*

Substrate (sub′-strayt). The substance that is acted upon or changed by an enzyme

Superinfection (sue′-per-in-fek′-shun). An overgrowth of one or more particular microorganisms; often microorganisms that are resistant to an antimicrobial agent that a patient is receiving

Surgical asepsis. The absence of microorganisms in a surgical environment (*e.g.,* an operating room)

Surgical aseptic technique. Procedures followed and steps taken to ensure surgical asepsis

Symbiont (sim′-bee-ont). One of the parties in a symbiotic relationship

Symbiosis (sim-bee-oh′-sis). The living together or close association of two dissimilar organisms

Symptomatic disease. A disease in which the patient experiences symptoms

Symptoms of a disease. Indications of disease that are experienced by the patient; subjective; examples include aches and pains, chills, blurred vision, nausea

Synergism (sin′-er-jizm). As used in this book, the correlated action of two or more microorganisms so that the combined action is greater than that of each acting separately; *e.g.,* when two microbes accomplish more than either could do alone

Systemic infection. See *generalized infection*

T cell. T lymphocyte; a type of leukocyte that plays important roles in the immune system

Taxa, sing. **taxon**. The names given to various groups in taxonomy; the usual taxa are kingdoms, phyla (or divisions), classes, orders, families, genera, species, and subspecies

Taxonomy (tak-sawn′-oh-me). The systematic classification of living things

Teichoic (tie-ko′-ick) **acids.** Polymers found in the cell walls of gram-positive bacteria

Temperate bacteriophage. A bacteriophage whose genome incorporates into and replicates with the genome of the host bacterium

Tetanospasmin (tet′-uh-noh-spaz′-min). The neurotoxin produced by *Clostridium tetani;* causes tetanus

Tetrad. A packet of four cocci.

Thermal death point (TDP). The temperature required to kill all microorganisms in a liquid culture in 10 minutes at pH 7

Thermal death time (TDT). The length of time required to kill all microorganisms in a liquid culture at a given temperature

Thermoduric (ther-mow-du′-rik) **organisms.** Organisms able to survive high temperatures

Thermophile (ther′-mow-file). An organism that thrives at a temperature of 50°C or higher; such an organism is said to be *thermophilic*

Tinea (tin′-ee-uh) **infections.** Fungal infections of the skin, hair, and nails ("ringworm"); named for the part of the body that is affected; *e.g.,* tinea capitis is a fungal infection of the scalp, tinea pedis is athlete's foot, tinea unguium is a fungal infection of the nails

Toxemia (tok-see′-me-uh). The presence of toxins in the blood

Toxigenicity (tok′-suh-juh-nis′-uh-tee) or toxinogenicity (tok′-suh-no-juh-nis′-uh-tee). The capacity to produce toxin; a microorganism capable of producing a toxin is said to be **toxigenic** (or toxinogenic)

Toxin (tok′-sin) As used in this book, a poisonous substance produced by a microorganism

Toxoid (tok′-soyd). A toxin that has been altered in such a way as to destroy its toxicity but retain its antigenicity; toxoids are used as vaccines

Transcription (tran-skrip′-shun). Transfer of the genetic code from one type of nucleic acid to another; usually, the synthesis of an mRNA molecule from a DNA template

Transduction (trans-duk′-shun). Transfer of genetic material (and its phenotypic expression) from one bacterial cell to another via bacteriophages; in *general transduction,* the transducing bacteriophage is able to transfer any gene of the donor bacterium; in *specialized transduction,* the bacteriophage is able to transfer only one or some of the donor bacterium's genes

Transfer RNA (tRNA). The type of RNA molecule that is capable of combining with (and thus "activating") a specific amino acid; involved in protein synthesis (translation); the anticodon on a tRNA molecule "recognizes" the codon on an mRNA molecule

Transformation (trans-for-may′-shun). In microbial genetics, transfer of genetic information between bacteria via uptake of "naked" DNA; bacteria capable of taking up "naked" DNA are said to be competent

Transient microflora. Temporary members of the indigenous microflora

Translation (trans-lay′-shun). The process by which mRNA, tRNA, and ribosomes effect the production of proteins from amino acids; protein synthesis

Transmission-based precautions. Safety precautions taken by healthcare workers, in addition to Standard Precautions, to protect themselves and their patients from infection via airborne, contact, or droplet routes

Tripeptide (try-pep′-tide). A protein consisting of three amino acids held together by peptide bonds

Tuberculocidal (too-bur′-kyu-low-sigh′-dull) **agent.** A chemical or drug that kills the bacteria that cause tuberculosis (*Mycobacterium tuberculosis*); a *tuberculocide*

Tyndallization (tin-dull-uh-zay′-shun). A process of boiling and cooling in which spores are allowed to germinate and then are killed by boiling again

Ubiquitous (you-bik′-wah-tus). Present everywhere

Ureteritis (you-ree-ter-eye′-tis). Inflammation or infection of a ureter

Urethritis (you-ree-thry′-tis). Inflammation or infection of the urethra

Vacuoles (vak′-you-oles). Membrane-bound storage spaces in the cell

Variolation (var′-e-oh-lay′-shun). The obsolete process of inoculating a susceptible person with material obtained from a smallpox vesicle; also known as variolization

Vasoconstriction (vay′-so-kon-strik′-shun). Narrowing of blood vessels

Vasodilation (vay′-soh-die-lay′-shun). An increase in the diameter of blood vessels

Vectors (vek′-tour). As used in this book, invertebrate animals (*e.g.,* ticks, mites, mosquitoes, fleas) capable of transmitting pathogens among vertebrates

Viable plate count. A laboratory technique used to determine the number of living bacteria in a milliliter of liquid; involves the use of plated media

Virion (veer′-ee-on). A complete, infectious viral particle

Viroid (vi′-royd). An infectious agent of plants that consists of nucleic acid, but no protein; smaller than a virus

Virologist (vi-rol′-oh-jist). One who studies or works with viruses

Virology (vi-rol′-oh-gee). That branch of science concerned with the study of viruses

Virucidal (vi-ruh-sigh′-dull) **agent.** A chemical or drug that inactivates viruses; a *virucide*

Virulence (veer′-u-lenz). A measure of pathogenicity; *i.e.,* some pathogens are more or less *virulent* than others

Virulence factors. Attributes or properties of a microorganism that contribute to its virulence or pathogenicity

Virulent (veer′-yu-lent). Pathogenic; capable of causing disease

Virulent bacteriophage. A bacteriophage that regularly causes lysis of the bacteria it infects

Viruses (vi′-rus-ez), sing. *virus.* Acellular microorganisms that are smaller than bacteria; intracellular parasites

Wandering macrophages. Macrophages that migrate in the bloodstream and tissues; sometimes called *free macrophages*

Zoonoses (zoh-oh-no′-seez), sing. *zoonosis.* Infectious diseases or infestations transmissible from animals to humans.

Chapter 1

Insight Figure 1–1ab: Volk WA, et al: ESSENTIALS OF MEDICAL MICROBIOLOGY, 4th Edition. Philadelphia: JB Lippincott, 1990.

Figure 1–9: Koneman EW, et al: COLOR ATLAS AND TEXTBOOK OF DIAGNOSTIC MICROBIOLOGY. 5th Edition, Philadelphia: Lippincott-Raven Publishers, 1997.

Figure 1–12: Volk WA, et al: ESSENTIALS OF MEDICAL MICROBIOLOGY, 5th Edition. Philadelphia: Lippincott-Raven Publishers, 1996.

Chapter 2

Figure 2–4, 2–13: Lechavalier HA, Pramer D: THE MICROBES. Philadelphia: JB Lippincott, 1970.

Figure 2–9, 2–11, 1–12: Volk WA, et al: ESSENTIALS OF MEDICAL MICROBIOLOGY, 5th Edition. Philadelphia: Lippincott-Raven Publishers, 1996.

Chapter 3

Figure 3–1, 3–6, 3–13: Volk WA, et al: ESSENTIALS OF MEDICAL MICROBIOLOGY, 4th Edition. Philadelphia: JB Lippincott, 1991.

Figure 3–2, 3–4, 3–18: Volk WA, et al: ESSENTIALS OF MEDICAL MICROBIOLOGY, 5th Edition. Philadelphia: Lippincott-Raven Publishers, 1996.

Figure 3–9a, 3–12: Lechavalier HA, Pramer D: THE MICROBES. Philadelphia: JB Lippincott, 1970.

Figure 3–10, 3–14: Davis BD, et al: MICROBIOLOGY, 4th Edition. Philadelphia: Harper & Row, 1990.

Chapter 6

Figure 6–8, 6–10: Volk WA, et al: ESSENTIALS OF MEDICAL MICROBIOLOGY, 4th Edition. Philadelphia: JB Lippincott, 1991.

Chapter 7

Figure 7–4: Volk WA, et al: ESSENTIALS OF MEDICAL MICROBIOLOGY, 5th Edition. Philadelphia: Lippincott-Raven Publishers, 1996.

Figure 7–5: Koneman EW, et al: COLOR ATLAS AND TEXTBOOK OF DIAGNOSTIC MICROBIOLOGY, 5th Edition. Philadelphia: Lippincott-Raven Publishers, 1997.

Chapter 8

Figure 8–3: Lechavalier HA, Pramer D: THE MICROBES. Philadelphia: JB Lippincott, 1970.

Chapter 10

Figure 10–1: Timby BK, Lewis LW: FUNDAMENTAL SKILLS AND CONCEPTS IN PATIENT CARE, 5th Edition. Philadelphia: JB Lippincott, 1992.

Figure 10–3: Taylor C, et al: FUNDAMENTALS OF NURSING, 2nd Edition. Philadelphia: JB Lippincott, 1993.

Figure 10–4, 10–5: Koneman EW, et al: COLOR ATLAS AND TEXTBOOK OF diagnostic microbiology, 5th Edition. Philadelphia: Lippincott-Raven Publishers, 1997.

Figure 10–9: Davis BD, et al: MICROBIOLOGY, 4th Edition. Philadelphia: Harper & Row, 1990.

Chapter 12

Figure 12–1: Lindberg JB, Hunter ML, Kruszewski AZ: INTRODUCTION TO PERSON-CENTERED NURSING. Philadelphia: JB Lippincott, 1988.

Figure 12–4, 12–18, 12–19, 12–20: Dobson RL, Abele DC: THE PRACTICE OF DERMATOLOGY. Philadelphia: JB Lippincott, 1985.

Figure 12–5, 12–14: Scherer JC: INTRODUCTORY MEDICAL-SURGICAL NURSING, 6th Edition. Philadelphia: JB Lippincott, 1991.

Figure 12–6, 12–13: Moffett HL: CLINICAL MICROBIOLOGY, 2nd Edition. Philadelphia: JB Lippincott, 1980.

Figure 12–9, 12–10, 12–15: Chaffee EE, Lytle IM: BASIC PHYSIOLOGY AND ANATOMY, 4th Edition. Philadelphia: JB Lippincott, 1980.

Figure 12–12: Davis BD, et al: MICROBIOLOGY, 4th Edition. Philadelphia: Harper & Row, 1990.

Figure 12–16: Koniak D: MATERNITY NURSING: FAMILY, NEWBORN, AND WOMEN'S HEALTH CARE, 17th Edition. Philadelphia: JB Lippincott, 1992.

Figure 12–21: Tortora GJ, Anagnostakos NP: ANATOMY AND PHYSIOLOGY, 4th Edition. Philadelphia: Harper & Row, 1985.

Figure 12–22, 12–23: DeVita VT, Jr, Hellman S, Rosenberg SA: **AIDS**. Philadelphia: JB Lippincott, 1985.

Figure 12–24a: Volk WA, et al: **ESSENTIALS OF MEDICAL MICROBIOLOGY,** 5th Edition. Philadelphia: Lippincott-Raven Publishers, 1996.

Figure 12–24b: Sauer GC: **MANUAL OF SKIN DISEASES,** 6th Edition. Philadelphia: JB Lippincott, 1991.

Figure 12–25: Davis BG, Bishop ML: **CLINICAL LABORATORY SCIENCE: STRATEGIES FOR PRACTICE**. Philadelphia: JB Lippincott, 1989.

Figure 12–28: Volk WA, et al: **ESSENTIALS OF MEDICAL MICROBIOLOGY,** 4th Edition. Philadelphia: JB Lippincott, 1991.

Appendix D

Figure D-1, D-4, D-5: Koneman EW, et al: **COLOR ATLAS AND TEXTBOOK OF DIAGNOSTIC MICROBIOLOGY,** 5th Edition. Philadelphia: Lippincott-Raven Publishers, 1997.

Figure D-2: Volk WA, et al: **ESSENTIALS OF MEDICAL MICROBIOLOGY,** 5th Edition. Philadelphia: Lippincott-Raven Publishers, 1996.

Color Plates

All Color Figures: Koneman EW, et al: **COLOR ATLAS AND TEXTBOOK OF DIAGNOSTIC MICROBIOLOGY,** 5th Edition. Philadelphia: Lippincott-Raven Publishers, 1997.

Index

Page numbers in *italics* denote figures; those followed by a 't' denote tables.

Abiogenesis, 11
Acellular microorganisms, 31, 32t
    (*see also* Prions; Viroids;
        Viruses)
Acellular vaccine, 326
Acetobacter, 12, 149
Acid-base reaction, 106, *107*
Acid-fast staining, of bacteria, 59,
    61
    in sputum, 391
Acidity, *106,* 107–108
    of common substances, 108t
    as factor in microbial growth, 182
    measurement of, 107–108
Acidophiles, 182
Acidophilic microorganisms, 182
Acids, *106,* 107–108
    characteristics of, 105–106
    common, and formulas, 106t
    fatty, 119–120
Acinetobacter, 214
Acne, 369–370
Acquired immunity, 323–328
    types of, 324t
Acquired immunodeficiency syn-
        drome (*see* AIDS; Human
        immunodeficiency virus)
Acquired resistance, 201
Actinomyces, mouth infections
        caused by, 378, 379
Active acquired immunity, 324–327
Active carrier, of pathogens, 253
Acute disease, 234
Adenine, 126–128
Adenosine triphosphate (ATP), 145
    in eucaryotes, 36
    in procaryotes, 41
    production of, 146–148, *147*
Adenoviruses, 375, 385, *385*
    enteric, 396–397
    model of, *85*
ADP (Adenosine diphosphate), 145
Aerobic bacteria, 62–64
Aerobic respiration, 146–150, *147*

Aerotolerant anaerobes, 63
African trypanosomiasis, 435, 442
Agammaglobulinemia, 331
Agar, 153
Agglutination, 347, 348t
Agricultural microbiology, 7,
    220–222
AIDS (Acquired immunodeficiency
        syndrome), 7, 422–425 (*see
        also* Human immunodefi-
        ciency virus)
Airborne transmission, of disease,
    255–256
    control of nosocomial, 271–272
    precautions for, 278, 279t
Alcohol, as antimicrobial chemical,
    192
Algae, 2, 31, 32t, 69–70
    characteristics of, 70t
    differences between protozoa
        and, 69
    diseases caused by, 4t, 69, 71t,
        407t (*see also* Phycotoxins)
    flagella of, 37
    inactivation of, 177 (*see also*
        Antimicrobial methods)
    phyla of, 70t
    typical, *72*
    uses of, 69–70
Algicidal agents, 177
Alkalinity, of solution, 182
Alkaliphiles, 182
Allergens, examples of, 341
Allergic response, 341–342 (*see also*
        Hypersensitivity)
Allergy shots, 344
Alpha-hemolytic bacteria, 240–
    241
Amebas, 72, *73,* 74
Amebiasis, primary, 410–411
Ameboid movement, 72
American trypanosomiasis, 436
AMI (Antibody-mediated
        immunity), 334

Amino acids, 121–122, *122*
    and transfer RNA, 131–132
Amino acyl-tRNA synthetase,
    131–132
Ammonification, 221–222
Amoebas (*see* Amebas)
AMP (Adenosine monophosphate),
    145
Amphitrichous bacteria, *43,* 43–44
    (*see also* Motility, bacterial)
Amplification procedure, for nucleic
    acids, 295
Anabolism, 144, 150–152
Anaerobic bacteria, 62–64
    discovery of, 63 (Insight section)
Anaerobic fermentation, 148–149
Anaerobic respiration,
    chemotrophic, 150
Anamnestic response, 330
Anaphylactic reactions, 339, 342,
    343
Anaphylaxis, localized and systemic,
    342–343
Angstrom (Å), 22
Aniline dyes, as antimicrobial chem-
    icals, 193–194
Animal cells, structures of, *33*
Animal vectors, of disease transmis-
        sion, 258–259 (*see also*
        Reservoirs)
Animalcules, 11
Animalia, kingdom of, 48
Anopheles mosquito, 434
Antagonism (*see* Microbial
        antagonism)
Anthrax, 370–371
Antibacterial agents (*see also* An-
        tibiotics; Antimicrobial
        agents)
    examples of, 200t
Antibiotics, 195
    broad spectrum, 197, 198t–200t,
        204
    examples of, 200t

Antibiotics—*Continued*
  mechanism of action of, 196–197
  precautions for, 203–204
  resistance of bacteria to, 164,
    201–202
Antibodies, 310
  blocking, 344
  classes of, 336, 337t, 338
  functions of, 337t
  in immune response, 329–331,
    336–339
  monoclonal, 338–339
  production of, *330*
  reasons for presence of, 346
  structure of, *337*
Antibody-mediated immunity
    (AMI), 334
Anticodon, 131
Antifungal agents, 197, 199t
Antigen-presenting cells, 334
Antigenic determinants, 329
Antigenic substance, 328
Antigenic variation, 324
Antigens, 310, 328–329
Antimicrobial agents, 195–205
  bacterial resistance to, 201–203
    (*see also* Superbugs)
    mechanisms of, 201–202
  characteristics of ideal, 195–196
  infectious disease treatment by,
    198t–200t
  mechanism of action of,
    196–197, 201
  precautions for, 203–204
Antimicrobial chemicals, 190–192
  action of, 192–194
  characteristics of, 192
  effectiveness of, 191
Antimicrobial methods, 183–194
    (*see also* Sterilization)
  chemical, 190–194 (*see also* An-
    timicrobial chemicals)
  chemotherapy as, 194–195 (*see
    also* Chemotherapeutic
    agents)
  effectiveness of, 183–184, *184*
  physical, 184–190
    cold as, 188–189
    drying as, 189
    filtration as, 190
    heat as, 184, *186*, 186–188
    radiation as, 189–190
    ultrasonic waves as, 190
Antimicrobial susceptibility testing
    (AST), 295–297, *296*
Antiprotozoal agents, 197
Antisepsis, 178

Antiseptic surgery, 16
Antiseptic technique, 178
Antiseptics, 177, 192
Antiserum, 326
Antitoxins, 326
Antiviral agents, 87, 89, 200t
  examples of, 201t
  mode of action of, 197
ANUG gingivitis, 380
Apoenzymes, 124, 143
Arachidonic acid, 120
Archaea, kingdom of, 49
Archaebacteria, 31, 32t, 68–69
  cell walls of, 42
  classification of, 49
  derivation of word, 68
  difference between eubacteria
    and, 49
Arthropod borne diseases,
    425–433, 465–466
Arthropod vectors, 65, 258,
    425–433, 465–466, 466t
    (*see also* Reservoirs)
Artificial active acquired immunity,
    324–327
Artificial passive acquired immunity,
    328
Ascomycetes, 76, 80
  antibiotic producing, 79
  characteristics of, 79t
Asepsis, 178, 274
Aseptate hyphae, 77, 79
Aseptic technique, 178
  medical, 274
  origin of, 16
  surgical, 274, *275*, 276
Aspergillus
  ear infections caused by, 382
  respiratory infections caused by,
    382
  toxicity of, 79
Aspirin, and inhibition of
    prostaglandins, 315
AST (Antimicrobial susceptibility
    testing), 295–297, *296*
Astroviruses, 396–397
Asymptomatic disease, 176, 235,
    247
Athlete's foot, 372
Atomic number, 100
Atomic weight, 100
Atoms, 99–102
  parts of, *100*
Atopic individuals, 341
ATP (*see* Adenosine triphosphate)
Attenuated vaccines, 325
Autogenous vaccine, 327

Autoimmune diseases, 340
  examples of, 344
Autolysis, 36
Autotrophs, 139–141
Avery, Oswald, 164
Axial filaments, 43–44, 61–62

B cells, 322, 329, 330
  development of, *332, 333*
  in humoral immunity, 334
B lymphocytes (*see* B cells)
Babesiosis, 433
Bacillariophyta, 69–70, *72*
Bacillary dysentery, 405–406
Bacilli (bacillus), 55, 56, 58 (Study
    Aid)
  sizes and shapes of, 57
Bacillus anthracis, 13, 370–371,
    450
  as biological warfare agent, 224
Bacillus cereus, enterotoxin produc-
    tion by, 242
Bacillus fastidiosus, cell wall of,
    40
Bacillus spp.
  in soil, 222
  spores formed by, 44–45
Bacteremia, 285, 421
Bacteria, 2
  atmospheric requirements of,
    62–64 (*see also* Aerobic bac-
    teria; Anaerobic bacteria;
    Capnophilic bacteria)
  biochemical activity of, 64 (*see
    also* Metabolism)
  cell morphology of, *55*, 55–58
  cell structures of, 38–45, *39*
  cell walls in, lack of, 57–58
  characteristics of, 54–65
  chromosome of, 159–160
  ciliates, 74, 75t, 457t
  colony morphology of, 62
  conjugation of, 44, 164–166, *165*
  diseases caused by, 4t (*see also*
    Bacterial infections)
    zoonotic, 251t–252t
  DNA of, 65, 159–160 (*see also*
    Deoxyribonucleic acid;
    Genetics)
  earliest, 68–69 (*see also* Archae-
    bacteria; Cyanobacteria)
  encapsulated, 320–321
  genetic constitution of, 159–160
    (*see also* Genetics)
    changes in, 160–166, *162*
  genetic engineering of, 166, 167
    (Insight)

genetic material of, 65, 159–160
(*see also* Deoxyribonucleic
acid; Genetics)
Gram-positive and Gram-negative, 58–59
cell wall of, *40,* 42
pathogens that are, 60t,
449
growth of, 138 (*see also* Culture
media; Growth, bacterial)
identification of, 55 (*see also* Identification)
inactivation of, 177–178 (*see also*
Antimicrobial methods)
agents for, 198t–199t, 200t
(*see also* Antibiotics; Bacteriophages)
mechanism of action of, 197,
202
indigenous (*see* Indigenous microflora)
L-forms of, 57–58
largest, 56
lysogenic, 91
medically important, categories
of, 449
metabolic activity of, 64 (*see also*
Metabolism)
motility of, 43–44, 61–62
nitrogen-fixing, 5, 68 (*see also* Nitrogen-fixing bacteria)
cycle of, 6
nutritional requirements of, 64,
139
nutritional types of, 139–141
pathogenic, 60t, 64, 450–452
(*see also* Bacterial
infections)
photosynthetic, 68 (*see also*
Cyanobacteria)
physiology of, 138–152 (*see also*
Physiology)
protein amino acid sequences of,
64
resistance to antimicrobial agents,
181, 201–203 (*see also* Superbugs)
rudimentary, 65–68 (*see also*
Chlamydia; Mycoplasma;
Rickettsia)
human diseases caused by,
66t
sizes of, 22–23, 38, 55–56
staining, 58–61 (*see also* Staining)
steroids produced by, 121
structures associated with virulence of, 238–239

taxonomic categories of, 46–49,
449
uses of, 5–8, 220–224
Bacterial infections
of circulatory system, 327–333
of ear, 381–382
of eye, 376–378
of gastrointestinal tract, 397–410
of genitourinary system, 417–420
of mouth, 379–381
of nervous system, 441–442
of respiratory system, 387–392
of skin, 367–372
Bactericidal agents, 177, 197
Bacteriocins, 232, 313
Bacteriologist, 6
Bacteriology, 6
Bacteriophages, 90–91, *91*
lysogenic conversion by, 161–162
transduction by, 162–163
Bacteriostatic agents, 177–178
Bacteriuria, 286
Bacteroides spp., 450
indigenous to humans, 217, 395
mouth infections caused by, 379,
380
Balantidium coli, 74
Barometric pressure, in bacterial
growth, 183
Barophiles, 183
Bartonella spp., 65, 374
Bases, 106, 107t
Basidiomycetes, 76, 77
characteristics of, 79t
Basophils, *316,* 317
Beard itch, 372
Benzene ring, *115*
Bergey's Manual of Systematic Bacteriology, 54–55
Beta-hemolytic bacteria, 240–241
Beta-lactamases, 202–203
Beta-lysin, 314, 318
Bifidobacterium spp., 218, 395
Binary fission, 38, 152, *153*
Binomial system, of classification,
46–47
how to use, 47–48
Biochemistry, 99, 102, 115–133
of carbohydrates, 115–119
of enzyme catalysts, 124–125
of lipid molecules, 119–121
of microorganisms, 113
of nucleic acids, 125–133
of protein molecules, 121–125
of protein synthesis, 129–133
Biocidal agents, 177
Biogenesis, 11

Biogeochemical cycling, 221
Biological theory of fermentation,
11–13
Biological warfare agents, 223–224
Bioremediation, 223
Biosynthetic energy metabolism,
150–152
Biotechnology, 222–223 (*see also*
Genetic engineering)
Biotherapeutic agents, 218
Bite infections, 374
Black death, 431–432
Blastomyces, 81
diseases caused by, 82, 384
Blocking antibodies, 344
Blood Bank, 291
Blood cells, types of, *316,* 316–317
Blood proteins, as defense mechanisms, 315–316
Blood specimens, collection of,
285–286
Bloodborne pathogens, standard
precautions for, 277
Body Substance Isolation, 276
Boils, 369
Bonds, chemical (*see* Carbon bonds;
Chemical bonds)
Book of Leviticus, 9
Bordetella pertussis, 392–393, 450
Bordetella spp., respiratory infections caused by, 392
Borrelia burgdorferi, 431, 450
Borrelia spp., 57, *57*
mouth disease caused by, 380
Botulism, 406–409, 441
Bovine spongiform encephalopathy,
92
Bradyrhizobium, 221–222
Brain (*see* Nervous system)
Brevibacterium, 214, 363
Brightfield microscope, 17
Broad spectrum antibiotics, 197,
198t–200t, 204
Brown algae, 69–70
Bubonic plague, 431–432
Budding
of fungi, 76, *77*
of yeasts, *80*
Burn infections, 373

Caliciviruses, 396–397
Calymmatobacterium granulomatis,
421
Campylobacter jejuni, 450
Campylobacter spp., 63
gastroenteritis caused by,
399–400

Cancer, caused by toxins of cyanobacteria, 68

Cancer drugs, mode of action of, 201

Candida albicans, 77, 80, 384
skin disease caused by, 82
superinfection by, 204

Candle extinction jar, atmosphere of, 64

Capnocytophaga spp., 374

Capnophilic bacteria, species of, 63–64

Capsid, of viruses, 83

Capsomeres, of viruses, 83

Capsular swelling, 349t

Capsules, bacterial, 42–43, 320–321
as virulence factor, 238

Carbohydrates, 115–119

Carbon bonds, 113–114, *114*

Carbon sources, in nutrition, 139–141, 140t

Carbuncles, 369

Cardiovascular system
anatomy of, *422*
infections of
bacterial, 427–433
ehrlichial, 429–430
protozoal, 433–436
rickettsial, 427–428
viral, 421–430

Carriers, of pathogens, 215, 234, 253

Catabolism, 144
of glucose, 146–148, *147*

Catalysts, enzymes as, 124, 141–145

Cations, 100

CCMS urine (Clean-catch midstream urine), 286

CD4 receptors, 90, 249

CDC (Centers for Disease Control and Prevention), 259, 298–299
historical highlights of, 299t

Cell counters, 155

Cell-mediated immunity (CMI), 334–335

Cell membrane
of bacteria, 39, 41
of eucaryotes, 33, *33, 34*
injury of, by antimicrobial chemicals, 192
of procaryotes, 41

Cell theory, 30

Cell types, relevance of, 32

Cell wall, 37, *37*

of bacteria, *40,* 42 (*see also* Gramnegative bacteria; Gram-positive bacteria; L-forms)
of eucaryotes, 37

Cells, 30
metabolism of, 30–31, *31*
structures of, 30–46 (*see also* Eucaryotes; Procaryotes)

Cellular microorganisms, 30–32, 32t
taxonomy of, 46–49
types of, 30–32 (*see also* Eucaryotes; Procaryotes)

Cellular secretions, as defense mechanism, 314–315

Cellular slime molds, 82–83

Cellulose, 45, 118
difference between starch and, *119*

Centers for Disease Control and Prevention (CDC), 259, 298–299
historical highlights of, 299t

Centimeter, 22, 23t

Central dogma, 130–131

Centrioles, 36–37

Cephalosporin, bacterial degradation of, 203, *203*

Cerebrospinal fluid, 437
collection of specimen of, 287

CFU (Colony-forming units), 155, 286

Chagas' disease, 436

Chain, Ernst Boris, 194

Chain, of carbon atoms, 114

Chain of infection, *231,* 231–232

Chase, Martha, 164

Chemical bonds, 102–104 (*see also* Carbon bonds)
conversion to cellular energy, 144–148, *145*

Chemical disinfection, 190–194

Chemistry, 99–108 (*see also* Biochemistry; Organic chemistry)
of acids and acidity, 105–108
of atoms, 99–102
of bases, 106
of chemical bonding, 102–104
of compounds, 99–102
of elements, 99–102
of molecules, 99–102
of pH, 107–108
of salts, 106–107
of solutions, 105
of water, 104–108

Chemistry Laboratory, 290

Chemoautotrophs, 140–141
chemosynthesis by, 151–152

Chemoheterotrophs, 140–141
chemosynthesis by, 151–152

Chemolithotrophs, 140–141
aerobic respiration by, 149–150
soil bacteria as, 221–222

Chemostat, for bacterial culture, 157, *158*

Chemosynthesis, 151–152
types of organisms using, 150

Chemotactic agents, 318, 321

Chemotaxis, 317–318

Chemotherapeutic agents, 194–205
action of, mechanism of, 196–197, 201
bacterial resistance to, 201–203 (*see also* Superbugs)
characteristics of ideal, 195–196
historical background of, 194–195
for infectious disease treatment, 198t–200t
precautions for, 203–205

Chemotrophs, 139–141
anaerobic respiration by, 150

Chickenpox, 364–365

Chitin, in fungi, 37, 76

Chlamydia spp., 65–66, 450
diseases caused by, 66, 66t, 384
genital infections caused by, 417–418
life cycle of, 66, *67*
method of transfer of, 65–66
urinary tract infections caused by, 413

Chlamydia trachomatis, 417–418, 421
conjunctivitis caused by, 376–377

Chlamydiasis, genital, 417–418

Chlorophyta, 69–70

Chloroplasts
of eucaryotes, 37
origin of, 36

Cholera, 400–401

Cholesterol, in metabolism, 121

Chromatin, of eucaryotes, 34–35

Chromosome(s), 31
of bacteria, 39, 41, 159–166
of eucaryotes, 34–35 (*see also* Deoxyribonucleic acid)

Chronic disease, 234

Chrysophyta, 70t

Cilia, of eucaryotes, 37–38, *38*

Ciliata, 74, 75t, 457t

Ciliates, 74, 75t, 457t

Ciliophora, 74, 75t, 457t

Circulatory system
anatomy of, *422*
infections of, 421–436
bacterial, 427–433
ehrlichial, 429–430
nonspecific, 421–422
protozoal, 433–436
rickettsial, 427–428
viral, 421–430
Cistrons, 129–130
mutations of, 132
Citric acid cycle, 147–148
Class category, in classification of
organisms, 48
Classification, of microorganisms
3-Kingdom, 49
5-Kingdom, 48–49
binomial system of, 46–48
Clean-catch midstream urine
(CCMS urine), 286
Clinical disease, 235
Clinical Microbiology Laboratory
(CML), 281–282
function of, 291
organization of, *291, 292*–293
procedures of, 467–472
quality control in, 297
relevance of findings of, 293
responsibilities of, 291–292
Clinical specimens, 283–297
collection of, 284–288
precautions for, 284–285
disposal of, 282
identification of pathogens in,
293–297, 467–472
processing, 289–297 (*see also*
Pathology Department)
role of healthcare professionals
in, 283–284
quality of, 283, 292 (Insight)
shipping, 288–289, *289, 290*
testing of, 293–297
types of, 285–288
Clostridium botulinum, 408–409,
450
as biological warfare agent, 224
neurotoxin production by,
241–242
temperature and pressure tolerance of, 183
Clostridium difficile, 450
enterotoxin production by, 242
superinfections caused by, 204
Clostridium perfringens, 409, 450
collagenase production by, 240
enterotoxin production by, 242
in focal wounds, 374

intolerance to oxygen of, 183
lecithinase production by, 241
Clostridium spp.
in focal wounds, 373–374
hyaluronidase production by, 240
indigenous to humans, 217
spores formed by, 44–45
Clostridium tetani, 374, *441,*
441–442, 450
booster shots for, 331
in focal wounds, 373
intolerance to oxygen of, 183
neurotoxin production by,
241–242
CML (*see* Clinical Microbiology
Laboratory)
CNA (Colistin-nalidixic acid) agar,
154
Coagulase, as virulence factor, 240
Cocci (coccus), *55,* 58 (Study Aid)
arrangements of, 56t
size of, 56
Coccidioides, 81
diseases caused by, 82, 384
Coccobacilli (coccobacillus), 57
Codons, 131
Coenzymes, 124, 143
Cofactors, of enzymes, 124
Cold, as antimicrobial method,
188–189
Colds, 385
Colicins, 313
Coliphages, 91
Colistin-nalidixic acid (CNA) agar,
154
Collagenase, as virulence factor, 240
Collection, of clinical specimens (*see
under* Clinical specimens)
Colony-forming units (CFU), 155,
286
Colony morphology, of bacteria, 62
(*see also* Identification)
Colorado tick fever, 425–426
Commensalism, 219, *219*
Common cold, 385
Communicable disease, 231
transmission of, 254–259
Community acquired infections,
267
Competence, bacterial, 163
Complement, 315
Compound lipids, 121
Compound microscopes (*see* Micro-
scopes, compound)
Compounds, chemical, 101
Condensation reactions, 105
Condenser, of microscope, 19

Condyloma acuminatum, 367, *416,*
416–417
Conidia, of fungi, 76
Conjugation, bacterial, 44,
164–166, *165*
Conjunctivitis, 375
bacterial, 376
chlamydial, 376
gonococcal, 377–378
inclusion, 376
Contact dermatitis, 341
Contact hypersensitivity, 341
Contact transmission, of infectious
disease
precautions for, 278, 279–280,
279t
Contagious disease, 231
Contagious microorganisms, 9
Contamination, 176 (*see also* Transmission)
Control of disease (*see* Environmental control; Infection control
procedures)
Convalescent carrier, of pathogens,
253
Coronaviruses, 385
Corynebacterium diphtheriae,
387–388, 450
exotoxin production by, 242
lysogenic conversion in, 91–92,
161–162
morphology of, 57
as opportunistic pathogen, 363
Corynebacterium spp., indigenous
to humans, 214–216
Coryza, acute, 385
Counts, of bacteria, 154–156
Covalent bonds, 102
formation of, *104*
Coxiella burnetii, 65
Crenation, of cells, 180, *180*
Creutzfeldt-Jakob disease, 92
Crick, Francis, 127
central dogma proposed by,
130–131
Cryptococcosis, 393
Cryptococcus neoformans, 384, 393
Cryptococcus spp., diseases caused
by, 82
Cryptosporidiosis, 247, 411
Cryptosporidium parvum, 75, 247,
411
testing for, in water supply,
302–303
Culture media, 152–154
acid-base balance in, 107
first developers of, 14

Culture media—*Continued*
  identification of bacteria by, 64
  inoculation and incubation of,
      467–468
  types of, 152–154
Cyanobacteria, 32t
  membranes of, 41
  toxins of, 68
Cyclic compounds, *114,* 114–115
Cystitis, 413
Cytokines, 313–315
  in cell-mediated immunity, 335
Cytology, 32
Cytoplasm, 30
  of bacteria, 41
  of eucaryotes, 35–37
Cytoplasmic membrane
  of bacteria, *39,* 41
  of eucaryotes, 33, *33, 34*
Cytoplasmic particles, of bacteria,
      41
Cytosine, 126–128
Cytosol, 35
Cytostome, 75
Cytotoxic reactions, 339

Darkfield techniques, 19–20
Death phase, of bacterial growth
      curve, 158
Decimeter, 22
Decomposers, of organic matter,
      140
Defense mechanism(s), 310–351
  blood proteins as, 315–316
  cellular secretions as, 314–315
  evasion of
      by encapsulated bacteria,
          42–43, 238, 320–321
      by parasites, 465
  fever as, 313
  first line of, *311,* 311–313
  inflammation as, 321–323 (*see
      also* Inflammatory
      response)
  iron balance as, 314
  nonspecific, 310–323, *311*
  phagocytosis as, 316–321
  second line of, 313–323
  specific, 323–351 (*see also* Im-
      mune response; Immunity;
      Immunology)
  third line of, 323–351
Degranulation, in allergic reaction,
      342, *342*
Dehydration synthesis reactions,
      105, 117
Dehydrogenation, 146

Delayed-type hypersensitivity
      (DTH), 340–341
Denatured proteins, 125
Denitrifying bacteria, 221–222
Dental caries, bacteria causing, 215,
      379
Deoxyribonucleic acid (DNA), 31,
      125–133 (*see also* Genetics)
  of bacteria, 65 (*see also under*
      Bacteria)
  bonding of bases in, *127,*
      127–128
  damage of, by antimicrobial
      chemicals, 193–194
  differences between RNA and, 129
  discovery of structure of, 127
  double helix of, 127–128, *128*
  extrachromosomal, 160
  mutations in, 132–133 (*see also*
      Mutations)
  protein synthesis controlled by,
      129–133
  replication of, 128–129
      visualizing (Study Aid), 129
  structure of, 127
  of viruses (*see* Bacteriophages;
      Viruses)
Deoxyribose, 126
Derived lipids, 121
Dermatomycoses, 372–373
Dermatophytes, 81–82, 372
Dermatophytoses, 372–373
Desmids, 69–70
Desiccation, of microorganisms, 179
Deuteromycetes, 77, 80
  antibiotic producing, 79
  characteristics of, 79t
Dextrose, 116
Diarrhea
  enterohemorrhagic, 401–403
  enterotoxigenic *Escherichia coli,*
      403
  traveler's, 403
  viral, 396–397
Diatoms, 69–70, *72*
Dick test, 351
Didinium nasutum, 74
Differential blood counts, 290
Differential media, for bacterial
      growth, 154
Differential staining, of bacteria, 61,
      61t
Dimorphic fungi, 80–81
  human diseases caused by, 81t,
      393–394
Dinoflagellates, 69–70
Dipeptides, 123, *123*

Diphtheria, 387–388
Diphtheroids, 363
Diplobacilli (diplobacillus), 57
Diplococci (diplococcus), *55, 56,* 56t
Disaccharides, 117
Diseases (*see also* Infectious diseases)
  acute, 234
  asymptomatic, 235
  chronic, 234
  clinical, 235
  communicable, 231
  contagious, 231
  germ theory of, 13, 14
  nationally notifiable (definition
      of), 363
  signs of, 235
  subclinical, 235
  symptomatic, 235
  zoonotic, 7 (*see also* Zoonoses)
Disinfectants, 177
Disinfection, 177
  chemical, 190–194
  discovery of, 16
Disk-diffusion testing, 295 (*see
      also* Immunodiagnostic pro-
      cedures)
Diversity, determination of, 49
DNA (*see* Deoxyribonucleic acid)
DNA helicase, 128–129
DNA polymerase, 128–129
DNA probes, 295
Domagk, Gerhard, 194
Double helix, of DNA, 127–128,
      128
Drinking water (*see* Water supply)
Droplet transmission, precautions
      for, 278, 279t
Dry heat, as antimicrobial method,
      186, *187*
Drying, as antimicrobial method,
      189
DT booster, 388
DTH (Delayed-type hypersensitiv-
      ity), 340–341
DTP (Diphtheria-tetanus-pertussis)
      vaccine, 326, 388, 392,
      442
Dysentery
  amebic, 410–411
  bacillary, 405–406

Ear
  anatomy of, *381*
  indigenous microflora of,
      215–216
  infections of, 381–382
Ear canal infections, 382

Ebola virus, epidemic of, 247
Ecology, 140
  microbial, 212 (*see also* Environmental microbiology; Indigenous microflora)
Ecosystem, 140
Ectoparasites, 219
Edema, of inflammatory response, 233
Ehrlich, Paul, 17t, 194
Ehrlichia spp., 65
  cardiovascular infections caused by, 429–430
Electrolytes, 106
Electron microscopes, 20–21
Electron transport system, 147–148
Electrons, 100
Elemental cycles, microorganisms in, 220–222
Elementary body, of infectious *Chlamydia*, 66, *67*
Elements, chemical, 100
  first twenty, 101t
  in periodic table, *101*
ELISA (Enzyme-linked immunosorbent assay), 347
Embden-Meyerhof-Parnas pathway, 147
..emia, meaning of suffix, 286 (Study Aid)
Encapsulated bacteria, 42–43, 238, 320–321
Encephalitis, 437
Endemic disease, 243–244
  factors influencing, *244*
  in U.S., 299, 299t
Endocarditis, 421
  infective, 430
Endocytosis, 318
Endoenzymes, 142
Endoparasites, 219
Endoplasmic reticulum (ER), 35
Endospore, bacterial, 44–45, *46*
Endosymbionts, 220–221
Endotoxins, 241
Energy metabolism, 144–152, *147*
  of bacteria, 41
  of cells, 144–145, *149*
  conversion of energy in, 150–152
  in eucaryotes, 35, 36
  in procaryotes, 41
Energy parasites, 65–66
Energy sources, for nutrition, 139–141, 140t
Enriched medium, for bacterial growth, 153
Entamoeba histolytica, 410–411

Enteric fever, 404–405
Enteritis, viral, 396–397
Enterobacter aerogenes, cell wall of, *40*
Enterobacteriaceae, 63
  indigenous to humans, 217, 395
  meningitis caused by, 438
Enterococcus spp., 450–451
  as biotherapeutic agents, 218
  multidrug resistance of, 181
  urinary tract infections caused by, 413
Enterohemorrhagic diarrhea, 401–403
Enteropathogenic organisms, 288
Enterotoxigenic *Escherichia coli* (ETEC), 403
Enterotoxins, 242
Enteroviruses, 439–440
  infections of eye by, 375
Environmental control, of disease, 297–303
  public health agencies in, 297–300
  sewage disposal in, 300–303
  standard precautions for, 277
  water supplies in, 300–303
Environmental microbiology, 8, 297–303
Enzyme-linked assays, 347, 349t
Enzymes, 124–125, 141–152 (*see also* Beta-lactamases; Polymerases)
  action of, 141–143, *142*
  associated with virulence, 240–241
  biotechnological use of, 223
  examples of, 125
  inactivation of, 143–145
    by antimicrobial chemicals, 192–193
  industrial use of, 223
  lysosomal, 318
  metabolic role of, 124–125, 141–143
  naming of, 124–125
  as protein molecules, 142–143
  toxins that interfere with, 125
  types of, 143
Eosinophilia, 317
Eosinophils, *316,* 317
Epidemic disease(s), 244–247
  childhood exposure and, 246
  cholera as, 400–401
  control of, 259, 300
  gonorrhea, in United States, *246,* 246–247

  graph of cases of, *245*
  in hospitals, 247
  influenza as, 247, 386–387
  introduction of, to native populations, 244, 246
  O157:H7 *Escherichia coli,* 401, 402 (Insight)
  prevention of, 300
  syphilis as, 419
  in United States, 244, 247, 419
Epidemiology, 243
  definitions of terms used in, 298, 298t
Epidemiology Service, 280
Epidermophyton spp., 372–373
Episomes, 161
Epitopes, 329
Epstein-Barr virus (EBV), 426
Epulopiscium fishelsonii, 56
ER (Endoplasmic reticulum), 35
Erhlichosis, 429–430
Erysipelas, 368–369
Erythrogenic toxins, 242, 389
Eschars, 370
Escherichia coli, 8, 451
  conjugation in, *165*
  ear infections caused by, 382
  enterotoxin production by, 242
  enterovirulent, 401–403
  meningitis caused by, 438
  O157:H7, 401, 402 (Insight)
  organic composition of, 113
  serotypes of, 401
  size of, 39
  testing for, in water supply, 302–303
Essential fatty acids, 120
ETEC (Enterotoxigenic *Escherichia coli*), 403
Eubacteria, 31, 32t
  kingdom of, 49
Eubacterium spp., of gastrointestinal tract, 395
Eucarya, kingdom of, 49
Eucaryotes, 3
  animal types of, 32, *33,* 45–46
  cell structures of, 32–38, *33*
  derivation of word, 32
  differences between procaryotes and, 45–46, 47t
  organisms included in, 31, 32t
  plant types of, 45–46
  ribosomal activity in, 131
  sizes of, 32
  transcription of DNA in, 131
Euglena, 69, *73*
Euglenaphyta, 69, *73*

Exfoliative toxins, 242

Exocytosis, 35

Exoenzymes, 142, 240

Exotoxins, 241–242

Expression, of gene products, 161

External otitis, 382

Extremophiles, 69

Exudates

of eye, 375

formation of, 234

inflammatory, 322

Eye

anatomy of, *375*

indigenous microflora of,
215–216

infections of, 374–378

Eyepiece, of microscope, 17

F+ gene, 164

F plasmid, 164, 166

Facultative anaerobes, 63

in gastrointestinal tract, 395

as opportunistic pathogens, 363

Familial insomnia, fatal, 92

Family category, in classification of
organisms, 48

Fastidious bacteria, 64, 152

Fat-soluble vitamins, 121

Father of Microbiology, 10, 11

Fats (*see* Lipids)

Fatty acids, 119–120

Feces specimens, collection of, 288

Fermentation

anaerobic, 148–149

biological theory of, 11–13

Fertility factors, bacterial transfer of,
164

Fever, as defense mechanism, 313

Fibrinolysins, as virulence factors,
240

Fibronectin, 314

Fibrous proteins, 123

Filtration, as antimicrobial method,
190

Fimbriae, bacterial, 44 (*see also* Pili)

5-Kingdom classification system,
48–49

Fixation of bacteria, purpose of, 58

Fixed macrophages, 317

Flagella, *319*

bacterial, 43–44, 61–62

as virulence factor, 238

of eucaryotes, 37–38

Flagellates, 74

Flagellin, 43–44

Flaming, of inoculating loop, 186,
187

Fleaborne typhus, 428

Fleming, Alexander, 194

Flesh-eating bacteria, 370 (Insight)

Floor manual, 284

Florey, Howard Walter, 194

Flu, 386–387

Fluorescent antibody technique,
348t

Fluorescent microscope, 19, 20

Focal infections, 373–374

Folliculitis, 369

Fomites, 253, *258*

disease transmission by, 259

nosocomial, 272–273

Food industry microbiology, 5, 5t,
223

Food poisoning, 406–410

etiologic agents of, 407t–408t

Food producers, 140

Foodborne infections, bacterial, 254
(Insight), 406–410

Formalin, as antimicrobial chemical,
193–194

Fracastorius, Girolamo, 9

Francisella tularensis, 374,
432–433, 451

Freeze-drying, of microorganisms,
180

Fructose, 116, *116*

Fungal infections

of respiratory system, 393–394

of skin, 372–373

of wounds, 374

Fungi, 2, 31, 32t, 75–82

characteristics of, 76, 79t

chitin in cell wall of, 76

commercial importance of, 79

differences between algae and,
76

differences between plants and,
76

dimorphic, 80–81, 393–394

diseases caused by, 4t, 81–82, 81t
(*see also* Fungal infections)

zoonotic, 252t

inactivation of, 177, 197 (*see also*
Antimicrobial methods)

agents for, 199t

kingdom of, 48

microscopic appearance of, *78*

molds, 77, 79

mushrooms, 77

pathogenic, 4t, 81–82, 81t (*see
also* Fungal infections)

reproduction of, 76

steroids produced by, 121

toxins of, 79, 407t

true, classification of, 76–80

yeasts as, 80

Fungicidal agents, 177

Furuncles, 369

Fusobacterium spp., 451

indigenous to humans, 217

mouth infections caused by, 379,
380

Gaia-A New Look at Life on Earth
(Lovelock), 141

Galactosamine, 118–119

Galilei, Galileo, 9

Gas tolerance, of microorganisms,
183

Gastritis, bacterial, 397–398

Gastroenteritis

caused by *Campylobacter,* 399

viral, 396–397

Gastrointestinal tract

anatomy of, *396*

indigenous microflora of,
217

infections of, 395–412

bacterial, 397–410

protozoal, 410–412

viral, 396–397

nonspecific defense mechanisms
of, 312

Gene products, 34

Gene therapy, 166–167

Genera (genus), in classification of
organisms, 47, 48

General infections, 234

General microbiology, 7

Gene(s), 129 (*see also* Deoxyribonucleic acid; Genetics)

Genetic diseases, mutations and,
132

Genetic engineering (*see also*
Biotechnology)

of bacteria and yeasts (Insight
section), 167

mutations in, 132

uses of, 166

Genetics, 159

of bacteria, 159–166

conjugation in, 164–166

lysogenic conversion in,
161–162, *162*

mutation in, 159–160

transduction in, *162,* 162–163

transformation in, 163–164

of eucaryotes, 34

gene therapy in, 166–167

and microbiology, 8

mutations in, 132, 159–160, *160*

recombination in, 163
of viruses, 84
Genitourinary system
anatomy of, *412, 414*
infections of, 412–421
bacterial, 417–420
protozoal, 420
viral, 413–417
Genotype, 159
Genus
in classification of organisms, 48
placement of, in binomial system, 47
Germ theory of disease, 13
proof of, 14, *14*
German measles, 366–367
Germicidal agents, 177
Germs, 2
Gerstmann-Strussler-Scheinker disease, 92
Giardia lamblia, 74, *74,* 411–412
testing for, in water supply, 302–303
Giardiasis, 411–412
Gingivitis, 379–380
acute necrotizing ulcerative, 380
bacteria causing, 215
Globular proteins, 123
Gloves, standard precautions using, 276
Glucosamine, 118–119
Glucose, 116, *116*
metabolism of, 146–149, *147, 149*
Glycocalyx, bacterial, 42–43
as virulence factor, 238
Glycogen, 118
Glycolysis, 146–148, *147*
Glycolytic pathway, 147
Glycosidic bonds, 102
of polysaccharides, 117
Golden Age of Microbiology, 16
Golden algae, 70t
Golgi complex, 35
Gonococcal conjunctivitis, 377–378
Gonorrhea, 418–419
asymptomatic, 235
epidemic, in United States, *246,* 246–247
Gonorrheal ophthalmia neonatorum, *377,* 377–378
Gram, Hans, 58
Gram-negative bacteria
culture media for, 154
endotoxins of, 241–242
medically important categories of, 449

as opportunistic pathogens, 363
pathogens that are, 60t
staining of, 58
Gram-positive bacteria
culture media for, 154
medically important categories of, 449
as opportunistic pathogens, 363
pathogens that are, 60t
staining of, 58
Gram staining, 58–59
negative and positive, 59t
preparation of smear for, 469–470
reaction of pathogens in, 60t
Gram-variable bacteria, 58–59
pathogens that are, 60t
Granules
in bacteria, 41
of eosinophil, *319*
in eucaryotes, 37
Granulocytes, *316,* 317
Green algae, 69–70
Green bacteria, 68
Griffith, Frederick, 164
Group A strep (*see Streptococcus pyogenes*)
Growth, bacterial, 152–159 (*see also* Growth, microbial)
culture media for, 152–154
meaning of, 152
population counts of, 154–156
population growth curve of, 156–159, *157*
rate of, 154–156
in vitro, 152
in vivo, 152
Growth, microbial, 137–173 (*see also* Growth, bacterial)
controlling, 174–214
importance of, 175–176
factor(s) influencing, 178–183
acidity as, 182
atmospheric gases as, 183
barometric pressure as, 183
moisture as, 179–180
osmotic pressure as, 180, 182
temperature as, 178–179
inhibition of, 183–194 (*see also* Antimicrobial methods)
Guanine, 126–128

Haemophilus ducreyi, 421
Haemophilus influenzae, 451
capsular types of, 42
conjunctivitis caused by, 376

ear infections caused by, 382
meningitis caused by, 438
multidrug resistance of, 181
respiratory infections caused by, 384
vaccine for, 204
Hair follicle infections, 369
Haloduric microorganisms, 182
Halogen compounds, as antimicrobial chemicals, 193
Halophilic microorganisms, 182
Handwashing
guidelines for, 273
standard precautions for, 276
Hanging-drop technique, for bacterial motility, 61–62, *62*
Hansen disease, 371–372
Hantavirus pulmonary syndrome (HPS), 385–386
Hantaviruses, 385–386
Haptens, 329
Hard measles, 365–366
HBIG (Hepatitis B immune globulin), 328
Heat sterilization, 184, *186,* 186–188
dry, 186
moist, 187
pressurized steam, 187–188
temperature and time in, 184
Helical proteins, 123
Helicase, DNA, 128–129
Helicobacter pylori, 397–398, 451
Helminths, 456–465
Hematology Laboratory, 290
Hemolysins, as virulence factors, 240
differentiation of hemolytic bacteria by, 240–241
Hemolytic-uremic syndrome (HUS), 401
Hepatitis, viral, 397
Hepatitis B immune globulin (HBIG), 328
Hepatitis viruses, 397, 398t
Herpes simplex viruses, 413, 415–416
eye infections caused by, 375
lip infections caused by, *415*
recurrent progenitalis, *415*
urogenital infections caused by, 417–418
Herpes viruses
anogenital, 413, *415,* 415–416
Epstein-Barr, 426
Herpes zoster, 365
Hershey, Alfred, 164

Hesse, Frau, 14
Heterotrophs, 139–141
Hexose, 116
HFr+ gene, 164
Hib vaccine, 42
High-dry objective, 18
High power objective, 18
Histiocytes, 317
Histones, in eucaryotes, 34–35
Histoplasma capsulatum, 384,
    393–394
Histoplasma spp., 80–81
  diseases caused by, 82, 393–394
Histoplasmosis, 393–394
HIV (*see* Human immunodeficiency
    virus)
Holoenzyme, 124
Hooke, Robert, 9
  naming of cells by, 30
Hospital-acquired infections,
    267–283 (*see also* Nosoco-
    mial infections)
Hospital infection control proce-
    dures, 280–282
Host, 212
Human immunodeficiency virus
    (HIV), 90, *90,* 424–425
  destruction of T cells by, 331
  as pandemic, 247–249, *248*
Human papilloma viruses (HPV),
    416–417
Humoral immunity, 333–334
Hyaluronidase, as virulence factor,
    240
Hybridoma, 339
Hydrocarbon compounds, 114, *114*
Hydrogen bonds, 103
Hydrogen ion concentration,
    107–108 (*see also* Acidity)
Hydrolases, 143
Hydrolysis, 105, 117, 143
  of starch, *118*
Hypersensitivity, 339–344
  allergic response in, 341–342
  allergy shots in, 344
  autoimmune diseases in, 344
  factors in development of, *341*
  latex allergy in, 343–344
  localized anaphylaxis in, 342
  skin testing in, 344
  systemic anaphylaxis in, 343
  types of, *339,* 339–341
Hypertonic solution, 180, *180*
Hyphae, of fungi, 76
Hyphal extension, of fungi, 76, *77*
Hypogammaglobulinemia, 331
Hypotonic solution, 182

Iatrogenic infections, 267
Identification, of pathogens, 55,
    293–297
  amino acid sequences in, 64
  antimicrobial susceptibility testing
    in, 295–297
  atmospheric requirements in,
    62–64
  biochemical activity in, 64
  cell morphology in, 55–58
  clinical microbiology laboratory
    procedures for, 467–472
  colony morphology in, 62
  genetic material in, 65
  immunodiagnostic procedures
    for, 339, 345–351, *346*
  metabolic activity in, 64
  molecular diagnostic procedures
    for, 294–295
  motility in, 61–62
  nutritional requirements in, 64
    (*see also* Culture media)
  pathogenicity in, 64
  phenotypic characteristics in, 294
  staining in, 58–61, 469–470
  types of tests for, 293–294
IDPs (*see* Immunodiagnostic proce-
    dures)
Illumination source, of microscope,
    19
Immobilization, 349t
Immune deficiency, 345 (*see also*
    Human immunodeficiency
    virus)
Immune response, 333–351 (*see also*
    Immune system; Immunity)
  antibodies in, 329–331,
    336–339, *337,* 337t
  antibody formation in, *330*
  antigens in, 328–329
  cell-mediated immunity in,
    334–335
  humoral immunity in, 333–334
  hypersensitivity as, 339–344
  K cells in, 335–336
  lack of, 345
  monoclonal antibodies in,
    338–339
  NK cells in, 335–336
Immune serum globulin (ISG),
    328
Immune system, 331–333
  cells involved in, 331–333
  development of, *332*
  primary function of, 331
  stimulation of, by indigenous mi-
    croflora, 217

Immunity, 323–328
  cell-mediated, 334–335
  humoral, 333–334
Immunocompetency, 345
Immunodiagnostic procedures
    (IDPs), 339, 345–351,
    346
  increasing the value of, 347
  reagents for, 347
  types of, 347–351
  visible reactions in, 347,
    348t–349t, 350
Immunofluorescence procedures,
    347
Immunogenic substance, 328
Immunoglobulin A (IgA), 336, 338
Immunoglobulin D (IgD), 338
Immunoglobulin E (IgE), 338,
    342
Immunoglobulin G (IgG), 338
  structure of, *337*
Immunoglobulin M (IgM), 338
Immunoglobulins (Ig), 329–331
  classes of, and functions, 337t
  location in body, 329, *330*
  production of, 329–331, *330*
Immunology, 328–331
  origin of, 13
Immunology Laboratory, 290
Impetigo, 367–368
in vitro microbial growth, 152
in vivo microbial growth, 152,
    195
Inactivated vaccines, 325
Inclusion bodies, viruses as, 89
Inclusion conjunctivitis, 376
Inclusions, in monocyte of AIDS
    patient, *423*
Incubators, atmosphere of, 64
Incubatory carrier, of pathogens,
    253
India ink preparation, 471
Indigenous microflora, of humans,
    2, 212–228 (*see also* Micro-
    bial antagonism)
  beneficial roles of, 217–218
  destruction of, effects of, 214
  exposure to, 212
  locations of, 212, 213t, 214–217,
    364
    ear, 215–216, 381
    eye, 215–216, 374–375
    gastrointestinal tract, 217, 395
    mouth, 214–215, 378
    respiratory tract, 216, 383
    skin, 214, 362–364
    urogenital area, 216, 412–413

symbiotic relationships of, 218–220

Industrial microbiology, 8, 223

Infantile paralysis, 438–440

Infection Control Committee (ICC), 280

Infection control procedures, 274–282
  hospital, 280–282
  medical and surgical asepsis in, 274, 276
  medical waste disposal in, 282
  reverse isolation as, 280
  standard precautions in, 276–277
  transmission-based precautions in, 277–280
    diseases requiring, 279t

Infection Control Professional (ICP), 280–281, 281 (Insight)

Infections (*see also* Infectious diseases)
  bacterial (*see* Bacterial infections)
  bite, 374
  burn, 373
  chain of, six components of, *231,* 231–232
  of circulatory system, 421–436
  development of, *233,* 233–234
    mechanisms of, 237–242
  difference between infectious disease and, 232 (Study Aid)
  of ear, 381–382
  by endoparasites, 219
  of eye, 374–378
  focal, 373–374
  foodborne, bacterial, 254 (Insight), 406–410
  fungal (*see* Fungal infections)
  of gastrointestinal tract, 395–412
  general, 234
  of genitourinary tract, 412–421
  iatrogenic, 267
  latent, 235
  local, 234
  meaning of, 175–176
  of mouth, 378–381
  of nervous system, 437–443
  nosocomial, 267–283 (*see also* Nosocomial infections)
  potential for, 231
  primary, 236
  protozoal (*see* Protozoal infections)
  reservoirs of, 249–254
  secondary, 236
  of skin, 362–374

systemic, 234
  viral (*see* Viral infections)
  of wounds, 373–374

Infectious agents, 2 (*see also* Prions; Viroids; Viruses)

Infectious disease process, 234–237
  four phases of, 236–237, *237*

Infectious diseases, 3, 230–243
  control of (*see* Environmental control; Infection control procedures)
  development of, 234–237
    mechanism of, 237–242
    potential for, 232
  endemic, 243–244
    in U.S., 299, 299t
  epidemic, 244–247
    control of, 259
  epidemiology of, 243–259
  four phases of, 236–237, *237*
  infections and, 230–231 (*see also* Infections)
    difference between, 232 (Study Aid)
  nonendemic, 249
  pandemic, 247–249
  reservoirs of infection in, 249–254
  sporadic, 249
  top ten notifiable, in U.S., 299t
  transmission of, 243–259 (*see also* Transmission)
    modes of, 254–259

Infestation, with parasites, 219

Inflammatory exudate, 322

Inflammatory response, *233,* 233–234
  physiological events of, 321–322, *322*
  purpose of, 321, *321*

Influenza, 386–387
  as pandemic disease, 247

Influenza viruses, 385, 386–387

Inorganic chemistry, 99

Insects (*see* Arthropod vectors)

Interferons, 314

Interleukins, 313, 315, 335

Interstitial plasma-cell pneumonia, 394

Intoxications, bacterial foodborne, 406–410

Intrinsic resistance, 201

Ionic bonds, 102
  formation of, *103*

Ions, 99

Iron balance, as defense mechanism, 314

ISG (Immune serum globulin), 328

Isotonic solution, 182

Isotope, 100

Janssen, Johannes, 9

Jenner, Edward, 13

Jock itch, 372

K cells, in immune response, 335–336

Kennel cough, 392

Keratitis, 375

Keratoconjunctivitis, 375
  chlamydial, 377

Ketone group, *116*

Kinases, as virulence factors, 240

Kingdom category, in classification of organisms, 48–49

Kirby-Bauer method, of AST testing, 295

Kissing disease, 426

Klebsiella pneumoniae, 451
  respiratory infections caused by, 384
  urinary tract infections caused by, 413

Koch, Robert, 14–16, 17t

Koch's postulates, *14,* 14–15
  exceptions to, 15–16

KOH preparation, 470–471

Krebs cycle, 148

Kuru, 92

L-forms, of bacteria, 57–58

Laboratory (*see* Clinical Microbiology Laboratory; Pathology Department)

Laboratory specimens (*see* Clinical specimens)

Lactobacillus spp., 451
  fermentation by, 149
  indigenous to humans, 217

Lactose, 117

Lag phase, of bacterial growth curve, 157

Latent infections, 235

Latex allergy, 343–344

Lecithinase, as virulence factor, 241

Leeuwenhoek, Antony (*see* van Leeuwenhoek, Antony)

Legionella pneumophila, 245, 388–389, 451

Legionellosis, 384, 388

Legionnaires' disease, 388

Leishmania spp., 465

Lens system, of microscope, 17

Leprosy, 371–372

Leukocidins, 242
Leukocytes, *316,* 316–317
Leviticus, Book of, 9
Lice, *429* (*see also* Arthropod vectors)
Lichens, 82
Light microscope, 17–20, *18*
   modified, 19–20
Light source, of microscope, 19
Limit of visibility, 23
Linens, standard precautions for,
   277
Linnaeus, Carolus, 46–47
Linoleic acid, 120
Lipids, 119–121
   classes of, 119–121
   compound, 121
   derived, 121
   simple, 119–120
   synthesis of fat, *120*
   waxes as, 120
Lipoteichoic acid, *40*
Lister, Joseph, 16, 17t
   development of antiseptic tech-
     nique by, 178, 193
Listeria monocytogenes, 451
   meningitis caused by, 438
   morphology of, 57
Lithotrophs, 139–141
Living organisms, distinguishing
   characteristics of, 113
Local infections, 234
Localized anaphylaxis, 342
Lockjaw, 441–442
Logarithmic growth phase, bacter-
   ial, 157
Lophotrichous bacteria, *43,* 43–44
   (*see also* Motility, bacterial)
Louse (*see* Lice)
Louse borne typhus, 428
Lumbar puncture, 287, *287*
Lyme borreliosis, 430–431
Lyme disease, 430–431
Lymphadenitis, 421
Lymphadenopathy, 421
Lymphangitis, 421
Lymphatic system
   anatomy of, *422*
   in inflammatory response, 322
   nonspecific diseases of, 421
   viral diseases of, 421–430
Lymphocytes, *316,* 317, 322
   retrovirus particle in, *424*
Lyophilization, of microorganisms,
   180
Lysis, by complement, 348t
Lysogenic conversion, 91–92
   by bacteriophages, 161–162

Lysogenic cycle, of viral infection,
   85
Lysogeny, 85, 91
Lysosomes, 35–36
   enzymes of, 318
Lysozyme, 35–36, 318
Lytic cycle, of viral infection, 85,
   88, 89t

MacConkey agar, 154
MacLeod, Colin, 164
Macrophages, *316,* 317, 322, 329
Magnification, by compound micro-
   scope, 17–19
Major histocompatibility complex
   (MHC), 334
Malaise, 362
Malaria, 433–434
Maltose, 117
Mannitol-salt agar (MSA), 154
Mastigophora, 74, 75t, 457t
McCarthy, Maclyn, 164
MDPs (Molecular diagnostic proce-
   dures), for pathogens,
   294–295
MDR (*see* Multidrug resistance)
Measles, 365–366
Measles-mumps-rubella (MMR)
   vaccine, 327, 366, 367,
   427
Measurement, units of, 22–23
   conversion of, 473–474
   representation of, 23t
Media (*see* Culture media)
Medical asepsis, 274
Medical microbiology, 7
Membrane (*see* Cell membrane; Nu-
   clear membrane)
Memory cells, 330
Memory response, 330
Meninges, 437
Meningitis, 437, 438
Meningoencephalitis, primary ame-
   bic, 442–443
Mercury salts, as antimicrobial
   chemicals, 193
Mesophiles, 179
Mesosomes, of bacteria, 41
Messenger RNA (mRNA), 126
   of eucaryotes, 35
   protein synthesis by, 129–132
   synthesis and distribution of, 132
   translation of, 132
Metabacteria, 56 (Insight section)
Metabolism, 144
   of bacteria, 64
   biosynthetic energy of, 150–152

   of cell, 30–31, *31*
   energy released during, 144–150
   energy used during, 150–152
   of microorganisms, 141–152
Metabolite, 144
Metchnikoff, Elie, 17t
Meter, divisions of, 22–23
Metric units, 22–23
   conversion of, 473–474
   representation of, 23t
MHC (Major histocompatibility
   complex), 334
Microaerophiles, 62–63
Microbe Hunters (De Kruif), 63 (Insight section)
Microbes, 2 (*see also* Microorganisms)
Microbial antagonism, 219, 232 (*see
   also* Indigenous microflora)
   in first line of defense, 312–313
Microbicidal agents, 177
Microbiologists, 7
Microbiology, 2
   areas of study in, 6–8
   contributors to, 9–16, 17t
   derivation of word, 2
   genetics and, 8 (*see also* Genetics)
   history of, 8–16, 17t
     advances of 1800s in, 11
   organisms studied in, 2 (*see also*
     Microorganisms, classifica-
     tion of)
   relevance of, 6
   tools of, 16–22 (*see also* Micro-
     scopes)
   units of measure of, 22–23 (*see
    also* Metric units)
Microbistatic agents, 177–178
Micrococcus spp., 214
   as opportunistic pathogens,
     363
Microcosmos (Margulis and Sagan),
   141
Microfilaments, in eucaryotes, 37
Microflora, indigenous (*see* Indigenous microflora)
Micrometer, 22, 23t
Microorganisms
   acellular, 31, *32,* 32t (*see also*
     Prions; Viroids; Viruses)
   agricultural uses of, 7, 220–222
   as biological warfare agents,
     223–224
   cell structure of, 30–46 (*see also*
     Eucaryotes; Procaryotes)
   cellular, 30–32, *32,* 32t
   chemistry of, 112–133 (*see also*

Biochemistry; Organic chemistry)
classification of, 46–49, 54 (*see also* Algae; Bacteria; Fungi; Protozoa)
contagious, 9 (*see also* Infections; Infectious diseases)
defense against, 310–351 (*see also* Defense mechanism(s); Immune response; Immunity)
difference between inanimate objects and, 113
discovery of, 9–16
diseases caused by (*see* Infections; Infectious diseases; Pathogens)
ecology of, 140, 211–228 (*see also* Indigenous microflora)
family tree of, *3* (*see also* Classification)
food industry uses of, 4, 5, 5t, 223
functions of, 2–6
growth of, 137–173 (*see also* Growth, bacterial; Growth, microbial)
control of, 174–214
inactivation of, 174–210 (*see also* Antimicrobial agents; Antimicrobial methods)
industrial use of, 8, 223
knowledge of, in Middle Ages, 9
measurement of, 22–23
metabolism of, 139–152 (*see also* Energy metabolism)
naming of, 46–49
nutritional requirements of, 139 (*see also* Energy metabolism)
nutritional types of, 139–141
obligate, 62–63 (*see also* Obligate microorganisms)
origin of, *3*, 8, 31
pathogen inhibition by, 4 (*see also* Antibiotics)
as pathogens (*see* Infections; Infectious diseases; Pathogens)
physiology of, 139–152
protected from phagocytosis, 320–321
pyogenic, 322
slime molds, 82–83
soil nutrients from, 5 (*see also* Nitrogen cycle)

structures of, 30–49 (*see also* Eucaryotes; Procaryotes)
study of, 6–8
taxonomy of, 46–49
ubiquitous, 3
uses of, 3–6, 222–223
virulent and avirulent, 230–231
visibility by microscope, 16–22
water in, importance of, 104–108
water purification by, 5–6
Microscopes, 16–22
brightfield, 17
characteristics of types of, 22t
compound, 9, 17–20, *18*
darkfield, 19–20
electron, 20–21
fluorescent, 19–20
light, 17–20, *18*
limit of visibility of, 23
magnification by, 17–19
modified light, 19–20
parts of, 17–20, *18*
resolution of, 19, 23
simple, 9, *10*
UV light, 20
Microsporum spp., 372–373
Microtubules, of eucaryotes, 37, 38
Middle Ages, knowledge of disease in, 9
Middle ear infections, 381–382
Milestones in Microbiology (Brock), 63 (Insight section)
Milk, bacteria in, 254 (Insight)
Millimeter, 22, 23t
Mini-systems, for identification of pathogens, 294
Mitochondria, 36
Mitosis, 37
MMR (Measles-mumps-rubella) vaccine, 327, 366, 367, 427
Moist heat, as antimicrobial method, 187
Moisture tolerance, of microorganisms, 179–180
Molds (*see* Fungi)
Molecular diagnostic procedures (MDPs), for pathogens, 294–295
Molecules, 100–101
Monera, kingdom of, 48
Monoclonal antibodies, in immune response, 338–339
Monocytes, *316*, 317
Mononucleosis, infectious, 426
Monosaccharides, 115–116
Monotrichous bacteria, *43*, 43–44 (*see also* Motility, bacterial)

Moraxella catarrhalis, 216
ear infections caused by, 382
Motility, bacterial, *43*, 43–44, 61–62
testing for, 61–62
Mouth
anatomy of, *379*
indigenous microflora of, 214–215, 378
infections of, 378–381
mRNA (*see* Messenger RNA)
MSA (Mannitol-salt agar), 154
Mucous membranes
collection of swab specimens of, 288
nonspecific defense mechanisms of, 311–313
Multidrug resistance (MDR)
of bacterial pathogens, 181, 202–203, 404, 405, 406
control of, 185
of ETEC, 403
of *Neisseria gonorrhoeae*, 418–419
prevention of, 185
Mumps, 427
Murein, 42
Murine typhus, 428
Mushrooms, 77
Mutagens, 159
Mutants, 159
Mutations, in bacterial genetics, 132, 159–160
agents that cause, *160*
genetic engineering and, 133
Mutualism, 218, *219*
Mycelium, of fungi, 76
Mycobacteriology Laboratory, 293
Mycobacterium leprae, 371–372, 451
Mycobacterium spp.
acid-fast staining of, 59, 61
Gram staining of, 58–59
Mycobacterium tuberculosis, 391–392, 451
generation time of, 157
multidrug resistance of, 181
waxes in cell wall of, 120
Mycologists, 7
Mycology, 7
Mycoplasma pneumoniae, *67*, 384, 389, 451
Mycoplasma spp., 66, 68
difference between L-form bacteria and, 68
diseases caused by, 66t, 68
lack of cell wall in, 42, 58

Mycoplasma spp.—*Continued*
  urinary tract infections caused by, 413
Mycoses, 79, 80–81
  subcutaneous, 82
  superficial and cutaneous, 81–82
  systemic, 82
Mycotoxins, 79, 407t
  physiological effects of, 438
Myelitis, 438
Myocarditis, 422

NADPH oxidase, 318
Naegleria fowleri, 442–443
Naegleriasis, 442–443
Nanometer, 22, 23t
National Center for Infectious Diseases (NCID), 298–299
National Committee for Clinical Laboratory Standards (NCCLS), 297
Natural killer (NK) cells, 335–336
Natural passive acquired immunity, 327
Navicula, 72
NCCLS (National Committee for Clinical Laboratory Standards), 297
Necrotoxins, 370, 373
Needle-stick injuries, precautions for, 277
Negative staining, of bacteria, 42
Neisseria gonorrhoeae, 63
  conjunctivitis caused by, *377*, 377–378
  infections caused by, 418–419, 451
    urinary tract, 413
  multidrug resistance of, 181, 204
  penicillin resistant, 247
Neisseria meningitidis, 438, 451
Neisseria spp., indigenous to humans, 216
Nervous system
  anatomy of central, *437*
  infections of, 437–443
    bacterial, 441–442
    protozoal, 442–443
    viral, 438–441
Neurotoxins, 241–242, 373
  of *Clostridium tetani*, 441–442
  physiological effects of, 438
Neutralism, 219
Neutrons, 100
Neutrophils, *316*, 317–318
Nightingale, Florence, 16
Nitrifying bacteria, 221–222

Nitrobacter, 222
Nitrogen cycle, 220–222, *221*
Nitrogen-fixing bacteria, 5, *6*, 68
  role of, in elemental cycle, 221–222
  types of, 221–222
Nitrogenous bases, 126–128
Nitrosococcus, 222
Nitrosolobus, 222
Nitrosomonas, 222
Nitrosospira, 222
NK cells, in immune response, 335–336
Nocardia spp., 451
Nodules, on legume roots, *222*
Nonendemic disease, 249
Nongonococcal urethritis (NGU), 216, 417–418
Nonpathogenic microorganisms, 2, 220
Norwalk viruses, 396–397
Nosocomial infections, 267, 269 (Insight)
  areas of transmission of, 269–270
  common types of, 269
  control of, 270–274
    contact contamination and, 273–274
  development of, 267–270
  increased incidence of, in U.S., 268
  pathogens involved in, 268, 384
  patient types vulnerable to, 270
  prevention of, 268
  viral, 270
  zoonotic, 271 (Insight)
Nostoc, 72
Notifiable diseases, nationally, 363
Nuclear membrane, 32–33
Nucleic acids, 125–133
Nucleolus, 34–35
Nucleoplasm, 30
Nucleotides, 126–133
  structure of, *126*
Nucleus, 33–35
Numerical aperture, of microscope, 19
Nutritional requirements, of microorganisms, 139
Nutritional types, of microorganisms, 139–141

Objective(s), of microscope, 18–19
Obligate microorganisms
  aerobic, 62
  anaerobic, 63, 151

of gastrointestinal tract, 395
  intracellular pathogens as, 65
Octads, 56, 56t
Ocular micrometer, 23
Oil-immersion objective, 18–19
Oncogenic viruses, 83
Oncoviruses, 83
One-gene/one protein hypothesis, 131
Oocystis, 72
Oomycetes, 76, 77, 79
  characteristics of, 79t
Opportunistic pathogens, 2 (*see also* Indigenous microflora; Microbial antagonism)
Opsonins, 315
Opsonization, 315
  in immunodiagnostic procedures, 348t
Oral cavity (*see* Mouth)
Order category, in classification of organisms, 48
Organelles, 30
Organic chemistry, 99, 113–115
  of cyclic compounds, *114*, 114–115
  of carbon bonds, 113–114
Orienta, diseases caused by, 66t
Oscillatoria, 72
Osmosis, 180, *180*
Osmotic pressure, 180, *180*
Osmotic tolerance, of microorganisms, 180, 182
Otitis externa, 382
Otitis media, 381–382
Oxidation, 146 (*see also* Aerobic respiration)
Oxidation-reduction reactions, 146
Oxidizing agents, as antimicrobial chemicals, 193
Oxygen holocaust (Insight section), 141

P-K antibody, 338
Pandemic disease, 247–249
  acquired immunodeficiency syndrome as, 247–249, *248*, 424–425
  cholera as, 400–401
  influenza as, 247, 386–387
Papillomatosis, genital, 416–417
Papillomaviruses, warts caused by, 367
  genital, 416–417
Parainfluenza viruses, 385
Paramecium, 73
Paramyxovirus, 427

Parapertussis, 392
Parasitism, of microflora, 219, *219*
Parasitology, 453–466
Paratrachoma, 376
Parotitis, infectious, 427
Passive acquired immunity, 323, 327–328
Passive carrier, of pathogens, 253
Pasteur, Louis, 11–13, *12*, 17t
   discovery of anaerobes by, 63 (Insight section)
Pasteurella haemolytica, 374
Pasteurella multocida, 374
Pasteurization, 12–13, 177
Pathogenicity (*see also* Virulence)
   factors determining, 242–243
   meaning of, 237
   and virulence, 242–243
   virulence as measure of, 230
Pathogens, *4* (*see also* Infections; Pathogenicity)
   avirulent strains of, 242–243
   bacterial, 60t, 64 (*see also* Bacterial infections)
      medically important, 450–452
      metabolic activity of, 64
   ciliates as, 74–75
   clinical identification of, 293–297 (*see also* Immunodiagnostic procedures)
      antimicrobial susceptibility testing in, 295–297
      molecular diagnostic procedures for, 294–295
      phenotypic characteristics in, 294
      types of tests for, 293–294
   destruction of, 4 (*see also* Antimicrobial agents; Antimicrobial methods; Multidrug resistance)
   in disease process, 229–259 (*see also* Infectious diseases)
   diseases caused by, 4t (*see also* Infections)
      nosocomial infections, 268, 271 (Insight)
   examples of, 4t
   first appearance of, 9
   flagellates as, 74
   fungi as, 81–82, 81t (*see also* Fungal infections)
   as indigenous microflora, 220
   inhibition of growth of, 4 (*see also* Antimicrobial methods; Multidrug resistance)

in kitchens, 254 (Insight), 257 (Insight)
mutation of, 132–133
in nosocomial infections, 268, 269 (Insight), 271 (Insight) (*see also* Nosocomial infections)
obligate intracellular, 65
opportunistic, 2 (*see also* Indigenous microflora; Microbial antagonism)
protozoa as, 71, 75 (*see also* Protozoal infections)
   parasitic, 455, *456*, 457t–458t (*see also* Parasitology)
relevance of, 6
standard precautions for, 277
transmission of, 254–259, *255* (*see also* Transmission)
   precautions for, 277–280
viruses as, 83–92 (*see also* Viral infections)
Pathologists, 290
Pathology, 290
Pathology Department, 289–297
   organization of, *290*
   responsibility of, 293
Patient care equipment, standard precautions for, 277
Patient placement, standard precautions for, 277
PCR (Polymerase chain reaction), 295
PEA (Phenylethyl alcohol agar), 154
Pediculus humanus, 429
Pellicle, of protozoa, 74–75
Penicillin, 196–197
   bacterial degradation of, 203, *203*
Penicillium notatum, 79
   antibacterial effect of, 194
   inhibition of *Staphylococcus aureus* by, *195*
Peptide bonds, 102
Peptidoglycan, *40*, 42
   penicillin interference in synthesis of, 196
Peptostreptococcus spp., 217, 395, 451
Pericarditis, 422
Periodic table, 100, *101*
Periodontal diseases, bacteria causing, 215, 379–380
Periodontitis, 379–380
Peritrichous bacteria, *43*, 43–44, *44* (*see also* Motility, bacterial)

Pertussis, 392–393
Petri, Julius, 14
Petri dish, 14
Pfiesteria piscicida, 69
pH (*see* Acidity)
Phaeophyta, 69–70
Phagocytes, in inflammatory response, 234
Phagocytosis, 35–36, *320*
   as defense mechanism, 316–321
   four steps of, 318
   by protozoa, 72
   by rat leukocytes, *318, 319*
   species protected from, 320
Phagolysosome, 318, *319*
Phagosome, 318
Phase contrast microscope, 19
Phenol coefficient test, 193
Phenolics, as antimicrobial chemicals, 192–193
Phenotype, 159
Phenylethyl alcohol agar (PEA), 154
..phile suffix, meaning and use of, 183 (Study Aid)
Photoautotrophs, 140–141
   obligate anaerobic, 151
Photoheterotrophs, 140–141
Photosynthesis, 150–151
   in absence of oxygen, 151
   in bacteria, 41, 68 (*see also* Cyanobacteria)
   in eucaryotes, 37
   organisms that utilize, 139–141, 150–151
Phototrophs, 139–141, 151
Phthirus pubis, 429
Phycologists, 6
Phycology, 6
Phycotoxins, 69, 71t, 407t
   physiological effects of, 438
Phylum category, in classification of organisms, 48
Physical antimicrobial methods, 184–190
   cold as, 188–189
   drying as, 189
   filtration as, 190
   heat as, 184, *186*, 186–188
   radiation as, 189–190
   ultrasonic waves as, 190
Physiology, of microorganisms, 138–152
Phytophtera, 77
Pili (pilus), bacterial, 44, *45* (*see also* Sex pilus)
   as virulence factor, 239

Pink eye, 376
Pinocytosis, by protozoa, 72
Plague, 431–432
Plantae, kingdom of, 48
Plantar warts, 367
Plasma cells, 330
Plasma membrane
  of bacteria, *39, 41*
  of eucaryotes, 33, *33, 34*
Plasmids, bacterial, 41, 160–161,
  164, 165
  in *Escherichia coli, 161*
  and resistance to antimicrobial
    agents, 202
  used in genetic engineering, 166
Plasmodial slime molds, 83
Plasmodium spp., 433–434
Plasmodium vivax, 75
Plasmolysis, 182
Plasmoptysis, 182
Plastids, of eucaryotes, 37
Pleomorphism, bacterial, 58
Pleuropneumonia-like organisms
  (PPLO), 66
PMNs (Polymorphonuclear cells),
  317
Pneumocystis carinii, 287–288
  pneumonia caused by, 384, 394
  silver staining of, *395*
Pneumocystis spp., human diseases
  caused by, 81t
Pneumonia
  caused by *Pneumocystis carinii,*
    384, 394
  interstitial plasma-cell, 394
  microorganisms causing,
    383–384
  mycoplasmal, 389
  primary atypical, 389
  viral, 384
Pneumonic plague, 431–432
Poliomyelitis, 438–440
Polioviruses, 439–440
  development of, in cells, *439*
Polymer, of nucleic acid, *127*
Polymerase chain reaction (PCR),
  295
Polymerases, 143
  DNA, 128–129
  RNA, 130
Polymorphonuclear cells (PMNs),
  317
Polypeptides, 123, *123*
Polyribosomes (*see also* Ribosomes)
  of bacteria, 41
  of eucaryotes, 35
Polys, 317

Polysaccharides, 117–119
  main functions of, 118
  structural purposes of, 118–119
Polysomes, of eucaryotes, 35
Pontiac fever, 388
Population counts, of bacteria,
  154–156
Population growth curve, of bacte-
  ria, 156–159, *157*
  stages of, 157–159
Porphyromonas spp., 452
  mouth infections caused by, 379
PPLO (Pleuropneumonia-like or-
  ganisms), 66
PPNG (Penicillinase-producing
  Neisseria gonorrhoeae),
  247
  treatment of, 378 (*see also* Super-
    bugs)
Precipitin tests, 347, 348t, *350*
Pressure tolerance, of microorgan-
  isms, 183
Prevotella spp., 452
  mouth infections caused by, 379
Primary amebiasis, 410–411
Primary atypical pneumonia, 389
Primary infections, 236
Primary response, 330
Prions, 7, 32t
  human diseases caused by, 92
Procaryotae, kingdom of, 48
Procaryotes, 3
  cell structures of, 38–45
  differences between eucaryotes
    and, 45–46, 47t
  organisms included in, 31, 32t
  ribosomal activity in, 131
  size of, 38
  transcription of DNA in, 131
Properdin, 315
Prophage, 91
Propionibacterium spp., 214, 363
  acne caused by, 369–370
Prostaglandins, 315–316
Prostatitis, 413
Protective equipment, personal, *275*
  standard precautions for,
    276–277
Proteins, 121–125
  amino acid structure of, 121–122
  configuration of, 123–124, *124*
  DNA control of synthesis of,
    129–133
    outline of, 129, *130*
  enzymes as, 124–125, 142–143
  globular, 123
  heat denaturing of, 125

quaternary structure of, 123,
    124
  secondary structure of, 123
  shape of, 123–124, *124*
  synthesis of, 129–133
    in bacteria, 41
    in eucaryotes, 35, 131
    outline of, 129, *130*
    in procaryotes, 41, 131
  tertiary globular structure of,
    123, *124*
Proteus spp., 452
  urinary tract infections caused by,
    413
Proteus vulgaris
  ear infections caused by, 382
  pili and flagella of, *45*
Protista, kingdom of, 48
Protons, 100
Protoplasm, 30
Prototheca, 69
Protozoa, 2, 31, 32t, 70–75
  amebas, 72
  cellulose digestion by, 118
  characteristics of, 75t
  cilia of, 37
  ciliates, 74–75, 457t
  classification of, 72, 73–75
    basis of, 72
  diseases caused by, 4t (*see also*
    Protozoal infections)
    zoonotic, 252t
  flagella of, 37
  flagellates, 74
  inactivation of, 199t (*see also* An-
    timicrobial methods)
  lack of cell wall in some, 74–75
  nonmotile, 75
  parasitic, 455, *456,* 457t–458t
    (*see also* Parasitology; Proto-
    zoal infections)
  pathogenic, 71, 75 (*see also* Pro-
    tozoal infections)
  pellicle of, 74–75
  representative pond, *73*
Protozoal infections, 410–412
  of cardiovascular system,
    433–436
  of genitourinary system, 420
  of nervous system, 438, 442–443
Protozoologists, 7
Protozoology, 7
Prusiner, Stanley B., 92
Pseudomonas aeruginosa, 373, 452
  blue-green pus of, 322
  ear infections caused by, 382
  multidrug resistance of, 181

Pseudomonas spp., 216, 364
  generation time of, 157
  in soil, 222
  urinary tract infections caused by, 413
Pseudomonicidal agents, 177
Pseudopodia, 72
Psychroduric organisms, 179
Psychrophiles, 179
Psychrotrophs, 178–179
Pubic lice, *429*
Public health agencies, in disease control, 297–300
Public health laws, earliest, 9
Public Health Service, of U.S., 298
Purines, 126–128, *127*
  difference between pyrimidines and, (Study Aid), 127
Purple bacteria, 68
Purulent exudate, 322
Pus, 322
  formation of, 234
Pyogenic microorganisms, 322
Pyogenic pathogens, 234
Pyrimidines, 126–128, *127*
  difference between purines and, (Study Aid), 127
Pyrogenic substances, 313
Pyrogens, 241, 313

Quality control, in clinical microbiology laboratory, 297
Quellung reaction, 349, 349t
  positive, *351*
  in *Streptococcus pneumoniae,* 239

R-colonies, 42
R-factor, 202
Rabbit fever, 432–433
Rabies disease, 440–441
Rabies immune globulin (RIG), 328
Rabies virus, *440,* 440–441
Radiation, as antimicrobial agent, 189–190
Radicals, neutrophilic production of, 318
Radioimmunoassay, 349t
Reagin, 341
Recombination, genetic, 163
Red algae, 69–70
Redi, Francisco, 17t
Reduction, 146
Reduviid bug, 436
Relatedness of microorganisms, 49
Replication, of deoxyribonucleic acid (DNA), 128–129

RER (Rough endoplasmic reticulum), 35
RES (Reticuloendothelial system), 317
Reservoirs, of pathogens, 249–254
  examples of, 250, *250, 253*
  inanimate, 253
  living, 249–253
Resistance factor, 202
Resolution, of microscope, 19
Respiration
  aerobic, 146–148, *147*
  anaerobic, by chemotrophs, 150
Respiratory system
  anatomy of, *383*
  indigenous microflora of, 216
  infections of, 382–394
    bacterial, 387–392
    fungal, 393–394
    nonspecific, 383–384
    viral, 385–387
  nonspecific defense mechanisms of, 312
Reticulate body, of *Chlamydia,* 66
Reticuloendothelial system (RES), 317
Retrovirus particle, from lymphocyte, *424*
Retroviruses, 90
Reverse isolation, 280
Reye's syndrome, 364
Rhinitis, acute viral, 385
Rhinoviruses, 385
Rhizobium, 221–222, *222*
Rhizopus, 77
Rhodophyta, 69–70
Ribonucleic acid (RNA), 125–133
  (*see also* Ribosomal RNA)
  differences between DNA and, 129
Ribose, 126
Ribosomal RNA (rRNA), 34, 126
  3-Kingdom classification system based on, 49
  differences between Archaebacteria and Eubacteria, 69
  of eucaryotes, 35, 131
Ribosomes (*see also* Ribosomal RNA)
  of bacteria, 41
  of eucaryotes, 35
Rickettsia prowazekii, 428–429
Rickettsia rickettsii, 427–428
Rickettsia spp., 65, 452
  difference between viruses and, 65

infections caused by, 65, 66t, 258
  cardiovascular system, 427–428
Rickettsia typhi, 428
RIG (Rabies immune globulin), 328
Ringworm infections, 81–82, 372–373
RNA (*see* Ribonucleic acid)
RNA polymerase, 130
Rochalimaea quintana, 65
Rocky Mountain spotted fever, 427–428
Rods (*see* Bacilli)
Room air, carbon dioxide in, 63
Rotaviruses, 396–397
Rough endoplasmic reticulum (RER), 35
Route of entry, in transmission of disease, 256t
rRNA (*see* Ribosomal RNA)
RSV (Respiratory syncytial virus), 385
Rubella, 366–367
Rubella virus, 366, *366*
Rubeola, 365–366
Rubeola virus, 365

S-colonies, 42
Saccharomyces, as biotherapeutic agent, 218
Saccharomyces cerevisiae, 80, 397
Salmonella enteritidis, 403–404
Salmonella spp., 452
  enterotoxin production by, 242
  peritrichous cells of, *44*
Salmonella typhi, 404–405
Salmonella typhimurium, 403–404
Salmonellosis, 403–404
Salts
  characteristics of, 106–107
  of heavy metals, as antimicrobial chemicals, 193
Sanitary microbiology, 8
Sanitization, 177
Saponification, 120
Saprophytes, 5
  cellulose digestion by, *5,* 118
  role in elemental cycles, 221–222
Sarcodina, 72, 75t, 457t
Saturated fatty acids, 120
Scalded skin syndrome, 367–368
Scalp itch, 372
Scanning electron microscopy (SEM), 20–21
Scarlet fever (scarlatina), 368–369
Scenedesmus, 72

Schick test, 350
Schleiden, Matthias, 30
Schröder, and von Dusch, 17t
Schultz-Charlton test, 351
Schwann, Theodor, 30
Scotch tape preparation, 471–472
Scrapie, 92
Secondary infections, 236
  to acquired immunodeficiency
      syndrome, 423–424
Secondary response, 330
Secretions, 35
Sedimentation coefficients, 35
Selective medium, for bacterial
      growth, 154
Selective permeability, 33
SEM (*see* Scanning electron mi-
      croscopy)
Semmelweis, Ignaz, 16
Sepsis, 178, 241
Septate hyphae, 79
Septicemia, 241, 286, 421
Septicemic plague, 431–432
Sequelae, 389
SER (Smooth endoplasmic reticu-
      lum), 35
Serologic procedures, 345
Serotype, determination of, 294
Serum sickness, 340
Sewage treatment, 5–6, *302,*
      302–303
Sex pilus, 44, 164–166
Sexually transmitted disease (STD),
      249, 413–431
Sharps, disposal of, 282
Shellfish poisoning, phytotoxins
      causing, 71t, 438
Shigella spp., 405–406, 452
  enterotoxin production by,
      242
Shigellosis, 405–406
Shingles, 364–365
Shipping clinical specimens,
      288–289, *289*
  label for, *290*
Shock, condition of, 241
Signs, of disease, 235
Silver salts, as antimicrobial chemi-
      cals, 193
Simple microscope (*see* Micro-
      scopes, simple)
Simple staining, of bacteria, 58, *59,*
      61t
16S ribosomal RNA, and related-
      ness of organisms, 49, 69
Skin, *363*
  growth of microbes on, 362

indigenous microflora of, 214,
      362–374
  infections of, 362–374
    bacterial, 367–372
    fungal, 372–373
    viral, 364–367
  nonspecific defense mechanisms
      of, 311–313
Skin testing, 344, 350
  for TB, 340–341
Sleeping sickness, African, 435,
      442
Slime layer, of bacteria, 42–43
Slime molds, 82–83
Small pox virus, as biological war-
      fare agent, 224
Smooth endoplasmic reticulum
      (SER), 35
Soaps, as surfactants, 192
Sodium chloride (NaCl), 102, *103*
Soil bacteria
  aerobic oxidation by, 149
  chemolithotrophic, 221–222
Soil microorganisms, 5, 222
Solute, 105
Solutions, 105
Solvent, 105
sp. and spp. abbreviation, how to
      use, 48
Spallanzani, Abbe Lazzaro, 17t
  discovery of anaerobes by, 63 (In-
      sight section)
Spanish flu epidemic, 386–387
Species
  abbreviation of names of, 48
  in binomial naming, 47–48
  in classification of organisms, 48
Specific epithet, in binomial nam-
      ing, 47
Specimens (*see* Clinical specimens)
Spectrophotometers, 155
Spermatozoa, flagella of, 37
Spinal tap, 287, *287*
Spirilla, 56
Spirillum minor, 374
Spirochetes, 43–44, 57
  medically important, 449
Spirogyra, 72
Spontaneous generation, 11
Sporadic disease, 249
Spores (*see* Sporulation)
Sporicidal agents, 177
Sporothrix, 81
  human diseases caused by, 81t
Sporozoa, 75, 75t, 457t–458t
Sporozoites, *454*
  of *Plasmodium,* 434

Sporulation
  bacterial, 44–45
  by *Clostridium* spp., 186
  fungal, 76, *77*
Sputum specimen
  acid-fast staining of, 391
  collection of, 287
Staining, of bacteria
  acid-fast, 59, 61
  differential, 61, 61t
  flagellar, 61
  Gram, 58–59, 469–470
  negative, 42
  simple, 58, *59,* 61t
  structural, 61, 61t
Standard Precautions, 276
Staphylococci (staphylococcus), *55,*
      56, 56t
Staphylococcus aureus, 363,
      367–370, 373, 452
  coagulase production by, 240
  ear infections caused by, 382
  enterotoxin production by, 242
  food poisoning by, 409–410
  image of
    light microscopic, *20*
    SEM, *21*
    TEM, *21*
  kinase production by, 240
  multidrug resistance of, 181
  respiratory infections caused by,
      384
  and toxic shock syndrome, 242
Staphylococcus epidermis, 181, 363
Staphylococcus pyogenes, 363,
      367–369, 373
  as flesh-eating bacterium, 370
      (Insight)
Staphylococcus spp., 214–217
  generation time of, 157
  hyaluronidase production by, 240
  urinary tract infections caused by,
      413
Staphylokinase, 240
Starch, 118
  difference between cellulose and,
      119
Stationary phase, of bacterial
      growth curve, 158
STD (Sexually transmitted disease),
      249, 413–431
Steam, pressurized, as antimicrobial
      method, 187–188, *188*
Stentor, 73
Sterile gown, *275*
Sterile technique, 178
  origin of, 16

Sterilization, 176
  developed by Pasteur, 11
  Lister's use of, 16
  methods of (*see* Antimicrobial methods)
Steroids, 121
Stomach flu, 396
Storage molecules, 118
Strep A (*see Streptococcus pyogenes*)
Strep throat, 389–390
Streptobacilli (streptobacillus), 57
Streptobacillus moniliformis, 374
Streptobacillus spp., 217
Streptococcal pharyngitis, 389–390
Streptococci (streptococcus), *55, 56,* 56t
  in liquid media, *390*
Streptococcus agalactiae, 452
  meningitis caused by, 438
Streptococcus pneumoniae, 452
  capsules of, as virulence factor, 239, *239*
  conjunctivitis caused by, 376
  ear infections caused by, 382
  meningitis caused by, 438
  multidrug resistance of, 181
  respiratory infections caused by, 384
  vaccine for, 204
Streptococcus pyogenes, 452
  complications caused by, 389–390
  ear infections caused by, 382
  respiratory infections caused by, 389–390
  skin infections caused by, 367–370
  toxic shock syndrome caused by, 242, 363, 390
Streptococcus spp., 215–217
  alpha and beta hemolytic, 215, 240–241
  differentiation of, 240–241
  fermentation by, 149
  hyaluronidase production by, 240
  mouth infections caused by, 378
  as opportunistic pathogens, 363
Streptokinase, 240
Structural staining, of bacteria, 61, 61t
Styes, 369
Subclinical disease, 176, 235
Subspecies category, in classification of organisms, 48
Substrate, of enzymes, 125, 142, *142*
Subunit vaccine, 326

Subunits of ribosomes, in eucaryotes, 35
Sucrose, 117, *117*
Suffixes
  ..emia, 286 (Study Aid)
  ..phile, 183 (Study Aid)
Sugar (*see* Disaccharides; Monosaccharides; Polysaccharides)
Sulfonamide drugs, 196, *196*
Superbugs
  control of, 185 (Insight)
  description of, 181 (Insight)
Superinfections, 204, 214, 217–218, 313
Surfactants, 192
Surgical asepsis, 274
Surgical infections, 373–374
Svedberg units (S), 35
Swimmer's ear, 382
Swine flu epidemic, 386–387
Symbionts, 218
Symbiotic relationship(s), of indigenous microflora, 218–220
  commensalism as, 219
  examples of, 218, *219*
  microbial antagonism as, 219
  mutualism as, 218
  neutralism as, 219
  parasitism as, 219, *219*
  pathogenic, 220
Symptomatic disease, 235
Synergism, 218
Syphilis, 419–420
  darkfield diagnosis of, *19,* 19–20
  epidemic of 1500s, 9
  primary, *419*
  secondary, *419*
  stages of, 235–236, *236*
Systemic anaphylaxis, 343
Systemic infections, 234

T cells, 322, 329
  destruction of, by HIV, 423–424
  development of, 331–333
  in humoral immunity, 333–334
T-dependent antigens, 330, 333–334
T-independent antigens, 330, 334
T lymphocytes (*see* T cells)
Taxonomy, 46
  basis of science of, 46–47
  binomial naming system in, 46–48
  classification systems of, 48–49
  of microorganisms, 46–49
TB (Tuberculosis), 391–392
  skin test for, 340–341

TCA (Tricarboxylic acid) cycle, 148
Teichoic acid, *40, 42*
TEM (*see* Transmission electron microscopy)
Temperate bacteriophages, 91
Temperature tolerance, 178–179
  bacterial categories based on, 178–179, 179t
  and time, in heat sterilization, 184
Tetanospasmin, 441–442
Tetanus, 441–442
Tetanus immune globulin (TIG), 328
Tetrads, *55, 56,* 56t
Tetrasaccharides, 117
Thayer-Martin agar, 154
Thermal death point (TDP), 184
Thermal death time (TDT), 184
Thermoduric organisms, 179
Thermophiles, 178–179
Three-dimensional proteins, 123
3-Kingdom classification system, 49
Thymine, 126–128
Tick fever, Colorado, 425–426
Tickborne diseases, 425–426, 427–433 (*see also* Reservoirs)
Ticks (*see* Tickborne diseases)
TIG (Tetanus immune globulin), 328
Tinea infections, 81–82, 372–373
Topoisomerase, DNA, 128–129
Total cell count, and viable cell count, 154–155
Total magnification, of microscope, 17–18
Toxemia, 421
Toxic shock syndrome (TSS), 242, 363 (*see also Streptococcus pyogenes*)
Toxins, 125, 232
  algal, 69, 71t, 407t, 438
  associated with virulence, 241–242
  of cyanobacteria, 68
  foodborne bacterial, 406–410
  fungal, 79, 407t, 438
  protozoal, 407t–408t
  viral, 408t
Toxoid vaccine, 326
Toxoplasma gondii, 434–435, 465
Toxoplasmosis, 434–435
Trachoma, 377
Transcription, of DNA, 129–132
  in eucaryotes and procaryotes, 131

Transduction, by bacteriophages, 162–163
Transfer RNA (tRNA), 34, 126
    amino acid activation by, 131–132
    synthesis and distribution of, 132
Transformation, bacterial, 163–164
Translation, of DNA, 131–132 (*see also* Proteins, synthesis of)
Transmission, of infectious disease, 254–259, 255 (*see also* Reservoirs)
    contact, 255
    nosocomial, 267–270
        control of, 270–274 (*see also* Infection control procedures)
        precautions for, 277–280
    precautions for, 277–280
    route of entry in, 256t
    vector, 257–259
Transmission electron microscopy (TEM), 20–21
Traveler's diarrhea, 403
Trench mouth, 380
Treponema pallidum, 19, 21, 57, 419–420, 452
    latency of infection by, 235–236, *236*
Treponema spp., mouth infections caused by, 379, 380
Tricarboxylic acid (TCA) cycle, 148
Trichomonas vaginalis, 74, 417–418
Trichomoniasis, 420
Trichophyton spp., 372–373
Tripeptide, 123
Trisaccharides, 117
tRNA (*see* Transfer RNA)
Trypanosoma brucei, 74, 435
Trypanosoma cruzi, 436, 465
Trypanosomiasis
    African, 435, 442
    American, 436
TSS (Toxic shock syndrome), 242, 363 (*see also Streptococcus pyogenes*)
Tuberculoidal agents, 177
Tuberculosis, 391–392
Tubulin, 38
Tularemia, 432–433
Turbidity, of solutions, 155
Tyndall, John, 11, 17t
Tyndallization, 11
Typhoid fever, 404–405
Typhoid Mary, 405
Typhus fever, tickborne, 427–428

Ubiquitous microorganisms, 3
Ulcers, 397–398
Ultrasonic waves, antimicrobial effect of, 190
Ultraviolet light, antimicrobial effect of, 189–190
Universal Precautions, 276
Uracil, 126–128
Ureaplasma spp., 66t
    in urinary tract infections, 413
Ureaplasma ureolyticum, 417–418
Ureteritis, 413
Urethritis, 413
Urinary tract
    infections of, 286, 413
    nonspecific defense mechanisms of, 312
Urine specimen, collection of, 286
Urogenital area, indigenous microflora of, 216
U.S. Public Health Service, 259, 298
UTI (Urinary tract infections), 286
UV light microscope, 20

Vaccination, 324–327
Vaccine(s)
    acellular, 326
    and acquired immunity, 324–327
    attenuated, 325
    autogenous, 327
    for *Bacillus anthracis,* 371
    for chickenpox, 365
    childhood, 327
    DTP (Diphtheria-tetanus-pertussis), 326, 388, 392, 442
    first preparation of, 13
    for *Haemophilus influenzae,* 42, 204
    for hepatitis viruses A and B, 397
    historical development of, 325t
    ideal, 324
    inactivated, 325
    MMR (Measles-mumps-rubella), 327, 366, 367, 427
    origin of word, 13
    for polioviruses, 439–440
    sources of, *326*
    for *Streptococcus pneumoniae,* 204
    subunit, 326
    toxoid, 326
    types of, 324–327, 327t
Vacuole(s)
    digestive, 318
    in eucaryotes, 37

van Leeuwenhoek, Antony, 9–11, 17t
    discovery of anaerobes by, 63 (Insight section)
Varicella-zoster virus, 365
Variolation, 13
Vasodilation, 321
    in inflammatory response, 233
Vaucheria, 72
Vectors (*see also* Reservoirs)
    animal, 258–259
    arthropod, 65, 258, 425–433
    types of, 465–466
Vehicular transmission, of disease, 259
    control of nosocomial, 272–273
Veillonella spp., 217, 395
Venereal warts, 367, 416–417
Verrucae vulgaris, 367
Veterinary microbiology, 7
Viable cell count, 155
    and total cell count, 154–155
Viable plate count, of bacterial growth, 155, *156*
Vibrio cholerae, 400–401, 452
    enterotoxin production by, 242
    shape of, 57
Vincent's angina, 380
Viral infections
    of circulatory system, 421–427
    of ear, 381–382
    of eye, 375
    of gastrointestinal tract, 396–397
    of genitourinary system, 413–417
    of nervous system, 438–441
    of respiratory system, 385–387
    of skin, 364–367
Virchow, Rudolf, 11
    cell theory of, 30
Viricidal agents, 177
Virions (*see* Viruses)
Viroids, 7, 32t, 92
Virologists, 7
Virology, 7
Virology Laboratory, 293
Virucidal agents, 177
Virulence, 230
    bacterial structures associated with, 238–239
    enzymes associated with, 240–241
    factors associated with, 237, *238*
    increase in, during epidemic, 243
    and pathogenicity, 242–243
    toxins associated with, 241–242

Virulence factors, 237, *238*
Virulent bacteriophages, 91
Viruses, 2, 83–92
  bacteriophages as, 90–91, *91,*
    161–162
  classification of, 31, 32t
    basis of, 83–84
  composition of, 83, *85*
  counting, in culture media,
    155
  difference between living cells
    and, 83
  diseases caused by, 4t, 86t, 89
    (*see also* Viral infections)
    zoonotic, 251t
  exclusion from living organisms,
    48
  genetic material of, 84
  genetic mutation by, 91–92
  important groups of, 86t
  inactivation of (*see also* Antimicro-
    bial methods)
    chemotherapeutic agents for,
      87, 89, 197, 201, 201t
  infectious process of, 85–86, *88,*
    89t
  nosocomial infections by, 270
  oncogenic, 83, 89
  origin of, 89–90
  parts of, *85*
  shapes of, 84t
  sizes of, 23, 83, 84t
    comparative, *87*

  tumors and cancers caused by,
    83, 89
  uses of, in gene therapy, 166–167
    (*see also* Biotechnology; Ge-
    netic engineering)
  visibility by electron microscope,
    16
Vitamins
  in metabolism, 121
  production of, by indigenous mi-
    croflora, 217
Volvox, 69
von Behring, Emil, 17t
von Dusch, and Schröder, 17t
Vorticella, 73

Wandering macrophages, 317
Warts, 367
  genital, 416–417
Waste disposal, medical, 282
Waste water purification (*see* Sewer-
  age treatment)
Water, 104–108
  essential characteristics of,
    104–105
  molecule of, 104, *104*
  percentage of cell that is, 104
  polarity of, 104–105
Water bacteria, aerobic oxidation
  by, 149
Water supply, 5–6
  bacteria in, 254 (Insight)
  contamination of, 300–301, *301*

Watson, James, 127
Waxes, 120
Wet mount, preparation of, 470
Whittaker 5-Kingdom System, of
  classification, 48–49
WHO (World Health Organiza-
  tion), 259, 297–298
Whooping cough, 392–393
Wilkins, Maurice, 127
Woese, Carl R., 49
Woolsorter's disease, 370–371
World Health Organization
  (WHO), 297–298
Worms (*see* Helminths)
Wound infections, 373–374

Yeasts, 80, 214
  budding, *80*
  genetic engineering of, 166, 167
    (Insight)
  human diseases caused by, 81t
  structure of, *34*
Yersinia pestis, 374, 431–432,
  452
  as biological warfare agent,
    224

Zoonoses, 7, 250
  examples of, 251t–253t
  nosocomial, 271 (Insight)
Zoonotic disease (*see* Zoonoses)
Zygomycetes, 76
  characteristics of, 79t